To ~~[illegible]~~

~~Thank you for many opportunities~~

~~that you and your management~~

~~team at Hester industries given me~~

in implementing Six sigma methodologies.

Kind Regards.

[signature]

8/14/13

MASTERING LEAN
SIX SIGMA

MASTERING LEAN SIX SIGMA
ADVANCED BLACK BELT CONCEPTS

SALMAN TAGHIZADEGAN

MOMENTUM PRESS, LLC, NEW YORK

Mastering Lean Six Sigma: Advanced Black Belt Concepts
Copyright © Momentum Press®, LLC, 2014.

First published by Momentum Press®, LLC
222 East 46th Street, New York, NY 10017
www.momentumpress.net

ISBN-13: 978-1-60650-404-8 (hardback, case bound)
ISBN-10: 1-60650-404-5 (hardback, case bound)
ISBN-13: 978-1-60650-406-2 (e-book)
ISBN-10: 1-60650-406-1 (e-book)

DOI: 10.5643/9781606504062

Cover design by Jonathan Pennell
Interior design by Exeter Premedia Services Private Ltd. Chennai, India

10 9 8 7 6 5 4 3 2 1

Printed in the United States of America

In memory of my parents who asked so little and gave so much,
God bless their soul
To my loving wife, Leila who makes my life so beautiful and our daughters Sara and
Setareh who bring so much joy to our lives.
I love you more than life itself

CONTENTS

PREFACE

In an environment of intense global economic competition in which competitors with the objective of lowering product manufacturing costs are embracing robust manufacturing, servicing, and technologies, a new method of thinking such as *Mastering Lean Six Sigma* is required in order to outperform the manufacturing techniques of the competition. A proactive way of meeting the increasing competition is to focus on maximizing productivity and achieving quality at the lowest manufacturing cost and at a faster rate than the competitors, in addition to building capacity to continuously introduce new ideas of quality/ reliability and process optimization methodologies. These measures lead to highest customer satisfaction and a robust bottom line.

Mastering Lean Six Sigma provides Advanced Black Belts concepts (ABC). This is one of the fundamental building blocks of any organization's Lean Six Sigma deployment, be it a manufacturing organization or performing transactional processes. The development of a Master Black Belt program is a critical component of corporate success in the strategic implementation of Lean Six Sigma objectives.

This book has been developed to help organizations deploy Six Sigma and support certified Master Black Belts with the necessary preparation so that they face the challenge of managing multiple Lean Six Sigma projects and lead a company-wide Six Sigma initiative.

Becoming a Master Black Belt involves a great deal more than just learning advanced statistical tools and methodologies. *Mastering Lean Six Sigma* provides students with a comprehensive Lean Six Sigma leadership and analytical tools, methodologies and road maps to drive successful implementation of Lean Six Sigma and other process improvement techniques within the organization.

The curriculum takes students beyond the tools and techniques that they practiced and mastered during the Black Belt projects and provides them with the techniques to manage and lead an overall Lean Six Sigma program. Various Lean Six Sigma examples and case studies are given throughout the text in addition to analysing what makes a successful Lean Six Sigma program, the pitfalls to avoid, and how these can be translated into success of the organization.

Mastering Lean Six Sigma provides an overview of various Lean Six Sigma tools, which are analyzed clearly in graphical forms with many examples from manufacturing or transactional practices. They are aimed at the following:

  i. Creating the need for organizational strategic goals
  ii. Launching the objectives and leading the efforts
  iii. Developing the Lean Six Sigma roadmap DMAIC (define, measure, analyze, improve, control) and DDVPC (define, design, verify, production, control). DDVPC is for new product in either manufacturing or services

iv. Complete and comprehensive analysis of DMAIC, DDVPC methodologies, statistical tools, with numerous examples and graphics

v. Various case studies with step-by-step DMAIC and DDVPC phases

Mastering Lean Six Sigma is aimed at preparing and coaching Black Belts in leading, teaching, training black belts/green belts in Six Sigma methodologies and statistical analysis skills to help them deploy full Lean Six Sigma initiatives.

KEYWORDS

Mastering Lean Six Sigma, Advanced Black Belt concepts, Lean Six Sigma mastering, Six Sigma Define Concepts and Strategies, Six Sigma Measure Concepts and Strategies, Six Sigma Analysis Concepts and Strategies, Six Sigma improve Concepts and Strategies, Six Sigma Control Concepts and Strategies, Lean Six Sigma Roadmap

ACKNOWLEDGMENTS

Much of my appreciation goes to my family. To my loving wife, Leila, and our daughters, Sara and Setareh, thank you for your long-lasting patience and continuous support. The completion of this project would have been impossible without your full support and for that I can't thank you enough. Further I would like to thank my colleagues and especially my students for their extensive support throughout the program.

Finally, my thanks to Joel Stein at Momentum Press for his support and patience throughout the publishing process and other members of the Momentum Press team for their support and assistance in making this book a reality.

ABOUT THE AUTHOR

Salman Taghizadegan, Ph.D. is one of the leading Lean Six Sigma masters who has extensive experience in Lean Six Sigma teaching, coaching, and training students academically through universities with industrial projects. Further, he teaches and practices plastics processing, design, control, and analysis. He received B.S. in Chemistry from Western Illinois University, B.S. in Chemical Engineering from the University of Arkansas, M.S. in Chemical Engineering from the Texas A&M University, and Ph.D. in Chemical Engineering with emphasis on plastics and control from the University of Louisville.

He has over 25 years of academic and full-time industrial experience in Lean Six Sigma, plastics engineering, chemical processing, design, and control engineering, primarily in injection molding industries. He has authored numerous technical publications along with his first book in Six Sigma entitled *Essential of Lean Six Sigma*. He has spent most of his professional career as an adjunct professor in engineering, as a highly technical specialist in Lean Six Sigma, in the plastics industry, as a leader in quality and process improvement, as well as managing of waste reduction in the manufacturing environment. Dr. Taghizadegan is certified Black Belt and Master Black Belt through the University of California at San Diego. Currently he is teaching Lean Six Sigma Green Belt, Black Belt, and Master Black Belt at the California State University in addition to his full-time industrial career.

PART I

DESIGN AND DEVELOP THE REQUIRED PROCESSES (THE NEED)

CHAPTER 1

Mastering Lean Six Sigma Principles

In an environment of intense global economic competition in which competitors with the objective of minimizing product manufacturing costs are embracing robust designing, product development, production, servicing, and technologies, *Mastering Lean Six Sigma* offers leadership skills, Six Sigma tools and methodologies to key players who aim to outperform the manufacturing techniques of their competition. A proactive way of meeting the increasing competition is to focus on maximizing productivity and achieving quality at the lowest manufacturing cost and at a faster rate than the competitors, in addition to building capacity to continuously introduce new ideas of quality and reliability engineering and robust process optimization methodologies. These measures lead to highest customer satisfaction and a robust bottom line. Virtually, *Mastering Lean Six Sigma* fuels top talent players with powerful ideas to create a breakthrough performance by arriving at solutions for unsolved problems.

1.1 LEAN SIX SIGMA: THEORY AND CONSTRAINTS

1.1.1 WHAT IS LEAN SIX SIGMA AND WHAT LEAN SIX SIGMA CAN DO FOR YOU?

If you want to know "what is Lean Six Sigma?" and "what it can do for you"? then you might ask yourself the following questions:

- Are you ready for world-class performance?
- Are you prepared to overtake your competition?
- Are you ready to experience the ultimate in process accuracy and speed?
- Do you want to improve your return on capital investment?
- Do you want to increase your market share?

What about just Lean and what it can do for you? Let's review some of the questions related to Lean as follows:

- Do you have a cost disadvantage against your competition?
- Is your process affected when an employee is absent or products are out of stock?

- Are you meeting the timelines for customer orders or request?
- Are people below par in productivity or resources underutilized?
- Do you have reinspection or perform lots of rework?

These are just a very few of the many questions related to Lean Six Sigma (LSS) process improvements. Now let's start reviewing the Six Sigma science of continuous improvement. Six Sigma simply means a measure of quality that strives for near perfection. It is a disciplined, data-driven approach and methodology for eliminating defects (driving toward six standard deviations between the mean and the nearest specification limit) in any given process, be it manufacturing or transactional, product or services.

As the Sigma level increases, quality improves and therefore costs go down. Customers become more satisfied as a result. Furthermore, working smarter not harder, with fewer mistakes at different stages such as manufacturing, filling out a purchase order, financial reporting, and employee turnover improve overall performance . In short, it brings world-class performance to a company and helps it overtake the competition by bringing in the ultimate in process accuracy and speed. It also improves the return on invested capital and increases market value.

There are four LSS success factors:

1. *Selecting the right project:* Six Sigma is all about selecting the right project that supports the business/engineering strategies and is linked to the goals of the organization. This should be the key issue that must be solved if the organization wants to be successful.
2. *Right people*: Six Sigma is all about selecting and training the right people to fill the key roles. Successful organizations select their best people to fill the key Six Sigma positions as sponsors, champions, black belts, and green belts.
3. *Project management and gate reviews*: Six Sigma is all about effectiveness. Management and steering committee for gate reviews are critical to the success of the company. Lack of management review reduces the impact of the Six Sigma effort.
4. *Sustaining the gain and improvement*: Any technique for maintaining the gains is an integral part of the Six Sigma approach. At a tactical level, this technique maintains the individual projects. At a strategic level, it is the continual identification of new projects for continuous improvements. The Six Sigma breakthrough model is shown in Figure 1.1 with three objectives: higher profits, maximized values, and minimized variation.

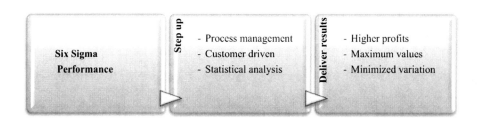

Figure 1.1. Six Sigma performance.

1.1.2 STATISTICALLY WHAT IS SIX SIGMA?

Sigma: By definition, Sigma (σ), a Greek letter, is the statistical quality measurement of standard deviation from the mean. Six Sigma describes how a process performs quantitatively. In other words, it measures the variation of performance.

Normal distribution curve: Normal distributions are probability curves that have the same symmetric configuration. They are mirror images with respect to the target such that sample data are more concentrated at the center of the curve than in the tails. The term bell-shaped curve is often used to describe normal distribution. The area under the distribution curve is unity. The height of a normal distribution can be expressed mathematically in two parameters: mean (μ) and standard deviation (σ). The mean is a measure of the center of the curve or the mean is the average of all the points on the curve, and the standard deviation is a measure of spread from the mean. The mean can be any value from minus infinity to plus infinity (in between $\pm \infty$), and the standard deviation must be positive. Thus, the probability of $f(x)$ (Equation 1.1) is equal to 1.

$$\int_{-\infty}^{+\infty} f(x)\mathrm{d}x = 1 = \text{Area under the normal distribution curve} \qquad \text{(Eq. 1.1)}$$

Suppose that x has a continuous distribution, then for any given value of x, the function must meet the following criteria:

$$f(x) \geq 0 \text{ for any event } x \text{ in the domain of } f.$$

Since the normal distribution curve (symmetric from the mean) meets the x-axis at infinity as shown in Figure 1.2, the area under the distribution curve and above the x-axis is assumed to be 1. This can be calculated by integrating the probability density on a continuous interval from minus infinity to plus infinity.

So, Six Sigma is a statistical measurement of process history (in the past as well as current). Its focus is on how good the company's product and service really is. Or, how far ahead or behind the company is from the industry standard. It allows the company to benchmark quality, and determine which direction should be taken to achieve the desired quality and how it can

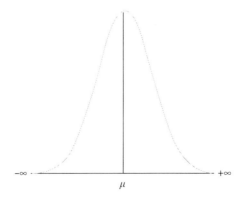

Figure 1.2. Normal distribution curve (symmetric) area equal unity.

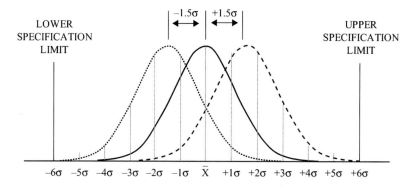

Figure 1.3. Normal distribution curve shifted by ±1.5σ to the right or left of the target.

be finally achieved. In a perfect world a process produces 99.9999998% defect-free work per million opportunities (or 0.002 defects per million). However, in the real world, the processes fluctuate and shift by ±1.5 Sigma. This translates into 99.99967% defect-free work per million parts or 3.4 defects per million opportunities (DPMO). This is visualized in Figure 1.3. For details refer to Chapters 7–11 that discuss Six Sigma application and tools.

1.1.3 WHAT IS LEAN CONCEPT?

The core concept is to maximize the customer value while eliminating the waste to near zero as possible. So Lean is a technique of reducing the cycle time and non-value-added work, resources, steps, and others. Furthermore, it is a business or engineering strategy that helps organizations to gain competitive advantage over the other players. An organization with a Lean culture establishes the ultimate goal of providing perfect value to the customer through a process that generates zero waste. One of the misconceptions is that Lean is applied only in the manufacturing environment. However, Lean applies to every process of a businesses or an organization. Some of the Lean goals are as follows:

1. Minimize lead time, process time (cycle time), and add value by removing waste (the non-value-added work). Simply, it is a strategy of removing waste from any process. The waste could be result of overdoing, delay, excess steps, unutilized talent, defects, variation, or quality issues. Anything that doesn't add value to the end user or customer is considered to be waste.
2. Apply value stream mappings with multiple process steps that run the process from start to finish. In other words, map out the boundaries of the end-to-end process. It provides a high level of extremely effective process view to ascertain the roadblocks and assists one to focus on improvement opportunities.
3. Continuously apply Kaizen process improvement. Kaizen is a Japanese term that stands for "improvement" or "change for the better." It refers to continuous process improvement.
4. Implement best practices and build in quality.
5. Put to use the talent of employees.

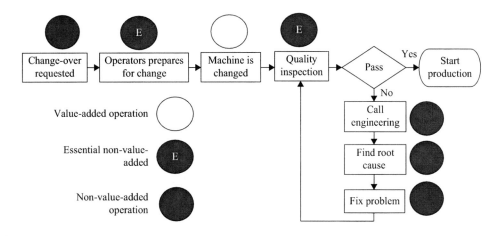

Figure 1.4. A Lean process with value-added and non-value-added steps.

Figure 1.4 is a Lean process mapping flow chart. The chart illustrates the automation operation change from product "A" to product "B." It includes the steps that add value, that are essential with no value added, and that are non-value-added.

See also Chapters 7 and 8 for a detailed discussion of Lean application tools.

1.2 LEAN SIX SIGMA MASTER BLACK BELT

Master Black Belts (MBB) is one of the foundational building blocks of any successful LSS organization. The development of MBBs is a critical success factor in the deployment of Six Sigma process improvement strategy. The tools in this book have been developed to design a successful LSS program. This will provide certified MBBs with the necessary preparation to meet the challenge of managing multiple projects and leading company-wide LSS initiatives.

Becoming an MBB involves a great deal more than just learning advanced statistical techniques. It provides the students of MBB with the comprehensive LSS leadership tools, methodologies, and road maps to drive successful implementation of LSS and other process improvement methodologies within the organization. The curriculum takes students beyond the tools and techniques that they practiced and mastered during the Black Belt project. It trains students with the techniques to manage and lead an overall LSS program. Various LSS case studies are included in Chapter 15 and they are analyzed to propose what makes a successful LSS program, the pitfalls to avoid, and how these can be translated into organizational success.

1.3 LEAN SIX SIGMA BLACK BELT OVERVIEW

An MBB program focuses on advanced LSS statistical methods used in LSS project phases, which are define, measure, analyze, improve, and control (DMAIC). In addition to Black Belt tools, we cover new tools in areas such as study of variation, multivariate experiments, nonparametric analysis, destructive testing, handling attribute responses (beyond Freeman-Tukey), practical experimentation, optimization experiments, handling multiresponse experiments, distributional analysis, advanced regression methods, advanced Statistical Process Control (SPC) methods, and much more. A brief review of the classification, goals, and objectives of the LSS in DMAIC follows. The detailed phases with practical examples are explained in Chapters 7–12.

1.3.1 DEFINE

There is no destination without a starting point. Every road or project has a beginning and an ending point. Define phase is the step one, the starting point of the project with strategic planning objectives. It is a road map to get to the destination and to achieve the deliverable results that is envisioned at the starting point. The elements of define phase are as follows:

1. Project charter (to include business case)
2. Project plans, boundaries, and timeline
3. List of deliverables
4. Stakeholder analysis
5. Voice-of-customer (VOC), critical-to-quality (CTQ)
6. Kano model analysis

Define also identifies the type of project that relates to either an existing process improvement in production or a new product/service development through the use of two Six Sigma submethodologies: DMAIC and DMADV.

Existing product or services — Process characterization and optimization

- Define the project

- Measure the current process in production or services

- Analyze the existing measurement using statistical tool

- Improve the process under study by deployment of world class benchmark and tools

- Control the improved process and continue to monitor

New product or services — Process characterization, development and verification

- Define the goals and objectives of the project

- Measure the identical process for selected criteria

- Analyze the measurement criteria

- Design the new process or product

- Verify the developed process or product

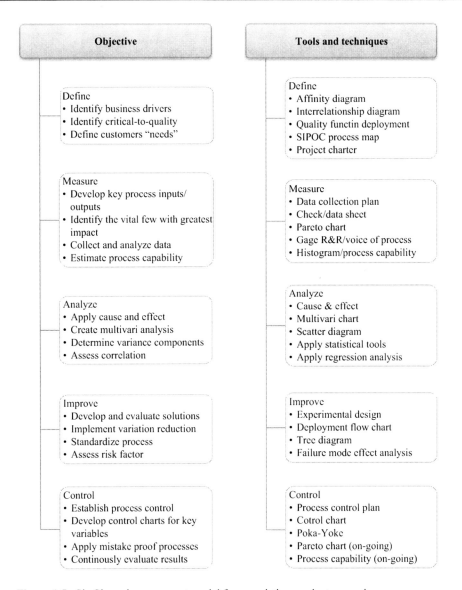

Figure 1.5. Six Sigma improvement model for an existing product or service.

A model of process improvement using Six Sigma tools and techniques for processes in the production cycle is shown in Figure 1.5.

A model of process improvement using Six Sigma tools and techniques for a new product or service is shown in Figure 1.6.

1.3.2 MEASURE

Measure phase is step two of the DMAIC or DMADV process. It focuses on numerical research and root cause analysis. The objective of this phase is to get as much information as possible

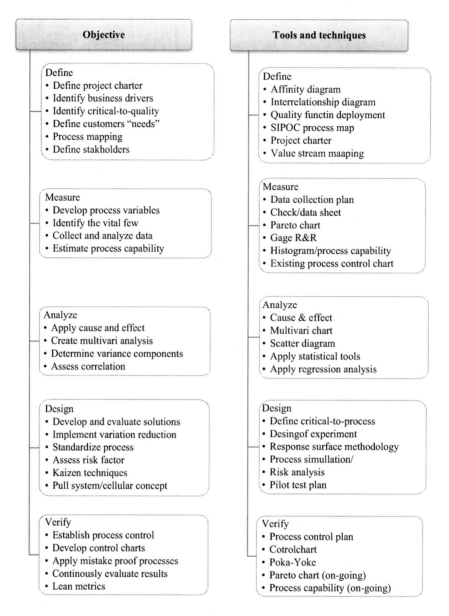

Figure 1.6. Six Sigma improvement model for a new product or service.

on the existing process or service so as to fully understand how well it operates. This entails the following key deliverable factors:

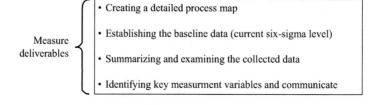

1.3.3 ANALYZE

The Analyze phase is the third step and the longest phase in the LSS methodology. This phase applies all the statistical analysis tools available for use. Almost all the essential measurement analysis is carried out in this step. The root cause inputs that affect the key outputs of the process or the system under investigation are identified and measures to eliminate them are proposed. Thus, the significance of inputs on the outputs is determined. Some of the deliverables in this phase are given below:

Analysis deliverables
- Analyzing the collected data
- Applying root cause analysis-relationship between input and output
- Determining the gap variations and improvement opportunities

1.3.4 IMPROVE

The Improve phase is the fourth step that requires knowledge of identifying and reducing the key process inputs that cause the effects (output). Thus, in this phase, the solution is identified, implemented, and the process variations reduced substantially. The return on invested time, planning, testing, optimization, and execution are achieved. The key deliverables in this phase are as follows:

Improve deliverables
- Optimize and test the improved process
- Identify the best possible process
- Design implementation and standardization plan
- Wastes removed, just-in-time flow established

1.3.5 CONTROL AND SUSTAIN

The Control phase is the final step and it concludes as well as sustains the gains made in the process. The process performance and capability are determined and documented.

Control deliverables
- Control and sustaining plan/ record process capability
- Operating procedures and document
- Transfer of ownership to production (project completion)

CHAPTER 2

LEAN SIX SIGMA AND MASTER BLACK BELT ROLES (WHO IS THE LEADER?)

"The past is gone,
The future isn't here yet,
What can we do now to change the future?"

Lean Six Sigma is all about technical leadership to engineer and solve the unsolvable problem, not through the classic approach, but using exceptional methods to make a breakthrough, that is, very quickly in the sense of "moving mountains" or by stimulating enormous changes in a complex global organization. The advancement of Lean Six Sigma applied to manufacturing and engineering and to service, health care, administrative, financial, transportation, and many other operations is widely recognized and can be taken for granted. Lean Six Sigma changes the world (Figure 2.1) that is not here yet. The role of Master Black Belts is to ensure that the Lean Six Sigma methodologies are deployed scientifically, efficiently, and effectively in achieving the ultimate customer satisfaction and bottom line in an organization.

2.1   MASTER BLACK BELT ROLES IN THE ORGANIZATION

The Six Sigma Master Black Belt (SSMBB), also known as quality manager, leader of improvement, or leader of change, possesses knowledge of advanced applied statistical analysis such as hypothesis testing, analysis of variance, design of experiment, response surface, business strategies, leadership training, and an extensive background in applying Lean Six Sigma methods. The Master Black Belts (MBB) is highly skilled in Lean Six Sigma (LSS) techniques such as define, measure, analyze, improve, control, and acts as a mentor in addition to teaching other black belt aspirants. So, MBBs have both managerial and technical skills and responsibilities. They must complete intensive training and oversee many projects before they earn this certification. Actually, they are process improvement scientists. This is a hands-on, full-time position.

13

Figure 2.1. Past, present, and future measures.

Normally, the MBB is selected from among the talented Black Belts (BB) in the company. The MBB also performs the role of a consultant to BBs in their projects to help, push, or direct if any hang-ups are needed to be resolved or cleared. Nevertheless, the company expects MBB to be an expert in advanced tools and management of Lean as well as Six Sigma with already proven projects. They should have completed several BB projects within a year. The LSS-MBBs are very efficient and effective individuals who utilize the knowledge they gained from training and what they have learned to improve the process performance of an organization.

2.2 MASTER BLACK BELT (MBB) QUALIFICATION

MBBs with the highest level of Six Sigma training are experts responsible for the strategic planning and implementation of LSS within an organization. They promote and support improvement activities in all business areas of their organization and at their suppliers and customers. Some of the selected criteria that all MBBs candidates should possess are the following:

1. completion of BB training and certification;
2. completed at least 10 BB projects related to commercial and engineering applications in business; and
3. necessary understanding of business, technical, or engineering applications. As mentioned above, a MBB has both leadership and technical activity roles. Their main roles and responsibilities are discussed in the following sections.

2.2.1 LEADERSHIP ROLES

An MBB helps create fundamental LSS infrastructure to ensure that LSS is established across the company vertically and horizontally. He or she provides, mentors, and supports the black belts, green belts, and LSS teams to use the continuous improvement tools relevant to a particular issue. MBBs assist in developing and utilizing organizational metrics or dashboards. They are held accountable for process improvement performance matrices that are implemented. The process charter and supporting data are available and up-to-date. This is very important for process improvement and project selection. Further, they can consider other benchmarks or world-class best practices for organizational process improvement. An MBB develops, provides, promotes, sustains, and modifies the LSS course of study, establishing program training and communicating with external agencies in the delivery of LSS training.

The daily roles of MBB as a LSS leader entails the following:

1. Offering a strategic planning vision for continuous improvement of product or services and working toward that goal.
2. Establishing a Six Sigma training program to enable employee career growth and impact in the future positions in the corporate.
3. Selecting projects and training BBs to be Six Sigma leaders.
4. Establishing project team to carry out Six Sigma project and monitoring the progress to ensure it will achieve the stated results in the project charter.
5. Continue to establish Six Sigma culture in the company.
6. Leading projects that has the biggest impact in the bottom line savings.
7. Contributing to increase in cash flow by implementing Six Sigma during new product development

Interacting with the senior management is one of the most important key strategic elements that is required for all MBB to be successful. Furthermore, to be effective and more influential in corporate strategic initiatives, MBBs need to report to the most senior leaders of the organizational division they are assigned to serve. They need to be role models for others in the organization and should align themselves with corporate culture. With this aptitude, MBBs can identify and take action that is of most tangible relevance to each division in the organization to result in effective corporate success. Thus, MBB must have an outstanding relationship and communication and diplomatic skills with business leadership and champions. They can earn it by

1. supporting leaders to focus on LSS with steady commitment,
2. continuously informing leaders with the latest improvements, breakthrough, and challenges,
3. offering the strategic vision and promoting to gain the support for the resources needed to make vision possible, and
4. providing simple method of report to other managers or staff to understand the progress.

Becoming an MBB is not just a matter of learning and training. It requires to be hands-on and demonstrating performance as a master in skill sets of leadership, teaching, coaching, LSS technology, facilitating, advising, consulting, project management, and mentoring as shown in Figure 2.2. Furthermore, they should have personal unmatched attributes. These involves athletic thinking; determination and dedication; strong technical skills; mission, vision and values; integrity; having a good understanding of business; engineering applications and their goals; and of course exceptional leadership skills. After a successful completion of MBB program, a candidate can choose one of the two areas of MBB career expertise.

2.2.1.1 MBB Instructor

An MBB should have excellent skills such as written, verbal, presentation, reporting, and interpersonal communication. Generally, not more than 20% of an organization's MBB time should

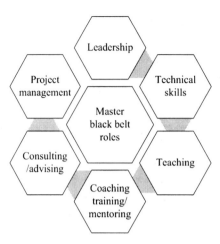

Figure 2.2. Master Black Belt roles.

be committed on training green belts (GBs) and BBs. Moreover LSS master requires one-on-one teaching and development of aptitude. The screening process of MBB selection for continuous process improvement, facilitation, lecturing, and coaching should be done under the supervision of a seasoned MBB coach. Originally, MBBs were chosen from super-BBs whose depth of experience and knowledge made them natural leaders. For instance, a BB may be certified by completing couple of projects. However, MBB may earn certification by coaching; mentoring many more projects than BB. Their diversity is impacted in the LSS success than any other role or management.

MBB will teach all levels of individuals from the most basic knowledge to most advanced. They train BBs how to lead projects, assist and motivate green belts, as well as how to communicate to apply the DMAIC (define, measure, analyze, improve, control)/DMADV (define, measure, analyze, design, verify) phases. The MBBs should become "process improvement scientists," continually monitoring, analyzing, and advancing business/engineering improvement methodologies for their organizations. As the time changes, improvement program also changes. So, the MBBs continually must challenge the theory of their existing processes. They must be aware of the changes

1. in the market,
2. technologies,
3. the competition's next move for advancement,
4. vision to design and verify, and
5. implement new concepts and techniques for future process improvements.

2.2.1.2   MBB Project Manager

An MBB must have experience in cross-functional project management and a strong ability to lead others without having direct authority. Mostly, MBBs manage large cross-functional design for Lean Six Sigma (DFLSS), new product development, or reengineering the existing projects already in progress. However, most of the organizations utilize MBBs' time for coaching BBs in the completing their projects either in business or engineering industries. The rank of MBBs is increasingly filled not only by engineers and statisticians but also by managers and health care professionals as well. This impressive diversity proves the expansion of LSS domain in business and engineering world in spite of the difference of opinion with regard to the philosophy and practice of LSS. One of the MBB process improvement scientist goals should be to understand the complexity of interrelationship between the organization, people, and technology issues involved and knowing that today's issues are yesterday's solution.

2.2.2 TECHNICAL ACTIVITY ROLES

Technical activity role may include the following disciplines:

1. Ensure all the critical-to-quality (CTQs), voice-of-customer (VOCs), and customer's "needs" and "wants" are fulfilled.
2. To carry out point 1, an MBB should possess knowledge and expertise in advanced analytical tools or LSS applications, i.e., analysis of variance (ANOVA), statistical process control (SPC), design of experiments (DOE), response surface analysis (RSA), and linear and nonlinear regression analysis.
3. Guarantee that corporate product and services are designed to be maximally robust.
4. Be able to train, teach, and develop workshops for GBs, BBs, and champions.
5. Coaching and advising BBs in their process improvement projects to confirm statistical tools and methodologies are properly used, results are achieved as expected, and the process of DMAIC is managed properly.
6. Supports technical direction for all corporate LSS projects.
7. Promotes main cross-functional LSS DMAIC and DMADV.

2.2.3 MBB JOB DESCRIPTION

The job description of an MBB may include some of the following management tasks:

1. Lead the effort to establish a LSS scorecard for monitoring the organization's business outputs (y's) and provide analysis of any detected defects to the level of process inputs (x's) for definition of DMAIC projects.
2. Educate and train organization leaders in establishing a portfolio of LSS projects in their business area. Figure 2.3 features the roles of different leaders in the organization.
3. Demonstrate a supervised instructorship for four or more BBs training program.
4. Support the management team in identifying the LSS projects that makes a big impact in their corporate strategic planning objectives.
5. Develop and sustain the project management role for assigned LSS projects.
6. Evaluate and identify qualified candidates for BB training in support of corporate strategic initiatives.

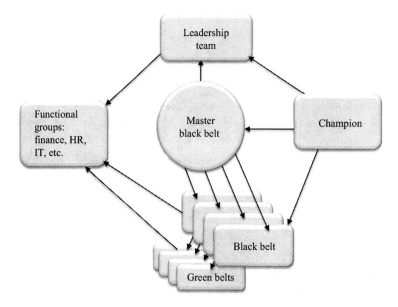

Figure 2.3. Lean Six Sigma roles and team structure.

2.2.4 COMPLETION OF CURRICULUMS

The complete curriculums for MBB are given as follows.

Leadership

    (a) Calculating the cost of poor quality
    (b) Project identification and prioritization
    (c) Change management leadership skills
    (d) Organizational structures for deployment of LSS
    (e) Competitive intelligence and industry benchmarks
    (f) Transitioning goals and objectives into actionable projects
    (g) Overcoming organizational resistance
    (h) Using financial measures to analyze performance

Design for Lean Six Sigma (DFLSS)

    (a) Pilot testing new products and services
    (b) Advanced failure mode effect analysis (FMEA)
        i. Design failure mode effect analysis (DFMEA)
        ii. Process failure mode effect analysis (PFMEA)
    (c) DFLSS—overview of DMADV (define, measure, analyze, design, and verify) and or DMADOV (define, measure, analyze, design, optimize, and verify)
    (d) Initiate project phase—This is the ideation phase. This is discussed in Chapter 7.
    (e) Define phase—Project must be defined in this phase. Again this is discussed in Chapter 7.

(f) Measure phase—This is the data collection phase (refer to Chapter 8).

(g) Analyze phase—The collected data are being analyzed in this phase. For details, refer to Chapter 9.

(h) Design or improve phase—The product is designed by using the lessons learned from analysis phase.

(i) Verify phase—The designed product or services will be tested and verified in this phase.

Optimizing Process Efficiency (Refer to Chapter 10)

(a) Applied value stream mapping

(b) Lean and Kaizen principles

(c) Advanced identification of capacity constraints

(d) The basics of theory of constraints

Measurement and Analysis (Refer Chapters 8 and 9)

(a) Force field analysis (FFA)

(b) Multivariate control charts

(c) GAP analysis (GA)

(d) Statistical normality tests

(e) Control charting continuous data

(f) Advanced interpretation of control charts

(g) Developing a process control plan

Advanced Statistics Theory and Practice (Refer to Chapter 9)

(a) Descriptive statistics

(b) Descriptive measures

(c) Probability statistics and distribution

(d) Discrete and continuous probability distributions

(e) Hypothesis testing, inferential statistics, and proportions

(f) Logistic, linear, and nonlinear regression and correlation analysis
   i. Simple linear regression
   ii. Multiple nonlinear regression

(g) Advanced analysis of variance (ANOVA)

(h) Review of other advanced distributions

(i) Confidence intervals and point interval estimates

(j) Mann-Whitney test

(k) Mood's median test

Design of Experiments (DOE) (Refer to Chapter 10)

(a) Design of experiments setup

(b) Selecting key process input variables x(s) and key process output variables y(s)

    (c) Selecting the optimal design
    (d) Design of experiments options and strategy
        i. Fractional factorial design (screening designs)
        ii. Full factorial design
        iii. Mixture screening design
    (e) Response surface methods
        i. Composite design
    (f) Nested designs and difficult to change variables
    (g) Attribute Gage study
    (h) Sampling strategies
    (i) Use of replication to estimate error
    (j) DOEs with two or more response variables
    (k) Developing optimal project metrics
    (l) Data mining

Statistical Software Analysis

    (a) Statistical analysis
    (b) Design of experimental analysis
    (c) Advanced analysis of variance (ANOVA)
    (d) Control charts for variables and attributes
    (e) Quality tools
    (f) Charts and graphs
    (g) Project application

Teaching, Training, Coaching, Mentoring Development

    (a) Resource allocation—managing black belts
    (b) Decision-making solutions—evaluating alternatives
    (c) Employee empowerment and motivation techniques
    (d) Rewards and recognition

Customer and Suppliers

    (a) House of quality or quality function deployment (QFD)
    (b) Customer retention and loyalty
    (c) Negotiation techniques
    (d) Supplier QA and performance
    (e) Cultural Issues in cross-cultural deployments of LSS

World-Class Industry Leadership

    (a) Advanced presentation, communication, and reporting skills
    (b) Career path for MBBs within the company

(c) International Standards Organization (ISO) 9001.
(d) Malcolm Baldrige quality criteria:
  i. Leadership, customer-focused, operation-focused, employee-focused, strategic planning, quality analysis and measurement, and results
(e) The Shingo prize model.

2.3   MBB PROGRAM DEVELOPMENT

The MBB program should begin with the screening of and selecting top talented BBs during their training or already trained MBB candidates. The process of choosing the best candidate who is capable and will meet the selection criteria for fulfilling the requirements of the position description entails four steps:

(a) For the BBs already employed by the company, the evaluation should be based on the job performance and suggestions for consideration as a MBB. To be selected for consideration, they must convince their division managers and possess the essential skills, aptitude, knowledge, teaching, and coaching capability to be a MBB.
(b) Each candidate must demonstrate a strong competence in technical skills of LSS, effective communication skills, an executive coaching, teaching, training, mentoring, and process facilitation.
(c) The upper management team should decide if the transition of the selected individual is aligned with the charter of program. Then necessary moves need to be made.
(d) After selection of the candidate, if the chosen person is a BB, then they must attend an MBB training program or an MBB already trained can start the program.

The LSS leadership characteristic and assessments can be evaluated by filling out Table 2.1 and rating from 1 to 5, 1 being very ineffective and 5 being very strong and effective.

2.4   DECISION-MAKING SOLUTIONS—EVALUATING ALTERNATIVES

This is the last step to go back and review the complete process data, findings, and conclusion. This final analysis helps to understand what can be done to achieve goals and objectives if

Table 2.1. Lean Six Sigma leadership and deployment assessments

Lean Six Sigma leadership characteristic	Effectiveness (1–5)	Strength	Opportunities to improve	Action planning
Committed Leadership				
Top Talent				
Supporting Infrastructure				

1 = Very ineffective, 2 = Ineffective, 3 = Marginally effective, 4 = Effective, 5 = Very effective

anything is missing. Here one may apply risk analysis tools to assess the possibility of failure or unexpected outcome from the final decision. This can be done using methods of decision analysis, uncertainty, mathematical modeling, statistical concept, and strength, weakness, opportunity, threats (SWOT) or force field analysis.

2.5 DEVELOPING AND UTILIZING A PROFESSIONAL NETWORK

Network promotes training and development. A professional network takes various forms and shapes. The online forums with journals, articles, and communications are very effective. Shared networks utilize a global competitiveness and innovation in a dynamic knowledge-based worldwide economy. This is visualized in Figure 2.4.

2.6 EMPLOYEE EMPOWERMENT AND MOTIVATION TECHNIQUES

Employee empowerment is a technique and concept that gives power or control to the employee in making decisions about their job function. In other words, it is a method of authorizing individuals to make a decision independently. Some of the empowerment and motivation techniques are as follows:

1. *Leadership role*: management should take the role of coaching and training.
2. *Career path*: improve the employee power of knowledge through constructive training.
3. *Responsibility*: give the employee higher responsibilities, and ability to make decisions in achieving their process improvement goals and objectives.
4. *Motivation*: motivate, mentor, encourage, and set goals for employees.

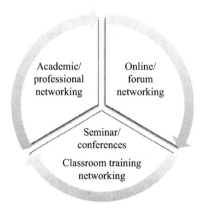

Figure 2.4. Roles of leaders.

2.7 EFFICIENT AND EFFECTIVE COACHING, TRAINING, AND MENTORING—SELF-DIRECTED

Everyone is born with a dream and ambition in life. Achieving dreams is not easy. It can be acquired with the help of a coach and with a bit of coaching. Effective coaching involves the practice of given proper direction, instruction, effective training, in the path of their goals and objectives. Although techniques of coaching vary from one person to another, there are some common techniques that include positive thinking, motivational speech, and training program.

1. Stimulate individuals and team progress toward the organization's strategic planning objective.
2. Create an atmosphere of trust that expedites the advancement of team members in contributing to organizational goals.
3. Sustain good eye contact with the audience and never ever turn back to read the slides.
4. Use effective communication skills that advance self-determination of others.
5. Apply SMART goals (see Chapter 7 for details).
6. Must have interest and enthusiasm to learn and serve as a trainer or coaching.
7. Capable of understanding and conveying the Six Sigma concepts to others.
8. Commitment and accountability to do high level of presentations.
9. Boost the team morale and be a good listener.
10. Manage your time and establish a positive image.

2.8 ADVANCED PRESENTATION SKILLS

Effective presentation and speaking skills are important in Six Sigma training, teaching, lecturing, coaching, and mentoring. Developing the confidence and ability to present a lecture in front of the audience and talking well are very important competencies. Like everything else, it takes a little preparation and practice. The style and goal of the presentation are different, for instance, spoken, media-visuals, audio, PowerPoint, training sessions, or simply just giving a talk on any subject.

Key essential elements of an effective speech and presentation are the following:

1. *Presentation material preparation*: Being prepared is an extremely important element of presentation. Use visual aids anywhere you can. Practice with sentence variety. For instance, every time you practice your presentations explain it differently. The objective is not to memorize the presentation rather keep it conversational. Use examples to support your points. The examples or stories must match your message.
2. *Time yourself*: Make sure that you stay within the boundaries of time and leave some time for questions and answers. Practice and organizing your timing will help you finish the speech without leaving any subject out. The practice should be at least 5 minutes less than actual talk depending on the questions and answers.
3. *Practice for an audience*: Rehearse in front of similar group. Know your audience's education level to explain things at their level. Be relaxed, prepared, and have full confidence in your speech. Most of the time audience will judge you in a couple of minutes.
4. *Opening and closing statement*: Practice your opening and closing statements. This is as

important as the entire speech. You got about 10 seconds to make a big impression on yourself by making a positive impact, thus come up with a very strong opening statement and practice as much as you can. You will perform better if you just be yourself.

2.9   REWARDS AND RECOGNITION

In LSS a rewards and recognition is as important as process improvement. It can play a very important role of bringing the future changes in the system. For example, an engineer can solve a problem faster than normal time if the individual encouraged and motivated by rewards or recognition. Rewards can be in any level possible, particularly if presented in front of colleagues. It could attract others to participate in process improvement, creativity, innovation, and so on if the rewards are presented in front of peers. Recognition of employees' achievements in front of peers is much more effective than if it presented individually.

Organizations can tie the salary or bonus of BBs and MBBs to their various process improvement projects, which involve cost reduction, improved productivity throughput, or other quality improvement factors to evaluate their rewards. Human resource (HR) can play a major role in this process. HR department along with higher management team can set up an appropriate system of rewarding and recognizing their employees.

CHAPTER 3

LEAN SIX SIGMA INFRASTRUCTURE: DESIGNING AND ENGINEERING (LEAN SIX SIGMA DEPLOYMENT)

The first question you may ask yourself is: how do I get started? When we talk about building an infrastructure of Lean Six Sigma (LSS), basically we are envisioning a project in itself, which follows DMAIC (define, measure, analysis, improve, control) steps and methodology. The purpose of define and measure phases of this technique is developing organizational strategic objective matrices as follows:

i. To align people with corporate strategic initiatives and use those initiatives as a guide to goal setting.
ii. To develop an effective way of communication for the company both horizontally and vertically.
iii. To provide information required in decision making, setting direction, and improvements.

The issues that most industries face is the lack of creating dashboard for define and measure phases. The dashboard enables organizations to set goals and deliver results. These concepts are discussed in Chapters 3–6 as envisioned in Figure 3.1.

3.1   INITIATE FINANCIAL GROWTH NEED PROJECTS

The financial goals and objectives need to be laid out for the foreseeable future. Several actions are required to do this.

Action point 1: The first action is that an organization should start doing self-assessment based on the world-class proven models such as those of Malcolm Baldrige. The assessment should measure the capability and advancement of the process, measurement system, and its results in the organization. This helps the organization to understand its current process status and identify the variables (causes) and opportunities for process improvements. The best evaluation to follow is the measure phase procedures and guidelines. After the completion of self-assessment (Table 3.1), the result should be shared with management and the employees who

Figure 3.1. Steps in building Lean Six Sigma infrastructure.

Table 3.1. Lean Six Sigma self-assessment/readiness review

Operational assessment questions	Effectiveness (1–5)
How does the organization measure success?	
Does the organization measure the right things?	
Does the organization have a few metrics that all employees understand and use?	
Are decisions based on data or assumptions?	
Have the critical process owners identified and communicated?	
Who owns the data?	
Where is the data?	
Where is the data stored and how accessible is it?	
Has the data been validated?	
Are the organization's reports written in simple, scientific, or financial terms, or free from using fuzzy language?	
Are there updated process maps of the most critical processes?	
Total Operational score	

1 = Very ineffective, 2 = Ineffective, 3 = Marginally effective, 4 = Effective, 5 = Very effective

are involved with the process. This becomes the turning point for the organization to make the cultural changes from the existing way of operation to LSS concepts and methodology. Some of the questions that the organization may consider during the process are as follows:

(a) How does the organization define and measure achievement?
(b) How frequently does the organization measure achievement?

(c) Has the organization established measurement metrics?

(d) How are the measurement data validated?

(e) Has the process map been developed?

Action point 2: Now that everything is clear, the company is ready for growth, increasing bottomline, and maximizing customer satisfaction. With the assessment results, the new strategic initiatives can be identified and developed.

Action point 3: Refer to what is stated in define and measure phases, which is discussed in Chapters 7 and 8. The process map is very powerful in defining and measuring even in decision making of strategic planning objectives. Therefore, we can utilize the same concept here to envision and identify the core strategic process map of LSS. Figure 3.2 illustrates the action that consists of four steps:

1. *Financial outlook*: First and foremost, the investor's financial goals and objectives should be clear.
2. *External (customer) outlook*: Value the customer strongly and trust in their recognition as your strength. Customer "needs" and "wants" must be met in order to achieve the investor's financial goals and objectives.
3. *Internal process outlook*: Internal process should be outperformed on customer expectations in order to achieve stockholder's financial goals.
4. *Teaching, training, coaching, and learning*: To deliver the strategic objective results, team and employees who own the processes must learn and implement the methodologies.

Action point 4: Upon the completion of action point 3, the organizational management can review the objectives of the stockholder's and customer's "need." Then, look for the cause-and-effect processes. Identify the measurement variables that will make a big impact on the corporate strategic objective and customer needs.

Action point 5: Develop a high-level process map such as SIPOC (supply, input, process, output, and customer). Using a process map, find the key process measures including Kaizen/Lean steps that will improve customer need and viewpoint. Use the defect per million opportunities (DPMO) as a baseline for improvement. Now, identify training, coaching, and

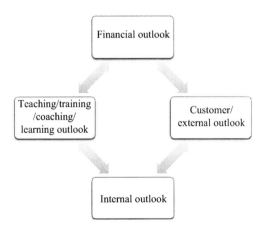

Figure 3.2. Envisioning the strategic initiatives.

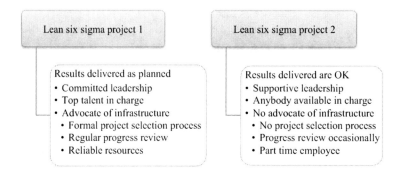

Figure 3.3. Similar attributes of successful project versus less successful project.

learning measures that link to these measures recognized in the process, strategic planning, and benchmark map. Make sure that organizational and LSS scorecard is implemented in each division to maintain the metrics as planned both vertically and horizontally.

Action point 6: Once the organization is committed to metrics, it is time to start collecting data on the metrics and determine the process off-sets. With the measurement variables in hand, LSS program should begin. According to Six Sigma, everything can be measured in the form of mathematics. In general, we do not question any process about which we do not know much in the form of mathematics. In other words, if we cannot measure the process in the form of mathematics metrics, we cannot even control it.

Similar attributes of a successful project versus a less successful project are summarized in Figure 3.3.

3.2 ELEMENTS OF SUCCESSFUL SIX SIGMA IMPLEMENTATION

The key success elements are arranged into three wide and related classifications.

3.2.1 MANAGEMENT SYSTEM SUPPORT AND COMMITMENT

Without support and commitment from top management of the organization, the program would not hold. Top management, functional strategic managers, and project managers should realize that their level of commitment and being on board with the program is required to make LSS implementation successful. As discussed in the previous chapters, if the leaders do not view Six Sigma as being strategically valuable and necessary, then they will not support it. The program loses its momentum and eventually due to lack of strategic deployment plan, it can never be established.

How do we know that leaders are supporting? They get involved with the game plan, program charter, tangible results, communications, recognition and rewards, and resources. So success starts at the top. The senior leaders personally should drive LSS implementation. Here are some key points:

  i. Senior and strategic leaders should ensure that leadership team is on board and they have a game plan.
 ii. They provide all the essential resources, funding, people, and support.

iii. They should ask for results to be delivered as planned.

iv. They are going to make cultural transition internally, through procedures, policies, and full support.

3.2.2 WELL-TRAINED BELTS

Without having the well-trained green, black, master black belts with full knowledge of methodologies, statistical analysis, communication, presentation skills, and a motivated mind, the program cannot deliver the results. Obviously the top trained talent delivers the results, and this talent pool is on the leadership pipeline. These individuals always take a more difficult and stressful path. On the other hand, weak leaders take the route of least resistance and less road block, even though it is not the way to achieve the goals and objectives. Successful companies follow the "no pain no gain" philosophy.

If the top-notch talent is selected to key essential roles in LSS, it proves the support and commitment of top management. This also makes others to compete for key positions in an LSS program. A step further, it is important to have future leaders with LSS background and their support.

3.2.3 WELL-DEFINED PROJECTS AND INFRASTRUCTURE

Since LSS is a project-based methodology, one of the essential factors is assembling well-defined projects (project charter) with end-to-end boundaries. Project selection is an essential component of success. A formal review of the process is necessary with the presence of a steering team to ensure that the projects are going to move in the right direction.

In addition, infrastructure support, resources, financial validation, and auditing infrastructure are critical components of success. Committed leaders provide the required infrastructure support.

3.2.4 LEAN SIX SIGMA SUCCESS MODELS

In mathematical terms, let the value of the dependent variable response y change as a function of independent variables (or adjustable variables) $x_1, x_2, x_3, ..., x_n$ of n quantitative factors such that

$$\text{Output} = f(\text{Input})$$
$$\text{or}$$
$$y = f(x) = f(x_1, x_2, x_3, ...x_{n-2}, x_{n-1}, x_n) \tag{Eq. 3.1}$$

where f is called response function and y is a dependent variable (in LSS terms strategic objective measurement values or symptom) and x is independent variable (or cause).

Simply, by definition, a process output is a function of process input. Thus, the product or services depend on the cause or process input changes as

$$\textit{Process output} = f(\textit{process input variables})$$

Or,
$$y = f(x)$$

This is also called the heart of Six Sigma.

Figure 3.4. Model of business success.

With the above definition, the success of the organization is a function of customer satisfaction and model of any corporate success. This is visualized in Figure 3.4.

$$Business\ success = f\,(customer\ satisfaction)$$

And the customer satisfaction is function of quality, cost, and delivery:

$$Customer\ satisfaction\ (delighted\ customer) = f\,(quality,\ cost,\ delivery)$$

Delighted customer is the result of exceeding customer's expectations. Furthermore,

$$Quality,\ cost,\ and\ delivery = f\,(process\ capability)$$

Likewise process capability is function of people's knowledge:

$$Process\ capability = f\,(knowledge\ of\ people)$$

And

$$Knowledge\ of\ people = f\,(training,\ education)$$

In summary, creating a world-class performance organization depends on three factors:

$$Lean\ Six\ Sigma\ world\text{-}class\ performance = f\,(well\text{-}trained\ belts,\ well\text{-}defined\ projects,\\ management\ system\ support)$$

In mathematical notation (Figure 3.5 and also refer Define phase in Chapter 7) we have

$$y = f(x_1, x_2, x_3)$$

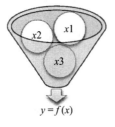

Figure 3.5. Model of world-class performance.

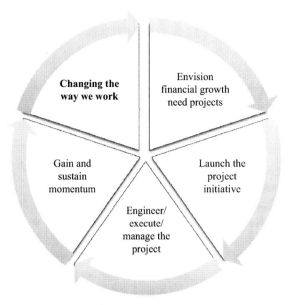

Figure 3.6. Overall high level of deployment process.

3.3   ROADMAP FOR DEPLOYMENT PHASE

The overall high level of deployment process for LSS includes five phases as shown in Figure 3.6. Of course, each phase is different from the other and has its own challenges in the process of deployment. One of the biggest obstacles would be cultural changes brought about by Six Sigma from the way operations are run presently to the new way using LSS methodologies. Additional improvement opportunities may even show up in the process of organizational transition.

3.3.1   ENVISION FINANCIAL GROWTH NEEDS PROJECTS

This topic was previously discussed in Section 3.1.

3.3.2   LAUNCH THE PROJECT INITIATIVE

The status of an existing process changes significantly in the organization. Many individuals would have honest and important questions. The overall vision and justification for implementation of Six Sigma should be made clear to all the contributors across the company.

   The following key elements need to be addressed:

(a) organization perspectives for a foreseeable future;
(b) lean Six Sigma program overall for short- and long-term objectives; and
(c) communication tools to convey the message to the organization.

It is extremely important that everyone understands the organizational direction. The top and strategic management should make sure that this is communicated.

3.3.3   ENGINEER, EXECUTE, AND MANAGE THE PROJECT

As mentioned before, success comes from top talent of the organization that will support all elements of Six Sigma infrastructure. The key essential success factors are as follows:

1. Strong trained leaders in LSS
2. Efficient and effective project selection team
3. Resources and budgets required
4. Effective communication process throughout the organization

3.3.4   CONTINUOUS PROGRESS AND MAINTAINING THE MOMENTUM

Momentum is defined as the force of movement, and the law of inertia states that a body or an object will maintain its velocity and direction as long as no other force acts on its motion's direction. Thus, developing and maintaining momentum is something that leaders should plan continuously. But, when it comes to global economic fluctuation, even a successful Six Sigma program may lose its momentum for a short period of time. However, with the top talent engaged in strategic planning, it is sure to gain its momentum again. Again, key components of supporting the program are as follows:

i. Well-trained green, black, and master black belts
ii. Well-recognized steering team who provide resources
iii. Establishment of the review team
iv. Auditing system for DMAIC

3.3.5   CHANGING THE WAY ORGANIZATIONS WORK

This is the last leg and turning point of the LSS implementation phase. The LSS methodologies must be implemented in order for this phase to be completed because the present systems are not working the way they should. Throughout the organizations, almost all the surveys indicate that most of the workforce does not feel like completely engaged at work or fully utilized, challenged, appreciated, and trusted on what they do best. Rather, employees feel distracted by many requests and as a result do not enjoy much from their job function. How can organizations build a culture with creative, motivated, and enjoyable employees to deliver results?

If the LSS is implemented fully and adequately then it will change how the leaders manage and the way work is executed. The investment on the employee core "needs" includes teaching, training, coaching, and motivating as required. So, they are fueled and inspired to make breakthrough and maximize customer satisfaction, which is core objective of LSS.

The details of the deployment phases are discussed in Chapters 4–8.

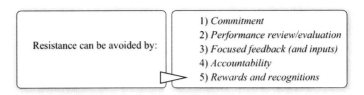

Figure 3.7. Avoiding cultural change resistance.

3.4 STRATEGIES TO OVERCOME ORGANIZATIONAL RESISTANCE TO CHANGES

It is human nature to resist any changes in the organization. Changes need new rules and methods to do the work while the old rules were just good enough to do the business. The transformation from old to new is necessitated by the changes outside the system that are out of the organization's control. Thus, it needs to be made clear that the changes are a necessity and not an option if the company wants to succeed.

Change is inevitable. As time progresses, change must happen. No one has control on it. But, how can we accept the fact of a matter that advancement must occur in order to move forward and even be ahead of the competition. This would not become a reality unless suitable changes are made in the organization. This is stated in Figure 3.7.

Commitment: Everyone in the organization from the top to bottom must be committed and be faithful to change. Obviously, as mentioned before, Six Sigma cannot be deployed unless top management completely backs the program and is willing to support it. It all starts at the top. Remember, all it takes one resistant in the leadership team to put an end to the changes.

Performance review/evaluation: The program transformation should be reviewed on a regular basis to confirm the progress as planned by the committee.

Focused feedback (*and inputs*): Individuals who are likely to be affected by the forthcoming transformation must be given a chance to voice their opinion. They should be able to address positive and constructive feedback about the changes.

Accountability: Every individual who will be affected by the changes must be held accountable for their job function, transformation activities, and responsibilities.

Rewards and recognitions: After successful deployment of the program, success should be recognized through compensation and rewards.

Leaders must build integrity, showing a commitment and support to their employees in the organization, while ensuring that they will get involved with providing feedback, listening, and encouraging process ownership by employees. If an individual fails to understand and accept the changes or strategic vision of the company for the foreseeable future, then transformation of infrastructure will allow the individual to see his role in the grand design.

3.5 CONVERTING GOALS/OBJECTIVES INTO ACTIONABLE PROJECTS

In general, strategic planning initiatives are tactical in nature. Every organization must follow the proper steps and procedures in achieving its corporate objectives. It is how you eliminate the gap between the goals and results. An operating plan must follow define and measure phases of LSS methodology, which is the roadmap for how to get to the maximum customer satisfaction. Leaders responsible for delivering the strategic initiative results must develop the plan.

Top management needs to come up with the program charter, resources, and financial means required to achieve the results they want. After deploying the LSS road map DAMIC (define, measure, analyze, improve, and control; see Chapters 7–11), all it takes from the top management is to fully support the infrastructure and demand weekly or bi-weekly updates and results.

PART II

LAUNCHING THE OBJECTIVES

CHAPTER 4

LAUNCHING THE LEAN SIX SIGMA PROJECT INITIATIVE: WHAT WORKS AND WHAT DOESN'T

A few years ago one of my engineering students asked the question, is it that once one earns a PhD degree then there is no more to study? The answer he got was that after a PhD you now realize how little you know about engineering and science. The higher you go the more you discover. It is a work in progress and you pass the knowledge on to the next generation along with the goal and objective you had set for yourself. But, you can't get there until you start. It all begins with the first step. Once again let's recall Figure 4.1 from Chapter 3.

Chapters 1–3 presented an overall deployment process for Lean Six Sigma (LSS) program once the initial preparations have been completed. Now it is time for the actual launch. Launching Six Sigma includes several steps.

Step 1: Start with team meeting for collection of information on project charter (initiation and define phases). Basically, this involves the investigation and identification of key root causes of the quality problems, high-cost processes, defects, key performance indicators (KPI), and the organization's SWOT (strength, weakness, opportunity, and threats) analysis.

Step 2: The next step is getting confirmation, involvement, and full support of management in the project initiatives. The goals and objectives should be supported by the stakeholders. To do this, a detail strategic planning objectives (SPO) and justification should be presented to the stakeholders.

4.1   SWOT ANALYSIS

What is SWOT analysis? How do you develop and analyze one?

SWOT is an acronym for Strength, Weakness, Opportunities, and Threats. By definition, strength and weakness are recognized to be internal factors that the organization has some control over its measurement. On the other hand, opportunities and threats are recognized as external factors that the organization has no control on its measurement.

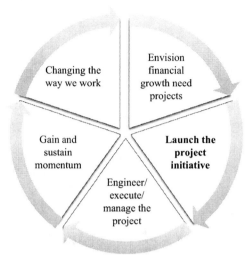

Figure 4.1. Steps in building Lean Six Sigma infrastructure.

SWOT analysis forms the basis of the infrastructure for developing your strategic planning initiative and tactics that then become the road map for maximum customer satisfaction in products and services. It is critical for full and partial deployment of LSS program of any organization.

A "SWOT" analysis details your organization's strength, weakness, opportunities, and threats in the form of matrix. It is a summary of the analysis of your company and is normally applied in identifying the overall strategic level projects and initiatives. Prior to defining the company's SWOT analysis, mission, vision, and values statement need to prepared. Then complete a thorough internal and external operation status and assessment. Figure 4.2 illustrates the understanding of the company's SWOT analysis.

4.1.1 STRENGTH

Strengths involve features that allow an organization to succeed in achieving the its strategic mission. They are foundations that continued achievement can be realized and maintained. Strength can be tangible or intangible. These factors show your expertise and assure your organization's accuracy and conformity. For instance, employee commitment, good credit, no debt, innovative product line, strong financial means, company name recognition and loyalty can be counted as strengths.

4.1.2 WEAKNESS

Weakness involve features that keep an organization from achieving its full strategic objective and potential. These weaknesses decay the momentum of company growth and success. Weaknesses of organizations do not meet the industry standards for instance, lack of a research, old machinery, or lack of variety product or services. The good news is that weaknesses are

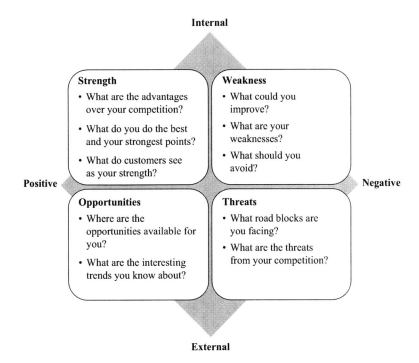

Figure 4.2. Understanding of company's SWOT.

causes and not symptoms. They are controllable variables. Other control variables are too much financial debt, workforce turn over, large product defects, scraps, warranty returns, and so on.

4.1.3 OPPORTUNITIES

Customers pay for quality product regardless of price. So, companies should seek and invest on opportunities that would improve their product quality. Thus, organizations can gain competitive edge by taking advantages of opportunities whenever they can. They include emphasis on new products, technologies, Lean and Kaizen operation, R&D, data validation/metrics, and advancing in Lean Six Sigma training, which creates world class manufacturing and services.

4.1.4 THREATS

Unlike weaknesses threats are uncontrollable. These include increasing competition leading to excess capacity, price wars and reduction in industry profits, or changing of required specifications for products or services. When a threat is present, the growth and success of the company can be unstable. Figure 4.3 lists elements to consider as threats.

What makes a SWOT analysis effort successful one? As stated before, the following are the ones:

    i. mission, vision, and value statements;
   ii. internal and external strategic planning and objectives;

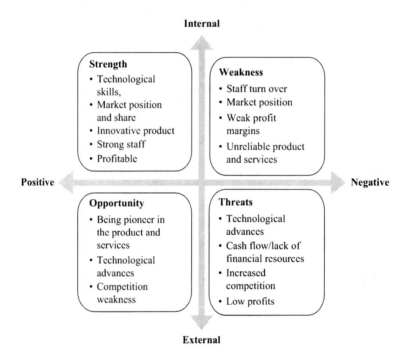

**SWOT analysis:
Threats**

Elements to consider:
• Voice of customer ("needs" and "wants")
• Competitive changes
• Technology transformation
• Environmental changes
• Financial changes, budget cuts, etc.
• Human resources rules and regulation changes

Figure 4.3. Elements to consider for threats.

Internal

Strength
• Technological skills,
• Market position and share
• Innovative product
• Strong staff
• Profitable

Weakness
• Staff turn over
• Market position
• Weak profit margins
• Unreliable product and services

Positive ← → **Negative**

Opportunity
• Being pioneer in the product and services
• Technological advances
• Competition weakness

Threats
• Technological advances
• Cash flow/lack of financial resources
• Increased competition
• Low profits

External

Figure 4.4. SWOT analysis sample.

iii. report specifics on strengths, weaknesses, and be sure to address specific and simple problems;
iv. establish strategic planning initiatives (SPI) and tactics that fit the SWOT analysis;
v. operation plan based on the SWOT analysis; and
vi. deployment of SPI and a result oriented operating plan.

Figure 4.4 presents a sample SWAT analysis.

4.2   PROJECT SELECTION CRITERIA

When selecting Lean Six Sigma project, there are two essential elements that should be considered.

1. Feasibility of the project
   (a) Resources needed
   (b) Expertise available
   (c) Complexity
   (d) Likelihood of success
   (e) Support or buy-in from the team and upper management
2. Organizational impact of the project
   (a) Learning benefits
   (b) Cross-functional benefits
   (c) Prioritization and chartering
   (d) Generate projects for financial, business growth, and market share.

Table 4.1 shows a template of project selection and prioritization and Table 4.2 is a sample preliminary project definition matrix.

Table 4.1. Project selection and prioritization

Process outputs	Complete within frame time	Support strategic engineering business goals	Meets customer needs	High return on investment (ROI)	Motivated team available	Data available	Meet budget constraint	
	Criteria and Weighting							
Level of importance	10	10	8	7	7	6	6	Total & Rates
Projects: Prioritized								
Standardized process for purchasing	9	9	3	9	3	1	9	348
Improve customer service	9	3	9	9	3	1	9	336
Improve on-time delivery	9	1	9	3	1	9	9	308
Reduce purchased labor and material costs	3	9	3	9	3	1	9	288
Supplier management	3	9	3	9	3	3	3	264
Improve employee training	1	3	3	9	9	1	3	214

Table 4.2. Preliminary project definition matrix

Project	Process owner	Sponsor	Black-belt (1x)	Green-belts (2x)	Problem statement	Key project metric and goal	Potential cost savings

4.3 MAKING THE OTHERS BUY IN AND SUPPORT FOR YOUR PROJECTS

In any organization, it's necessary to support your colleagues and to rely on them to execute work. Absolutely, by applying leadership principles and working as a team, it is possible to arrange your workload and encourage your colleagues to directly buy-in to your goals. Here are some of the basic principles for convincing the others to buy-in for your projects.

Convince them it's their idea and they're the owners of the project: It's difficult to convince anyone to support your project. It's easy for them to say no. What you need to do is make the goals and objectives clear by presenting metrics, ROI (return on investments), and make sure that your colleagues understand these key goals are what are best for all team members, and they'll positively support those objectives.

Keep in mind that it's the end result and objective that's important, not whose idea it was. If you are a strategic leader, allowing your colleagues take credit for their efforts now and then is part of that. The top management would recognize that your colleagues seem to have been developing a lot of great ideas since you joined the organization.

Talk Less and Listen More: Part of getting your colleagues to be interested in your project or idea is to be open to their ideas. In the course of time, it's all about the end results. If you can trust in your colleague's contributions toward your goals, then they must have some good ideas that will have an impact on your project. In any case, get the team, and your shared objectives before your own pride, and be proficient at the organization. Of course, this is the team that it has to go both ways by helping the others in exchange for their contributions. The main people who need to be on your team are the stakeholders. So, we need to define project stakeholders.

4.3.1 IDENTIFY PROJECT STAKEHOLDERS

Stakeholders are a person, a group, organization, or system who affect or get affected by a person or organization's action (such as Table 4.3). In other terms, an individual, group, or organization who has a direct or indirect "stake" (or vested interest) in the outcome of your project or corporate. Either you may lose or acquire something as a consequences of others' action. You may provide something they want or vice-versa. Commonly known stakeholders are the following:

Table 4.3. Lean Six Sigma project stakeholders analysis chart

Project stakeholder analysis										
Project Stakeholder/group	Relationship to project					Communication/Involvement strategy				
	Is effected by outcome	Can influence outcome	Has useful experience	Provides resources	Has decision authority	Meet with regularly	Invite to team meetings	Send copy of meeting minutes	Speak with informally as required	Others (describe)

   i.   Project champion
  ii.   Project sponsor
 iii.   Process owner
 iv.   Project team members
  v.   Individuals who contribute directly to the project or process

     Other stakeholders are individuals who are responsible for the system that supports the process, person who manages the process, and person who is directly involved with the process.

4.3.2 ANALYZE PROJECT STAKEHOLDERS

One should minimize the stakeholder list with key players you want to further analyze such as the stakeholder's belief that you don't know, higher stakeholder who will resist in supporting the project, and stakeholders that will support your project and will bring other contributors as well.

4.3.3 CREATE PROJECT STAKEHOLDER PLAN

One strategy is that to come up with a plan how to get stakeholders involved with the project and how to communicate the progress of the project to them. Table 4.3 is one sample that may help to brainstorm in planning.

4.4 SIX SIGMA TEAMING

Having excellent technical skills and the best technical solution is not enough to ensure successful completion of Lean Six Sigma projects. So what other tools we need to make it happen? Let's review some of the road blocks that may slow down the project progression.

4.4.1 BARRIERS TO A SIX SIGMA CULTURE

Barriers to Lean Six Sigma culture and progress come from a variety of sources.

    Management: Lack of leadership involvement, unclear direction, wanting immediate results, not tracking the right metrics, continued emphasis on the old way of doing business, and reluctant to support the new methods.

Time: Too busy firefighting, people are weary from long hours of work, improper time management, too many arguments without facts and data, too many unproductive meetings, timeline pressure, and too busy reorganizing.

Communication: Culture not right for sharing knowledge, conflicting message, information politics, tools for proper communicating not known or used, and inability to properly present something.

Training: Management last to be trained, lack of funding, training is not just in time (JIT), not enough emphasis on return on investment (ROI), people not trained properly, and lack of application.

Resources: Limited resources, insufficient internal experts to assist others, and improper allocation of resources.

Rewards: Promotes firefighting, people in groups against each other, no consequences for lack of implementing, and of course people not being held accountable for their job function.

Motivation: Rush to result without planning, don't perceive current condition as threatening, resistant to change, of course easier to blame others than to take responsibilities in other words easier to blamestorming than brainstorming , fear of failure using new methods, and unwilling to share knowledge with others.

4.4.2 WHY TEAM?

It's not as easy as it seems: Getting team members to show up for a project meeting, maintaining team momentum, and keeping the team focused on the project direction is a challenge. Further, getting data from the process owners, and gaining the cooperation and support of various stakeholders is a hard work and time-consuming.

Project can't be completed by an individual: One need to lead the project team and participate in project team's discussion. Keep in mind that you can't force the team to cooperate with you but can only request. Thus, you may need the influencing tools to get the others help your project (see Section 4.7).

The four stages of teaming are forming, storming, norming, and performing.

4.5 SIX SIGMA TEAMING: FORMING/STORMING/NORMING/ PERFORMING MODEL

Effective and efficient teamwork is a must in today's competitive global economy, but as you have been directing a team for some time, you can't expect them to perform at the exceptional level. As a strategic team leader, your goal is to direct your team reach and maintain high performance at the fastest time possible. As Warren Bennis said, managers are people who do things right, while leaders are people who do the right thing. To achieve this, you will need to change your approach at each phase and steps. The steps below (Figure 4.5) show how to make the right decision in guiding a team.

Continuously review the performance of your team and align the team with your direction that would place them in the phase they belong to become a leader. This is visualized in Figure 4.6.

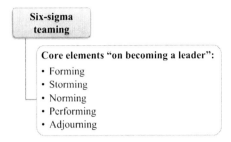

Figure 4.5. Core elements of a leader.

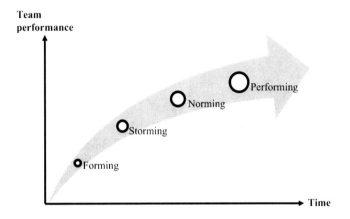

Figure 4.6. Core elements: Steps of becoming a leader.

4.5.1 FORMING OR ORIENTATION

This is relatively *introductory* easy phase. In this phase, team members are introduced. They are polite to each other and trying to figure out what is the team concept. Normally team is positive for most of the meetings. Team leader directs the team by developing a "team charter and tasks" (strategic objective, scope, etc.) and begins with what needs to be done effectively and efficiently to get the team to the performing stage. Such topics could include: tasks to be accomplished and a summary of concepts and issues. Further details related to forming are the following.

1. Understanding

When a team first forms, team members are like hesitant swimmers standing by the side of the pool and dabbling their toes in the water.

2. Teams Feelings

Team members are excited, anticipated, optimistic, and have pride in being selected for the team; there is also suspicion, fear, and anxiety about the job ahead.

3. Behaviors

Team members attempt to define the task and decide how it will be accomplished so they can determine acceptable behavior and find out how to get along with the team problem. They make decisions on what information is needed and come with an abstract discussions of concepts, issues, problems, or symptoms not relevant to task. It becomes difficult to identify relevant issues. They continue to complain about the corporate roadblocks to get the job done.

4. Leadership Roles

Leadership can help the team members get acquainted and get to know each other; guide them in clear path and goals; get them included in developing plans, focusing on becoming team players; and support the team with the knowledge that they need to begin.

4.5.2   STORMING OF DISSATISFACTION

In this phase, the team is transformed from "as is leader" to "to be true leader." This is a challenging phase. Team might clash with each other, disagree with the concepts of charter, develop an infrastructure process, remove all the ambiguity, and build strong relationship among team members. In the case of being challenged by team members to your charter goals, stay focused and positive. To avoid the conflicts you may explain the tooling of forming, storming, norming, and performing concept. So, the team members might understand the reasons for the possibility of conflict. Assertiveness as well as conflict resolution techniques are needed in this phase. At the end of this phase, team members should understand each other and start to work together. Some techniques may help to ease this phase are the following:

(a) Management need to train and coach the differences among the members
(b) One-on-one focused feedback may help to come to the same terms
(c) Possible challenges may include
   i. Resisting the concept and methodology
   ii. Defensiveness and competition among the members
   iii. Tension, disunity, even jealousy

Let's review understanding, feeling, and behavior of team as we did in the forming.

1. Understanding

As team members begin to learn the amount of work that lies ahead of them, it is normal for them to almost panic. Now they are like people who jumped into the pool and don't know how to swim, think they are about to sink, and their heart starts beating faster.

2. Team Feelings

Team is resisting tasks and techniques of work difference from what each individual member is comfortable using. Team shows sharp fluctuations in attitude about their chance of success.

3. Behaviors

Team members argue among themselves and even they agree on the real problems. They become defensive and create competition among themselves, even form factions and start choosing sides. They are concerned about the objectives and develop unrealistic goals, even question the knowledge of the individuals who came up with the project. Creation of disunity, jealousy, and tension is normal in this phase.

4. Leadership Roles

Leadership can assist in resolving issues of power and authority or establish an agreement plan about how decisions are being made and who makes them. Further providing a leadership role so they can become independent, i.e., let them to take additional responsibilities, helps.

4.5.3   NORMING OR RESOLUTION

The team members agree on "*to be leader*" process and has turned around from the "storming" phase. Creates a social and team building environment. It leaves the team alone so they will take responsibilities and work out the differences to agree on the agenda. Some may have to give up their own beliefs and honor the others in order to keep the team unity and its functionality. Characteristics of the norming are understanding, team feeling, behaviors, and leadership roles.

1. Understanding

As team members get along to play key roles in working as a team, their primary resistance fades away. They begin to cooperate with one another and stay unified rather than competing against each other.

2. Team Feelings

Norming consist feelings of

- (a) Agreeing to be a team player
- (b) Being trained to avoid conflict
- (c) Being friendly to the team members
- (d) Being able to have or express constructive feedback
- (e) A sense of team cohesion and a common objective

3. Behaviors

Almost everyone tries to avoid conflict by being friendly to each other and talking about individual personal problems or team dynamics. They learned how to express their objection with constructive feedback by respecting ground rules and boundaries.

4. Leadership Roles

Leadership can utilize team knowledge, skills, and experience to promote respect and honor among the team members.

4.5.4   PERFORMING OR PRODUCTION

The team members have come to a decision on their relationships and expectations. Once the team has achieved their high-performing skills, they will function as a group and find the ways to get the job done effectively and efficiently without having any outside supervision. By now, the team is self-motivated and educated in decision making process. Team leader (or coach) can focus on the other goals and objectives.

   (a) Members have better understanding of each other's strength and weaknesses.
   (b) Team members are united due to completion of projects or tasks.
   (c) Individuals complete development of leadership thinking.

1. Understanding

As team members have good feelings to one another and becom team players or realize the expectations, then they perform more effective, efficient, and united with each other.

2. Team Feeling

Team members understand each other's strengths and weaknesses and gain satisfaction as they progress.

3. Behaviors (Constructive Self-Change)

They are now capable of working through their group issues.

4. Leadership Roles

Leadership should provide team how to manage the change, should monitor work progress and recognize the progress, and update team's procedures and rules to support the team work.

4.5.5 ADJOURNING

The team shares the process improvements with other members. In this phase team breaking up after the task has been achieved. The team briefs and shares the improvements during this phase. The relationship formed among the team will continue long after the team dismissed.

4.6 CONFLICT MANAGEMENT: THE FIVE CONFLICT HANDLING MODES

Conflict involves any place of activity where your concerns or interests differ from another person's. The conflict environments are manpower resources such as equipment and facilities, capital expenditure, costs, technical options and trade-offs, priorities, administration procedures, scheduling, responsibilities, and personality clashes.

4.6.1 AVOIDING

This is an unassertive and uncooperative mode. Try to avoid the conflict, concerns and opinions are not shared by stepping aside. Sometimes just listening might be a good thing to cool down the emotions.

4.6.2 ACCOMMODATING

Accommodation is essential when you want to get a quick solution, keep the harmony, relationship, and basically to gain temporary agreement. Thus, you need to focus on relationship and even give up your personal goals to please the interest of others.

4.6.3 COLLABORATING

Collaborating is opposite of avoiding. Thus, attempt to be assertive and cooperative to work with the other person to come up an agreeable resolution. As a team you diagnose the problem and find the cause to improve the effect. It will even help further if you understand the "needs" and "wants" of each other in the product or process advancement.

4.6.4 COMPETING OR CONTROLLING

This mode is recognized when someone who follows their beliefs at somebody else's cost, using any means to gain other individual's case. It is very assertive and uncooperative; basically it is all about the controlling and competing. Even though this mode might be considered as a negative, sometimes it is the most effective way of resolving a conflict.

4.6.5 COMPROMISING

This is a give-and-take mode as pictured in Figure 4.7. So it is in the middle of assertiveness and cooperativeness. You find an overall common ground that brings both of you to an agreement. Basically, you withdraw most of your requests when you are in the competitive mode but much less when you are in accommodating mode. You talk about the issues more openly and directly rather than avoiding without paying attention to its detail and analysis like you do when collaborating.

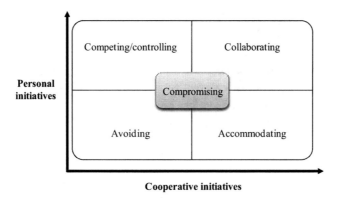

Figure 4.7. Five conflict handling modes.

4.7 CONFLICT RESOLUTION

Your conflict mode is equal to your skill and the place of activity.

4.7.1 EFFECTIVE CONFLICT RESOLUTION BEHAVIORS

The key elements are

(a) Effective communication
(b) Patience and composure
(c) Sense of timing and rhythm
(d) Being specific about your "needs" and "wants"
(e) Behaviors that indicate that creative solutions can be achieved
(f) Empathy with regard to the discomfort of the other party
(g) Constructive dialog and feedbacks
(h) Help the other party understand your intent and importance of your situation

4.7.2 KEY CONFLICT POINTS TO CONSIDER

The issues, as initially presented, are oftentimes not the real issues and the disputants are often blind to the systemic influences of a conflict. Sometimes, very few individuals see conflict as healthy. Another point is that there is a bias to label conflict as simply a matter of style. We all know that it is almost impossible to solve conflicts in the past.

4.7.3 CONFLICT AND POWER

It is interesting all conflicts are struggles for power. Knowing that it is not the only or even the highest force but it is always there. First question should be, "What is it that each party wants, but doesn't have"? Conflict is always about wanting something.

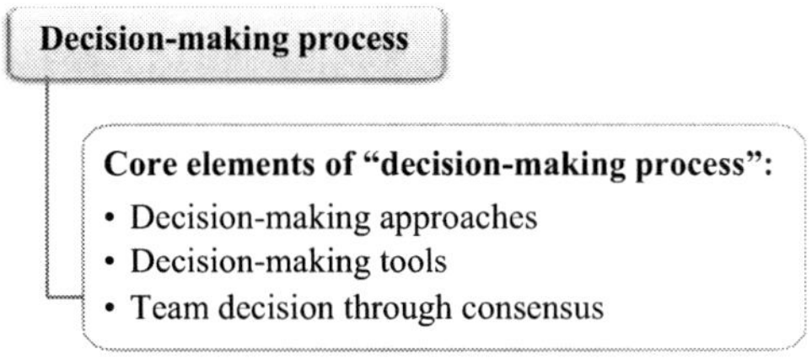

Figure 4.8. Core elements of decision-making process.

4.8 LEADERSHIP DECISION-MAKING PROCESSES AND TOOLS

Effective decision making is a necessary and most important leadership skill. Often it can be difficult and nerve breaking. It is all about time. If you can train yourself to make a decision on time with a well-considered judgment then you can lead your team to success. In other terms, the poor decisions will represent that your time as a manager were short. Therefore decision-making is a balance between time available and amount of buy-in needed of people who are affected by the decision. Oftentimes there is a conflict or give-and-take to include everybody in decision making versus the time needed to make a concrete decision. This process is shown in Figure 4.8.

4.8.1 DECISION-MAKING APPROACHES

Unilateral decision: One individual alone makes the decision. The decision should be made fast and the individual who makes that call has the right expertise to make highly acceptable and thorough decision. In this case there is very low need for the opinion of others to accept the decision.

Consultative decision: One individual alone makes the decision after consulting or brainstorming with others. It makes sense to request others' opinion due to the fact that their acceptance for the decision is required. So, the right people with the specific expertise who will provide the highly qualified decision are being advised.

Group decision: All the team members are included in the decision making. There is ample time to get everyone's opinion, particularly the right individuals who have the expertise to make a high-quality decision since their acceptance of the decision is needed.

4.8.2 DECISION-MAKING TOOLS

1. Voting

The team decides to make a decision through a voting system, so the majority rules. Again, the decision needs to be made fast. The team members are sure that people won't feel animosity and get angry at.

2. Polling

The team members vote their choices but in spite of the voting still no decision being made. So, first you need to determine how far off the team members are from understanding and complying with the decision. Second, check for achievable agreement without any opposition to the decision.

3. Multi-Voting

Team members are given permission to vote on multiple concepts. Basically, you are putting efforts to reduce number of the promising or achievable solutions and convincing the team that their ideas are part of the solution without breaking the team unity.

4. Consensus

The problem is discussed in detail among the members of the team taking in each member's input so they can have a chance to influence on the other. The decision is modified with everyone's input. All the members wait until they come up with a decision that they all support it. These are the factors a consensus decision looks for:

   i. The problem is highly essential.
  ii. Agreement on the decision a must.
 iii. There should be ample time for review and discussion.
  iv. All the team members view point should be included in the solution to be decided.

The Lean Six Sigma has some of the most efficient and effective decision-making tools that can be utilized in presenting the data-driven points and analysis. Details of the following tools are described in Chapters 7–9.

1. *Pareto analysis*: It is also called as "the 80/20 rule." Twenty percent of the cause affects 80% of the process.
2. *SWOT analysis*: SWOT analysis is a business and engineering strategic planning technique to determine the Strengths, Weaknesses, Opportunities, and Threads associated with a decision making of a project, process, or any critical plan of action. Any leadership decision should be made after SWOT analysis has been performed. This would provide assuring goals and objectives to be carried out for the organization.
3. *Force field analysis*: It assists leaders to make a decision by analyzing the weight of forces for and opposite of changes and it proposes the proof in supporting your decision making. This is the principal dynamic force that energizes people in achieving organizational goals.
4. *Statistical analysis*: Statistical analysis will help you to remove any source of errors in the risky decision-making problems or objectives. It will utilize the past and current history analysis in decision making of the future forecasting and processes or strategic planning.

Figure 4.9 represents the decision making with qualitative and quantitative tools.

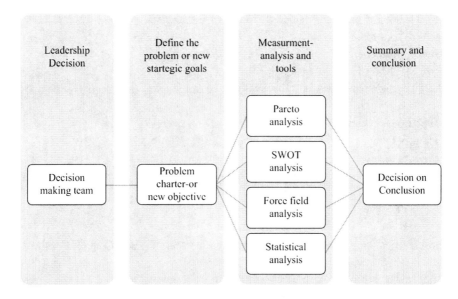

Figure 4.9. Decision-making tree.

4.8.3   TEAM DECISION THROUGH CONSENSUS

Key principle: If used properly, consensus is a decision-making method that results in the highest quality decision so that all the team members agree on it. However, if used improperly, time would be wasted without any achievement, which results in disappointment.

Hints in reaching consensus: Team members should share their concept, ask questions, as well as sell their ideas. In addition, they need to include all the core issues, concerns, win/win opportunities, strong points, weaknesses, and threads. Further, team needs to listen, be open minded, not dominate in their discussion, and be open for continuous discussion if needed.

4.9   PROJECT AND PROCESS ASSESSMENT MATRIX

A project assessment matrix must be designed to analyze the risk factors in question before the work begins. The financial, resource, and workforce impact must be evaluated as well. Project assessment assists you in understanding the progress, success, efficiency, and effectiveness of a project. You can use the outcome of the project assessment matrix to do the following:

(a) Recognize ways to analyze and improve project plan of action.
(b) Determine the end-to-end time boundaries.
(c) Learn more about the SWOT analysis of organization.
(d) Keeping track of project while in progress normally called process assessment.
(e) Process assessment lets you to identify the challenges at early stages and make course actions:
  • How accurately the project is implemented according to strategic plan
  • What barriers are experienced

(f) End results of project will enable you and shareholders to assess whether your project worked or didn't work

(g) Prepare project reports

The following steps allow you to assess matrix and identify the improvement factors:

1. state project charter;
2. identify the measurement and risk factors (apply affinity);
3. identify the assessment tools;
4. develop assessment improvement, risk analysis, and other factors that may prevent the project from being successful; and
5. begin the plan of action.

Different members of the team can assess the project matrices (Tables 4.4 and 4.5) and record their input.

Table 4.5 shows project prioritization assessment based on the ratings.

The matrix can be summarized in a project assessment report.

Project Assessment Template

Names of team members: _____

Project charter: _____

Type of project-in details: _____

Cost of completion of the project (use the matrix to estimate the completion of the project):

_____

Approximate date of completion: _____ dd/mm/yy

Approximate date of review: _____ dd/mm/yy

Source of financial support:

4.10 SIX SIGMA FINANCIAL REPORTING (USING FINANCIAL MEASUREMENT TO ANALYZE PERFORMANCE)

Lean Six Sigma financial reporting and measurability are essential to present the bottom line on the profit and loss of a project in the engineering business unit, production, manufacturing, and entire company projects. Any successful organization is required to define the rules and guidelines for measurability, accuracy, accountability, integrity, and consistency.

4.10.1 PLAN OF ACTION

1. Financial statement
2. Actual savings versus budgeted
3. Standard rules and procedures for Six Sigma financial performance

Table 4.4. Project assessment and measurement matrix

	A	**B**	**C**	**D**
Ideas	Useful idea	Poor statement	Well thoughts	Potential idea
Analysis tools	Well developed	Not utilized	Used all tools	Well analyzed
Financial	Analytically performed	Need improvement	Some factors missing	Optimistic values
Performances	Well performed	Clarity lacks	Fair performance	Well prepared
Achievements	Excellent work	Achieved little	Fair results	Quality work

Table 4.5. Project assessment matrix (Rating are based on high = 9; medium = 3; low = 1)

Process outputs	Complete within frame time	Support strategic engineering business goals	Meets customer needs	High return on investment (ROI)	Motivated team available	Data available	Meet budget constraint	
				Criteria and Weighting				
Level of importance	10	10	8	7	7	6	6	Total & Rates
Projects: Prioritized								
Standardized process for purchasing	9 = (10)(9) = 90	9 = (10)(9) = 90	3 = (8)(3) = 24	9 = (7)(9) = 63	3 = (7)(3) = 21	1 = (6)(1) = 6	9 = (6)(9) = 54	348
Improve customer service	9	3	9	9	3	1	9	336
Improve on-time delivery	9	1	9	3	1	9	9	308
Reduce purchased labor & material costs	3	9	3	9	3	1	9	288
Supplier management	3	9	3	9	3	3	3	264
Improve employee training	1	3	3	9	9	1	3	214

4. Prompt way of reporting, auditing financial, and validating savings
5. Rules and accountabilities for all process owners
6. Reporting timelines

4.10.2 FINANCIAL ACCOUNTABILITIES

1. Determine the savings created by the Six Sigma project
2. Find out if the savings are based on hard savings or soft savings
3. Decide how to report the balance sheet reduction, normally not hard savings
4. Be able to audit the financial benefits
5. Confirm that higher management agreed with the savings
6. Reflect the financial in the future budgeting
7. Demonstrate the return on investment (ROI)

PART III

LEADING THE EFFORT

CHAPTER 5

LEADING AND ENGINEERING MULTIPLE LEAN SIX SIGMA PROJECTS

Leading and engineering multiple successful projects require proper planning, execution, verification, financial validation, and approval before advancing to the next phase. Review of Lean Six Sigma project milestones and measurement of Key Performance Indicators (KPIs) are essential for the success of a project. If the key players or the project owner does not apply Lean Six Sigma efficiently and intelligently, the methodology will not be effective. Effective leaders ensure that the project is progressing as planned. Thus, all the key elements of leading a Lean Six Sigma effort are discussed in this Chapter.

5.1 MANAGING MULTIPLE PROJECT AND PROJECT REVIEWS

5.1.1 PROJECT MANAGEMENT AND REVIEWS

Lean Six Sigma is about efficient and effective project management, involving the project selection, planning, and management reviews. Management reviews are essential for success of a project and default in management reviews considerably reduces the impact of Lean Six Sigma effort.

5.1.2 WHY REVIEW?

Evaluating the progress report and accomplishment of the project according to the plan and assessing the changes on the implementation system are needed. Likewise, prioritizing the tasks based on the changes to the strategic planning as well as planning for successful achievement and system for managing the changes are crucial. Environmental changes and its consequences may also modify the strategies on the continuation of the project or process improvement.

5.1.3 HOLDING REVIEWS

Do not be flexible when reviewing actions, responsibilities, obligation, and make it a hard rule that commitments must be met. If for whatever reasons the strategies and planning are not giving results, better change them now than later. And utilize the review meetings to come up with new concepts to implement the strategy. Table 5.1 is a sample template which presents overall single project scorecard and status review.

5.1.4 LEAN SIX SIGMA BLACK BELTS: THE CRITERIA FOR SELECTION

Black Belts are Lean Six Sigma team leaders accountable for deployment of process improvement projects DAIMAC (Define, Measure, Analyze, Improve, and Control) or DFLSS (Design, for Lean Six Sigma) within the business and engineering or manufacturing operations. The goal is to maximize customer satisfaction levels and product or service productivity. Black Belts are experts and skilled in the application of the Lean Six Sigma techniques and tools.

Black Belts typically complete up to four weeks or 160 hours of Lean Six Sigma training and have demonstrated mastery of the subject matter through the completion of projects.

Customer advocacy: Understanding customer "needs" (Critical-to-Quality) and "wants" is the key to process improvement. Thus, a Black Belt candidate must speak clearly about how eliminating process variation or defect is a key to manufacturing or business improvement.

Passion: Black Belts must be self-motivated, positive thinker, display initiative, and affection to solve the unsolvable problems. From time to time they are expected to be a cheerleader, to pick the team and guide them as well as motivate them to move forward and increase their productivity.

Change leadership: Black Belts have shown performance as a change agent in the past, regardless of their job functions. Changing the corporate culture and how business is accomplished may agitate the employees and create roadblocks among them; change executers and change leaders have a way of achieving positive and constructive change while developing and establishing support for the change.

Communication: Black Belts are effective communicators, which is necessary for the many roles they play: as trainers, coaches, or mentors; Black Belts must have self-confidence and speak to all kinds of audiences. Accepting the identified various "needs" of audience and tailoring the content of the lecture to express their concerns is the sign of an effective communicator.

Project management: Lean Six Sigma is achieved one project at a time. We must not lose the view of the fact that Black Belts must manage projects from scope, requirements, timeline, resources, and various perspectives. Understanding the science of project engineering management fundamentals and experience in managing projects are necessary.

Technical aptitude: Black Belts do not have to be an engineering or statistical graduate, but in some cases this qualification becomes important and handy. However, in all cases a Black Belt is required to define, collect, and analyze data for establishing a strategic continuous improvement.

Team player and leader: Black Belts must possess the capability to lead, manage, work with teams, be a team player, and understand team dynamics (forming, storming, norming, and performing). In order to effectively and efficiently lead a team, a Black Belt must be friendly, get along with people, have good influence skills, and positively motivate the other members.

Table 5.1. Holding reviews: single project scorecard and status review

	Lean Six Sigma single project scorecard status review				
Define	Measure	Analyze	Improve	Control	Cumulative percent complete (%)
Process owner	Develop existing process flow	Validate project scope and process flow	Identify/prioritize root causes	Develop advanced process control charting concepts	10
Project charter	Implement Kaizen and 5S housekeeping	Identify cause and effect	Screen potential causes/ Design of Experiment	Develop and implement plan	20
Develop stakeholder analysis	Identify areas of the 7 wastes	Cause-and-effect diagram	Taguchi/full factorial design	Issue revised procedures	30
Project selection/ prioritization (QFD)	Develop data collection plan/gage R&R and Cpk study	Analysis of mean/ Box plot of x and or y variable	Identify and prioritize improvement actions	Revalidate cause and effect of X's to achieve Y's	40
Establish CTQ/SIPOC	Measurement system analysis	Multivariate analysis	Develop implementation plan	Implement mistake proofing-Poka-Yoke	50
Project metrics/baseline performance	Plot output (y) data over time (box plot & time series)	Histogram of input (x) variables	Cost–benefit analysis/ quality and delivery analysis	Monitor process performance (Control chart)	60
Establish goals (% of actual)	Pareto high-impact cause/ process input	Regression analysis/ scatter diagram/ correlation	Risk assessment/failure mode effect analysis	Develop before/ after summary table of performance improvements	70

(Continued)

Table 5.1 (*Continued*)

Lean Six Sigma single project scorecard status review					
Define	**Measure**	**Analyze**	**Improve**	**Control**	**Cumulative percent complete (%)**
Develop value stream map	Histogram of existing data/apply descriptive statistics	Scale and probability analysis: discrete and continuous	Establish operating tolerance for input variables	Identify opportunities to leverage improvements	80
Develop value-added analysis	Establish process capability of existing data	Hypothesis testing concept and Chi square	Develop new process flow (development flowchart)	Process and Capability study (C_p, C_{pk}, P_p, P_{pk})	90
Potential project cost savings	Determine sigma level of existing process	Analysis of variance (ANOVA)	Porter's five forces	Risk analysis	100
Review with sponsor	Review with sponsor	Review with sponsor	Review with sponsor	Review with sponsor	
		Overall Project Completion Percentage%:			
Sponsor signature	**Sponsor signature**	**Sponsor signature**	**Sponsor signature**	**Sponsor signature**	

Results oriented: Black Belts are expected to highly perform and produce tangible financial gains for the business operating in a corporate environment. They must be hard working and demonstrate success in a short period of time.

Table 5.2 demonstrates the template for multiple Six Sigma projects progress review scorecard.

Table 5.3 introduces Black Belt and Green Belt Six Sigma project scorecard template.

Table 5.2 Holding reviews: multiple project scorecard and status review

	Project #1 status	Project # 2 status	Project #3 status
Lean Six Sigma multiple project scorecard status review			
Define phase			
Identify process owner			
Establish project charter			
Develop stakeholder analysis			
Project selection/prioritization (QFD)			
Establish CTQ/SIPOC			
Project metrics/baseline performance			
Establish goals (percentage of actual)			
Develop value stream map			
Develop value added analysis			
Potential project cost savings			
Review with sponsor			
Measure phase			
Develop existing process flow			
Implement Kaizen an 5S housekeeping			
Identify areas of the seven wastes			
Develop data collection plan/gage C_{pk} study			
Measurement system analysis			
Plot output (y) data over time (box plot & time series)			
Pareto high impact cause/process input			
Histogram of existing data/apply descriptive statistics			
Establish process capability of existing data			
Determine sigma level of existing process			
Review with sponsor			

(Continued)

Table 5.2 (*Continued*)

	Project #1 status	Project # 2 status	Project #3 status
Lean Six Sigma multiple project scorecard status review			
Absolute Score			
Possible Points			
Percent on Schedule			
Analyze phase			
Validate project scope and process flow			
Identify cause and effect			
Cause and effect diagram			
Analysis of mean/box plot of x and or y variable			
Multivariate analysis			
Histogram of input (x) variables			
Regression analysis/scatter diagram/correlation			
Scale and probability analysis: discrete and continuous			
Hypothesis testing concept and Chi square			
Analysis of variance (ANOVA)			
Review with sponsor			
Improve phase			
Identify/prioritize root causes			
Screen potential causes/Design of Experiment			
Taguchi/full factorial design			
Identify and prioritize improvement actions			
Develop implementation plan			
Cost–benefit analysis/quality and delivery analysis			
Risk assessment/failure mode effect analysis			
Establish operating tolerance for input variables			
Develop new process flow (development flowchart)			
Porter's five forces			
Review with sponsor			
Absolute Score			
Possible Points			
Percent on Schedule			

(*Continued*)

Table 5.2 (*Continued*)

	Project #1 status	Project # 2 status	Project #3 status
Lean Six Sigma multiple project scorecard status review			
Control			
Develop advanced process control charting concepts			
Develop and implement plan			
Issue revised procedures			
Revalidate cause and effect of x's to achieve y's			
Implement mistake proofing-Poka-Yoke			
Monitor process performance (Run chart/Control chart/s chart)			
Develop before/after summary Table of performance improvements			
Identify opportunities to leverage improvements			
Process and Capability study (C_p, C_{pk}, P_p, P_{pk})			
Risk analysis			
Review with sponsor			
Absolute Score (i.e., 48, 42, 35)	48	42	35
Possible Points (i.e., 48, 48, 42)	48	48	42
Percent on Schedule	100.00%	87.5%	83.33%

Table 5.3. Lean Six Sigma project scorecard template

Scoring: Not started = 0 Started but not complete = 1 Complete = 2	Project black belt/Green belt			
	Black belt #1	Black belt #2	Green belt #1	Green belt #2
Control				
Develop process control plan	1	1	2	1
Implementing control plan	1	1	1	1
Issue revised procedures	1	2	2	0
Revalidate cause and effect of X's to achieve Y's	2	2	1	2
Implement mistake proofing	1	2	1	1
Monitor process performance (run chart/control chart)	1	2	1	1

(*Continued*)

Table 5.3. (*Continued*)

Scoring: Not started = 0 Started but not complete = 1 Complete = 2	Project black belt/Green belt			
	Black belt #1	**Black belt #2**	**Green belt #1**	**Green belt #2**
Develop before/after summary table of performance/metrics	2	2	2	1
Identify opportunities to leverage improvements	2	1	1	1
Hand-off to process owner	2	2	2	2
Review with sponsor	2	2	2	2
Total Score	15	18	14	12
Strengths: Kanban, Inventory process, Structure approval	Identification	Analysis	Commitment Determination	Measurement of process data
Opportunities: Mistake proofing	Improve/ more analysis/control area	Organization	Process analysis	Organization clear path/ emphasis on more analysis

5.2 HOW TO MASTER THE SKILLS OF LEAN SIX SIGMA FACILITATION

5.2.1 HOW TO BECOME AN EFFECTIVE FACILITATOR

Being an effective facilitator means getting educated, being capable, and trained to become a skillful individual who has a different approach that can distinguish you from other organizers or leaders. The goal is to get the team to focus on self-development and continual learning. To be a good organizer in Lean Six Sigma and achieve tangible results, you must be master in facilitation. One of the most powerful skill tools for an effective facilitator is communication skills.

Communication skills set: These skills count: communicating the past, current, and next steps of the process or project under the study, specified duration of meetings, the team activity, reduction of side-bar submeetings, listening, and being open for questions and responding to them. Other factors are decision making, team accountability, preview and review strategy, commitments, and conflicts as discussed in Chapter 4.

5.2.2 STRATEGIC ROLES OF THE FACILITATOR IN THE ORGANIZATION

As a team leader: Any team without a direction will produce no results. Effective team must have a direction from the facilitator or its leader. A good facilitator should tell what is the team's direction and expectation. Based on the record, they also need to know even where they are at. Some of the following items that should be in the review.

  i. *Task*—Analysis and plan of action.
 ii. *Process*—The process flow that will take the team to their goal.
iii. *Time bound*—The end-to-end timeline to complete the project.
 iv. *Roles and accountability*—Roles of team members and their responsibilities to be efficient and productive. In fact, even team should know where they have been. As a master Black Belt and facilitator, you need to motivate the team to plan their objective, teach them to concentrate on real-time project, and have them to discuss ideas.

5.2.3 EFFECTIVE ELEMENTS OF COMMUNICATION STRATEGIES AND SKILLS

Everyone involved in communication should want to and develop their talent to be proficient in listening, observing, and note taking. These are some of the basic skills for improving the communication. When you are getting to know and work with different stakeholders, teams, or process owners, listening skills are among the most important of all skills you need. As a listener and observer, it is the first step toward strengthening your ability in communication. Listening and observation skills are necessary throughout the projects and during any communication they will help you in monitoring the process.

5.2.3.1 Listening, Paraphrasing, and Questioning

As you listen to the words of a speaker, pay your undivided attention and recognize the speaker's nonverbal communication. Put aside all the external and internal distraction and look at the speaker directly. You can show your own body language that you are interested in speaker's point of view. You may also provide feedback by paraphrasing "what I heard you are saying...," or "so you mean that ..." and "as I understand it...." Asking effective questions are an essential skill of facilitator. Be candid, open, and honest in your questions. Express your thoughts and opinions respectfully. When asking questions be prepared for closed questions (requires only one-word answers) and open-ended questions that need more than yes or no answer.

5.2.3.2 Observation

In any research, teaching, training, coaching, lecturing, and community work, well-established skills are necessary. Observation is important if you are facilitating a team, or if you are liable for taking notes on any process or activities. In this case, it is essential to document the action and any observation on how such action might be defined. Keep in mind that majority of the communication is nonverbal. Particularly when you are communicating with people of variety cultural experiences where the language is an issue, interpreting observation is the key. The type of behavior you may observe can be yawning, smiling, nodding, frowning, bored, and moving hand or feet.

In a group discussion, training, teaching, multistakeholder meetings, and observation would focus on who is participating, who is leading the discussion, and who is quite. This makes the facilitator to get people who are quite involved with the discussion activities.

5.2.3.3 Effective Note Taking

Listening and note taking is an essential skill of a facilitator or a leader. Listening requires you not only hear the speaker but also to understand as well. Comprehending requires three types of actions: dynamic listening, paying attention, and focusing.

5.2.4 TIME YOUR TIME FROM TIME TO TIME

In other words, manage your time. Lean Six Sigma projects time metrics are as important as project itself. So, let's review this.

What really is time management? There are so many definitions. In the Six Sigma context, it is the process of planning and control over the amount of the time spent on specific activities, particularly to maximize the efficiency, productivity, and effectiveness of the process.

How do you measure the time? By seconds, minutes, hours, days, weeks, months, years, or light (and distance) years. Time has different values in various areas but they all have their own special values in achieving an objective. An objective requires a limited time frame that has a start and an end. For instance, time on the operating table depends on the type of surgery and postoperation recovery process. In sports, in basketball, running, or swimming, even one-tenth of a second makes the difference. Other examples are time to travel, time to spend with family, and time to complete a project in a given period of time.

Why time is so important? Because there isn't enough time to begin with and passes by so fast. You can't slow it down. Once you have tackled the concept of time management, you will find it possible to get more done. To manage time (future time), the current and past time is required to be evaluated. This is also called time auditing.

5.2.5 BUILDING TEAM COMMITMENT AND INTERACTIONS

Team building is an idea of activities designed to advance team performance. Normally it relates to organizational development, sports teams, and school activities. It concentrates on bonding and making the team stronger and coherent to bring out the best in a team and assure development, leadership, communication, team dynamics (relationships, problem solving, and leadership), and capability of working together toward the same goals and objectives in producing the best results.

5.3 COMMUNICATION PLANNING

5.3.1 SIX SIGMA PROJECT COMMUNICATION

When embarking on a project, Six Sigma Black Belts and project champions need to be aware of the importance of establishing a communication plan during developing and validating a team charter. A carefully executed Black Belt project can suffer disappointing results if an efficient mechanism is not already in place to ensure that vital information relayed to those members who need it.

Typically, team charters include such deliverables such as a business case, problem and goal statements, scope, milestone, and roles. What should be added, perhaps in the team charter or as a separate define phase deliverable, is a plan or strategy for communicating information that is related to Six Sigma project to its appropriate recipients.

At a minimum, Black Belts should give a thought on how the proceedings of team meetings and project work will be communicated, so that others in the organization who are on a need-to-know basis will be assured that they remain in the loop. A simple table could be constructed that would display what will be communicated, who will do the communicating, when the communication will take place, to whom the communication will be delivered, how the communication will be delivered, and finally where the information will be stored.

5.3.2 COMMUNICATION PLAN CONSIDERATIONS

Who—The person who is accountable for delivering the communication, e.g., project champion, Master Black Belt, Black Belt, Green Belt, quality analyst, process owner, team member, and so on.

What—The type of communication (e.g., team meeting, minutes/meeting action items, project status report, end-to-end timeline, gate reviews, success story, story committee, etc.)

Why—A course of action for communication plan, such as to organize and enforce a contract for communication.

Where—A place where the beneficiary will find the communication, if decided definitively.

When—The time at which the communication is conveyed, for instance, every Monday at 10:00 AM, daily, monthly, within 48 hours, or every other day.

How—The delivery mechanism that will facilitate the communication, such as voice mail, electronic mail, video conference call, and video presentation.

To Whom—The beneficiary of the communication, that is, who will be communicated like leadership team, higher management, project sponsors, champion, and team members.

Table 5.4 is a sample template for communication tool that could be implemented in a Lean Six Sigma project that addresses all of the considerations mentioned above.

Normally, team charters consist of such deliverables as engineering or business case, issues and objectives, scope, milestones, and roles. What should be added perhaps in the team charter or as a separate define phase deliverables is a strategic plan for communicating information that is connected to Lean Six Sigma projects to its true beneficiaries.

The team members who need information the most are generally subject matter experts who are not part of the project, but whose knowledge on specific factors of a process, or a subprocess, is valuable to the team or to the organization. These experts are required to be informed as a project progresses and when key meetings are to take place. They should be informed and they can take part in activity of the discussions when important business or engineering decisions are being made. A well-accepted and understood communication plan helps to ensure that promising contributors are not left out of the loop.

Table 5.4 Project charter communication plan

	Lean Six Sigma project charter communication plan					
What	**To Whom**	**When**	**Who**	**How**	**Where**	**Comments**
Project team meetings	Project team, Invitees	Weekly (every Wednesday at 9 AM)	Black Belt	Notices, agendas sent out 1 week ahead via e-calendar	Six Sigma conference room	
Meeting minutes	Distribution list	By next day COB	Black Belt or team scribe	Via e-mail	MS word file on shared drive	
Team work/action items	Project team, champion	TBA	Black belt	Via e-mail	Nature of file TBD, placed on shared drive	
Status reports, including timeline	Project team, champion, customer/client	Weekly (every Friday at COB) TBD	Black belt	Via e-mail	MS word file on shared drive, e-mail to customer representative	
Project budget	Champion, project financial analyst, quality department head	TBD (To be decided)	Black Belt or project financial analyst	Via e-mail	MS Excel file on Six Sigma database	
Project reviews	Project team, champion, quality dept. head	TBD (monthly)	Black Belt	Notices sent out 1 week ahead via e-calendar	Six Sigma conference room	
Project storyline	Deployment champion, quality dept. head, senior management	TBD	Black Belt or team members	Gallery walk notices sent out 2 weeks prior	Six Sigma gallery room	

5.4   PROJECT CLOSURE

Review the lessons learned from the process of project.

    i. About your results
   ii. About the work process
  iii. About the team's process
  iv. Share across Lean Six Sigma network

When you finish your project, it should contain your final results and conclusions. Then, present the completed presentation to the guidance team, the people whose jobs are changing as a result of the work, customers of the change, and other interested people. Collect, catalog, and make the documentation available to others. Have your team discuss the following issues and compile recommendations that you will submit to your sponsor or the guidance team.

1. Are there opportunities for replicating in other areas?
2. What are your recommendations for sustaining the gains put in place?
3. What roles would you like your team to play?
4. How much improvement is still needed to achieve the business, engineering, or manufacturing goals?
5. What portions of the problem are still left to address?
6. Which of the remaining problems are the most urgent to address?
7. What would you and your team like to work on next, if approved by management?
8. Where do you think management should devote resources next?

Communicating the team's results is a joint task for the project team and its sponsor or guidance team. Communicate results and closure to the people who were involved. If not done already, identify the people who will be involved in implementing the improved methods.

1. Which employees could benefit from your lessons learned?
2. How can you communicate to management and to the rest of the organization?
3. How can the end of this project sow the seeds for future projects?

5.5   LEAN SIX SIGMA MASTER BLACK BELT DEPLOYMENT PLAN

This Lean Six Sigma deployment and implementation plan builds on current efforts and presents a breakthrough steps forward by aligning the organization to a culture of continuous process improvement with a Six Sigma standards, data-driven disciplined approach that will achieve efficient and effective results. So, here are the contents of deployment plan.

1. Strategy and goals for Lean Six Sigma
2. Performance metrics (overall program)
3. Project selection criteria
4. Project identification and prioritization
   (a) Import completed project prioritization matrix

Table 5.5. Lean Six Sigma: Typical problem and its recovery (control phase)

Problem/Symptom	Probable cause	Recovery

5. Organization structure and roles
   (a) Summarize how the organization will support the effort
   (b) Who are the players in the key roles and any changes in the organizational structure?
   (c) Import completed project definition matrix
6. Training requirements
   (a) Summarize training requirements for leadership, "belts," and others
7. Management review process
   (a) Frequency and attendees for the management review process
8. Communication plan
   (a) Develop communication plan from the "who," "what," "why," etc.
9. Rewards and recognition system
   (a) Summarize how project participants will be recognized for successful completion and any proposed changes to the reward system

Table 5.5 is the template for a typical problem in Lean Six Sigma and its recovery phase.

5.6 CASE STUDY: LEAN SIX SIGMA DEPLOYMENT PLAN

Organization: SXTX Corporation
  Lean Six Sigma Master: ST

5.6.1 STRATEGY AND GOALS FOR SIX SIGMA

Engineer a strategy, technique, and an infrastructure that would enable all the managers to work as a team in improving the performance of the SXTX Corporation in achieving a higher market share through customer satisfaction. Further create an environment that will gain shareholder support in achieving these goals.

Strategy: List your strategy of deployment

(a) Apply Six Sigma in manufacturing and engineering and then continue on in the administration and transactional processes.
(b) Pick top talent in the company for Green or Black Belts to be trained.
(c) Select top-performing Black Belts for Master Black Belt training.
(d) Start training Black Belts and Green Belts for the next 6–9 months.

Goals for Six Sigma: Set the Six Sigma financial goals

(a) Five million dollars in productivity improvements, achieving customer satisfaction, and adding bottom line.

5.6.2 PERFORMANCE METRICS (OVERALL PROGRAM)

Metrics for financial gain are based on the following criteria:

(a) Scrap reduction
(b) Cost reduction
(c) Capacity improvement
(d) Downtime reduction

5.6.3 PROJECT SELECTION CRITERIA

All the projects are based on end-to-end timeline within maximum 6–9 months

1. Material 1: Cost reduction project
   (a) Areas of improvement: properties, process ability, and strength
   (b) Effect on customer satisfaction: a better and improved functionality
   (c) Effect on the bottomline: more than $300 K per year
2. Material 2: Cost reduction
   (a) Areas of improvement: product properties and strength
   (b) Effect on customer satisfaction and product longer lifetime reliability
   (c) Effect on the financial bottomline more than $350 K per year
3. Molded parts scrap and defects per million reduction
   (a) Areas of improvement are process, quality, and reduction in defect rate of assembled products
   (b) Effect on customer satisfaction and reduction on warranty return
   (c) Effect on the financial bottom line more than $600 K per year
4. Product 1: capacity improvement, cycle reduction, mold, and machine justification
   (a) Areas of improvement are capacity, productivity, and machine downtime
   (b) Effect on customer satisfaction are quality and longer lifetime
   (c) Effect on the financial bottom-line is more than $250 K per year
5. Product 2: downtime reduction and mold/machine justification
   (a) Areas of improvement are reduction in downtime, quality, and properties.
   (b) Effect on customer satisfaction are improved product and reduction in product warranty returns
   (c) Effect on the bottom-line is more than $200 K per year

5.6.4 PROJECT IDENTIFICATION/PRIORITIZATION

Table 5.6 prioritizes the project by their importance and possibility.
    Table 5.7 illustrates another example of Table 5.6.

5.6.5 ORGANIZATION STRUCTURE/ROLES

Table 5.8 prioritizes the project by their potential cost savings.

Table 5.6. Project identification and prioritization

Process outputs →	Complete within a defined timeframe	Support strategic business goals	Meets customer requirements	High return on investment (ROI)	Available resources	Data available	Top/Bottom line	Rating: High=10 Low = 1 ← Total
Level of importance → Projects	9	10	9	8	7	7	10	Total
Material 1: Cost reduction	9 "9 × 9" = 81	9 "9 × 10" = 90	9 "9 × 9" = 81	9 "9 × 8" = 72	9 "9 × 7" = 63	3 "3 × 7" = 21	9 "9 × 10" = 90	Sum = 408
Material 1: Cost reduction	9	9	9	9	9	1	9	394
Molded parts Scrap/dpm Reduction	9	9	9	9	9	1	9	394
Product 1: Capacity Improvement/ Cycle reduction	3	9	9	9	3	3	9	312
Product 2: Molds failure, machine downtime reduction	9	3	9	9	3	3	3	306

Rating Criteria: High = 9, medium = 3, low = 1

Example of total: Total (material-1 cost reduction) = $(9 \times 9) + (9 \times 10) + (9 \times 9) + (9 \times 8) + (9 \times 7) + (3 \times 7) + (9 \times 10) = 408$

Table 5.7. Project identification and prioritization

Process outputs →	Complete within timeframe	Support strategic business goals	Meets customer requirements	High return on investment (ROI)	Available resources	Data available	Top/Bottom line	Total Rating: High=10 Low = 1 ←
Level of importance								
Projects	8	9	9	8	7	7	10	Sum = 246
Manufacturing by location	"3 × 8"=24	"9 × 9"=81	"3 × 9"=27	"9 × 8"=72	"3 × 7"=21	"3 × 7"=21	"9 × 10"=90	Sum = 246
Smaller mold versus larger cavity mold	3	3	1	3	3	3	3	126
Developing international market	9	9	3	9	9	3	9	336
Time to market from design to production	9	9	3	9	9	3	9	336
Outsource mold fabrication	9	3	3	9	3	3	9	240

Rating Criteria: High = 9, medium = 3, low = 1
Example of total: Total (manufacturing by location) = (3×8) + (9×9) + (3×9) + (9×8) + (3×7) + (3×7) + (9×10) = 336

Table 5.8. Project identification and prioritization

Project	Process owner	Sponsor	Black belts (1x)	Green belts (2x)	Problem statement	Key project metrics and goal	Potential cost savings
Material-1 cost reduction	Molding	Quality process improvement manager	Process engineer	Molding technician	High cost	Financial impact	More than $300 K
Material-2 Screen cost reduction	Molding	Quality process improvement manager	Senior process engineer, Quality engineer	Molding technician	High cost	Financial impact	More than $350 K
Molded parts scrap/ PPM reduction	Production	Vice president of manufacturing	Quality engineer	Molding technician	Defect per million opportunity	Defect reduction/ Financial impact	More than $600 K
Product 1: Capacity improvement (Cycle reduction)	Molding	Quality process improvement manager, Manufacturing engineering manager	Quality process improvement manager	Molding technician	Longer cycle time: mold design and machine issues	Process improvement/ Financial impact	More than $250 K
Product 2: capacity improvement/ molds failure	Molding	Quality process improvement manager, Mfg. engineering manager	Process: Quality manager	Molding technician	Longer cycle time: mold design and machine issues	Process improvement/ Financial impact	More than $200 K

5.6.6 *TRAINING REQUIREMENTS*

  i. Who should be trained are listed as follows.
    Champions: Senior Managers
    Black Belts:
    (a) Manufacturing Engineers
    (b) Product/Sustain Engineers
    (c) Process Engineers
    (d) Purchasing Manager
    (e) Quality Engineers
    Green Belts:
    (a) Molding Technicians
    (b) Testing Technicians
  ii. What are the training criteria and how long:

Champion training:
  1. Day One
    (a) Overview and introduction to Six Sigma
    (b) Overview of company's prioritization of projects
    (c) Overview of define and measure models
  2. Day Two
    (d) Overview of analyze phase
    (e) Overview of improve phase
    (f) Overview of control and sustain phase
  3. Day Three
    (g) Define elements of project team members
    (h) Define and evaluate project charter
    (i) Project selections and prioritization

Black Belts training:
Absolutely all the DMAIC phases along with a thorough analysis of statistical tools will be carried out. Project selections and prioritization are done first.

  1. Week 1: Define and Measure (includes software, i.e., statistical software review)
    (a) Business process and project management/project charter/performance metric/measurements/continuous improvement tools
    (b) Process mapping (SIPOC)/data collecting and summarizing/project capability
  2. Week 2: Analyze and Explore
    (a) Descriptive statistics (mean and variance)/probability and statistics/normal and binomial distribution/central limit theorem/
    (b) Graphical statistics (histograms/box plots/stem-and –leaf diagram)/multivariate analysis/correlation and regression analysis/
    (c) Hypothesis testing: one and two samples means known and means unknown/paired t-test, variance/ANOVA
  3. Week 3: Improve
    (a) DOE: full factorial [2–3], fractional factorial, response surface techniques/steepest ascent methods/central composite design
    (b) Failure mode effects analysis/RPN

4. Week 4: Control
   (a) Application of control charts/analysis of control charts/process control planning/Lean tools for control 5S/, visual control and Kanban, TPM/mistake proofing/Poka-Yoke
   (b) Value stream mapping, set-up reduction/quick change over/SMED/design for Lean Six Sigma

Green Belt training: Green Belts will attend the Black Belt class only half the time of Black Belts. The more advanced topics such as multivariate analysis, response surface, and composite design/steepest ascent are not part of the Green Belt training program.

5.6.7 MANAGEMENT REVIEW PROCESS

Management team will review the improvements of the Six Sigma initiative on a quarterly basis with respect to all elements of the deployment plan as illustrated in Figure 5.1.

Review team	Review timeline
Process Technicians	Daily
Sustain Eng./mfg. Eng./Testing/Process improvement engineering	Weekly
Manufacturing/quality/molding/purchasing/Process improvement/sustain Engineering managers	Monthly
Senior managers	Quarterly

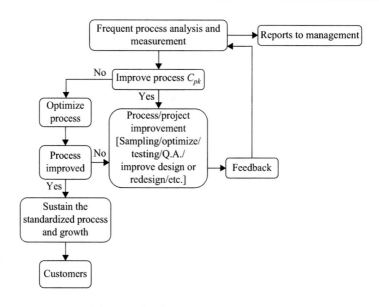

Figure 5.1. Brief schematic of the process.

5.6.8 COMMUNICATION PLAN

Tables 5.9 through 5.13 are the communication plans for Tables 5.6 and 5.8.

Table 5.9. Material 1: Cost reduction project communication plan

What	To whom	When	Who	How	Where	Comments
Project team meetings	Project Team, Invitees: product/Sustain engineer, manufacturing engineer, molding, purchasing mgr., testing, Q.A.	Weekly (every Thursday at 9:00 AM)	Black Belt	Notices, agendas sent out every Fridays via e-calendar	Building A Conference Room #1	
Meeting minutes	Project team list	By next day	Black Belt	Via Lotus Notes	MS Word, Excel, MS Project files, on share drive	
Project review and new actions Items	Project Team List	Weekly (every Thursday at 9:00 AM)	Black Belt	Via e-mail	MS Word, Excel, MS Project files, on share drive	
Project status reports and timeline	Champion [Mfg. Eng., Quality Sys., Production, Contract Mfg. Manager]	Monthly (every first week of the month) on Monday 10:00 AM	Champion [Mfg. Eng., Quality Sys., Production, Contract Mfg. Manager]	Via e-mail	MS Word, Excel, MS Project files, on share drive	
Project budget	Financial Analyst	Monthly (every first week of the month) on Tuesdays 10:00 AM	Project Financial Analyst	Notices, agendas sent out every Fridays via e -calendar	Building A Conference Room #1	

Table 5.10. Material-2 screen cost reduction communication plan

What	To whom	When	Who	How	Where	Comments
Project team meetings	Project Team, Invitees: Product/sustain Eng., mfg. Eng., molding, purchasing mgr., Tooling	Weekly (every Thursday at 10:00 AM)	Black Belt	Notices, agendas sent out every Friday via e-calendar	Building A Conference Room #1	
Meeting minutes	Project Team List	By next day	Black Belt	Via Lotus Notes	MS Word, Excel, MS Project files, on share drive	
Project review and new action Items	Project Team	Weekly (every Thursday at 10:00 AM)	Black Belt	Via e-mail	MS Word, Excel, MS Project files, on share drive	
Project Status reports and timeline	Champion [Mfg. Eng., Quality Sys., Production, Contract Mfg. Manager]	Monthly (every first week of the month) on Monday 10:00 AM	Champion [Mfg. Eng., Quality Sys., Production, Contract Mfg. Manager]	Via e-mail	MS Word, Excel, MS Project files, on share drive	
Project budget	Financial Analyst	Monthly (every first week of the month) on Tuesdays 10:00 AM	Project Financial Analyst	Notices, agendas sent out every Friday via e-calendar	Building A Conference Room #1	

Table 5.11. Molded parts scrap/PPM reduction communication plan

What	To whom	When	Who	How	Where	Comments
Project team meetings	Project Team, Invitees: Product/ Sustain Eng., Mfg. Eng., Molding, Tooling	Weekly (every Wednesday at 9:00 AM)	Black Belt	Notices, agendas sent out every Friday via e-calendar	Building A Conference Room #1	
Meeting minutes	Project Team List	By next day	Black Belt	Via Lotus Notes	MS Word, Excel, MS Project files, on share drive	
Project review and new action items	Project Team	Weekly (every Wednesday at 9:00 AM)	Black Belt	Via e-mail	MS Word, Excel, MS Project files, on share drive	
Project status reports and timeline	Champion [Mfg Eng., Quality Sys., Production, Contract Mfg Manager]	Monthly (every first week of the month) on Monday 10:00 AM	Champion [Mfg Eng., Quality Sys., Production, Contract Mfg Manager]	Via e-mail	MS Word, Excel, MS Project files, on share drive	
Project budget	Financial Analyst	Monthly (every first week of the month) on Tuesdays 10:00 AM	Project Financial Analyst	Notices, agendas sent out every Friday via e-calendar	Building A Conference Room #1	

Table 5.12. Product-1 capacity improvement/Cycle reduction and mold/machine justification communication plan

What	To whom	When	Who	How	Where	Comments
Project team meetings	Project Team, Invitees: Product/ Sustain Eng., Mfg. Eng., Molding, Tooling	Weekly (every Wednesday at 10:00 AM)	Black Belt	Notices, agendas sent out every Friday via e-calendar	Building A Conference Room #1	
Meeting minutes	Project Team List	By next day	Black Belt	Via Lotus Notes	MS Word, Excel, MS Project files, on share drive	
Project review and new actions Items	Project Team	Weekly (every Wednesday at 10:00 AM)	Black Belt	Via e-mail	MS Word, Excel, MS Project files, on share drive	
Project status reports and timeline	Champion [Mfg Eng., Quality Sys., Production, Contract Mfg Manager]	Monthly (every first week of the month) on Monday 10:00 AM	Champion [Mfg Eng., Quality Sys., Production, Contract Mfg Manager]	Via e-mail	MS Word, Excel, MS Project files, on share drive	
Project budget	Financial Analyst	Monthly (every first week of the month) on Tuesdays 10:00 AM	Project Financial Analyst	Notices, agendas sent out every Fridays via e-calendar	Building A Conference Room #1	

Table 5.13. Product-2 downtime reduction and mold/machine justification communication plan

What	To Whom	When	Who	How	Where	Comments
Project team meetings	Project Team, Invitees: Product/ Sustain Eng., Mfg. Eng., Molding, Tooling	Weekly (every Wednesday at 11:00 AM)	Black Belt	Notices, agendas sent out every friday via e-calendar	Building A Conference Room #1	
Meeting minutes	Project Team List	By next day	Black Belt	Via Lotus Notes	MS Word, Excel, MS Project files, on share drive	
Project review and new action items	Project Team	Weekly (every Wednesday at 11:00 AM)	Black Belt	Via e-mail	MS Word, Excel, MS Project files, on share drive	
Project status reports and timeline	Champion [Mfg Eng., Quality Sys., Production, Contract Mfg Manager]	Monthly (every first week of the month) on Monday 10:00 AM	Champion [Mfg Eng., Quality Sys., Production, Contract Mfg Manager]	Via e-mail	MS Word, Excel, MS Project files, on share drive	
Project budget	Financial Analyst	Monthly (every first week of the month) on Tuesdays 10:00 AM	Project Financial Analyst	Notices, agendas sent out every Friday via e-calendar	Building A Conference Room #1	

CHAPTER 6

DESIGN AND DEVELOP ORGANIZATIONAL LEAN SIX SIGMA ROADMAP: DELIVERING CONTINUOUS BREAKTHROUGH PERFORMANCE

The fundamental objective of the Lean Six Sigma Roadmap DMAIC (Define, Measure, Analyze, Improve, and Control) is the implementation of data-driven strategy that concentrates on process improvement and minimizing variability. So, the direction of the Roadmap is "Lean = waste reduction" and "Six Sigma = defects elimination." The DMAIC aims at continuous breakthrough performance, which is what the product end users are looking for. A well-designed and developed organizational roadmap if deployed without skipping any step will help the organization outperform the competition at all levels.

6.1   ROADMAP FOR SUCCESSFUL CORPORATE RESULTS

In today's global competitive market, new product development (NPD) is the key to a company's long-term tangible growth and success. In fact, companies perceive NPD as the cornerstone for continuous success in their business and market share. Without NPD, corporations tend to deteriorate. Nowadays, corporations are placing increased emphasis on innovations, breakthrough technologies, gaining new market share, and increasing customer satisfaction. To achieve this goal, NPD process needs to be promoted and supported by implementing Design for Lean Six Sigma (DFLSS) methodologies.

6.2   DESIGN FOR LEAN SIX SIGMA PROCESS

The DFLSS is an engineering and business process that is applied to the company's existing and future product development in designing as well as manufacturability either in production or transactional services. It takes product or services from ideation to production, market, using data-driven approach in analyzing processes, offering high-end quality, and scrap-free products. DFLSS concentrates on the design and process factors based on customer requirements.

6.3   VISION OF LEAN SIX SIGMA PROCESS

The success of any organization depends on the capability to assure the highest product quality at the lowest cost and maximum customer satisfaction. Six Sigma not only ensures resolving the existing process issues, but it also removes the cause from the original source. The technique analyzes the status quo measurement, predicts the future issues, and takes action to eliminate the possible failure before it takes place. No other theory has such a powerful scientific and engineering approach to maximize improvement by eliminating defects.

6.3.1   WHERE AND WHEN DO WE USE LEAN SIX SIGMA?

Lean Six Sigma is deployed in the case of unknown independent variables (cause), undefined problem, unclear approach, failed methods, unsolvable or complex problem, and many more difficult cases. It can bring benefit to any company that has product or services (e.g., manufacturing, health care, medical devise, transportation, chemical companies, technologies, electronics, automotive, pharmaceuticals, polymer and plastics, administrations, and the list goes on). It focuses on the project scope and deliverables in delivering the results.

6.3.2   WHY USE LEAN SIX SIGMA?

Because it is very clear and marketable, Six Sigma approach delivers results based on scientific techniques, changes the corporate culture, when applied correctly produces better result than any other practice, has defined approach to problem solving, brings consistency with the result, is focused on maximizing customer satisfaction and bottom line that makes the leadership team to be attracted and support the projects.

On the other hand, Six Sigma deployment requires cultural change and patience. Six Sigma is all about achieving the right results, not just any result. It is always true that more ideas or course of actions fail to succeed if there is fear for cultural changes, lack of Six Sigma understanding, and support of the leadership team.

6.4   DESIGN FOR LEAN SIX SIGMA ROADMAP

The DFLSS roadmap consists of six phases with gate reviews in between phases. The gate reviews are designed to review progress in each phase and agree on the results before moving to the next phase. This is illustrated in Figure 6.1.

The objectives and tools associated with the basic processes of Six Sigma roadmap (phases) are described in the following sections. Some of the tools may overlap across the phases.

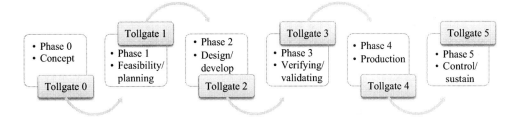

Figure 6.1. Design for Lean Six Sigma phases.

6.4.1 PHASE 0: CONCEPT AND IDEATION

Phase 0 provides various methods, tools, and steps to generate new ideas.

(a) Identify the concept and project charter.
(b) Define, develop the business case, and project scope using brainstorming activities.
(c) Align the project strategically with organizational objectives.
(d) Conduct market analysis, competition profile, economics, benchmarking, forecasting analysis, and customer target (Voice of Customer).
(e) Use the knowledge-based engineering techniques and utilize mathematical modeling, graphical, statistical, prototyping for scientific and engineering analysis.
(f) Design the product modular and platform.
(g) Carry out the house of quality (quality function deployment) analysis to determine key factors in designing the product.
(h) Finally, evaluate the preliminary financial analysis on the quality, cost, time, and profitability of the design.

6.4.2 PHASE 1: DEFINE, FEASIBILITY, AND PLANNING

Phase 1 supports and provides different tools and techniques to complete the define analysis.

(a) Preliminary concept is discussed and identified and a scope is developed.
(b) Strategic fit to organization objectives is decided.
(c) Market formation, survey, competitors, and target customers are discussed.
(d) Define customer technical requirements.
(e) Discuss and confirm the basic information on technical feasibility.
(f) Improve performance of product proposal with more business case justification.
(g) Develop and finalize initial ideation along with engineering requirements.
(h) Create introductory manufacturing and course of action for qualification.
(i) Layout preliminary design plan.

6.4.3 PHASE 2: DESIGNING AND DEVELOPING

Phase 2 provides tools and steps in designing and developing in product and services.

(a) Develop the KJ (Kawakita Jiro) or affinity diagram analysis.
(b) Exercise statistical analysis tools on the status quo design or product.
(c) Carry out design failure mode effect analysis (DFMEA).
(d) Apply experimental design and response surface methods.
(e) Determine the essential design variables and processes that will affect the product functionality.
(f) Benchmark your product against the best-in-class designs in the market.
(g) Confirm and finalize customer specification limits (Voice of Customer) for design.
(h) Design for Lean Six Sigma manufacturability and assembly production.
(i) Finalize engineering design and verify using software technology.
(j) Create detailed manufacturing and qualification course of action.
(k) Develop reliability modeling and testing if necessary.

6.4.4   PHASE 3: VERIFYING AND VALIDATING THE DEVELOPED DESIGN

Phase 3 confirms the design by implementing various tools and methods.

(a) Complete the production tooling and qualification part.
(b) Verify, validate, qualify, test, and confirm that customer specifications are met.
(c) Assure the manufacturability and processability of the product/services.
(d) Optimize any production noise factors from the system; apply analysis of variance (ANOVA), analysis of mean (ANOM), measurement system analysis (MSA), tolerance analysis, DOE, and response surface or Taguchi's methods for robust design if necessary.
(e) Finalize the design capability, life testing, and reliability assessment studies.
(f) Lay out the details of preproduction plans, and procedures.

6.4.5   PHASE 4: PRODUCTION AND COMMERCIALIZING

Phase 4 presents the launch phase for commercialization by introducing various methods. This is also the last phase to confirm if anything missed in phase 3.

(a) Introduce the full-scale launch plans including technical staff and sales training.
(b) Confirm the quality control plans, course of action, and process support engineering.
(c) Implement marketing, commercializing, and on-time delivery system strategies.
(d) Follow-up product capability, manufacturability, and process ability.
(e) Collect customer's feedback.
(f) Prepare for process monitoring system.

6.4.6   PHASE 5: CONTROL AND SUSTAINING

Phase 5 confirms and oversees that project objectives will sustain product quality based on the customer "needs and wants" and continue maintaining the end-user satisfaction.

(a) Assure the manufacturing processes, procedures, and system are in line, as planned to continue with product quality and repeatability.

(b) Continue monitoring the product production that is already started and through its life-cycle.

(c) Apply Six Sigma continuous improvement tools as needed.

(d) Analyze product performance and quality data as needed.

(e) Review the lessons learned.

The above phases "design for Lean Six Sigma roadmap steps and tools" is summarized in Table 6.1.

Thus, design for Lean Six Sigma can be used in new product development process or for the modification of the existing product. It applies data-driven methodologies to measure and decide on the product specification limit for the intended use. Products are developed based on the knowledge obtained from measurement studies of components, finished products, and manufacturability factors related to the performance of the finished goods along with customer's

Table 6.1. Design for Lean Six Sigma (DFLSS) roadmap phases

Ideation (Concept)	Feasibility and Planning (Developing Technical Requirment)	Designing and developing	Verifying and Validating	Manufacturing, Optimization, Production Introduction and Commercializing	Controlling, Sustaining, and Production
Market analysis	Process owner	Develop existing process flow	Validate project scope and process flow	Identify/ prioritize root causes	Develop advanced process control charting concepts
Economic forcasting	Project Charter	Implement Kaizen and 5S housekeeping	Identify cause and effect of similar process	Screen potential causes/ Design of Experiment	Develop and implement plan
Bench-marking	Develop stakeholder analysis	Identify areas of the 7 waste	Cause and effect diagram	Taguchi/ full factorial design	Issue revised procedures
Customer's "needs"	Scope, metrics, Probelm statement, baseline	Develop collection plan/gage Cpk study	Analysis of mean/box plot of x and or y variable	Identify and prioritize improvement actions	Revalidate cause and effect of X's to achieve Y's
Concept generation	Project selection/ prioritization (QFD)	Measurement system analysis	Multi-vari analysis	Develop implementation plan	Implement mistake proofing-Poka-Yoke
QFD analysis	Establish CTQ/ SIPOC	Plot output (Y) data over time (box plot & time series)	Histogram of input (x) variables	Cost benefit analysis/quality and delivery analysis	Monitor process performance (Run chart/Control chart/ s chart)
Engineering methods	Establish goals (% of actual)	Pareto high impact cause/ process input	Regression analysis/ scatter diagram/ correlation	Risk assessment/ failure mode effect analysi	Develop before/after summary table of performance improvements
Math grafical, modeling	Develop value stream map	Histogram of existing data/ apply descriptive statistics	Scale and probability analysis: discrete and continuous	Establish operating tolerance for input variables	Identify opportunities to leverage improvements
Generate product proposal	Develop value added analysis	Establish process capability of existing data	Hypothesis testing concept and Chi square	Develop new process flow	Process and Capability study (Cp, Cpk, Pp, Ppk)
Generate research proposal	Potential project cost savings	Determine sigma level of existing process	Analysis of variance (ANOVA)	Portter's five forces	Risk analysis
	Quality				
	Cycle time				
	Cost				

feedback on the functionality. New products can be designed, built, tested, established, manu-factured, and marketed with Lean Six Sigma high-quality reliability meeting customer "needs" and "wants," advancing in product innovation, uniformity, achieving maximum customer sat-isfaction, and profitability.

6.5 LEAN SIX SIGMA CONTINUOUS PROCESS IMPROVEMENT ROADMAP

Once the product hits the production line, the process requires continuous monitoring system. Now it is transformed into continuous process improvement in the production line. The pro-duction variations become the cause of the effect where the independent variables begin shift by 1.5 Sigma due to the wear and tear of the equipment. The Lean Six Sigma (LSS) roadmap consists of six phases with gate reviews in between phases as illustrated in Figure 6.2. Again, the purpose of the gate reviews is to review the deliverables of each phase and sign off before moving to the next phase.

The basic processes of roadmap for breakthrough steps and tools are described in the fol-lowing sections. Some of the tools may overlap across the phases.

6.5.1 PHASE 0: CONCEPT

Phase 0 provides various tools, techniques, and steps to generate new ideas for existing product or services process improvement.

(a) Initiate the concept, vision, philosophy, project charter, method, tool, and project scope using brainstorming activities.
(b) Align the project strategically with organizational initiatives such as cost reduction, scrap or NCMR reduction, and improve productivity.
(c) Conduct competition benchmarking, and customer target (Voice of Customer).
(d) Adopt knowledge-based engineering techniques.
(e) Apply mathematical modeling, graphical, statistical, and engineering analysis.
(f) Evaluate the preliminary financial analysis on the quality, availability, cost, and profit-ability.

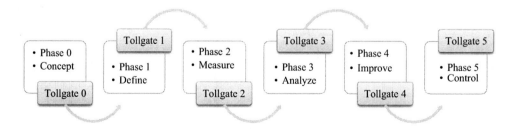

Figure 6.2. Lean Six Sigma phases.

(g) Develop and finalize initial ideation along with engineering requirements.

(h) Carry out the Strengths, Weaknesses, Opportunities, and Threats (SWOT) analysis.

6.5.2 PHASE 1: DEFINE

Phase 1 utilizes various methods in defining the project.

(a) Define the project charter.

(b) Identify customer technical requirements.

(c) Conduct KJ analysis (the affinity diagram).

(d) Create introductory manufacturing, sustaining, and qualification course of action.

(e) Carry out the house of quality (quality function deployment) analysis to determine key factors in designing the product.

(f) Determine the project timeline end-to-end boundary.

6.5.3 PHASE 2: MEASURE

Phase 2 evaluates the past history and current activity of the process by employing Six Sigma measurement tools.

(a) Apply seven quality control tools (see Chapter 8).

(b) Determine a base line Sigma (σ) level for the existing process.

(c) Measure process performance variables and establish performance capability.

(d) Determine the essential process variables that will affect the product functionality.

(e) Create a process flow chart and identify where Lean manufacturing applies.

(f) Use statistical analysis tools on the status quo design or product or services.

(g) Conduct process failure mode effect analysis (PFMEA).

(h) Benchmark your product, process, and service against the best in class in the market.

(i) Finalize process design and course of actions using software technology.

(j) Apply cause and effect interrelationship diagram to determine the biggest drivers.

(k) Develop reliability and confidence modeling if essential.

6.5.4 PHASE 3: ANALYZE

Phase 3 applies all the analysis tools used to evaluate and understand the data collected in measurement phase.

(a) Use Pareto (80/20 rule) analysis to find variables that make biggest impact on the process.

(b) Utilize statistical tools: statistical measurement, inferences, hypothesis testing, analysis of variance, linear and nonlinear regression.

(c) Confirm and finalize customer specification limits (Voice of Customer) for process design.

(d) Assure the manufacturability and processability of the product or services.

(e) Finalize the process design capability and reliability assessment studies.

6.5.5 PHASE 4: IMPROVE

Phase 4 designs or redesigns the current process by applying various improve tools and methods.

(a) Optimize any production noise factors from the system; apply analysis of variance (ANOVA), analysis of mean (ANOM), and measurement system analysis (MSA), tolerance analysis, DOE, response surface, or Taguchi's methods for robust design if necessary.

(b) Design the process for Lean Six Sigma manufacturability and assembly production.

(c) Follow-up product capability, manufacturability, and processability.

(d) Verify, validate, qualify, test, and confirm if customer specifications are met.

(e) Prepare for the process monitoring system.

6.5.6 PHASE 5: CONTROL AND SUSTAINING

Phase 5 monitors that project objectives will sustain the product quality based on customer specifications and continue maintaining the end-user satisfaction.

(a) Assure the manufacturing processes, procedures, and systems are in line as planned to continue with product quality and repeatability.

(b) Confirm the quality control plans, course of action, and process support engineering.

(c) Continue monitoring the product production already started and through its lifecycle.

(d) Apply Six Sigma continuous improvement tools as needed.

(e) Analyze product performance and quality data as needed.

(f) Review the lessons learned.

The above phases of "Lean Six Sigma continuous process improvement roadmap steps and tools" are summarized in Table 6.2.

So, what are the benefits of Six Sigma breakthrough for an organization?

 i. Maximum customer satisfaction, gross income, and profitability

 ii. Focused on high quality, cost reduction, time (cycle time), and growth

iii. Consistent product, processes, and speed to market

iv. Effective and efficient communication, operation, and management

 v. Changes the culture of thinking, team work, increased motivation, and on-time project completion

6.6 LEADING THE EFFORTS

By now, the Lean Six Sigma objectives are in progress. A deployment and implementation course of action are all set and prepared. The projects are identified and Black Belts training is

Table 6.2. Lean Six Sigma continuous process improvement roadmap

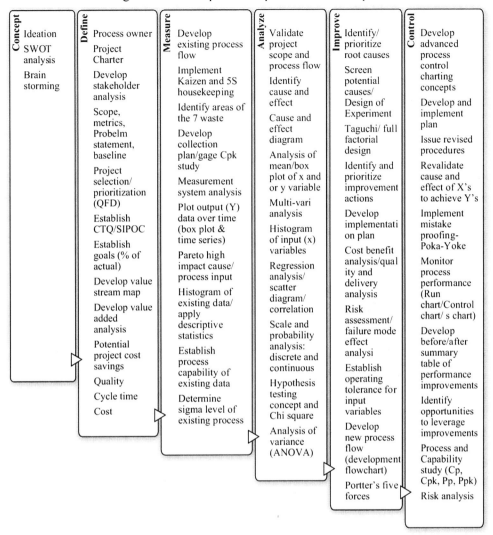

Concept	Define	Measure	Analyze	Improve	Control
Ideation	Process owner	Develop existing process flow	Validate project scope and process flow	Identify/ prioritize root causes	Develop advanced process control charting concepts
SWOT analysis	Project Charter	Implement Kaizen and 5S housekeeping	Identify cause and effect	Screen potential causes/ Design of Experiment	Develop and implement plan
Brain storming	Develop stakeholder analysis	Identify areas of the 7 waste	Cause and effect diagram	Taguchi/ full factorial design	Issue revised procedures
	Scope, metrics, Probelm statement, baseline	Develop collection plan/gage Cpk study	Analysis of mean/box plot of x and or y variable	Identify and prioritize improvement actions	Revalidate cause and effect of X's to achieve Y's
	Project selection/ prioritization (QFD)	Measurement system analysis	Multi-vari analysis	Develop implementati on plan	Implement mistake proofing-Poka-Yoke
	Establish CTQ/SIPOC	Plot output (Y) data over time (box plot & time series)	Histogram of input (x) variables	Cost benefit analysis/qual ity and delivery analysis	Monitor process performance (Run chart/Control chart/ s chart)
	Establish goals (% of actual)	Pareto high impact cause/ process input	Regression analysis/ scatter diagram/ correlation	Risk assessment/ failure mode effect analysi	Develop before/after summary table of performance improvements
	Develop value stream map	Histogram of existing data/ apply descriptive statistics	Scale and probability analysis: discrete and continuous	Establish operating tolerance for input variables	Identify opportunities to leverage improvements
	Develop value added analysis	Establish process capability of existing data	Hypothesis testing concept and Chi square	Develop new process flow (development flowchart)	Process and Capability study (Cp, Cpk, Pp, Ppk)
	Potential project cost savings	Determine sigma level of existing process	Analysis of variance (ANOVA)	Portter's five forces	Risk analysis
	Quality				
	Cycle time				
	Cost				

underway. Once the initial projects are completed, everything will move so fast that you will need a strong set of talented people and a solid infrastructure to lead the efforts.

6.6.1 PROJECT REPORT AND REVIEWING PROGRESS

As noted in Chapter 5, it is extremely important to review the progress at each phase to make sure that the project is moving forward. Also, the other management levels should be informed of the progress and delivered results of the initiatives. A key factor of the reporting system is financial auditing system. Leadership team always needs to know how the bottom line savings by Lean Six Sigma projects is improving. So, they can communicate it to the shareholders and the organization. This is discussed in details in the "Tollgate Review" section.

6.6.2 COMMUNICATION

As previously defined, communicating the vision, progress, and course of actions will have a big impact on the organization's viewpoint and acceptance of the program. Organizations should require Lean Six Sigma activities for advancement of financial gains, and openness for answering questions, Who? What?, When?, Where?, Why?, and How?, through an effective communication.

6.6.3 AWARDS AND APPRECIATION

The question is what creates motivated, energetic, contributing people? To keep the employee momentum at the highest level, they need to be recognized. Rewarding and appreciating them will keep them inspired, positive, proactive, above-the-line performers, with high morale, result oriented, and enthusiastic. Below-the-line performing people will never produce the result to impact the bottom line.

6.7 MAINTAINING AND GAINING THE MOMENTUM

By this time, you have completed the launch initiatives and trying to organize a management system that will sustain the achievements of deployment effectively. The momentum defined as "physical energy or the force of movement." It is a language used by leaders to maintain and gain the momentum and motivation in achieving their vision. The concept is not achieving the goal at once. The vision is keeping the momentum for continuous process advancement. The momentum of Lean Six Sigma can be sustained by maintaining the improvement (see Chapter 11 for details). The progress gained from Lean Six Sigma projects are maintained by implementing the control plan, making sure that all the process team and owners associated with the process are trained.

Along with the control, the financial tracking system and process auditing must be installed. The financial tracking will evaluate whether the predicted benefits are being realized or not. As noted in Chapter 5, the best strategy is to review the key process elements in each phase, quarterly, and an overall annually.

6.8 TOLLGATE REVIEW

As discussed before, a Tollgate Review is a necessity. Tollgate Review is like a gate review, or a checkpoint in a Lean Six Sigma project, where all the team members meet and evaluate if the work criteria have been executed and performed as outlined in the project plan and whether the goals and strategic objectives have been accomplished successfully. Tollgate members or steering team also called gate keepers. These members could be Master Black Belts, Black Belts, Green Belts, quality council members, project champions, or sponsors such as vice president, senior manager, and leaders.

In general, each Tollgate Review occurs at the end of each phase in the DMAIC or DMADV process. That means that the steering team or gate keeper team members will get together at the end of each of the Define, Measure, Analyze, Improve, and Control phases. Likewise a Tollgate Review happens at the end of design for Six Sigma phases Define, Measure, Analyze, Design, and Optimize/Verify.

So the gate keeper must approve the project phase success criteria before a project can proceed to the next level or phase x. Figures 6.3 and 6.4 indicates how a Tollgate Review takes place at the end of each phase.

Tools: A number of tools can be used to present the phase outcome in the Tollgate Review such as

  i. Check sheet
 ii. Project deliverables for each phase
iii. RAM (responsibility assignment matrix) also called RACI chart
 iv. List of milestone.
  v. Project evaluation and scoring criteria: to approve or request for reexamination/ change and even eliminate the project due to the lack of requirements such as
   (a) Strategic alignment with the organizational initiatives
   (b) Competitive advantages
   (c) Process and product attractiveness
   (d) Core competencies
   (e) Technical and financial feasibilities

The project manager must prepare an acceptable presentation with all the check-list requirements for the attendees and gate keepers such that all the subject matter should be clear to the attendees. The projects that are found lacking to demonstrate the deliverables of each phase (of DMAIC or DMADV) are the ones that fail and are unsuccessful. A Tollgate Review will make secure this doesn't take place and make the project more efficient, productive, and profitable.

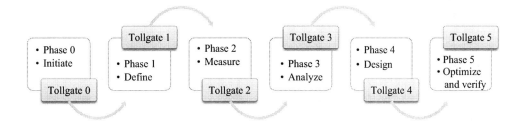

Figure 6.3. Design for Lean Six Sigma (DFLSS) projects: Tollgate review with prior similar product history.

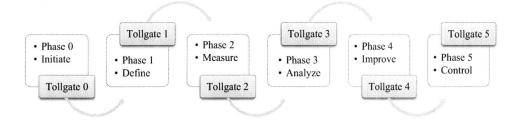

Figure 6.4. Lean Six Sigma process improvement projects: Tollgate review.

In addition to gate meeting a RACI (Responsible, Accountable, Consulted, and Informed) chart must be developed for each phase to make sure that each member of the project team will complete their assigned deliverables.

6.8.1 DEVELOP A RACI MATRIX

RACI is a method utilized to identify roles and responsibilities during an organizational process improvement, product development, and/or making a business or engineering decision. This will assist in visualizing the roles and responsibilities and helps to identify if anything missing. RACI stands for Responsible, Accountable, Consulted, and Informed. Each of this term is defined as

  i. *Responsible*: Those who are responsible to carry out the work or perform an activity that deliver the results.
 ii. *Accountable*: Those who are ultimately accountable for the satisfactory execution and completion of the deliverables.
iii. *Consulted*: Those whose assessment and analysis are requested, the person assigned the responsibility provides feedback, and can contribute ideas prior to decisions are made. This will make sure that all the implication of the end decision is accordingly well analyzed. This opens up a two-way conversation.
 iv. *Informed*: Those are informed of the project progress such as one-way conversation, and status updates.

For instance a RACI model is cited as shown in Table 6.3.

Table 6.3. RACI Chart

Lean Six Sigma project process improvement phases	Organizational roles							
R = Process improvement responsibility A = Process accountability C = Consult the process improvement I = Inform the process improvement	VP of manufacturing	VP of engineering	Project champion	Project sponsor	Master black belt	Black belt	Green belt 1	Green belt 2
Fulfill the Six Sigma phases	**RACIs**							
1.1 Define: Projection selection	I	I	I	C	C	A	R	R
1.2 Project charter-Scope	I	I	I	C	C	A	R	R
... .	..	..	..	..	..	..	..	..
2.1 Measure: data collection	I	I	I	I	C	A	R	R
2.2 Cause and effect identification	I	I	I	I	C	A	R	R
.. ..	..	..	..	..	..	..	..	..
3.1 Statistical analysis	I	I	I	I	C	A	R	R
3.2 Phase 3 deliverables	I	I	I	I	C	A	R	R

6.9   LEAN SIX SIGMA CULTURE AND THE WAY IT WORKS

The cultural transition (the game changer) is a must. Either you take in control of your corporate culture or it will control you. What is culture? It is the way people think and behave.

It is defined as "A pattern of shared basic assumptions created, discovered, or developed by a given group throughout the time as it learns to get along with its problems of external adaptation and internal integration." In addition, what has worked well enough to be considered valid should be taught to new members as the correct way to see, think, and feel in relation to issues solved or those facing them.

Every corporation has a culture but in the direction of corporate goals or against it. The corporate culture can make the difference between success and failure. Basically, the cultural transition becomes the game changer of the failure to extraordinary success along with Six Sigma methodology implementation. To continue improving performance, a cultural change is a must if the culture does not produce the results you are looking for. Consider a Results Target as envisioned in Figure 6.5. Mathematically, result is function of efforts and effort depends on beliefs. Likewise, the belief is function of experience.

Table 6.4 is a template to cultural assessment that one can answer the questions and evaluate the effectiveness.

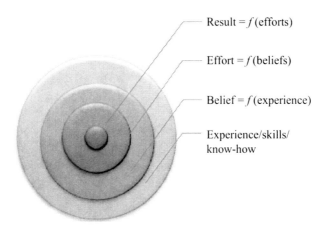

Result = f (efforts)

Effort = f (beliefs)

Belief = f (experience)

Experience/skills/know-how

Figure 6.5.   Organizational culture transition.

Table 6.4.   Lean Six Sigma self-assessment/readiness review

Cultural assessment questions	Effectiveness (1–5)
How are important decisions made?	
How fast are the decisions implemented?	
How does the organization recognize successes and failure?	
How does the organization handle failures?	
Does anyone in the firm understand the mission, vision, and strategy?	

(*Continued*)

Table 6.4. (*Continued*)

Cultural assessment questions	Effectiveness (1–5)
Is everyone in the organization aware of the critical customers and their needs, revenue and operating expense issues across the organization?	
How does the organization set up their objectives?	
How clear are these organization objectives?	
Are the organization objectives measurable?	
How much important information is released in the organization?	
Total operational score	

1 = Very ineffective, 2 = Ineffective, 3 = Marginally effective, 4 = Effective, 5 = Very effective

CHAPTER 7

DEFINE CONCEPTS AND STRATEGIES

Define begins at the end. In this phase, the leaders of the Lean Six Sigma project set the vision, "project charter," and "project boundaries" based on the customer's needs and wants (or Critical-to-Quality). Primarily, the target improvement of the Sigma level is normally based on the impact of increasing and improving the effect(s) or reducing and eliminating the cause(s). The define phase can be further expanded once the analysis phase is completed. The project statement must be strategically aligned with organizational initiatives, that is, the problem taken up for solution should be high priority and should have a high impact.

Initiation Phase
Define the process
Determine customer requirements
Identify key process output variables

7.1   CONCEPTS, VISION, AND IDEATION PHASE

In an era of intense global economic competition pressure, countries with lower product manufacturing costs quickly advance with robust manufacturing and technologies. New methods of thinking are required to outperform the manufacturing or transactional techniques of the competition. A *can-do* and dedicated way of meeting the increasing competition is to focus on maximizing the productivity and quality at the lowest manufacturing cost and at a faster rate than the competitors. This will enable the organizations to continuously introduce new ideas of quality/reliability and process optimization methodologies. These measures in turn lead to highest customer satisfaction and profitability.

Lean Six Sigma with precision engineering entails the following challenges.

1. Maximize manufacturing productivity, process quality, and reliability by reducing product warranty returns through increasing life cycle of products using advanced reliability concepts. Further, establish consistent and repeatable process for uniform product functionality. Apply principles of Lean Six Sigma, robust design, and processes in maximizing customer satisfaction.

2. Improve availability by reducing cycle time, which maximizes productivity and process efficiency. Sustain the optimized cycle time, which leads to consistent manufacturability.

3. Minimizes costs by controlling manufacturing or business process through implementing efficiency, lean manufacturing principles, implementing most cost-effective and efficient solution to a problem or design for a robust process.

4. Ensures the highest product performance by applying engineering optimization techniques.

The concepts for Lean Six Sigma projects may come from customer, marketing, employee surveys, suggestions, benchmarking, or existing products. The management or sponsors may identify potential process improvement variables and develop a process to create and prioritize project ideas. The project selection process needs to be aligned with quality focus and the strategic objectives of the organization . These elements should be evaluated when project opportunities are prioritized.

Any corporate strategic change is accompanied by a plenty of requirements. An initial concept must have justification that it is worth pursuing, most probably along with world class success stories and benchmarks. Of course, an idea absolutely requires senior management support to evolve into an effective project, which leads to long-term process improvement. All the abovementioned activities should be taken into consideration for designing a Lean Six Sigma project.

Define Phase (problem definition)
Project Selection Engineering and Strategy

7.2   WHAT IS SIX SIGMA "DEFINE PHASE"

It is extremely important and necessary that enough time is spent on planning the tasks in this phase to give the Lean Six Sigma project the best possible key results and success. The define phase is where the problem statement, goals, and scope of the project are defined and finalized. Define phase provides a fundamental understanding of why launch the project activities and what problem serves as a trigger for the launch. Therefore, to develop the project charter, the following steps along with essential components are necessary in identifying and measuring the selected project.

1. Project title
Finalize process improvement name or title (also see phase zero).

2. Summary and overview of project
   (a) Define "phase," stages within "phase," metrics, process changes, subprojects, target time, communication, and assign a team leader.
   (b) Define project and business case including a description of the business process problems, specify goals, and values such as
      i. Assign a project with high-impact variations, resources and expertise available, process complexity, chance of success, and get support or buy-in from higher management.

      ii. Select process parameter input (independent or causes) and output (dependent or symptoms) variables.

     iii. Design the process flowchart, or SIPOC (supply, input, process, output, and customer).

     iv. Compile data on process output.

      v. Prioritize opportunities.

3. Objectives

   (a) Develop a statement of the objective.

      i. Understand why the project is required by the company.

     ii. Determine who, how, and when.

   (b) Develop metrics to measure the progress. Measurability is a characteristic of the Objectives.

      i. Establish standardization by lean and cost reduction.

   (c) Increase productivity of the process.

   (d) Identify business financial drivers—the areas that will bring the highest rate of profit by preventing or eliminating waste.

   (e) Determine Critical-to-Quality (CTQ) processes through Voice-of-Customer (VOC), Voice of Business (VOB), and customer requirements. Establish communication with customer and define customer's lower (LSL) and upper specification limits (USL) and any other needs.

4. Project charter

   (a) Develop the business case (financial analysis).

   (b) Draft a statement of the scope.

   (c) List roles and responsibilities of team members.

      i. Clarify what is expected from team members.

     ii. Aligning the team with organizational priorities.

   (d) Identify the vital few milestones and deliverables. The scope needs to be tied to deliverables.

   (e) Decide process boundaries and constraints or Voice-of-Process (VOP), for example, what is included and what is excluded? What is start and stop point?

      i. Identify the customer who receives the process output either internal or external.

     ii. Think, describe, rank, and prioritize customer's expectation and specify tangible or intangible deliverables linked to these expectations.

     iii. Recognize Critical-to-Quality measurable attributes that have the greatest impact on customer satisfaction. Select these variables by mapping a detailed flowchart of the existing process.

5. Assumptions, concerns, and main constraints of the project

6. Stakeholders analysis

   (a) Customers (i.e., business unit manager, division president)

   (b) Sponsors (vice presidents, directors, IT team)

   (c) Engineering team leaders

   (d) Select fully trained project team members. The success of the Lean Six Sigma project will be obtained by giving responsibilities to those who are actually involved in the process.

      i. Define skills and knowledge needed for the project team

     ii. What is the commitment expected of team members?

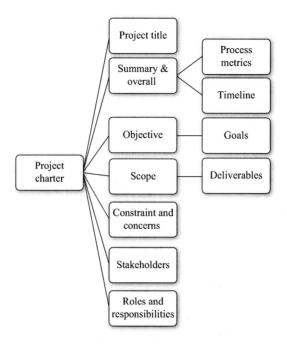

Figure 7.1. Project charter tree diagram.

7. Gain corporate top management commitment
8. Set up launch plan, gate review meetings with steering committee that governs the project and a member of the project customer should be in the committee.
   (a) Design the project charter
   (b) Approve the project charter

Lean Six Sigma follows the projects in which solutions are not known. It is about solving a business or engineering problem by improving processes. A Lean Six Sigma project should contain at least one of the following criteria:

1. Process with lower Sigma level.
2. Higher return on investment (ROI). The higher the ROI the higher the impact.
3. Quantitatively measurable.
4. High probability of success.
5. Create a timeline. Be able to complete within 3–9 months.
6. Strategically high on the company priority list or an important project for the corporate future growth.

The project or process improvement results should have an impact on the company's strategic business plan, competitive position, customer's needs, "core competencies," financial results, urgency, trend, sequence, and dependency. Figure 7.1 categorizes the above concepts.

7.3 LEAN SIX SIGMA VARIATION

Regardless of process specification any departure or deviation from the targeted value has a quality cost to company. It includes cost of inspection, test, rework, and increasing customer

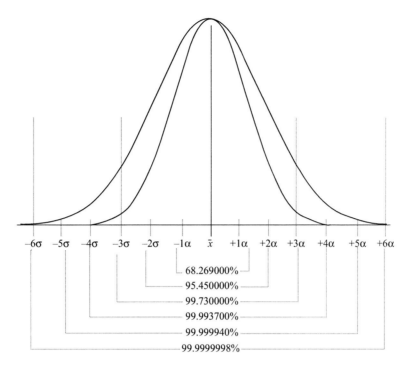

Figure 7.2. Short-term process capability.

dissatisfaction. Thus, if we put it in the mathematical model, the variation and reproducibility are inversely related to each other such that as variation increases, producibility decreases due to increase of nonconformance (defect) probability. The empirical rule is only valid for bell-shaped (normal) distributions. The following charts are true. Figure 7.2 features the process capability for short term when the process is centered on the target and process distribution shift is present.

Table 7.1 illustrates the information from Figure 7.2 in terms of defects per million for 1.0 sigma to 6.0 sigma when the process capability distribution centered at the target.

Table 7.1. Mathematical comparison of sigma capability concepts for short term

Sigma capability	Defect free per million (%)	Defects per million (%)	Quality/profitability
1.0 Sigma	68.269000	31.731000	Loss
2.0 Sigma	95.450000	04.550000	
3.0 Sigma	99.730000	00.270000	An industry average
3.5 Sigma	99.953500	00.465000	
4.0 Sigma	99.993700	00.006300	Above average
4.5 Sigma	99.999320	00.000680	
5.0 Sigma	99.999940	00.000060	
6.0 Sigma	99.9999998	00.0000002	World class performance

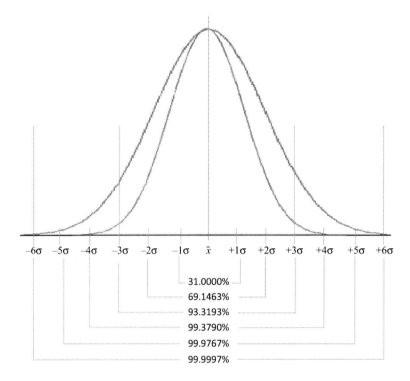

-6σ   -5σ   -4σ   -3σ   -2σ   -1σ $\bar{x}$ +1σ   +2σ   +3σ   +4σ   +5σ   +6σ

31.0000%
69.1463%
93.3193%
99.3790%
99.9767%
99.9997%

Figure 7.3. Long-term process capability.

Figure 7.3 illustrates the process capability for long term when the process distribution shifted by 1.5 sigma to right or left of the target value.

Table 7.2 presents the information from Figure 7.3 in terms of defects per million for 1.0 Sigma to 6.0 Sigma when the process capability distribution shifted by 1.5 Sigma from the center or target.

The three major types of variations in Six Sigma are discussed below.

7.3.1 POSITIONAL VARIATION

Variations from within the units or unit family, variations within a single unit: left side versus right side or top versus bottom, variations across a single unit: thickness of a printed circuit board, variations from plant cell to plant cell, machine to machine, operator to operator, mold to mold, and cavity to cavity. An example of this is muli-variate chart in Figure 7.4. The characteristic of interest, the "width," varies with the position. It is used to identify and reduce possible independent variables $x_{(s)}$ or group of causes of variation to a much smaller group of control factors.

7.3.2 CYCLICAL VARIATION

Variations as shown in Figure 7.5 from piece to piece or unit to unit family, variation between consecutive units in the same time frame, batch-to-batch variation, lot-to-lot variation, changes of economic activity as an outcome of recurring causes, and time-based variations with a

Table 7.2. Empirical rules of Lean Six Sigma when process is offset by 1.5 sigma (Long term)

Sigma range	Percent defect-free per million (%)	Defects per million	Quality/profitability
1.0 Sigma	31.0000	690,000	Loss
1.5 Sigma	50.0000	500,000	
2.0 Sigma	69.1463	308,300	Noncompetitive
2.5 Sigma	84.1350	158,650	
3.0 Sigma	93.3193	66,807	Average industries
3.5 Sigma	97.7300	22,700	Entering above average
4.0 Sigma	99.3790	6,220	Above average
4.5 Sigma	99.8700	1,350	
5.0 Sigma	99.9767	233	Below maximum productivity
5.5 Sigma	99.9968	32	
6.0 Sigma	99.9997	3.4	Near-perfection

Lean Six Sigma (6σ) contains 99.9997% of all values.

Note: Table 7.1 shows that the short-term process produces no more than 0.002 nonconforming per million when Sigma is equal to 6. In the long term, this value increases to 3.40 when Sigma is equal to 6.

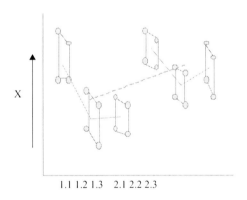

Figure 7.4. Multivariate chart.

repeating pattern. Therefore, the cycle occurs more than once and a noncyclic does not occur more than once.

Figure 7.5 also shows that the cycle length can be measured from X_1 to X_2, from Y_1 to Y_2, or from Z_1 to Z_2.

The length of the cycle is the period of that cycle which can be measured from one peak to the next, one valley to the next, or from the time value at which the cycle crosses the horizontal line to the value where it completes the cycle and returns to this point.

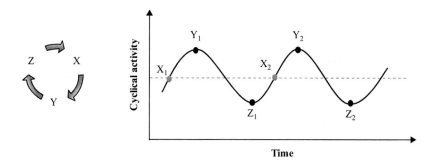

Figure 7.5. Cyclical variation.

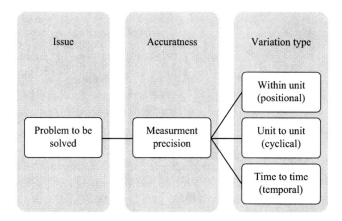

Figure 7.6. Temporal variation.

7.3.3 TEMPORAL VARIATION

Variations (Figure 7.6) from time to time, year to year, month to month, week to week, shift to shift, day to day, and hour to hour.

Other variations such as direct variation are that the ratio of two variables is constant. Inverse variation is when the product of two variables is constant. For instance, direct variation is $y = ax$ and indirect variation would be $y = a/x$ or $a = x \cdot y$.

7.4 LEAN SIX SIGMA PROJECT SELECTION PROCESS

The objective of Lean Six Sigma (LSS) is the methodology for an organization to discuss the unsolved problem that affects corporate bottom line and customer satisfaction. Therefore, implementing LSS tools in achieving ultimate operational performance is important. This will assure the success of the company's LSS initiatives and project goals. The study shows that the projects on the higher priority list and focused on the organization's core strategic operations achieve the highest financial gain at the completion of the project. Consequently, the proposed methodology can prioritize the financial gain and Voice of Customer, which are the key performance indicators (KPI) for a LSS project. Project selection criteria and subcriteria for sustainable DMAIC (define, measure, analyze, improve, and control) project results are based on business strategy, financial impact, and operational strategies.

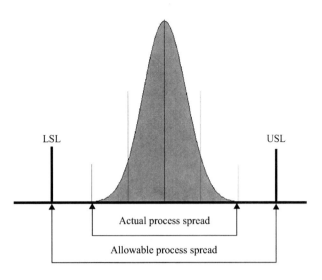

Figure 7.7. Process capability and specification limits.

7.4.1 BUSINESS STRATEGY

Lean Six Sigma is a process quality standard. Businesses that implement a Lean Six Sigma strategy are aiming for near-perfect (defect free) quality control in their products. The core competencies of Lean Six Sigma are made up of key concepts.

Design for Lean Six Sigma (*DFSS*): design exactly to customer needs, wants, and process capability

Cristal-to-Quality (*CTQ*): extreme attributes to the customer

Defect: not delivering customer needs

Process Capability (C_{pk}): a process where almost all the measurements fall inside the specification limits is a capable process. This can be represented pictorially by the plot shown in Figure 7.7.

Variation: Product inconsistency that customers receive and feel or see. In other terms, customers experience variations and not the mean. Simply suppliers are always being judged on the product variations. For this purpose, Lean Six Sigma focuses on the reduction of process variation to product quality as well.

Stable process: Any process that can produce a product or services based on the customer needs. Customers always honor products with the world class standards and quality. This is what LSS competes to meet higher standards.

7.4.2 FINANCIAL IMPACT ANALYSIS

Financial impact analysis (FIA) is a tool that allows the financial representatives, executives, or senior managers to review the financial impact of the proposed projects prior to allocating resources, and for making valid justifications and decisions for those projects. Projected financial impacts must be analyzed by a leading project manager or a Black Belt in agreement with the corporate finance representative. Black Belt/Green Belt must understand the financial measures, variance, and estimated impact of the project. The reported outcome of the project may be audited. Thus, the accuracy of the process improvement is necessary.

On the other hand, strategically, in some cases, a project does not need to have financial impact on the revenue or on return on investment (ROI) in order to be qualify as a LSS project. Such a project may perform as risk avoidance, customer satisfaction, loyalty, customer score-card improvement, cost prevention, and so on. These projects are considered to bring in qualitative rather than quantitative benefits.

7.4.3   OPERATIONAL ENGINEERING

Projects are the essential product of the Lean Six Sigma efforts because they achieve improvements in customer value in addition to operational and financial impact.

7.4.3.1   Resources Needed to Implement Six Sigma

One of the alarming realities is that it is not only employees who are unaware of such complexity, but also top management as well. Therefore, it is essential for senior management to understand the importance of Six Sigma training and the need to implement it at all levels of the organization. Other aspects of Six Sigma training is the future reengineering that helps companies move from doing what is most effective today to meeting tomorrow's challenges. It is essential in meeting the needs of organization and customers.

Roles and responsibilities for resources are normally assigned among Six Sigma champion, Master Black Belts, Black Belts, Green Belts, and a financial analyst to assess and audit the project impact on the key corporate results. These necessary roles and responsibilities are often determined by a LSS steering committee. The primary role descriptions are as follows.

Master Black Belts (MBB), also known as quality manager, with the highest level of training possesses knowledge of advanced applied statistical analysis, business strategies, and act as teachers and mentors. Responsible for delivering results and handling multiple projects, MBBs have strong leadership skills and must be recognized in the organization in order to influence decisions. The MBB also perform as a consultant to BBs in their projects to help, push, or direct if any hang-ups are needed to be resolved or cleared.

Black Belts are the technical leaders of the Lean Six Sigma team with an extensive level of training. These individuals should possess the following capabilities:

1. Mentoring other leaders in achieving the Lean Six Sigma goals.
    i. Selling the idea and philosophy of the concept.
    ii. Developing, setting a direction, facilitating, and leading the team to a higher level.
2. Teaching, training, and coaching the Lean Six Sigma tools as well as new techniques, case studies to project leaders in groups, and one-on-one cases.
    i. Must have the ability to convert the concept to a highly successful project. They should pass the Lean Six Sigma skills and tools to their peers and to the customer-oriented team.
    ii. Technically oriented with an analytical approach. Develop an in-depth knowledge of Lean Six Sigma statistical tools and techniques to improve key processes.
    iii. Develop and create techniques and short cuts to achieve objectives.
3. Think like senior management, i.e., time, financial results, performance, and corporate dynamics.

Green Belts are Lean Six Sigma project leaders, with part-time responsibilities. Spread across an organization, they must have statistical knowledge and be trained in basic Lean Six Sigma concepts. This individual is involved in problem solving in daily operations, in addition to contributing to more specific improvement efforts.

Financial auditor analyzes bottom-line results as a third party to avoid distrust. This person, rather with Black Belt or Green Belt certification, could also support in analyzing the cost of a defect and accepting projects.

7.4.3.2 Likelihood of Success

The basic principle of successful Lean Six Sigma projects are being in alignment with corporate strategic planning initiative, key financial results, and the support of corporate management.

7.4.3.3 Support or Buy-in

If there is no buy-in, then that means no project. Therefore, to avoid project from dying, Black Belts should pick a project from pipeline to gain the stakeholders' buy-in. As previously mentioned, projects must support corporate's global initiatives and objectives. This means achieving financial or strategic (non-financial) but measureable targets in key results.

Figure 7.8 illustrates the Six Sigma project selection process and tools that are required to achieve design for Six Sigma and continuous process improvement.

7.5 LEAN SIX SIGMA PROCESS MANAGEMENT AND PROJECT LIFE CYCLE

7.5.1 BUSINESS PROCESS MANAGEMENT

Business process management (BPM) is a scientific approach to make an organization's workflow more productive, effective, efficient, and capable of adapting to an ever-changing environment. So, today global economy leaves no room for organizations to remain complacent. They must constantly increase their efforts to improve their competency, which has a positive impact on their financial performance. Corporates looking to grow revenue and decrease operating costs are increasingly using BPM to deliver their results.

A business process is the complete set of "end to end" activities required to fulfill a transaction that provides value to a customer. In other words, process management is the complete and dynamically coordinated set of collaborative and transactional "end-to-end" activities that deliver value to a customer. An overall process management (Figure 7.9) involves the supplier, input, process, output, and customer.

BPM is focused on aligning all aspects of an organization with the "needs" and "wants" of customers. It promotes business effectiveness and efficiency while striving for innovation. Studies have shown that BPM helps organizations gain higher customer satisfaction. End-to-end processing refers to a system that performs a business process from "beginning to end," including all intermediate steps, such as data capturing, data processing, analyzing, and the generation of output. A process is an end-to-end work that involves everything you do to conduct your business.

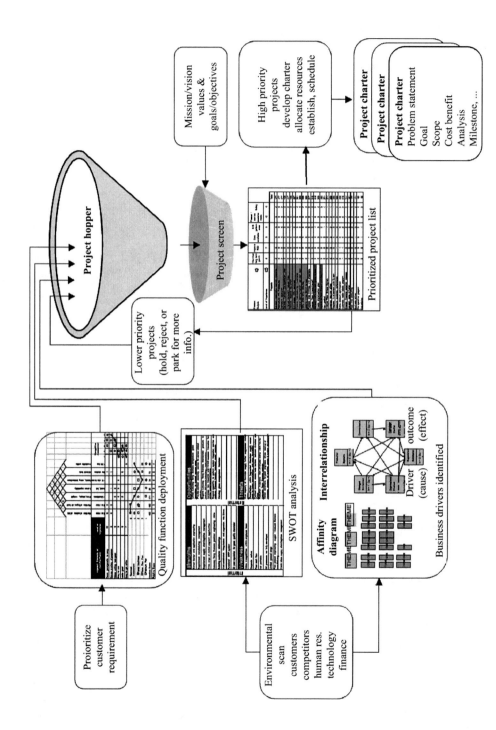

Figure 7.8. Six Sigma project selection process.

Figure 7.9. Process management.

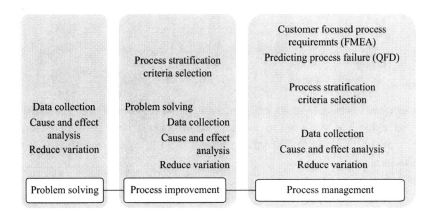

Figure 7.10. Quality planning management.

Every business process has a direct impact on the organization's efficiency, cost structure, and customer satisfaction. But how can one tell what's working and what's not? Thus, studying a business process will help one to understand the following:

1. What to do and what not to do for the process "end-to-end"
2. Business strengths for sustainable competitive advantage
3. Business weaknesses and look for process improvements
4. Simulation of business change before implementing the change
5. Measurement and monitoring business process performance in real time

With powerful analysis, visualization, and data filtering tools, process mining can help one discover, monitor, and improve real processes. LSS process management is a combination of problem-solving techniques that make use of quality improvement tools, process improvement using quality management tools, and process management using advanced quality planning tools as shown in Figure 7.10.

7.5.2 BPM PROJECT LIFE CYCLE

Any developed specialty is most often organized in the form of a lifecycle, with phases that are logically separate from each other and gate reviews in between but have well-defined hand-over points to move from one phase to the next.

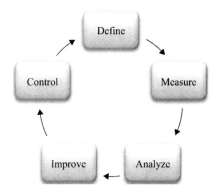

Figure 7.11. DMAIC process development.

Business process management or engineering product development life-cycle activities can be grouped into five, six, or more categories depending on the organizational process flow. The main activities that can be covered in a typical BPM lifecycle consisting of well-defined phase are as follows:

1. BPM: Ideation, design, modeling, execution, monitoring, and optimization
2. Defining, modeling, simulation, activation, execution, analyzing, and optimization
3. New product development: Concept, feasibility, planning, design, verify, launch, and production
4. Lean Six Sigma process development in Figure 7.11: Initiation, define, measure, analyze, improve, and control

Each phase includes a list of functions. Each function of phase is responsible for certain sets of processes made up of tasks and activities that are to be executed and reported as planned. Multiple processes are required for functional operations, and multiple operations are combined to achieve organizational goals.

7.6 WHO IS A CUSTOMER?

In common terms, a customer is someone or organization that a marketer thinks will profit from the process, products, and services proposed by the marketer's organization. In other terms, the person or group who actually uses the product or direct recipient of service. Customer is generally categorized into three types:

  i. *Current customers* include those who are and continue to purchase an organization's products or are using services for some period of time. For some organization, the current customer's timeframe could be discrete continuous process.
  ii. *Previous customers* consist of those who have previously purchased an organization's products or used its services, normally within a specific period of time in the past.
  iii. *Potential customers* are those who have yet to purchase or use an organization's product or services. But they possess the financial means to buy what the marketer thinks are the needs to eventually become future customers.

7.7   VOICE OF CUSTOMER

Whether you are an engineer, manager, or just an employee in any organization the chances that you will work with customers on a daily basis are high. It may be on an hourly basis, even though it may involve interaction with customers.

Customers are at the heart of all Lean Six Sigma initiatives, and this focus on customers is what makes LSS a superior organizational continuous process improvement strategy. A robust Voice-of-Customer (VOC) strategy is a major element in successful Lean Six Sigma program. Simply, one objective of Lean Six Sigma is to meet the needs and the wants of the customer. Voice of the Customer is the methodology used to find out the customer needs and wants. It is also an essential input at every phase of the Lean Six Sigma DMAIC process, specifically at the define phase.

The need is the real or true requirement to which the customer should be putting the product. For instance, the "need" of basic living is essential water, food, shelter, work, or money. This, if achieved, leads to customer satisfaction. Statistically it is customer's specification limits. The "want" is the perceived needs the customer feels should be met. So, "want" is something you feel you need or would like to have, i.e., second home, fancy clothes, nice car, and so on.

Further, VOC starts with defining LSS goals for collecting and analyzing customer requirements. The LSS project team members must identify various customer needs, especially the ones most recognized with the project. Then the team members must identify and select the most effective technique for collecting customer feedback and requirements. Using the collected information, VOC is converted into measurable, actionable project objectives and used to define the problem, which will set the direction of its LSS efforts to deliver the results projected.

7.8   KANO MODEL OF QUALITY

The Kano model of quality is a tool that can be used to prioritize the Critical-to-Quality (CTQ) characteristics, as defined by the Voice of Customer (or customer's needs) and the three categories identified by the model of quality are as follows:

i.   *Must Be*: Whatever the quality characteristic is, it must be present, such that if it is not, the customer may opt for an alternative.
ii. *Deliver Performance*: The better the needs are met, the happier and more satisfied the customer is.
iii. *Delighter*: Those qualities that the customer was not expecting but received as a bonus.

The model includes two dimensions on an x–y axis graph as shown in Figures 7.12 and 7.13. The horizontal (x-axis) represents how good the organization is in achieving the customer's need (s) or CTQs and is called achievement. It begins with the supplier "not delivered" the service to "fully delivered." The vertical (y-axis) corresponds to the customer's level of satisfaction the customer should have, as a result of organizational level of achievement. It is also called customer satisfaction, which starts with "total dissatisfaction" with the service or product to "total delighted."

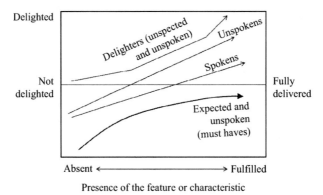

Figure 7.12. Sample 1: Kano model of quality.

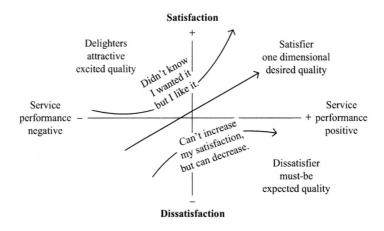

Figure 7.13. Sample 2: Kano model of quality.

7.9   SWOT (STRENGTH, WEAKNESSES, OPPORTUNITIES, THREATS) ANALYSIS

A "SWOT" as shown in Figure 7.14 analysis details the organization's strengths, weaknesses, opportunities, and threats in the form of a matrix. It is a summary of the analysis of the organization. It is often used in identifying strategic level projects and initiatives. Depending on the organization's reasons for using SWOT, the questions can be changed accordingly. The strength and weakness correspond to internal factors within the organization. The opportunity and threats apply to factors outside the company from competition.

7.9.1   STRENGTH

Strengths are characteristics of the business or project team that give an advantage over others. Strengths are helpful in achieving the initiatives.

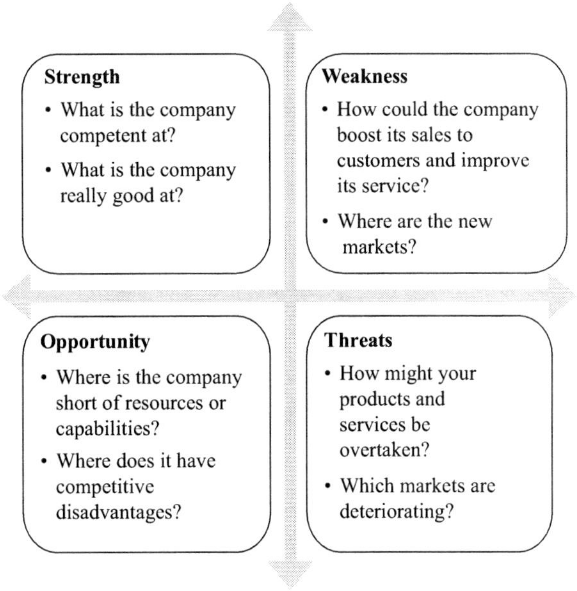

Figure 7.14. SWOT analysis.

7.9.2 WEAKNESS

Weaknesses are characteristics that place the team at a disadvantage relative to others. Weaknesses are harmful in achieving the initiatives.

7.9.3 OPPORTUNITIES

Opportunities are external chances to improve performance (e.g., make greater profits) in the environment. Opportunities are helpful in achieving the goals.

7.9.4 THREATS

Threats are external elements in the environment that could cause trouble for the business or the project. Threats are harmful to achieving the objectives.

Understanding of company's SWOT analysis:

Figure 7.15 is an example of strategic SWOT analysis for any Lean Six Sigma project.

7.10   PROJECT SCOPE, CHARTER, AND GOALS

Project scope: The project scope is defined as specific project with a beginning and end points, the budget and time that have been allocated to achieve the objectives. The more specific the details (what is in the scope and what is out of the scope), the less a project may experience "scope creep." Project scope and work effort required to complete project must also be

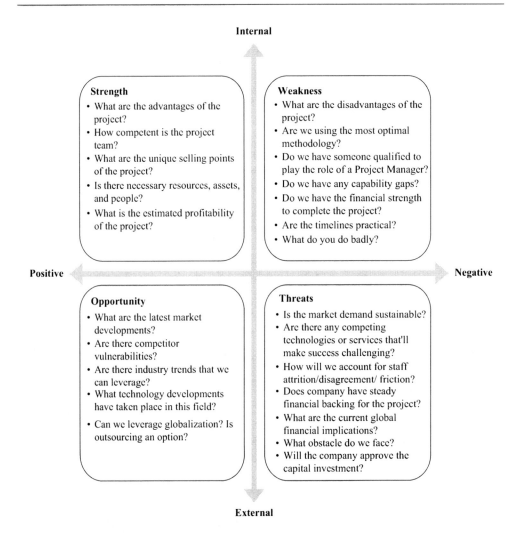

Figure 7.15. Six Sigma SWOT analyses.

monitored as the project proceeds to ensure that all changes are properly managed. The scope consists of two main elements: Sum of the products (deliverables) and services provided as a project.

Deliverables: any measureable, tangible, and verifiable items that must be produced to complete the project.

- Milestone: a significant event in the project, usually the completion of a major deliverable.
- Project phase: a collection of logically related project activities, usually maximizing the completion of a major deliverable.

Boundaries: establish the boundaries for the projects, which are described in the project statement, which is helpful to separate the items that support the scope of the project from those out of the scope. If the scope of the project is not clearly defined, then there is a chance that project may fail.

Project charter: The Lean Six Sigma Project Charter is the most basic Six Sigma tool. It outlines summary of the project and includes statement of the problem, the scope of project, the metrics involved, the return on investment, and the project team members are all selected in the project charter. Table 7.3 illustrates Lean Six Sigma project launch plan charter. The define phase is completed with the completion of all the areas identified in the project charter. The project would not start unless all the requirements in the project charter are fulfilled. However, project charter does not have to be precise once it completed. Minor change may occur throughout the phases as the project progresses. Additionally, some of the assumptions may not be accurate at the start of the project. This is because as one goes through the measurement and analysis phase the process data become more accurate and concrete. If one concluded something along the way that proved prior thinking was incorrect, that is already a huge success.

Project goal: The project objectives are the outputs that one is trying to optimize using the input variables of the process. The output could vary from one to many. All the outputs should be measureable. The statistical tools allow one to measure all the variables. The baseline, which is simply current measurement (performance of existing process) and targets the optimum process, should be recorded in the project charter. The question is what the process measurement would be if the process was perfect. So, the goal should be how this process would behave in a near defect-free situation. The SMART (Specific, Measurable, Attainable, Relevant, and Time-bound) goals along with other Six Sigma methodology would enable one to achieve these goals.

Furthermore information regarding roles and responsibilities in the project charter may be helpful in the progress of the project.

Beginning point: project identification (define phase)

Ending point: project completion (completion of control phase) and standardization of process

Process Owner

1. Identifies project opportunity through use of pipeline reports, business plans, process dashboards, and so on.
2. Quantifies financial benefit of the project
3. Identifies the project lead
4. Determines if Lean Six Sigma Kaizen training is required
5. Supports the lead in completing the project launch plan and the project charter
6. Supports the lead in project charter reviews
7. Administers regularly scheduled project tollgate reviews
8. Drives overall project implementation or elimination

Project Lead Committed to Lean Six Sigma Resource

1. Completes project launch plan and project charter; reviews with process owner
2. Leads project from concept (define) through deploy
3. Applies proper Six Sigma tools and methodology to their project
4. Plans and performs data collection as well as analysis

Table 7.3. Lean Six Sigma project launch plan

Project Name/Number		Expected Net Gain (profit) ($)	
Process Owner		Sponsor (Champion)	
Project Lead		Business unit	
Black Belt coach		Target completion date	

1. **Business Case** • How would you describe the project to your manager? • What are the financial benefits? • Why is it urgent that you do this project now? • How does this link to your functions or key initiatives? • What would happen to the organization if you don't address this?	
2. **Process Definition** The process in which the opportunity exists.	
3. **Problem Statement** • Who is impacted by the problem? • What is actually happening? • Where is it occurring in the process? • When is it occurring in the process? • How much and how extensive is the problem?	

4. **Project Objective**	**Goal Statement** • By when do you want to have the improvement in place? • How much of an improvement are you aiming for? (x% error reduction, x% cycle time reduction) • What measurable business impact will the improvement have? Type? Magnitude?	*Specific, Measurable, Attainable, Relevant and Time-bound (SMART)*			
	Define: *replace definitions below with definitions appropriate for this project* • **Key Output Signal (KPI or Metric)**—*a key measure (key performance indicators) that indicates the performance of the business process*	**Baseline**			
		Errors/ Defects (D)	Units Processed (N)	Opportunities per Unit (O)	DPMO *(Defects per million opportunities)*

		Future State	
	• **Error/Defect**—*any part of a product or service that does not meet customer specifications or requirements, or causes customer dissatisfaction, or does not fulfill the functional or physical requirements.* • **Unit**—*something that can be quantified by a customer. It is a measurable and observable output of your business process. It may manifest itself as a physical unit or, if a service, it may have specific start and stop points.* • **Opportunity**—*the total number of chances per unit to have a defect.*	**Output Signal Goal**	**Target Improvement (%)**
	Data source and frequency of measurement		
5. Voice of the Customer (VOC) Who are the customers of the output product; what are their key measures; what is important to them?			
6. Project Scope What processes, systems, products, services, channels, etc. will you consider in this project? What will be excluded? *Scope to allow for completion of Define through Analyze in 60–90 days*			
7. Team Members Names and roles of team members			
8. Principal Stakeholders Who are the approvers of team decisions, resource of subject matter experts, or interested party who needs to be kept informed?			

9. Project Timeline Key milestones/ dates for each phase	**Project Start**	**Define Complete**	**Measure Complete**	**Analyze Complete**	**Improve Complete**	**Control Complete**	**Implement Complete**

10. Additional Support Required Are any special capabilities, hardware, trials, etc. needed?	

5. Follows process owner direction and support when necessary
6. Plans meetings with the Black Belt coach to review progress and receive just-in-time coaching
7. Reports project status at project reviews as needed

Lean Six Sigma Kaizen Black Belt Coach/Mentor (Dedicated Six Sigma Resource)

As required, provides the following skills to support the Project Lead:

1. Assists the lead in completing project launch plan prior to Six Sigma training.
2. Assists the lead in structuring or breaking down tasks into manageable assignments to be done between their meetings.
3. Assures the rigor and discipline of the Six Sigma methodology are applied appropriately by the lead.
4. Provides just-in-time training of quality tools and helps guide lead's efforts when statistical expertise is needed.
5. Helps lead decide what data will be useful and how best to gather the data.
6. Advises the lead on project progress, deliverables/milestones, and next steps.
7. Reviews the lead's project template for completeness and proper use of quality tools.
8. Helps the lead for project gate reviews.

7.11 LEAN SIX SIGMA METRICS AND PERFORMANCE MEASURES

What is Lean Six Sigma metrics? Why measure the process? It is often said that, "What gets measured gets done." Metrics are a set of measurements that quantify results. A measurement in the form of metrics communicates the values and creates priorities for management to allocate the resources available. Priorities may depend on the complexity of the process. The information collected from the process may be measured in different ways depending on the type of the process. That is the continuous process that will produce variable data with more information than discrete process with attribute data like small or large, good, or bad. Project metrics tell us whether the project is meeting its goals and what is its limitation. Metrics are generated using process measurement into various operational and financial categories like the following:

Productivity: Depends on the industry type, i.e., manufacturing or services. Productivity measures may include labor cost, product per unit cost, up time or down time of machine asset productivity.

Personnel: This measures information about the people's skill, habits, employee turnover, missing time, sick time, and experience.

Safety: In any industry safety measures can be critical in identifying issues such as OSHA (Occupational Safety and Health Administration) lost workday incidents, injuries resulting from duties, incident rate, and workman's compensation cost per employee.

Assets: The measurement provides information about organization resources, value-added cycle time, product inventory, number of days inventory supply, capital assets utilizations, and maintenance cost.

Maintenance: Maintenance as well as preventive maintenance play a big role in the performance of product production or services. Some of the common measurements are time to repair, machine down time, and duration of production after fixes.

Quality: Quality is extremely essential in Lean Six Sigma transformation. It proves how well customers' "needs" and "wants" are met based on their requirements. Some of the measures include warranty claims per unit, warranty percent of sales, C_p (potential process capability), C_{pk} (actual process capability), first pass yield (units required no rework), defects per million opportunities, on-time delivery rate, and errors per delivery.

Transactional service: Although quality measures in a manufacturing industry focus on products, customer satisfaction is the quality indicator (i.e., key performance indicator) for service environments. Whatever the organization may be, common measures include customer complaints per unit and the rate of fully satisfied customers.

7.11.1 CRITICAL TO QUALITY

Critical-to-Quality (CTQ) measures characteristics of a product, process, or services in an effort to meet performance standards or specification limits (USL/LSL). In other terms, CTQ is the customers' "needs" and "wants" that must be performed fully. This is a necessary part of customer satisfaction and is highly important to the overall success of an organization. The typical features of some product or service can be as clear as a word or phrase that is descriptive. It does not have to characterize the whole product or service, but must cover at least one critical function. Measures will define how the features of the product or service should be quantified. It is essential to understand what factors or input is applied to design the product/process or services and what is the result. The information obtained from output will be used in the analysis of improving overall customer satisfaction. As the outputs are measured and customer needs are achieved, then it will help company to grow stronger. That is why CTQ has become an important factor of customer fulfillment and why it is often measured using Lean Six Sigma methodology. Figure 7.16 illustrates a tree diagram of Voice of Customer (VOC).

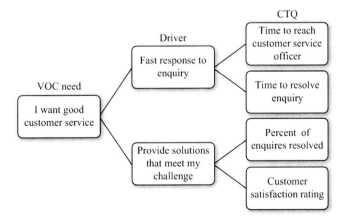

Figure 7.16. Illustratration of a tree diagram of Voice of Customer.

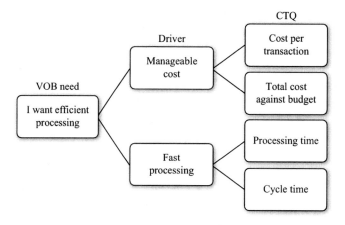

Figure 7.17. A tree diagram of Voice of Business.

7.11.2 CRITICAL TO BUSINESS AND VOICE OF BUSINESS

The Voice of Business (VOB) is derived from financial outcome of the business operations. It encompasses the primary "needs," "wants," expectations, both the spoken and unspoken, of the shareholders, executive officers, or other people who involved in the organization's management. It may be used to develop the strategic planning objectives of the company. From the corporate strategic initiatives information, the Voice of Business will support the qualified projects in moving the organization closer to its goals and objectives. Figure 7.17 illustrates a tree diagram of Voice of Business.

7.11.3 COST OF QUALITY

The term "cost of quality (COQ)" refers to the price of producing products that do not match with the customer's needs. Simply, it means cost of producing defects. This includes every element that contributed to defects in the production of products or services. Cost of quality (Figure 7.18) can be estimated using the following equation:

$$Y_{qc} = \frac{(X_{md} + X_{ld} + X_{rc}) + (X_{dpc} + X_{ac})}{X_{cmc}} \qquad \text{(Eq. 7.1)}$$

where

Y_{qc}	Cost of quality
X_{md}	Material defect cost
X_{ld}	Labor defect cost
X_{rc}	Rework cost
X_{dpc}	Defect prevention cost
X_{ac}	Appraisal cost
X_{cmc}	Complete manufacturing cost

Some of the quality cost categories are described below.

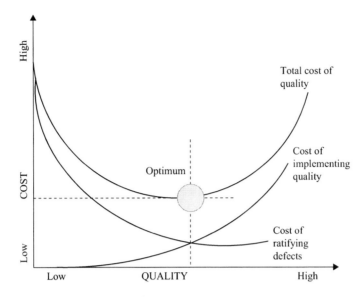

Figure 7.18. Cost of quality.

7.11.3.1 Defect Prevention Cost

This means costs of all the activities designed to prevent the poor quality in products or services such as process quality control, process capability evaluation, quality planning and training, data analysis and corrective action, improvement of communication, and new product design reviews.

7.11.3.2 Appraisal Cost

Appraisal costs are associated with testing instruments (capital and maintaining), testing and inspection, process and product quality system audit, calibration of measuring testing instruments, data collection cost, and process or service audit.

7.11.3.3 Internal Defects

Internal defects costs occur ahead of product shipment/delivery, completing a service to customer. For instance rework, scrap, reinspection, retesting, material review, cost of correcting defects, and conforming them to the customer specification limits, material, labor, and overhead defects on unfixable components, equipment downtime, and failure analysis.

It is fact, there is a cost associated with building a higher quality for applications, but there is also a cost associated with a less quality. If you sum up both cost functions you'll end up with the "total cost of quality" as shown in Figure 7.18. As the graph clearly indicates, both ends of the spectrum—less quality or none and too much quality—are undesirable. So, the goal must be to work toward an optimum point where costs generally balance each other out.

7.11.3.4 External Defects

Failure costs occurring after shipment or delivery of the product, and during or after providing of a service to the customer, for example warranty returns, customer complaint correction, material returns (receipt and replacement), cost to customer, and product recalls.

7.12 SPECIFIC, MEASUREABLE, ATTAINABLE, REALISTIC, TIME-PHASED

A SMART goal is a process monitoring system to detect and report on various key performance indicators (KPI) of reliability.

7.12.1 SPECIFIC

A specific goal has much higher chance to achieve than a general one. To plan a specific goal one need select the team members, goal and objective, process location to be improved, timeline, requirements and constraints, and benefits as well as justification for conducting the project.

7.12.2 MEASURABLE

Develop a solid criteria to measure the process progress toward the achievement of each step of the goal being set. This will allow you to measure the progress within the given time to meet the target date and feel the excitement of continued process improvement in achieving your goal.

7.12.3 ATTAINABLE (ACHIEVABLE)

The goal must be realistic and achievable. Once it is set, then you can begin to find out the ways for achieving it. You may have to develop motivation, skills, abilities, and financial means to get there. However, slowly you will find yourself closer to the achievement of your goals. The more goals you have the more you can get as long as you plan right and follow the steps within the time frame. Even if the goals are hard and too far away, eventually through the time and effort it moves closer and becomes achievable.

7.12.4 REALISTIC

To be practical, a goal must be realistic and achievable toward which you are willing to work hard. You are the only one who can decide how important and high priority your goal is. The higher priority goals are faster to reach than low-priority ones, which reduce motivation and positive thinking. Every project, high or low priority, must show progress in its different steps.

7.12.5 TIME-PHASED

A goal should be completed within a time frame. Without a time table plan for the project there's no urgency to meet the target. There is no sense of achieving project performance and delivering result without a time frame. Upon experience if one believes that the project and goals are realistic then it can be accomplished.

7.13 FORCE FIELD ANALYSIS

Force field analysis (FFA) is a useful technique for analyzing all the forces for and against a decision such as a manufacturing process improvement. It illustrates process problems in terms of positive forces that support change in the desired direction (driving forces) and the negative forces that promote the status quo (restraining forces). A positive force can be people, resources, positive attitudes, traditions, regulations, value adds, needs, desires, and so on.

Driving positive forces are forces that impact the decisions that push to improve the process; they tend to initiate a change and keep moving on. In terms of improving productivity, pressure from managers and recognitions may be examples of driving forces. So, FFA can help one to improve the probability of a project success in two ways: To reduce the strength of the forces opposing a project and to increase the forces pushing a project.

Restraining negative forces are forces acting to reduce the driving forces. Lack of interest, confusion, disorganization, mismanagement, and poor preventive maintenance of equipment may be examples of restraining forces against increased production. Final decision at current level of productivity is achieved when the score of the driving forces equals the score of the restraining forces. This is featured in Figure 7.19.

7.13.1 DEFINE THE CURRENT PROCESS PROBLEM

What is the nature of the existing process that is causing problem and requires changes? It is important to distinguish the process problems from those that are performing well.

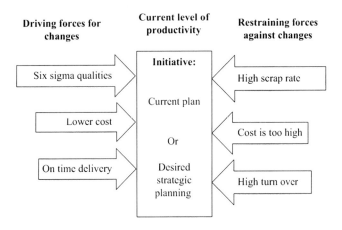

Figure 7.19. Force field analysis.

7.13.2 DEFINE THE IMPROVEMENT GOALS

What is the desired direction that would justify improving the process? It is necessary to state as clear as possible.

7.13.3 DEFINE THE DRIVING FORCES

Define the forces that promote improvement in the desired direction. And define forces that approximate strengths of these forces. Then record these driving forces on the chart of FFA diagram as labeled arrows with the length of the arrow reflecting the approximate strength or weight of each force. Finally for the purposes of analysis, define the relationships among the driving forces.

7.13.4 DEFINE THE RESTRAINING FORCES

Define the forces that resist the proposed improvement and sustain the current situation. Place these forces on the diagram as you did those for the driving forces. Define the relationships among the restraining forces.

7.13.5 ESTABLISHING THE COMPREHENSIVE CHANGE STRATEGY

Any possible change can occur in the middle line by strengthening any of the driving forces, adding new driving forces, or removing any of the restraining forces. Therefore, the total strength of driving forces must be bigger than restraining forces for the transformation to take place.

7.13.6 FORCE FIELD ANALYSIS EXAMPLE

Describe your plan or proposal for change in the current level of productivity (Figure 7.20). List all driving forces for change in one column, and all restraining forces against change in another column. The score is based on the strength of the force and the degree to which it is possible to influence the force. Assign a score to each force, from 1 (weak) to 5 (strong). Calculate a total score for each of the two columns. Decide if the goal or change is feasible. If so, brainstorm a course of action that strengthens positive forces, weakens negative forces, and creates new positive forces.

Table 7.4 is an example of force field analysis template.

7.14 TOLLGATE REVIEW AND CHECKLIST FOR DEFINE PHASE

This is the last chance before moving to the next phase (measure). Make sure that you have completed and prepared all the requirements for the *Define Phase*. You may tailor Figure 7.21

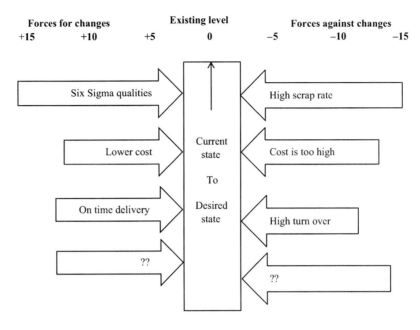

Figure 7.20. Force field analysis for existing productivity.

Table 7.4. Force-field analysis

<hr>

Force-field analysis template

<hr>

Force-field analysis template definition
Use this template to analyze two opposing forces of a problem or potential solution.

Potential uses:
- Identify pros and cons of options prior to making decisions.
- Explore what is going right and what it going wrong.
- Explore any two diametrically opposed items.

Type the question or problem being analyzed here

<hr>

Positive Forces (+)	Negative Forces (−)
•	•
•	•
•	•
•	•
•	•
•	•

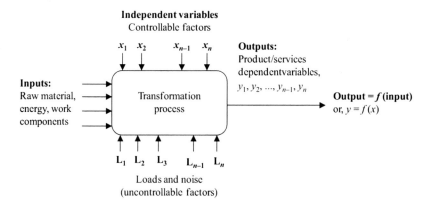

Figure 7.21. Lean Six Sigma process improvement projects: Tollgate review "define the process."

to your own process and come up with a solid project definition. That includes, but is not limited to, the define phase deliverables and checklists.

7.14.1 DEFINE PHASE DELIVERABLES AND CHECKLISTS

(a) Concept and ideation brainstorming
(b) Apply as many as define tools and methodologies such as
    i. Project charter and scope
       • Is the project scope justifiable?
       • Project success criteria
       • Identify project risk factors
    ii. VOC (Voice of Customer)
    iii. Customer requirments
    iv. CTQ (Critical-to-Quality) tree diagram
    v. Affinity diagram
    vi. Kano Model
(c) Lean Six Sigma project selection
(d) End-to-end project timeline.
(e) Complete the Lean Six Sigma Scorecard
(f) Feasibility and planning
(g) Project strategic alignment with organizational business direction.
(h) Define the problem statement, which includes what is the problem, when did it happen, and where was it occur and how serious it is?
(i) Identify the deliverables to be achieved with end-to-end timeline and target.
    i. Identify the essential milestone.
(j) Identify the financial business cases and the impact of that in organization future market share and growth.
(k) Create and define RACI-Matrix. Include team members, gate keepers, and anyone who should be part of RACI-chart.

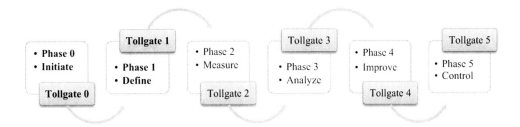

Figure 7.22. Lean Six Sigma process improvement projects: Tollgate review "Define phase."

(l) Identify the customer and their "needs" and "wants," timeline, requirements, specification limits, quality needs, etc.

(m) Determine list of key shareholders and if they are going to be included in the project. Or, how the information will be conveyed to them.

(n) Find out the roadblocks and obstacles that are required to be eliminated for project to continue

(o) Apply all the Define tools as much as needed in describing the project as discussed in this chapter.

Once the checklist has been completed, the project needs to go through the gate review (Figure 7.22) as discussed in Chapter 6 for gate review team approval.

CHAPTER 8

MEASURE CONCEPTS AND STRATEGIES

CROSS FUNCTIONAL CORE COMPETENCIES "ADVANCED MEASUREMENT SYSTEM"

Measure phase is all about the assessment of the current state. The goal is to locate the source of the problem as accurately as possible by understanding the existing system, which can be either manufacturing or transactional processes. If you can't measure the defects in a process strategically, you can't take steps to eliminate them. Simply if you can't measure them in a process you can't control it, manage it, or even improve it. Thus, systematic measurement of a process determines how the organization's product and services are meeting the needs of the customer. It offers an efficient and effective tool in collecting data and establishing a matrix that demonstrates whether the organization is achieving its objective with its current status quo. It involves numerical data and the Voice of Process (VOP).

8.1   THE SEVEN QUALITY CONTROL TOOLS FOR MEASUREMENT

The essential seven quality control measurement tools are described in the following sections.

8.1.1   CAUSE-AND-EFFECT DIAGRAM (FISHBONE OR ISHIKAWA) $y = f(x)$

It is also called fishbone or Ishikawa diagram. It identifies the causes for an effect or process problem. Here cause means "why it's happening" and effect means "what is happening." Thus, cause and effect matrices are used to study a problem or improvement opportunity by identifying root causes. It focuses on causes, not symptoms. In other words, the cause and effect diagrams are used to identify all the potential or real causes (inputs) that result in a single effect (output). Causes are laid out based on their level of importance or detail. This can help one to search for root causes and identify areas where there may be problems. A simple example of cause and effect diagrams tool for pizza delivery processes, as shown in Figure 8.1, consists of three sections: baking, packaging, and delivery. Each section is caused by different categories of variations.

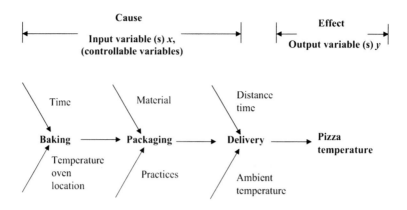

Figure 8.1. Cause and effect for pizza delivery.

Cause and effect is a key concept that one must fully understand and is a fundamental principle of Six Sigma methodology. In its simplest form, it is called "cause and effect." In its more robust mathematical form, it is called "y is a function of x." Another way of saying this is that the Output = dependent (inputs). Since the output is dependent on the inputs we cannot directly control it, but we can only monitor it. Only independent variables can be controlled.

Let's review which process variables (causes) have critical impact on the output (effect). Since mathematically $y = f(x)$ is a transfer function tool to determine which input variables (x's) affect the output responses (y's). Therefore, here we have output "pizza temperature" being a function of input baking, packaging, and delivery or in the form of function

$$Pizza\ temperature = f\ (baking,\ packaging,\ delivery)$$

or $$y = f(x) \tag{Eq. 8.1}$$

Additional cause-and-effect diagram is illustrated in Figure 8.2 for the employee's lack of communication with managers or vice versa. The cause focuses on the understanding, clarity, and delivery of messages as well as frequency of communication.

Now, the question is, *if we are so good at the x's why are we constantly testing and inspecting the y*? Various methods are reviewed in this chapter for resolving the output issues.

8.1.1.1 Establish Cause (Why It's Happening) and Effect (What Is Happening)

Ask *why* five times. Identify root causes by analyzing potential causes as long as one can ask *why* and get an answer. The potential cause was not the root cause. For example, ask the "5 *whys*" questions; in this case "Pizza didn't taste good." This is shown in Figure 8.3.

1. Why didn't the pizza taste good?
    It was due to cold temperature.
2. Why was the temperature cold?
    It was due to late delivery.

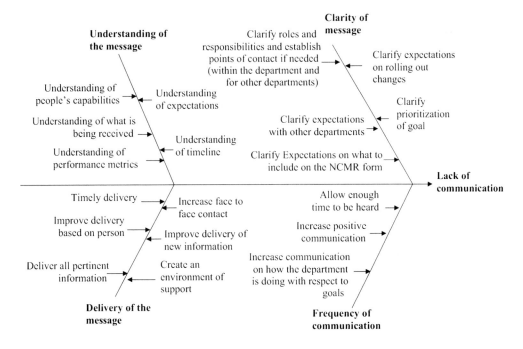

Figure 8.2. Cause and effect for "lack of communication."

3. Why was the pizza late?
   It was due to accident.
4. Why did accident happen?
   It was due to heavy traffic.
5. Why was the traffic heavy?
   It was due to lunch time.

Normally, the third question will explore the majority of the problems.

Defect

	Answer why pizza didn't taste good	Answer why the machine failed
1st Why		
2nd Why		
3rd Why		
4th Why		
5th Why		

Figure 8.3. Five whys.

8.1.2 DATA COLLECTION (PROCESS MEASUREMENT AND CHARACTERIZATION): VOICE OF CUSTOMER (VOC)

In many organizations, Six Sigma simply means a measure of quality that strives for near perfection. Thus, the complete evaluation and measurement of the ongoing process is necessary for redesign or future designs of process or product. The process measurement must be recorded without any manipulation. Data collected (continuous or discrete) represents the actual real time capability of the process. For instance, data collected in a process should be relevant to output variables (dependent or response and or effect) such as mechanical properties as well as input variables (independent or cause) of the equipment, operator, and or process (e.g., pressure, volume, temperature, velocity, and so on). The data should be selected randomly with a large sample size or population.

8.1.2.1 Process Measurements and Data Collection

The highlights of the process measurement steps are as follows:

1. Define and clarify key process measures which make big impact on the outcome of the process.
2. Identify the vital few causes that have the greatest effect on the process.
    i. Identify process parameters [process input x's (causes) and process output y's (effects)] and correlation between them: $y = f(x)$.
    ii. Develop a detailed process map (or SIPOC of existing process) to identify possible measures. For details of SIPOC, refer to Section 8.3.1.
3. Analyze the collected data.
    Identify what is being observed and keep the data collection process as easy as possible. Simple check marks are the easiest.
    i. Collect data on key process response and convert any transactional data to numerical data.
    ii. Develop operational definitions and procedures.
    iii. Validate measurement system and methods. Collected data should be grouped in a way that makes the data valuable and reliable.
    iv. Start data collection and continue to improve the measurement consistency. Similar problems must be in similar groups. And create a format that will give you the most information with the least amount of effort.
4. Measure output variable's performance.
5. Perform process capability analysis (C_p, C_{pk}, P_p, P_{pk}).
6. Apply failure analysis such as five *whys* method (Section 8.1.1).
7. Continue to ask five *whys* that eventually will lead to the root cause of the problem.

Example 8.1: Root cause analysis

A leading book store located in a major shopping mall gradually had lower sales per day than budgeted. The owner noticed that quite a few customers came into the store to browse, but left without buying anything. When considering this problem (not tapping the customer base potential), a wide range of possible causes surfaced, which included the fact that the customers did not find what they were looking for. The staff did not offer the necessary help.

1. Sought items were temporarily sold out.
2. There was no customer help.
3. Interested items were not carried by the store.
4. Cost was too high.
5. There were too long lines at the checkout counter.
6. Some types of the credit cards were not accepted.
7. Store lighting was poor in some areas.
8. There were no places to sit and look through books before deciding to purchase them.

8.1.2.2 Root Cause Analysis

The challenge in identifying the actual issue and how often it occurred make it difficult for the store staff to implement any changes. During a two-week period, many of the customers leaving without making purchases were courteously surveyed why this happened. The responses were recorded in a check sheet shown (Table 8.1), which gives a much clearer idea of where to start to improve the situation.

8.1.3 PARETO CHART

It was named after Vilfredo Pareto, a 19th-century Italian economist who proposed "the vital few and trivial many rules" and predictable imbalance 80:20 rules. Vilfredo Pareto studied distribution of wealth in Italy and found that 20% of the Italian population owned 80% of Italy's wealth. He then noticed that 20% of the pea pods in his garden accounted for 80% of his pea

Table 8.1. Check sheet/data collection

Cause of no. purchase	Week 1 survey	Week 2 survey	Total number of occurrences
Could not find item	||||| ||||| ||||| ||	||||| ||||| ||||| |||||	37
No customer help	||||| ||||| ||	|||||		20
Items sold out	||	|||	5
Items not carried	|||||	|||||		9
Cost was too high				1
Line too long			|||	4
Wrong credit cards	||		2
Poor lighting	||||| ||	||||| ||||| ||	19
No place to sit	||	|||||	6
Total number of customers who didn't purchase	49	54	103

crop each year. Thus that brings us to 80/20 rule: If we work on the correct 20% we will use our time and resources to the maximum benefit. A small number of causes are responsible for a large percentage of the effects, usually a 20% to 80% ratio. This basic principle translates well into quality problems: most quality problems result from a small number of causes. One can apply this ratio to *almost anything*, from the engineering science of management to the physical world. The following examples illustrate this concept.

1. Addressing the most troublesome 20% of the problem will solve 80% of it.
2. Within your process, 20% of the individuals will cause 80% of your headaches.
3. Of all the solutions you identify, about 20% are likely to remain viable after adequate analysis.
4. 80% of the work is usually done by 20% of the people.
5. 80% of the quality can be obtained in 20% of the time; perfection takes five times longer.
6. 20% of the defects cause 80% of the problems.

Project engineers or managers know that 20% of the work (the first 10% and the last 10%) consume 80% of the time and resources. A Pareto chart is a useful tool for graphically depicting these and other relationships. It is a simple *histogram style* graph that ranks problems on the order of magnitude to determine the priorities for improvement activities. The goal is to *target the largest potential improvement* area and then move on to the next, then next, and in so doing addressing the area of most benefit first. Figure 8.4 illustrates where the company should allocate the time, human, and financial resources that will yield the best results.

Pareto diagram: The contribution, the causes of whatever is being investigated, is listed across the bottom (*x*-axis) and a percentage is assigned for each (relative frequency) to total 100%. A vertical bar chart is constructed, from left to right, in order of magnitude, using the percentages for each category. A 80% improvement in quality or performance can reasonably be expected by eliminating 20% of the causes of unacceptable quality or performance. This is envisioned in Figure 8.5.

Break point: The percentage point on the line graph for cumulative frequency at which there is a significant decrease in the slope of the plotted line.

Vital few: Category contributions that appear on the left of the break point account for the bulk of the effect.

Trivial few: Category contributions that appear to the right of the break point account for the bulk of the effect.

In addition to selecting and defining key quality improvement programs, Pareto chart is used to prioritize problems, goals, objectives, identifying root causes, selecting key customer

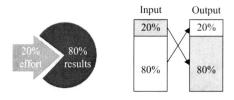

Figure 8.4. 80/20 rule.

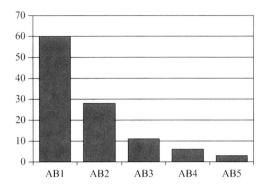

Figure 8.5. Pareto chart.

relations and service programs, selecting key employee relations improvement programs, and defining key performance improvement programs. Further, it maximizes research and product development time, verifies operating procedures, and manufacturing processes.

8.1.4 HISTOGRAM

The purpose of a histogram is to graphically summarize the distribution of causes or relationship data set. Histogram illustrates frequency data in the form of a bar graph that is bell-shaped. It is a highly effective tool in identifying the mean and capability of the process. Normally the frequency represents the dependent variable (y-axis) and independent variable (x-axis). The histogram in Figure 8.6 graphically illustrates the following:

1. Center (i.e., the location) of the data
2. Spread (i.e., the scale) of the data set
3. Skewness of the data set
4. Presence of outliers
5. Presence of multiple modes in the data.

These features provide strong indications of the proper distributional model for the data. The probability ploy test can be used to verify the distributional model.

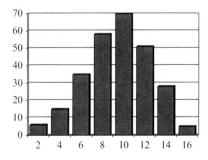

Figure 8.6. Histogram.

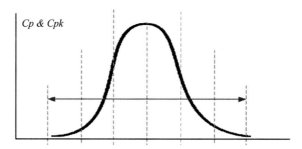

Figure 8.7. A bell-shaped histogram.

(a) Construct and interpret histogram, i.e., bell shape (Figure 8.7), binomial, skewed, truncated, isolated peak, comb, and so on.
(b) Summarize data variation in the process.
(c) Visually and graphically communicate process data.
(d) Apply process capability tools (C_p, C_{pk}, P_p, P_{pk}) and calculations.

8.1.5 SCATTER DIAGRAM AND CORRELATION

Scatter diagram in Figures 8.8 through 8.10 (Pearson's R value) illustrates the visual view of the qualitative relationship, either linear or nonlinear, that exists between two variables (of input and output) using an x and y diagram. Furthermore

  i. It provides both a visual and statistical means to test the strength of a potential relationship.
 ii. It supplies the data to confirm that two variables are related.
iii. The steps to accomplish this are the following:
    1. Collect paired data samples.
    2. Draw the x- and y-axis of the graph.
    3. Plot the data.
    4. Interpret the outcome of the plot, such that values can be only between $(-1, +1)$. For example:
        (a) Strong positive linear slope in $x - y$ coordinate means "strong positive correlation," as shown below:

$$Slope = \frac{\Delta y}{\Delta x} = +1.0.$$

        (b) Strong negative linear slope in $x - y$ coordinate means "strong negative correlation," as shown below:

$$Slope = \frac{\Delta y}{\Delta x} = -1.0.$$

        (c) Scatter (non-linear) data all over the $x - y$ coordinate means "no correlation," e.g.,

$$Slope = \frac{\Delta y}{\Delta x} = 0.0.$$

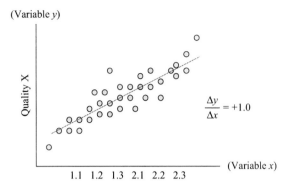

Figure 8.8. Scatter diagram: Strong positive correlation.

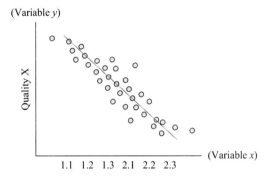

Figure 8.9. Scatter diagram: Strong negative correlation.

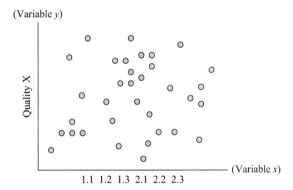

Figure 8.10. Scatter diagram: No correlation.

Pearson correlation: Pearson is another method of distinguishing the different levels of correlation as visualized in Figure 8.11.

8.1.6 CONTROL CHARTS

Originally this control chart was introduced by Dr. Walter A. Shewhart in 1920s. Often it is referred to as "Shewhart chart." The objectives of control charts (see Figure 8.12) are to enable

Pearson correlation

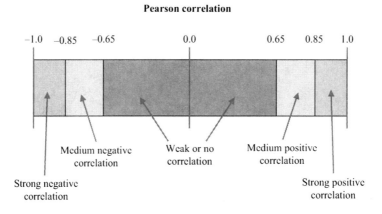

Figure 8.11. Pearson correlation.

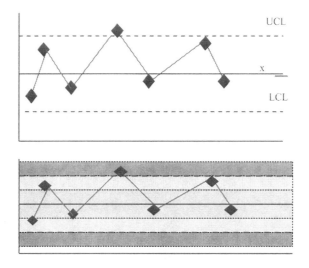

Figure 8.12. Pearson correlation.

the front-line worker to distinguish between random variation and variation due to an assignable cause and to monitor process performance over time for stability. It also helps to identify opportunities, understand, and control variations.

8.1.7 STRATIFICATION (TRENT, FLOW, OR RUN CHART)

Stratification describes the way in which different data are divided into groups. When measurement from a variety of sources or categories have been lumped together, the meaning of the data can be impossible to see. Stratification technique separates the data so that trends can be clearly seen. It divides the data into groups and detects a pattern that identifies a problem by looking at why the frequency of impact varies between time, location, or condition.

The ways to stratify factors are who (department, individual, customer type), what (complaint type, defect category, incoming call reason), when (month of the year, quarter of the year, week of the month, day of the week, time of the day), and where (region, city, specific location on product).

8.2 THE DESIGN OF SEVEN MANAGEMENT/PLANNING TOOLS

8.2.1 AFFINITY DIAGRAM

Purpose and Selection: You need to generate a high volume of ideas by tapping the creative side of the brain. In addition, capture ideas not available in typical meeting discussions. Creatively brainstorm and organize a large number of ideas or issues. Summarize natural groupings among ideas to understand the essence of a problem. Figure 8.13 illustrates these concepts.

Example 8.2: Creating Affinity Diagram

What do we need to learn about human body bones?

First we need to brainstorm ideas on what do we need to learn from human body bones. Then, list the ideas on a separate sheet. Figure 10.14 illustrates this concept.

Then organize Figure 8.14 into like groups and adding a header for each class. This is shown in Figure 8.15.

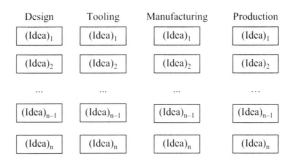

Figure 8.13. Affinity diagram template.

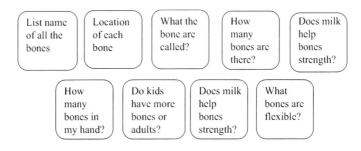

Figure 8.14. Affinity diagrams of human body bones.

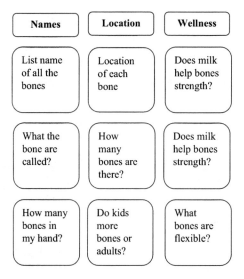

Figure 8.15. Categorized affinity diagrams for human body bones.

Others examples for students to work on are as follows:

1. What do we want to know about the Lean?
2. What do we want to know about the Six Sigma?
3. How can we make students not to miss the class?
4. What are the barriers to getting students to class on time?
5. How can we better communicate with high school students to focus on their school work?
6. What are the barriers of obtaining results from the field test on time?
7. What are the barriers to increasing the turnover time for requisitions of new employees?
8. What are the barriers to getting people to fitness center?

8.2.2 INTERRELATIONSHIP DIAGRAM

The purpose of interrelationship diagram is to identify, analyze, classify the cause-and-effect relationship matrix among critical issues and determine key drivers or outcomes for effective problem resolutions. The key steps in creating an interrelationship diagram are to agree on the issues or problem statement and look for cause-and-effect relationship. By creating graphical representation of all the factors in a complicated problem, system, or situation and the final interrelationship diagraph is drawn and the largest factors (drivers) and outcome are identified (see Figure 8.16). This tool can be also used in identifying root causes, even when the data is unavailable.

Step six begins with "B" and repeats the same questions for all the remaining combinations. Draw only one-way arrows in the direction of the stronger cause or influence (drivers and outcomes). A high number of outgoing arrows illustrates that an issue is a driver or possible root

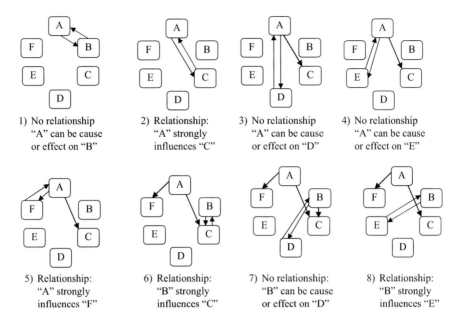

1) No relationship
   "A" can be cause
   or effect on "B"

2) Relationship:
   "A" strongly
   influences "C"

3) No relationship
   "A" can be cause
   or effect on "D"

4) No relationship
   "A" can be cause
   or effect on "E"

5) Relationship:
   "A" strongly
   influences "F"

6) Relationship:
   "B" strongly
   influences "C"

7) No relationship:
   "B" can be cause
   or effect on "D"

8) Relationship:
   "B" strongly
   influences "E"

Figure 8.16. Interrelationship diagram.

cause. In other words, a high number of incoming arrows indicates that an issue is an outcome. Make a decision on the stronger direction.

Interrelationship Diagram Exercise

- Create a work-related problem statement (possibly within your organization).
- Lay out all of the ideas/issue cards that have either been brought from other tools or brainstormed.
- Look for cause/influence relationships between all of the ideas and draw relationship arrows.
- Choose any idea as a starting point and work through them in sequence.
- Determine the Final "Driver (Cause)" and "Outcome (Effect)."
- Draw the final interrelationship diagram.
- Draw a conclusion.

Figure 8.17 illustrates interrelationship diagram example on automation machine efficiency improvement.

An another example of cause and effect from an automotive survey regarding tire sensor variation is shown in Figure 8.18.

8.2.3   TREE DIAGRAM

Tree diagram breaks any broad goal into levels of detailed action plans. It also visually displays connectivity between goals and action plans. This is shown in Figure 8.19.

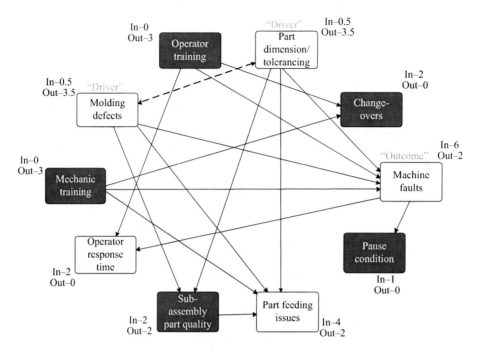

Figure 8.17. Automation machine efficiency improvement.

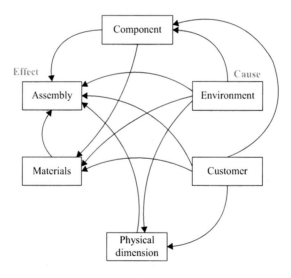

Figure 8.18. Automotive sensor variation reduction.

8.2.4 THE MATRIX DIAGRAM OR MATRIX CHART

The matrix diagram illustrates the relationship between two, three, or more groups of information (Table 8.2). It also can present the knowledge about the relationship (e.g., its strength and the roles played by various individuals or measurements). A good example of matrix is house of quality or quality function deployment (QFD) in that the roof-shaped matrix diagram and internal process metrics are formed in different shapes.

Figure 8.19. Tree diagram.

Table 8.2. Matrix diagram (chart)

	Title # A	Title # B	Title # C
Item # 1			
Item # 2			
Item # 3			
Item # 4			
Item # 5			
Item # 6			
Item # 7			

8.2.5 MATRIX DATA ANALYSIS

Matrix data analysis is also called prioritization matrices. It is a complex mathematical methodology for analyzing of matrices. One of the most rigorous, careful, and time-consuming of decision-making tools, a prioritization matrix is an L-shaped matrix (Table 8.3) that uses pairwise comparisons of a list of options to a set of criteria in order to choose the best option(s).

Table 8.3. Matrix data analysis

	Type #1	Type #2	Type #3
Item # 1			
Item # 2			
Item # 3			
Item # 4			
Item # 5			
Item # 6			
Item # 7			

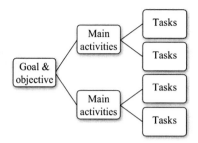

Figure 8.20. Process decision program chart.

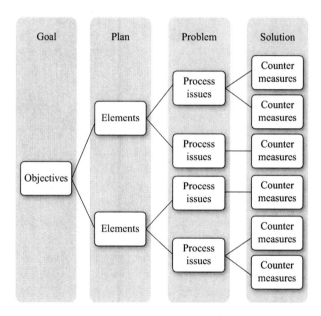

Figure 8.21. Process decision program: expanded chart.

8.2.6 *PROCESS DECISION PROGRAM CHART (PDPC)*

It is used when the plan is large, complicated, and too costly. In addition, it must be accomplished on time and on schedule. As shown in Figure 8.20, a tree diagram is used with goals and objectives. Thus a tree describes the main activity and defined tasks that support the main objectives.

A useful method of strategic planning is to break down activities into a hierarchy using a tree diagram in Figure 8.21. PDPC simply expands this chart a couple of levels to identify risks and countermeasures for the lower level tasks, as in Figure 8.21. Different shaped boxes can be used to highlight the risks and countermeasures.

8.2.7 *ARROW DIAGRAM (THE ACTIVITY NETWORK DIAGRAM)*

It is also called network diagram, activity network diagram, activity flow chart, node diagram, and critical path method chart (CPMC). The arrow diagram illustrates the required order of

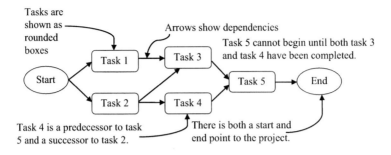

Figure 8.22. Arrow diagram.

tasks in a strategic planning project or process, the best schedule for the overall project, potential planning, issues, and their solutions. The arrow diagram lets you calculate the "critical path" of the project. The flow chart shown in Figure 8.22 is the essential step where interruption will affect the timeline of the project. On the other hand, additional resources can speed up the project process.

8.3 PROCESS MAPPING

Process mapping is a process flowchart, which displays an accurate and detailed picture of the process. You have probably heard of the term "a picture worth a thousand words." In any organization a process map or a process flow chart describes a process workflow. But what exactly is a process map? Are there different kinds of process maps? The answer for these questions is examined here for different kinds of process maps and how they are used to define a process. After all, the foundation of all businesses is a common set of core processes.

In general, a "process mapping" visually illustrates the flow activities of a process that transform inputs into outputs. A process flow can be described as the sequence and interactions of related process steps, levels, activities, or tasks that make up an individual process, from start to finish (the boundaries). A process map is a flow from left to right or from top to bottom. One may get confused if the arrows go from right to left or bottom to top. Thus, backflow is preferred to be minimized. All processes should be measurable with clear key performance indicators (KPI). Processes are strategic assets of an organization that if managed well deliver a competitive result. It also helps us in defining roles and responsibilities, internal controls and communication, and establishes standards for compliance, consistency, and performance.

Steps to consider when mapping a process are as follows: identify the flow or sequence of events in a process that product or service follows, the opportunities for improvement, the boundaries, the steps, sequence of steps, draw appropriate symbols, validate the flowchart, and finalize the flowchart. This is visualized in Figure 8.23.

It also helps to explore more improvements, i.e., SIPOC, FMEA, Process-FMEA, Design FMEA C_p, C_{pk}, P_p, P_{pk}, etc. One of the most commonly used flow chart is SIPOC. Other examples are shown in Figure 8.24 for manufacturing and "DMAIC" process in Figure 8.25.

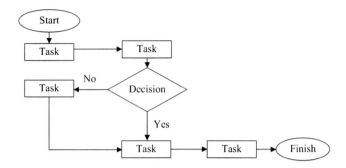

Figure 8.23. Process flow chart.

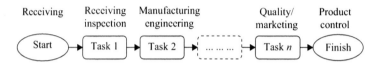

Figure 8.24. Process flow charts for manufacturing.

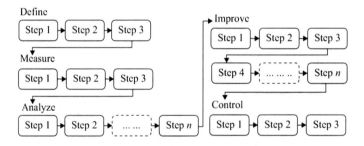

Figure 8.25. "DMAIC" model process map.

8.3.1 SIPOC CHART (SUPPLY, INPUTS, PROCESS, OUTPUT, CUSTOMER)

The acronym SIPOC (Figure 8.26) stands for suppliers, inputs, process, outputs, and customers and is a high-level process map that "maps out" its basic overall steps. It is used at the start to define the scale and scope of the project, which also distinguishes boundaries (X = process inputs, and Y = process outputs) of the process. Through the process, the suppliers (S) provide input (I) to the process. The process (P) that your team is assigned to improve adds value, resulting in output (O) that meets or exceeds the customer (C) expectations. This also allows you to identify the opportunities for improvement. A SIPOC is mapped out most easily by starting from the right (customer's expectation) to identify all relevant elements of a process improvement and working toward the left. SIPOC also called overall business process map.

The following SIPOC is an example of manufacturing production in which SIPOC analysis for machine maintenance cost reduction is shown in Figure 8.27.

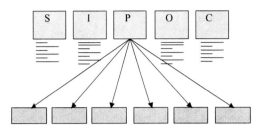

Figure 8.26. Process flow chart.

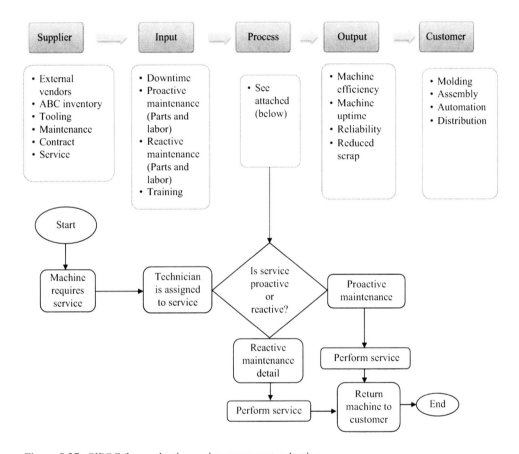

Figure 8.27. SIPOC for production maintenance cost reduction.

Figures 8.29 illustrates a high-level snapshot that captures information critical to a project and provides definition and agreement on project scope and boundaries. It also describes potential areas of data collection and project measurement. At the highest level, proactive maintenance activities are value-added and all reactive activities are unplanned activities and considered non-value-added. When reactive maintenance situations arise, they require immediate attention regardless of available resources. This has a direct impact on the completion of value-added activities due to the fact both reactive and proactive resources are shared.

8.3.2   VALUE STREAM MAPPING

Value stream mapping (VSM) is a lean manufacturing methodology used in Six Sigma process to analyze and design the flow of materials and information required to deliver a product or service to a consumer at the highest speed. It provides maximum value to the customer with the minimum waste in the process of

- design concept to meet or exceed customer "needs" and "wants";
- build and deliver on time and within the boundaries; and
- maintain through life cycle to service.

8.3.2.1   Value-Added versus Non-Value-Added Activities

Value added: Activities for which an external customer is willing to pay for quality, cost, and shipment. Or, an activity that transforms or shapes raw material or information to meet customer needs and wants.

   Non-value-added: The activities that take time, resources, or space, but do not add value to the product or services. So, where does the non-value come from? It comes from top seven wastes of operations.

1. *Overproduction*: Products or services are being produced in more quantity than customer demand. Excess products may be sold at discounted prices.
2. *Delay and wait time*: Employee or parts/services that wait for work cycle to be finished. This would include processing delays, machine or system downtime, response time, or signature required for approval wait-time.
3. *Transportation*: Unnecessary transportation of parts/services between processes or locations, such as wasted floor space, travel distance, and poor communications.
4. *Processing and complexity*: Over processing the product/services beyond the customer requirement, i.e., out of innovation, outdated standards, and lack of continuous improvements.
5. *Excess inventory*: Storing excess products with no orders in the warehouse and having excess work in progress (WIP), i.e., inaccurate forecasting, excessive downtime, push in place of pull.
6. *Wasted motions/Unutilized talent*: Motions that may lead to injury in the manufacturing environment resulting in process delay.
7. *Errors and defects*: A defect is a component that will add additional rework, inspection, or waste to the product and process.

8.3.2.2   Visual Representation of Production

Value stream mapping creates a visual representation of a production process as shown in Figure 8.28 to identify and eliminate waste and maximize efficiency. In other words, this includes all the activities that are required to satisfy a customer need from order to delivery. For instance, a paper took 210 seconds to obtain 3 signatures versus a wait time of 24 days.

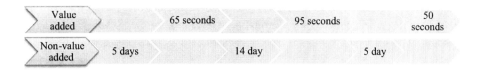

Figure 8.28. Visual representation of value and non-value added time.

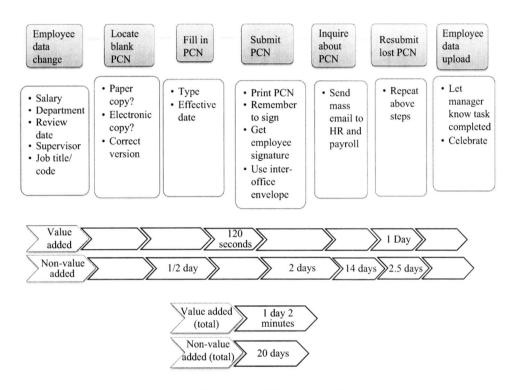

Figure 8.29. Value stream mapping. Who has the PCN (Personnel Change Notice?).

Figure 8.29 represents the real-time value stream mapping in a manufacturing environment for "who has the personnel change notice (PCN)" for employees.

Thus, value-added analysis approach are as follows:

1. Draw the process flowchart.
2. Identify the activities that add value.
3. Identify the non-value-added activities.
4. Decide whether activities should be kept, combined, or eliminated.
5. Determine appropriate actions.

8.4   KAIZEN EVENTS: PLANNING AND EXECUTION

Kaizen is a Japanese term meaning "improvement" or "change for the better." In other words, "kai" means to change and "zen" means make better. Event refers to rapid improvement. Rapid improvement processes are commonly recognized to be the building block of lean manufacturing methodology. Its core focus is eliminating waste, improving productivity, attaining and maintainability of continuous improvement processes in manufacturing, engineering, and business management.

Rapid continuous improvement normally requires a corporate culture where employees are allowed to identify root causes and solve the issues. Organizations that are applying *kaizen* improvement processes have very well-developed methodologies along with training systems communicated through rules and techniques. *Kaizen* events are very effective and very powerful because they are carefully selected and planned. Most of the *kaizen* power comes from the people who are being developed around the lean culture. If a lean culture is implemented, *kaizen* events will be successful. The basic steps for "kaizen events" are outlined in the following and visualized in Figure 8.30.

Phase 1: Planning. The first priority is to identify an opportunity for rapid improvement events. Such areas might be composed of work-in-process (WIP), administrative process, production delays, machine down time, product or service performance, financial impact projects, areas that create a big mess in the factory or office, or anything that can be improved quickly. Generally it is focused on three to four-day breakthrough events and include the following activities:

1. Preparing and training the team
2. Defining the problem/goal
3. Recording the existing state
4. Analyzing the present methods
5. Brainstorming, developing future state and verifying ideas.

Phase 2: Execution. The *kaizen* team must have a clear understanding of the existing issues of the targeted process that everyone working toward the solution. Couple of popular methodologies for finding the current manufacturing waste is "five whys" and value stream mapping. Toyota has developed a technique of asking "why" five times and responding with an answer to uncover the root cause (see Section 8.1.1 for details). The second method is "value stream mapping" (see Section 8.3.2). Value stream mapping identifies all the non-value-added activities in the targeted process that *kaizen* team is trying to improve. In the *kaizen* event, it is important to collect data on the targeted process issues such as source of process scrap, scrap rate, product quality, frequency of production change over time, amount of WIP (work-in-progress),

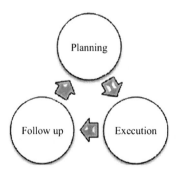

Figure 8.30. Kaizen events.

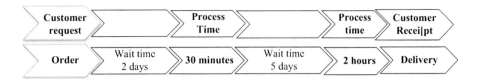

Figure 8.31. Customer request and received order lead time.

or just seven popular types of waste (see Section 8.3.2.1). For instance, one of the important non-value-added activities is wait time. For example, we look at when a customer places an order and receives the order after how long. Figure 8.31 illustrates the "lead time" for customer request and shipment of order.

Furthermore, we know that lead time is "Lead-time = Process time + Delays" and activity ratio (AR)—the percentage of time spent on work to the product flowing through the process

$$AR = PT/LT \times 100$$

$$100 - AR = Idle\ time.$$

The following steps can be utilized in the *kaizen* rapid process improvement event.

Days 1 and 2: Analyze the existing process and perform root cause analysis. Then design the future state.

Days 3 and 4: Design and verify improvements and check by applying Lean tools and align the team to gain the buy-in.

Day 5: Standardize improvements; train workforce.

Kaizen team presents its ideas weekly and leadership team makes sure that project team has taken into consideration all the options and all the implications. Leadership's role during proposal is that not to say "no" but to say or ask questions such as "Have you considered this?" "How would we handle…?" "What if…?" What this means is that corporate workforce grows stronger as they learn how to identify and eliminate waste. Leadership grows stronger as they're able to shift their focus to strategic planning initiatives, performance monitoring, and coaching or teaching staff.

Phase 3: *Follow-up.* A key part of a *kaizen* event is that present results and follow-up activity ensure continuous improvement and not just improvement for a short period of time. *Kaizen* event team routinely follows the key performance indicators (i.e., matrices) to document the process improvement gains. *Kaizen* events, however, cannot target and solve every problem within an organization. There are certain types of improvements for which other methods such as Six Sigma should be utilized.

8.5   LEAN: IMPROVES EFFICIENCY/SIX SIGMA AND IMPROVES EFFECTIVENESS

Lean manufacturing is based on the idea of *kaizen* or continuous improvement. At its core, *kaizen* refers to a process of continuous improvement that establishes a system to focus on eliminating all kinds of manufacturing or transactional process waste. Thus, by removing waste processes, efficiency and speed improve. The advantage of Lean is that it includes employees from multiple functions who may have a role in a given process, and strongly motivates them to participate in eliminating waste activities. Lean is a powerful tool for uncovering the hidden wastes or waste-creating process and removing them.

Six Sigma is a philosophy of identifying defects in product or services and eliminating them. In other words, it is a measure of quality that strives for near perfection. Thus Lean Six Sigma improves process accuracy and speed.

8.6 QUALITY FUNCTION DEPLOYMENT

Quality function deployment (QFD) was developed in Japan in the late 1960s by Professors Shigeru Mizuno and Yoji Akao to assure quality in new product development. In 1972, the first QFD matrix, the "house of quality," was added to QFD as a new tool. Following that, it was implemented at the Mitsubishi Heavy Industry's Kobe shipyard. QFD was introduced to the U.S. industries from Japan in 1984, and in 1988, John Hauser and Don Clausing described the use of a simple subset of QFD by U.S. automotive parts suppliers in the classic *Harvard Business Review* article "The House of Quality" (May–June 1988).

8.6.1 WHAT IS QFD QUALITY?

Quality function deployment (QFD) is a design planning process or communication tool that is driven by customer requirements. QFD deploys "the Voice of the Customer" throughout the organization and uses planning matrices called the "The House of Quality." It is applicable to any product or process manufacturing and service or transactional processes. The customer dictates the attributes of a product, which is the driving force behind QFD. So, the main focus should be on meeting or exceeding customer expectations, which must be taken literally and not translated into what the organization defines them as. When collecting customer's information one may ask the following questions. What does the customer really want? Are customer's expectation used to drive design process? What can design team do to achieve or exceed customer satisfaction?

8.6.2 BUILDING A "HOUSE OF QUALITY"

Begin with listing customer requirements (What's) and technical descriptors (How's). Then develop a matrix relationship between "What's" and "How's", develop interrelationship (How's), and competitive and technical assessments. Finally, prioritize customer requirements and technical descriptors (see Figures 8.32 and 8.33).

1. **Create focus on customer requirements (the What's)**
   (a) Gather customer attributes (in the words of the customer).
   (b) Identify customer need (the What's): Expected, spoken, unspoken, exciters, needs, preferences, and problems.
   (c) Group attributes logically and assess relative importance of the attribute.

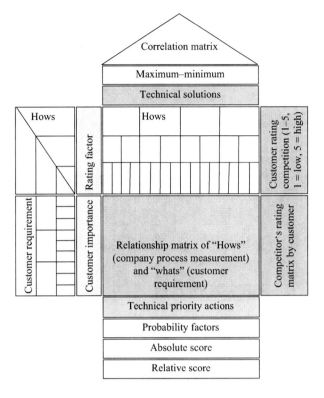

Figure 8.32. Quality function deployment (QFD).

2. **Develop improvement activities (How to deliver What's)**
   (a) Direction for improvement and how to get to goals.
   (b) Tool: tree diagram structure used to document the sub-group of *How's*.
3. **Relationship matrix of How's) and What's**
   (a) How will each *how* fulfill each *what*
   (b) Two-stage analysis
      i. Is there a relationship?
      ii. If yes, is it a low (score = 1), medium (score = 3), or high (score = 9)
4. **How's interrelationship (correlation matrix) conflicts or supports**
   (a) Interrelationship between technical descriptors
   (b) Illustrates positive and negative relationships among the *how's* (How's versus How's)
   (c) Rating scores are strong positive (+9), positive (+3), negative (−3), strong negative (−9)
   (d) Focuses on complimentary and conflicting *How's*
5. **Technical assessment**
   (a) Assess competitive performance on the attributes
   (b) Investigates competitive benchmarking on *How's*
   (c) Evaluate weaknesses and strengths of product or service
   (d) Find out about the drawbacks in competitors' product or service
   (e) Achieves targets and specification to outperform the competitors' market
6. **Describe product in terms of engineering characteristics**
   (a) Detail the influence of engineering characteristics on customer attributes
   (b) Detail the interaction between engineering characteristics

O Negative influence
P Positive influence
X Strong influence

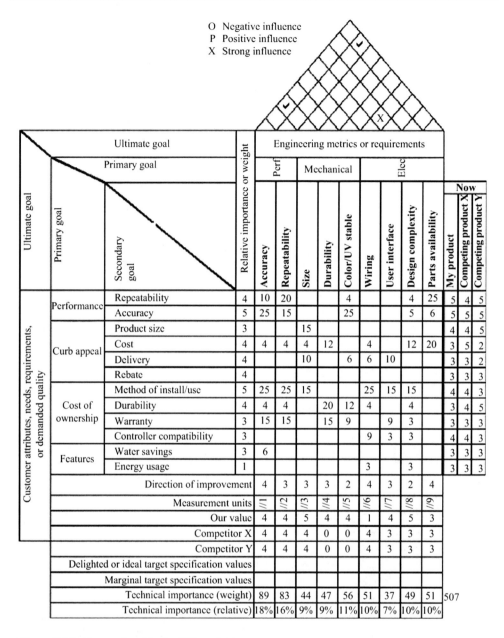

Primary goal	Secondary goal	Relative importance or weight	Accuracy	Repeatability	Size	Durability	Color/UV stable	Wiring	User interface	Design complexity	Parts availability	My product	Competing product X	Competing product Y
Performance	Repeatability	4	10	20			4			4	25	5	4	5
Performance	Accuracy	5	25	15			25			5	6	5	5	5
Curb appeal	Product size	3			15							4	4	5
Curb appeal	Cost	4	4	4	4	12		4		12	20	3	5	2
Curb appeal	Delivery	4			10		6	6	10			3	3	2
Curb appeal	Rebate	4										3	3	3
Cost of ownership	Method of install/use	5	25	25	15			25	15	15		4	4	3
Cost of ownership	Durability	4	4	4		20	12	4		4		3	4	5
Cost of ownership	Warranty	3	15	15		15	9		9	3		3	3	3
Cost of ownership	Controller compatibility	3						9	3	3		4	4	3
Features	Water savings	3	6									3	3	3
Features	Energy usage	1						3		3		3	3	3
	Direction of improvement	4	3	3	3	2	4	3	2	4				
	Measurement units	/1	/2	/3	/4	/5	/6	/7	/8	/9				
	Our value	4	4	5	4	4	1	4	5	3				
	Competitor X	4	4	4	0	0	4	3	3	3				
	Competitor Y	4	4	4	0	0	4	3	3	3				
	Delighted or ideal target specification values													
	Marginal target specification values													
	Technical importance (weight)	89	83	44	47	56	51	37	49	51	507			
	Technical importance (relative)	18%	16%	9%	9%	11%	10%	7%	10%	10%				

Figure 8.33. House of quality for PF1 product.

7. **Target value for technical descriptors**

   At this step in the process, the QFD team starts to draft target values for each technical descriptor. Target values represent "how much" for the technical descriptors, which can then act as a baseline to compare against.

8. Calculate absolute weight and percent using equation:

$$a_j = \sum_{i=1}^{n} R_{ij} c_i$$

where R is relationship matrix and c is customer importance.

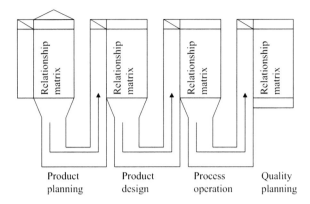

Figure 8.34. Steps of product development.

9. Calculate *relative weight* and percent can be calculated using the following relation:

$$b_j = \sum_{i=1}^{n} R_{ij} d_i$$

where R is relationship matrix and c is customer absolute weight.

Full QFD for all steps of product development in product planning, product design, process operation, and quality planning is illustrated in Figure 8.34.

8.7   MEASUREMENT SYSTEM ANALYSIS (MSA)

A measurement system analysis is an essential factor of quality improvement initiatives. It is an important tool to assess the capability of measurement system to identify the standard deviation (variation from target) in the process variable. Then, take action on minimizing these factors that could excessively contribute to the variation in the data. For example, in an experimental design, the level of variation may be attributed to an operator, machine parameters, and interaction effects. All these variations can be analyzed and assessed during the measure phase of the define-measure-analyze-improve-control (DMAIC) cycle. The goal is to confirm if the measurement system used to obtain data is valid. There are statistical methods available to estimate the repeatability and reproducibility called gage R&R. That is a study of variability in a product or process.

Repeatability means taking the same measurement more than once on the same components with the same instrument (gage) to get identical results. In this case, normally there are more than 10 parts in a homogenous batch.

Reproducibility means different people measuring the same part with the same instrument (gage) will get identical results. However, there are some parts that can't be measured more than once such as parts in destructive tests. There are statistical methods available to estimate the repeatability and reproducibility (R&R). This is pictured in Figure 8.35.

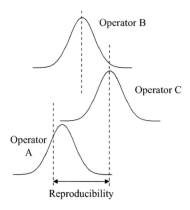

Figure 8.35. Gage R&R.

8.8   PROCESS MEASUREMENT

8.8.1   DATA COLLECTION

Any organization's success depends upon how it delivers on its processes. Before the process can be improved, it must be measured with the appropriate tools. The essential steps of data collection and process measurement lead to process improvement in any Lean Six Sigma initiative. A successful data collection begins with careful planning; a knowledge of various data types, sampling strategies, and measurement methods. Only reliable and suitable data will lead to dependable analyses that translate into required process improvements. Lean Six Sigma project leaders, such as Black Belts, will oversee data collection efforts during the measure phase of the Lean Six Sigma process. They will determine what should be measured and the kind of techniques used in collection of data and must ensure the integrity and accuracy of data. Furthermore, the following steps will improve the process of data mining.

1. Decide the type of data that needs to be collected in a given process.
2. Recognize the advantages of automated data collection.
3. Be able to convert data into a different data type (i.e., transactional or attributes).
4. Match measurement scales to associated statistical analysis tools.
5. Identify the use of best practices for ensuring data accuracy and integrity in data collection.
6. Select the correct measurement tools that matches category description.
7. Follow the order of steps in the data collection process.
8. Determine the classification of engineering tests as destructive and nondestructive tests.

8.8.1.1   Identification of the Key Measurement Variables (x's)

Develop operational definitions, procedures, and validate the measurement system. Then, begin data collection (data collection is time-consuming). Understanding the variation in Y's

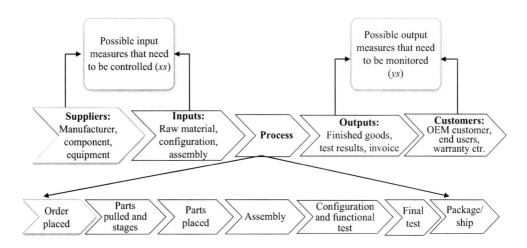

Figure 8.36. SIPOC steps for process measurement system.

(output) requires data about the X's (process inputs). Use the SIPOC map to identify possible measures. The process measurement step for a particular process system is illustrated in Figure 8.36.

Other funneling tools are Critical-to-Quality (CTQ) tree, prioritization matrix (PM), quality function deployment (QFD), interrelationship diagraph (ID), cause-and-effect (CE) or fishbone diagram (FD), failure mode effect analysis (FMEA): Ask five *Whys*, design FMEA, process FMEA, and continue to improve the measurement consistency. Identify key measurements using a tree diagram as pictured in Figure 8.37.

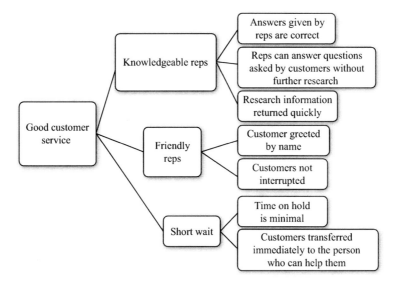

Figure 8.37. Key measurements using a tree diagram.

8.8.1.2 Basic Measurement Concepts

Observe first and then measure. Know the difference between variable and attribute data. *Variable data* is what you would call quantitative. There are two types of data: discrete data and continuou) data. *Attribute data* is always binary and unusable for the purpose of quantification. Examples include good or bad, yes or no, and high or low. Once you convert it to discrete data by counting the number of good or bad, it becomes discrete variable data. *Or,* Attribute data represents the absence or presence of characteristics. Go or no-go gaging and the presence or absence of a component yield attributes data. Measure for a reason and have a measurement system in process. This concept is outlined in Figure 8.38.

8.8.2 PRINCIPLES OF VARIATION

Everything varies. Individual things are unpredictable. Groups of things from a constant system of cause are predictable, i.e., every time you write the letters of your name, it is different than the last one. You don't know how the next letters of your name will be different. There is something about your name that distinguishes it from if someone else writes letters of your name. Variation exists in all the processes. Understanding and reducing variation remains key to success.

8.8.3 TYPE OF VARIATION

Attributes or discrete variables (also called qualitative variables) can be scored, but not be counted into a continuous. Variable data (or quantitative) are measured on a continuous and infinite divisible numeric scale. Continuous variable data falls along a continuum, i.e., age, weight, and height. Quantitative (independent variable) sets along a measurable axis and can be measured to see their position with respect to others. Discrete variables falls among discontinuous variables such as number of basketball player in each team and door size.

8.8.4 TYPE OF DATA

Attribute data are discrete (discontinuous) data. These data types measure includes blue, red, automobile type BMW, Benz, Honda, injection molding process, molded parts, integer number, counting numbers, blood type, tree type, apple color, pass or fail, leak or no leak, gender, small, medium, or large, go or no-go and yes/no tests, and positive or negative options. Continuous data consist of extrusion process, pipe or plastics rope, real number, and numbers with decimal.

8.8.5 SCIENCE OF STATISTICS

Statistics is the science of collecting information from data, organizing, analyzing, understanding, and translating data for decision making under uncertainty. So statistics involves the study of how numerical data or facts are collected, how they are analyzed, and how they are interpreted.

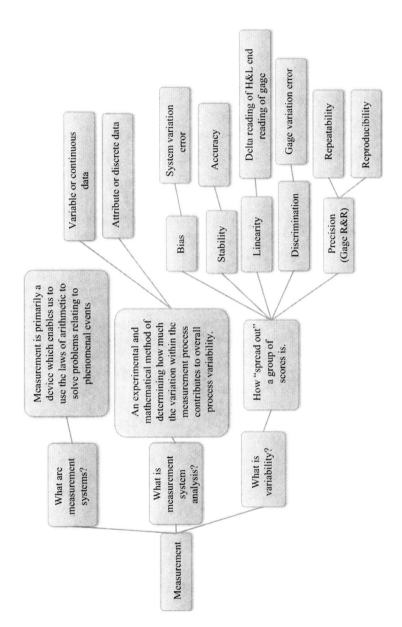

Figure 8.38. Measurement system classification.

1. Descriptive statistics simply collects, organizes, or summarizes a particular set of measurements and describes the same set of data (e.g., histogram, bar chart, pie chart, and so on).
2. The inferential statistics analyze the measurement collected from a sample to make inferences or decision about the population from which the sample data was obtained. In other words, inferential statistics will estimate or predict about a population based on sample information.
3. Population includes the set of all possible measurement of elements.
4. Sample: A set (or portion) of data drawn from the population or a subset of population. It is less costly to obtain data for a selected subset of a population, rather than the entire population. Observations are easier to measure and analyze with a sample set than with a complete count.
5. Parameter: A number, i.e., x-bar ($\bar{x}$) average of a given population.

8.8.6   CLASSIFICATION OF NUMERICAL DATA

Measurement is the process of observing and recording the observations. However, the methods of measuring differ from one to another. It depends on what you are measuring. Normally, when one hears measurement, the first thing to come to mind is in the form of physical world. Either it is measuring today's temperature, or the dimensions of a house, height, length, or even weight, speed, distance, and so on. Basically, the information has to convey properties of some attributes that are of interest to us. Scales of measurement refer to ways in which independent (input) variables/numbers are defined and categorized. The four scales of numerical data are nominal, ordinal (or discrete), interval, and ratio (can be discrete or continuous).

8.8.7   QUALITATIVE DATA (NOMINAL OR ORDINAL)

The discrete numerical data is referred to as integer or counting number (e.g., 3, 5, 2, 4, 4, 3, 5, 5, 1, 2, 4) or collected from marketing survey such as strongly agree, agree, uncertain, disagree, and or strongly disagree. Furthermore, using code for female as 1 and male as 2 or vice versa are examples of nominal data. Thus, one can assign numeric code to the number 1 and 2 to represent categories of data. Rating a process performance on a scale of 1 to 10, with one being very poor and 10 being excellent represents ordinal or ranked data. So, measurements can be arranged in order such as worst to best, low to high, 1= fastest to 10 = slowest and level of customer satisfaction.

8.8.8   QUANTITATIVE DATA (INTERVAL OR RATIO)

Interval is a measurement scale that represents quantity and has equal units and does not have a "true" zero point. For example, if human body temperature was measured on an interval scale, then a delta between a score of 30°C and a score of 31°C would illustrate the same difference in temperature as would a difference between a score of 40°C and a score of 41°C. Thus it is impossible to make statements about how many times higher one score is than another. Ratio measurement scale is similar to the interval scale representing quantity and has equality of units. No negative scale ratio exists, e.g., there is no negative length or weight. It always has

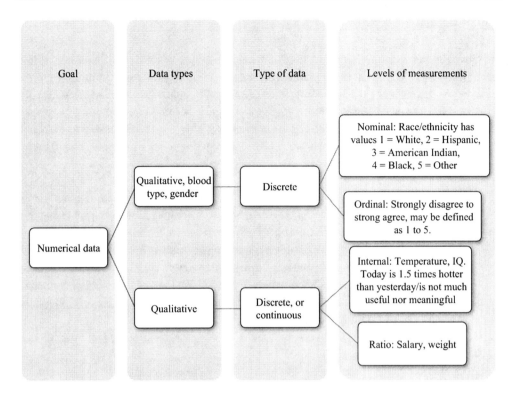

Figure 8.39. Continuous and discrete numerical data measurement.

to be positive or none. Continuous numerical data are defined as real numbers (e.g., weight 5.5, 5.503, 5.50372…, or age (in yr) 2.1, 2.50, 2.75). Other examples are 100° F twice as 50°F ratios. Figure 8.39 clarifies the forgoing discussion further.

Another example of variable and attribute data is given in Table 8.4.

The advantages of attribute data outcome are easier to measure (pass or fail) versus actual value, interpretation-satisfied versus unsatisfied customer, determining a sigma performance level. Basic sigma calculations are required on the number of defects. On the other hand, the disadvantage of attribute data outcome is a loss of precision, the need to collect more data-looking for pattern in the data categories, and the likelihood of missing important data due to "either/or" nature.

Table 8.4. Example of variable and attribute data

Type of Data for the Same Measurement	
Variable	**Attribute**
Hold time per incoming call	Number of calls on hold > 30 seconds
Average temperature per hour	Hours with temperature > 25°C
Cost per unit	Units exceeding target value
Quantity of gasoline in tank	Tank empty or not empty

8.8.9 SAMPLING STRATEGY

Before sampling, one needs to ask the following questions. What type of data (variable or attribute) is expected? What is the plan for the data? How confident do you want to be? What trade-off or risk between the detail, precision, cost, and sample size is expected? Which factors determine the size of sample the degree of confidence, the maximum allowable error, and the variation of the population? None of the above factors has direct relationship to the size of the population. Sampling approaches are random, stratified random, systematic, and subgroup sampling. They are defined as follows:

- *Random*: Each unit has the same chance of being selected. A simple random sample (SRS) of size n is produced by a scheme that ensures that each subgroup of the population of size n has an equal probability of being chosen as the sample.
- *Stratified random*: Randomly sample a proportionate number from each group.
- *Systematic*: Sample every nth one (e.g., every 10th, or 20th, etc.). It is a method of selecting sample members from a larger population according to a random starting point and a fixed, periodic interval. Typically, every nth member is selected from the total population for inclusion in the sample population.
- *Subgroup*: Sample n units every nth time (e.g., 10 units every hour) and then calculate the mean for each subgroup.
- Continuous data is based on a continuum of numbers (e.g., currency, time, weight, and height). Sample size can be calculated using the following equation:

$$n = \left(\frac{zS}{E} \right)^2 \qquad \text{(Eq. 8.2)}$$

where n is the sample number, z the standard normal distribution (z score) associated with the degree of confidence, S is sample standard deviation, and E is the allowable error.

- Discrete (attribute/discontinue) data is based on counts or classes (e.g., flipping a coin to yield a head or a tail). Sample size can be calculated using the following equation:

$$n = \frac{(z^2)pq}{E^2} \qquad \text{(Eq. 8.3)}$$

where $q = p - 1$, p is the estimated proportion of correct observation, and q is the estimated proportion of defects (scraps). Then Equation 8.3 becomes 8.4.

$$n = \frac{(z^2)p(p-1)}{E^2} \qquad \text{(Eq. 8.4)}$$

Since Lean Six Sigma is statistically data driven, it follows that accuracy of process measurement is necessary in Lean Six Sigma. In this step, the nature of the ongoing process will be analyzed. The result will represent the actual real-time capability of the process for the products under ongoing production.

8.8.10 DATA ANALYSIS

For data analysis, the following are needed.

1. Understand the relationship between quality and variation.
2. Differentiate between common and special cause variation. Common cause occurs naturally in every process and fluctuation is caused by unknown factors; for example, the variation in the moisture content of a resin, particle size distribution in a powder, and wax thickness in a roll of heat transfer labels. On the other hand, special cause occurs by known factors that result in a nonrandom disruption of output. It is unavoidable in almost every process, i.e., earthquake or cartons near the door of a warehouse are exposed to rain and degradation.
3. Interpret control charts, time plots, Pareto charts, and histogram; for example, bar chart can be used to focus on the specific causes of a problem or opportunity and identify those that have the greatest impact. It is used only for attribute data. Histogram/frequency chart used only for variable data. Run chart is a measure of a process over a specified period of time used to identify trends or patterns.
4. Differentiate and understand between control limits (process capability C_{pk}) and specification limits (customer requirements, USL and LSL).

8.9 TOLLGATE REVIEW AND CHECKLIST FOR MEASURE PHASE

Again this is the last chance before moving to the next phase (analyze). Make sure that you have achieved and ready for all the gate meeting requirements for the *Measure phase*. You must complete all the essential activities that are already assigned. You may tailor Figure 8.40 to your own process and come up with a solid project measurement mapping. This includes but is not limited to the measure phase deliverables and checklists.

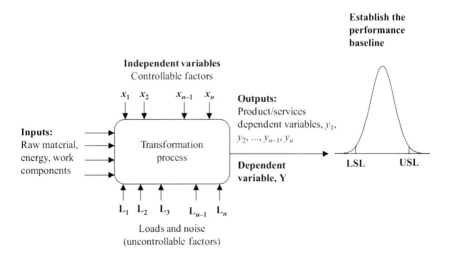

Figure 8.40. Lean Six Sigma process improvement projects: Tollgate review "measure the process."

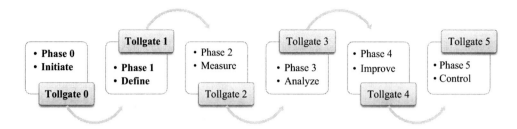

Figure 8.41. Lean Six Sigma process improvement projects: Tollgate review "measure phase."

8.9.1 MEASURE PHASE DELIVERABLES AND CHECKLISTS

(a) Apply data collection tools and system.
  i. Key measures identified and approved.
(b) Apply as many as measure tools and techniques such as
  i. Pareto charts, histogram, scatter diagram, affinity diagram, interrelationship diagram, matrix data analysis, process mapping, variation analysis, etc.
  ii. Gage R&R study
(c) Complete the Lean Six Sigma Scorecard.
(d) Compute the baseline process performance (*Sigma* level, Capability study) for the measurements.
(e) Identify the cause and effect for the Sigma level base line.
(f) Communicate process variation analysis using charts, graphs, plots.
(g) Does Sigma level base line changes anything in the project charter?
(h) Conduct value-added and non-value-added analysis.
(i) Identify the deliverable to be achieved with end-to-end timeline and target.
(j) Identify the financial business requirements and resources for the completion of this phase.
(k) Create and define RACI-Matrix. Include team members, gate keepers, and anyone who should be part of RACI-Chart.
(l) Identify the essential milestone.
(m) Determine Voice of the Process (VOP).
(n) Focus on the Voice-of-Customer (VOC).

Once the checklist has been completed, the project must to go through the gate review (Figure 8.41) as discussed in Chapter 6 for gate review team approval.

CHAPTER 9

ANALYSIS OF CONCEPTS AND STRATEGIES: ADVANCED STATISTICAL ANALYSIS—ACHIEVING ULTIMATE PERFORMANCE SCIENTIFICALLY

The goal of data analysis is the thorough examination of the components of a process from measure phase, in order to identify a pattern, relationships between input and output, opportunities, causes, and issues for improvement. It applies statistical knowledge, which is a science of data collecting, organizing, analyzing, describing, summarizing, and interpreting numerical data. Thus, the statistical concepts are used in analyzing and identifying the root causes of a process. The most dominant factors and source of variations that contribute to the Sigma mean will be recognized in the analysis phase.

9.1 DESCRIPTIVE STATISTICS

Descriptive statistics is the most basic form of statistics that is used to analyze, describe, and or summarize a given data (samples or measurements) in a simple and clear form of tabular, graphical, and numerical summaries, which include the commonly used forms as frequency distributions, histogram, mean, median, mode, range, variance, standard deviation, box plot, stem, and leaf plots, random number generators, and coefficient of variation (CV). It does not allow us to make a conclusion. Simply, it is a way to describe our data. Let's say we are interested in graphical summaries that show the spread of the data, and numerical summaries that either measures the central tendency (a "typical" data value) of a data set or that describe the spread of the data. To help explain descriptive statistics, suppose the following tornadoes has been occurred in the past two years as shown in Table 9.1. We will use this data to create a descriptive analysis.

Histograms are one kind of graphical summary, which combines data into groups or classes as a way to generalize the details of a data set while at the same time illustrating the data's overall pattern. On a histogram, the x-axis represents the data values arranged into classes while the

Table 9.1. The total number of recorded tornadoes, arranged alphabetically by state

State	Number of tornadoes	State	Number of tornadoes
Alabama	50	Montana	15
Alaska	0	Nebraska	57
Arizona	0	Nevada	3
Arkansas	35	New Hampshire	0
California	20	New Jersey	0
Colorado	55	New Mexico	7
Connecticut	15	New York	6
Delaware	0	North Carolina	22
Florida	1	North Dakota	29
Georgia	35	Ohio	27
Hawaii	0	Oklahoma	39
Idaho	59	Oregon	4
Illinois	75	Pennsylvania	6
Indiana	25	Rhode Island	0
Iowa	65	South Carolina	23
Kansas	93	South Dakota	19
Kentucky	43	Tennessee	32
Louisiana	72	Texas	159
Maine	5	Utah	5
Maryland	9	Vermont	0
Massachusetts	2	Virginia	17
Michigan	5	Washington	6
Minnesota	35	West Virginia	5
Mississippi	36	Wisconsin	19
Missouri	29	Wyoming	9

y-axis shows the number of occurrences in each class. So, how can we summarize and describe Table 9.1? One of the central tools of descriptive statistics is to construct a frequency distribution for each of the variables. The frequency distribution for the data given in Table 9.1 is shown below. To establish a continuous data frequency distribution, complete the following steps.

1. Each class interval must have an upper and a lower limit. The rule of thumb is to use 5–20 classes depending on the volume of the data.
2. The interval length is equal to its upper limit minus its lower limit; i.e., 60 – 40 = 20; 80 – 60 = 20; 100 – 80 = 20. That is classes should have equal length and at least one observation.

3. Find the minimum class width using the given equation and round the class width up to a convenient value.

$$Class\ Width = C.W. = \frac{Rang}{k} = \frac{Highest\ value - Lowest\ value}{Number\ of\ class} = \frac{159 - 0.0}{8} = 19.88$$

$$\approx 20.0$$

4. Calculate the number of values in each class as given in Table 9.2 to construct a histogram.

Horizontal axis show all values of the measurement in the population and vertical axis show number of observations.

Notice, in the Figures 9.1 and 9.2, histogram of each class has the same width along the x-axis. The decision to set the number of classes (k) is arbitrary. A different number of classes

Table 9.2. Histogram class calculations

Number of tornadoes	Number of frequency	Relative frequency	Cumulative	
			Cumulative frequency	Relative frequency
00 and under 20	28	28/49 = 0.571	28	0.571
20 and under 40	11	11/49 = 0.224	39	0.796
40 and under 60	05	5/49 = 0.102	44	0.898
60 and under 80	03	3/49 = 0.061	47	0.959
80 and under 100	01	1/49 = 0.020	48	0.980
100 and under 120	00	0/49 = 0.00	48	0.980
120 and under 140	00	0/49 = 0.00	48	0.980
140 and under 160	01	1/49 = 0.020	49	1.000
	$\Sigma = 49$	$\Sigma = 1.00$		

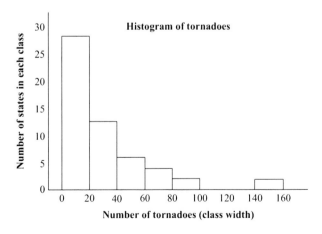

Figure 9.1. A histogram showing the tornado data for the last two years.

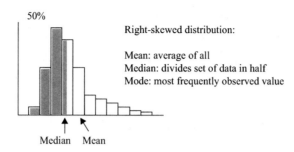

Figure 9.2. A histogram showing the median is greater than mean.

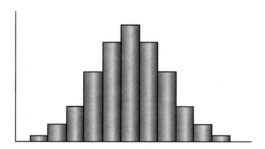

Figure 9.3. A histogram showing symmetric normal distribution.

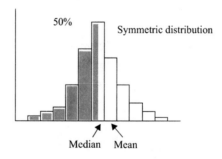

Figure 9.4. A symmetric histogram showing mean equal median.

could easily be used and would likely change the overall appearance of the histogram. In Figure 9.1 there are two empty classes between 100–120 and 120–140. Whenever the histogram data concentrates mostly to one side or the others, i.e., to the left or right, the shape of the histogram is called as "skewed" (nonsymmetric). In Figure 9.1 the tornado data are distributed on the lower or left-hand side, which is known as positive skew. The data is tailed to the positive side due to the single outlier between 140 and 160 classes.

Figures 9.3 and 9.4 illustrate histogram of a symmetric (not skewed) normal data set due to the absence of outliers clustered around the mean. A symmetric distribution is one in which the two "halves" of the histogram appear as mirror-images of one another with respect to the mean.

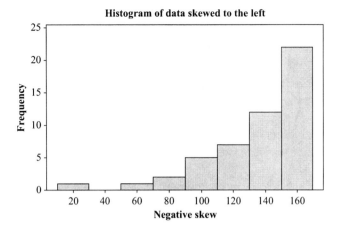

Figure 9.5. A histogram showing negative skew.

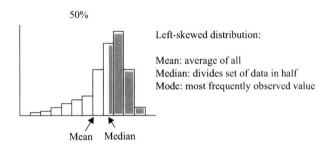

Figure 9.6. A histogram showing mean is less than median.

Figures 9.5 and 9.6 illustrate the different degrees of skewness that are typical of data sets. Data sets that have a greater number of high values, with outliers on the low end of the data scale (data that "tail" to the negative side), are said to be skewed to the left. Histogram in Figure 9.5 is an example of data having a negative skew.

Discrete and Continuous Data Type

Process information in the form of numerical data (in visible form) falls into two types of categories: discrete and continuous. Discrete data is high, low and fast, slow, measured by counting numbers or integer that can take only certain values; however, continuous data is measured both by integer or real numbers that can take any numbers and is not restricted. An example may clarify this concept.

Example 9.1

A 10-question exam is given in a Statistics class. The number of correct answers would have to be one of the following: 0, 1, 2, 3, 4, 5, 6, 7, 8, 9, or 10. There are not an infinite number of values, therefore this data is *discrete*. Further, if we were to draw a number line and place each possible value on it, we would see a space between each pair of values.

The height of trees at a nursery is an example of *continuous data*. Is it possible for a tree to be 76.2" tall? Sure. How about 76.29"? Yes. How about 76.2914563782"? Yes. The possibilities depend upon the accuracy of our measuring device.

Let's review the above discussion by classifying more cases as discrete or continuous:

1. *Discrete.* The number of suitcases lost must be a whole number.
2. *Continuous.* The height of corn plants can take on infinitely many values (any decimal is possible).
3. *Discrete.* The number of ears of corn must be a whole number.
4. *Discrete.* The number of white chocolates must be a whole number.
5. *Continuous.* The amount of time can take on infinitely many values (any decimal is possible).
6. *Continuous.* The weight of the tomatoes can take on infinitely many values (any decimal is possible).

9.1.1 DESCRIPTIVE STATISTICS TECHNIQUES AND GRAPHING: STEM AND LEAF

One of the quickest methods for performing a preliminary analysis of quantitative data is called *stem and leaf diagram*. The stem and leaf diagram is similar to histogram with one difference that actual individual data values are visible. For example, if a mathematical tests are marked out of 50 and the marks for the class are 6, 35, 45, 37, 25, 23, 22, 19, 27, 32, 14, 15, 33, 16, 7, 10, 11, 47, 49, 50, 43, 34, 26, 15, 41, 31, 25, 12, 13, 5. This data can be more easily interpreted if we represent it in a *stem and leaf diagram*.

Solution

A stem and leaf diagram (Figure 9.7) is formed by splitting the numbers into two parts: in this case, tens and units. The tens form the "stem" and the units form the "leaves." It is usual for the numbers to be ordered such as the row **2:** 2 3 5 5 6 7 shows the numbers 22, 23, 25, 25, 26 and 27, respectively, in order. The numbers are listed from lowest to highest. The forgoing two examples 9.2 and 9.3 will explain this idea furthermore.

Stem	Leaf
0	5 6 7 ◄——— 7 means 07
1	0 1 3 4 5 5 6 9
2	2 3 5 5 6 7
3	1 2 3 4 5 7
4	1 3 5 7 9 ◄——— 9 is actually 49
5	0

Figure 9.7. Stem and Leaf diagram.

Example 9.2

According to the latest report the top movies for the last week are

Theater movie	Earnings in millions of US dollars
1. Future sports	15.4
2. Football	13.4
3. LA Highway	8.6
4. Chicago night	7.3
5. Robotic	5.9
6. Mars	5.5
7. Africa	4.8
8. Return to Asia	3.1
9. Road to the Future	2.5
10. The last era	2.4

(a) Form a stem-and-leaf display. Does any outlier appear?
(b) Form a frequency distribution, starting with the class 0 and under 5.
(c) Interpret the shape of the distribution.

Solution

To make steam-and-leaf diagram as shown in Figure 9.8 we will use digits after decimal as a leaf and before the decimal point as a steam. The steam in this case will be one or two digits.

Steam-and leaf display

Stem	Leaf	
2	4 5	← 4 and 5 represent 2.4 and 2.5
3	1	
4	8	
5	5 9	
6		
7	3	
8	6	
9		
10		
11		
12		
13	4	
14		
15	4	← 15 space 4 means 15.4

Figure 9.8. Stem and Leaf diagram for movie earnings.

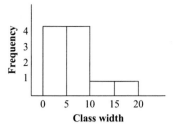

Figure 9.9. A histogram showing
positive skew.

Frequency distribution classes can be calculated as

$$Class\ Width = C.W. = \frac{Rang}{k} = \frac{Highest\ value - Lowest\ value}{Number\ of\ class} = \frac{15.4 - 2.4}{4} = 3.25 \approx 4$$

Class width	Frequency	Cumulative frequency
0.0 and under 5.0	4	4
5.0 and under 10.0	4	8
10.0 and under 15.0	1	9
15.0 and under 20.0	1	10
	Sum = 10	

The shape of distribution in Figure 9.9 indicates positive skew. The data shows that a majority of the movie earned $5.9 million or less.

Example 9.3

Many apartment rental companies offer special monthly rates on medium apartments. A sample of U-rent homes monthly medium rate at 18 locations is shown below.

Monthly medium rates (in US dollars):

207, 194, 191, 190, 190, 184, 179, 170, 169, 169, 145, 144, 142, 137, 137, 133, 131, 120

(a) Construct stem-and-leaf diagram using the last digit as the leaf.
(b) Construct a frequency histogram with nine class intervals.
(c) Compare the shape of the stem-and-leaf in part (a) and the frequency histogram diagram in part (b).

Stem-and-leaf diagram

Stem	Leaf
12	0
13	1 3 7 7
14	2 4 5
15	
16	9 9
17	0 9
18	4
19	0 0 1 4
20	7

Figure 9.10 Steam and leaf diagram for U-rent.

Solution

(a) To make a stem-and-leaf (Figure 9.10) plot, we divide each value into two parts: a stem and a leaf. We will use the last digit as the leaf. The steam will contain one, two, or more depending on the observation value. Here all the observation values are three digits. Thus, the steam will be two digits only.

(b) Frequency distribution classes is computer as

$$Class\ Width = C.W. = \frac{Rang}{k} = \frac{Highest\ value - Lowest\ value}{Number\ of\ class} = \frac{207 - 120}{9} = 9.67 \approx 10$$

Frequency histogram class width	Frequency	Cumulative frequency
120 and under 130	1	1
130 and under 140	4	5
140 and under 150	3	8
150 and under 160	0	8
160 and under 170	2	10
170 and under 180	2	12
180 and under 190	1	13
190 and under 200	4	17
200 and under 210	1	18
	18	

Figure 9.11 is frequency distribution for U-rent.

(c) Diagram comparisons of part (a) and (b) indicate that the frequency histogram distribution is much easier to interpret than data.

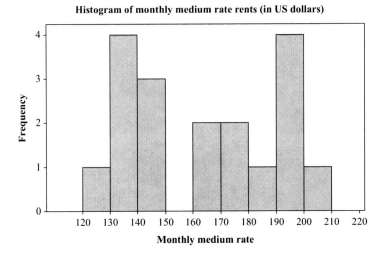

Figure 9.11. Histogram for monthly medium rate.

Example 9.4

Ten days annualized yields for 15 biggest mutual funds opens to investors soon are listed as follows:

Given	Money market fund (order of asset size)	Week of 5/15/20X1 (In %)	Week of 5/15/20X2 (In %)
1	Merrill Lynch Retirement Fund-II	5.24	5.34
2	Fidelity Spartan-ZX	5.30	5.33
3	Smith Barney Class AX	5.07	5.31
4	Fidelity Cash Reserves and Saving	5.28	5.28
5	Paine Webber XM	5.11	5.26
6	Merrill Lynch CXMA Money Fund	5.11	5.25
7	Prudential Money Mart Assets XG	5.05	5.17
8	Chicago Liquid Asset Fund-NX	5.15	5.13
9	Vanguard Money Market Fund	5.33	5.12
10	Merrill Lynch Ready Assets-II	5.09	5.10
11	Centennial Money Saving Trust	5.03	5.10
12	Schwab Money Market Fund	4.98	5.10
13	Schwab Value Ultimate Advantage	5.32	5.08
14	Prudential/ Insurance Command Money	5.18	5.08
15	Dean Witter-II/Active Assets-I	5.25	4.99

(a) Construct relative frequency distributions for the funds yields on April 15, 20X1. Assume five classes.

(b) Repeat (a) for April 15, 20X2.

(c) Explain the results of (a) and (b).

Solution

Figure 9.12 is the frequency distribution for the Week of 5/15/20X1.

Class width	Frequency	Cumulative frequency
4.9 and under 5.0	1	1
5.0 and under 5.1	4	5
5.1 and under 5.2	4	9
5.2 and under 5.3	3	12
5.3 and under 5.4	3	15
	15	

Figure 9.13 is the frequency distribution for the Week of 5/15/**20X2**.

Class width	Frequency	Cumulative frequency
4.9 and under 5.0	1	1
5.0 and under 5.1	2	3
5.1 and under 5.2	6	9
5.2 and under 5.3	3	12
5.3 and under 5.4	3	15
	15	

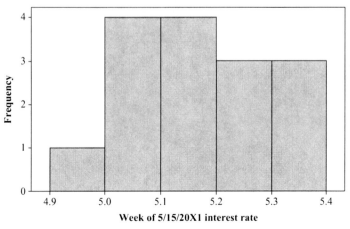

Figure 9.12. Frequency distribution for the week of 5/15/20X1.

Histogram of money market funds week of 5/15/20X2

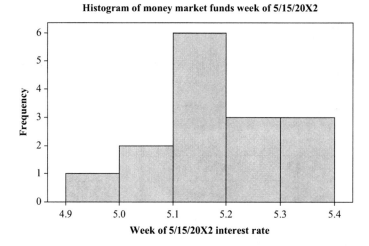

Week of 5/15/20X2 interest rate

Figure 9.13. Frequency distributions for the week of 5/15/20X2.

Comparisons between A and B above show both were within the same range and the "A" data is much more evenly distributed while the "B" data is much more centered.

9.1.2 HISTOGRAM

In stem-and-leaf plots each data point has a view. As the data gets immense stem-and-leaf becomes clumsy and awkward to work with. Therefore, the graphical representation of the data-set such as *histogram* becomes proficient, which is a visual description of the data-set. For instance, suppose for about five days and each day samples of 15 parts were collected, measured, and plotted for a critical dimension concerning a shrinkage issue as shown in Figure 10.14. In this case the Figure 10.14 indicates that data is symmetrically distributed.

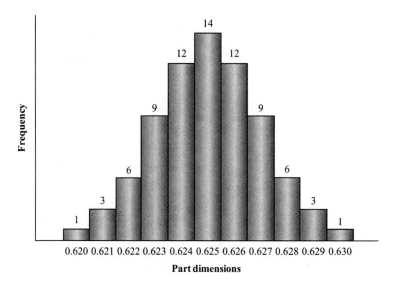

Part dimensions

Figure 9.14. A performance histogram of parts dimension.

In histogram the variable under study is measured on the horizontal (x) axis while the frequency is labeled on the vertical (y) axis. Bar height illustrates frequency of classes.

9.1.3 MEASURE OF CENTER TENDENCY

The mean: A population mean is the sum of the entire set of objects, observations, or scores that have something in common. This can be expressed as Equation 9.1.

$$\mu = \sum_{i=1}^{N} X_i / N = \frac{1}{N}\sum_{i=1}^{N} X_i, \qquad \text{(Eq. 9.1)}$$

where μ is the population mean and
    X_i = Part measurement
    N = Total number of parts (population)

A sample is a set of observations drawn from a population. Since it is usually time-consuming and too costly to test every one of the population, a sample from the population is normally the best approach to ascertain the population outcome. Sample mean is expressed in Equation 9.2.

$$\bar{x} = \sum_{i=1}^{n} x_i / n = \frac{1}{n}\sum_{i=1}^{n} x_i = \frac{x_1 + x_2 + \cdots + x_{n-1} + x_n}{n} \qquad \text{(Eq. 9.2)}$$

where $\bar{x}$ = the sample mean and
    x_i = sample measurement
    n = total number of sample parts

The median – Median is the middle value of the relevant set of data or the value that divides the set of data in half.
    Data set (1):

$x_{(1)}$	$x_{(2)}$	$x_{(3)}$	$x_{(4)}$	$x_{(5)}$	$x_{(6)}$	$x_{(7)}$	$x_{(8)}$	$x_{(9)}$
5	3	7	9	6	−3	2	−5	3

Rearranging the data set in the order of decreasing (high to low) or increasing (low to high)

$x_{(1)}$	$x_{(2)}$	$x_{(3)}$	$x_{(4)}$	$x_{(5)}$	$x_{(6)}$	$x_{(7)}$	$x_{(8)}$	$x_{(10)}$
−5	−3	2	3	3	5	6	7	9

The 50% of the measurements is above the median and the 50% of the measurement is below the median point. Thus in this case the median is equal to 3. Since the sample number (n = 9) is odd number thus the median falls in the middle $x_{(5)}$ = 3.
    Data set (2),

$x_{(1)}$	$x_{(2)}$	$x_{(3)}$	$x_{(4)}$	$x_{(5)}$	$x_{(6)}$	$x_{(7)}$	$x_{(8)}$	$x_{(9)}$	$x_{(10)}$
0.2	9.3	11.3	10.4	6.6	2.3	5.7	2.3	8.9	14.2

Rearranging the data set in the order of decreasing (high to low) or increasing (low to high):

$x_{(1)}$	$x_{(2)}$	$x_{(3)}$	$x_{(4)}$	$x_{(5)}$	$x_{(6)}$	$x_{(7)}$	$x_{(8)}$	$x_{(9)}$	$x_{(9)}$
2.3	5.7	6.6	8.9	9.3	10.2	10.4	11.3	12.3	14.2

If "even" number of elements is in your data set then the mean of the middle pair ($x_{(5)}$ and $x_{(6)}$) is the median as shown in Equation 9.3.

$$\text{Median} = \frac{x_{\left(\frac{n}{2}\right)} + x_{\left(\frac{n}{2}+1\right)}}{2} = \frac{x_{(5)} + x_{(6)}}{2} = \frac{9.3 + 10.2}{2} = 9.75 \qquad \text{(Eq. 9.3)}$$

And using the Equation 9.2 the mean equals to

$$\bar{x} = \sum_{i=1}^{n} x_i / n = \frac{1}{n} \sum_{i=1}^{n} x_i = \frac{x_1 + x_2 + \cdots + x_{n-1} + x_n}{n}$$

$$\text{(Eq. 9.2)}$$

$$\bar{x} = \frac{2.3 + 5.7 + 6.6 + 8.9 + 9.3 + 10.2 + 10.4 + 11.3 + 12.3 + 14.2}{10} = 9.12$$

9.1.4 MEASURES OF VARIABILITY

Variation measurement (also called dispersion or spread measurement) is a measure that explains how the data values are spread out around the measure of central tendency as presented in the Figure 9.15.

Thus, a basic function in statistical analysis is to distinguish the *dispersion*, or variability, of a data set. Measures of variation are used to estimate this variability. When determining the variability of a data set, there are couple essential elements to consider:

1. How spread out are the data set values around the mean?
2. How spread out are the tails?

Various analysis results will produce different measurement to the above elements. The decision of measurement estimator is normally based on which of the two elements

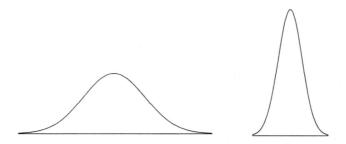

Figure 9.15. Variability of normal distribution curve.

you prioritize. Common examples of variation measurement are the *range, variance and standard deviation, coefficient of variation* (CV), *coefficient of skewness*, and *interquartile range* (IQR).

9.1.4.1 Range

The *range* is simplest measure of variation to calculate. Basically, it is by taking the difference between the highest value of data-set and the lowest value.

Example 9.5

A sales person travels six days a week. The distance traveled by the sales person in a week (Monday through Saturday) is listed below. Find the range of distance traveled by the sales person in miles.

25   35   42   9   28   5

Solution

As it is evident, the largest and shortest distance traveled in a week is 42 and 5 miles. Thus, the range is the highest value minus lowest value.

$$\text{Range} = \text{Greatest value} - \text{Least value} = 42 - 5 = 37 \text{ miles}$$

9.1.4.2 Variance and Standard Deviation

9.1.4.2.1 POPULATION VARIANCE AND STANDARD DEVIATION

The population variance (Equation 9.4) is a measure of variability (spread out) in a population data set or distribution curve.

$$\sigma^2 = \frac{\sum_{i=1}^{N}(X_i - \mu)^2}{N} \qquad \text{(Eq. 9.4)}$$

And mathematically the population standard deviation is the square root of population variances:

$$\sigma = \sqrt{\frac{\sum_{i=1}^{N}(X_i - \mu)^2}{N}} \qquad \text{(Eq. 9.5)}$$

where σ^2 is the variance and σ is the standard deviation of the population.

9.1.4.2.2 SAMPLE VARIANCE AND STANDARD DEVIATION

The sample *variance* is a measure of variability (spread out) in a sample data set or distribution curve.

$$S^2 = \frac{\sum_{i=1}^{n}(x_i - \bar{x})^2}{n-1} = \frac{\sum x^2 - \frac{(\sum x)^2}{n}}{n-1} \qquad \text{(Eq. 9.6)}$$

And sample standard deviation (Equation 9.7) is the square root of variances:

$$S = \sqrt{\frac{\sum_{i=1}^{n}(x_i - \bar{x})^2}{(n-1)}} \qquad \text{(Eq. 9.7)}$$

where S^2 is the variance and S is the standard deviation of the sample. The accuracy of standard deviation and normal distribution increase with the higher sample size.

Example 9.6

Calculate the difference between each data value and the mean for data set

$$46 \quad 54 \quad 42 \quad 46 \quad 32$$

Solution

The above data set is selected from long list of data. For the data the mean is

$$\bar{x} = \sum_{i=1}^{n} x_i / n = \frac{1}{n}\sum_{i=1}^{n} x_i = \frac{x_1 + x_2 + \cdots + x_{n-1} + x_n}{n}$$

$$\bar{x} = \frac{46 + 54 + 42 + 46 + 32}{5} = 44$$

Using the information from Table 9.3 and Equations 9.6 and 9.7, we can calculate the variance and standard deviation as follows:

$$S^2 = \frac{\sum_{i=1}^{n}(x_i - \bar{x})^2}{n-1} = \frac{256}{4} = 64$$

$$s = \sqrt{64} = 8$$

Table 9.3. Deviation and squared deviation for sample variance

x	$\bar{x} = 44$	$x_i - \bar{x}$	$(x_i - \bar{x})^2$
46		$46 - 44 = 2$	$(2)^2 = 4$
54		$54 - 44 = 10$	$(10)^2 = 100$
42		$42 - 44 = -2$	$(-2)^2 = 4$
46		$46 - 44 = 2$	$(2)^2 = 4$
32		$32 - 44 = -12$	$(-12)^2 = 144$
$\sum x = 220$		$\sum(x - \bar{x}) = 0$	$\sum(x_i - \bar{x})^2 = 256$

9.1.4.3   Coefficient of Variation (CV)

The *coefficient of variation* (CV) is the ratio of the standard deviation (measure of the spread of data) to its mean (Equations 9.8 and 9.9)

$$\textbf{Population coefficient of variance: } CV = \left(\frac{\sigma}{\mu}\right)(100\%) \qquad \text{(Eq. 9.8)}$$

$$\textbf{Sample coefficient of variance: } CV = \left(\frac{s}{\bar{x}}\right)(100\%) \qquad \text{(Eq. 9.9)}$$

Example 9.7

Assume that there are two types of stocks A and B. The average price for stock "A" is equal to $50 and standard deviation is $5. On the other hand, the average price for stock "B" is equal to $100 and standard deviation is $5. Which one has less variation to its price?

Solution

Using Equation 9.9 we have

$$CV_A = \left(\frac{s}{\bar{x}}\right)(100\%) = \left(\frac{\$5}{\$50}\right)(100\%) = 10\%$$

$$CV_B = \left(\frac{s}{\bar{x}}\right)(100\%) = \left(\frac{\$5}{\$100}\right)(100\%) = 5\%$$

Both stocks have the same standard deviation. But stock B is less variable relative to its price.

9.1.4.4   The Mode

The most frequently observed value of the measurements in the relevant set of data. A data set may have no modes (10, 12, 14, 16), or more than one mode (**10**, **10**, 12, **14**, **14**, 16, 18, 21). In this case the data set has two modes 10 and 14.

9.1.4.5   Coefficient of Skewness (CS)

In probability and statistics, skewness is a measure of the asymmetry of the probability distribution of a random numbers of a real data set. The skewness values can be either negative or positive, or even undefined. The skewness for normal distribution is zero and any symmetric data set should have a skewness near zero.

Negative skewness: A negative skewed means that the *left tail* is long relative to *right tail* and most of the data values are concentrated on the right of the mean.

Positive skewness: A positive skewed means that the *right tail* is long relative to *left tail* and most of the data values are concentrated on the left of the mean.

The skewness as shown in Figure 9.16 can be calculated using Equation 9.10:

$$skewness = \frac{\sum_{i=1}^{N}(Y_i - \bar{Y})^3}{(N-1)s^3} \qquad \text{(Eq. 9.10)}$$

where $\bar{Y}$ is the mean, s is the sample standard deviation, and N is the number of data set points.

9.1.4.6   The Interquartile Range

From the past, we know that the outlier in the dataset do not impact the median. Thus, the interquartile range is a measure of spread, which is resistant to outliers similar to median that is resistant to minimum and maximum value. The interquartile range value is the difference between the 75th percentile value (the largest) and 25th percentile value (the smallest) in the middle 50% of the data set. On the contrary, the range uses the difference between the two most extreme values and interquartile uses middle portion only.

The interquartile range is obtained by the third quartile value minus first quartile value.

$$IQR = Q_3 - Q_1$$

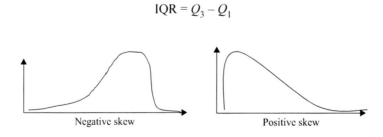

Figure 9.16. Negative and positive skewed distribution.

Example 9.8

The monthly starting salaries for a sample of 12 business school graduates are shown below:

2050, 2150, 2250, 2080, 1955, 1910, 2090, 2330, 2140, 2525, 2120, 2080

Determine the three quartiles and the interquartile range for the given data set.

Solution

To find the three quartiles for the data set, first we arrange the 12 values in increasing orders as follows:

1910, 1955, 2050, 2080, 2080, 2090, 2120, 2140, 2150, 2250, 2330, 2525

Since there is an even numbers of data set thus, the value of the middle term is determined by the average of 6th and the 7th values as discussed in Section 9.1.3:

$$Q_2 = \text{median} = \frac{x_{\left(\frac{n}{2}\right)} + x_{\left(\frac{n}{2}+1\right)}}{2} = \frac{x_{(6)} + x_{(7)}}{2} = \frac{2090 + 2120}{2} = 2105$$

There are six values in the dataset below $Q_2 = 2105$, and since there are even number below Q_2 thus we use Equation 9.3 again to find Q_1.

$$Q_1 = \frac{x_{\left(\frac{n}{2}\right)} + x_{\left(\frac{n}{2}+1\right)}}{2} = \frac{x_{(3)} + x_{(4)}}{2} = \frac{2050 + 2080}{2} = 2065$$

Likewise, there are six values in the data set above $Q_2 = 2150$, and since there are even number above Q_2 thus we use Equation 9.3 once more to find Q_3.

$$Q_3 = \frac{x_{\left(\frac{n}{2}\right)} + x_{\left(\frac{n}{2}+1\right)}}{2} = \frac{x_{(3)} + x_{(4)}}{2} = \frac{2150 + 2250}{2} = 2200$$

The value of interquartile is given by the difference between the values of the third and the first quartile.

$$\text{IQR} = Q_3 - Q_1 = 2200 - 2065 = 135$$

9.1.4.7 Box and Whisker Plots (Box Plots)

A box plot is used to show the distribution of a single variable in a data set. In other words, it provides a visual summary of the quartiles (four quartiles) of a data set or it is a graphical representation of dataset using five measures:

1. Smallest value.
2. First Quartile (Q_1): 25% of values are less than Q_1 and 75% are larger than Q_1.
3. Second Quartile (Q_2): 50% of values. Q_2 divides data set into two equal parts, hence Q_2 = median.
4. Third Quartile (Q_3): 75% of values are less than Q_3 and 25% are greater than Q_3. Difference between Q_3 and Q_1 is called interquartile range (IQR).
5. Largest value.

This helps to visualize the center, the spread, and the skewness of a data spread as shown in Figure 9.17.

Example 9.8

Let's use the same data set from Example 9.9, gives monthly starting salaries for a sample of 12 business school graduates:

2050, 2150, 2250, 2080, 1955, 1910, 2090, 2330, 2140, 2525, 2120, 2080

Construct a box plot.

Solution

To create a box plot, we apply five measurements summary results: the lowest value, the first quartile, the second quartile (median), the third quartile, and the highest value in the data set under study. By taking the results from Example 9.7 we have

1910, 1955, 2050, 2080, 2080, 2090, 2120, 2140, 2150, 2250, 2330, 2525

Lowest Value	Q_1	Q_2	Q_3	Highest Value
1910	2065	2105	2200	2525

Figure 9.17. Boxplot Q_1 to Q_3 spread.

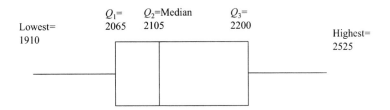

Figure 9.18. Box plot of monthly starting salaries.

To draft a box plot, the five values need to be used. Of course it is easier to use any statistical software to plot the given data set.

In the box plot of Figure 9.18, the two lines on the both sides of the box are called "whiskers." Due to this, most of the time box plot is also called "Box and Whisker Plot." The whisker on the left side of the box starts at 1910, which is the lowest value in the data set. The box starts at 2065, which is first quartile, the vertical line inside the box represent second quartile 2105 or median, and the box ends at the third quartile 2200. The whisker on the right side of the box ends at the highest value of the data set 2525.

Therefore, the box plot is made of four sections. First, the whisker 25% of the dataset values prior to first quartile, second the first section of the box between first quartile and median (second quartile) and last section of the box, which is between median and third quartile, and finally the fourth section the largest 25% of the values.

9.1.4.8   The Outliers Analysis

Outliers are data set values that are much larger (mild or extreme) than the rest of the data set values. In some cases outliers may be caused by unknown cause or malfunction of the process. There are some measurements that we can use to find out if a data set contains any outliers. Thus, we need to calculate the *inner* and *outer fence* for the data set.

$$\text{Lower inner fence} = Q_1 - 1.5\ (\text{IQR})$$

$$\text{Upper inner fence} = Q_3 + 1.5\ (\text{IQR})$$

$$\text{Lower outer fence} = Q_1 - 3.0\ (\text{IQR})$$

$$\text{Upper outer fence} = Q_3 + 3.0\ (\text{IQR})$$

Any response or process outputs that falls between the two lower fences (inner and outer) and between the two upper fences (inner and outer) are called mild outliers. The response of observation that falls under lower outer fence and above upper outer fence is called extreme outliers.

Example 9.9

We will continue Example 9.8 and determine the inner and outer fences.

Given dataset:

1910, 1955, 2050, 2080, 2080, 2090, 2120, 2140, 2150, 2250, 2330, 2525

Solution

The values of Q_1, Q_3, and IQR are given in Example 9.8. Thus, we can calculate the lower, upper, inner, and outer fence as follows:

$$\text{Lower inner fence} = Q_1 - 1.5(\text{IQR}) = 2065 - 1.5(135) = 1862.5$$

$$\text{Upper inner fence} = Q_3 + 1.5(\text{IQR}) = 2200 - 1.5(135) = 2402.5$$

$$\text{Lower outer fence} = Q_1 - 3.0(\text{IQR}) = 2065 - 3.0(135) = 1660$$

$$\text{Upper outer fence} = Q_3 + 3.0(\text{IQR}) = 2200 + 3.0(135) = 2605$$

By examining at the 12 data set values, we notice there is no data set value that falls within mild and extreme outlier's fence. On the other hand, the number 2525 highest value of data set falls between upper inner fence and upper outer fence, which is a mild outlier.

9.1.4.9   The Percentile

The **pth** percentile of a distribution is a value such that at least (approximately) p percent ($p\%$) of the numbers in the distribution are equal to this value or less than that number and at least $(100 - p)$ percent of the numbers equal to this value. Thus, if "35" is the 60th percentile of a big batch of numbers, 60% of those numbers are less than or equal to 35.

To determine the value of the pth percentile, where p is an integer from 1 to n, arrange the data set in increasing order so that x_1 is the smallest value, and x_n is the highest value, assuming n being the total number of observations (Equation 9.11). Then compute i and if i is not an integer, then round up. However, if i is an integer, then take the average *of "i and i+1"* and that is the pth percentile (i.e., for instance average $x_i + x_{i+1}$)

$$i = \left(\frac{p}{100}\right)n + \frac{1}{2}; \; p\,(percentile), \, n\,(number \; of \; items; n = 1, 2, 3, \ldots, n-1, n) \quad \text{(Eq. 9.11)}$$

Example 9.10

Calculate the 25%, 50%, and 75% percentile value using the following data set with $n = 7$

$$3 \quad 4 \quad 6 \quad 7 \quad 9 \quad 10 \quad 11$$

By applying Equation 9.11 we have

$$i = \left(\frac{p}{100}\right)n + \frac{1}{2} = \left(\frac{25}{100}\right)7 + \frac{1}{2} = 2.25 \; (round \; 2.25 \; to \; 2)$$

So the 25th percentile is equal to 4 ($i = 4$). Likewise the 50th percentile = 7, and 75th percentile = 10. Now, if x_i is the p_i^{th} percentile of the data set then we have Equation 9.12:

$$p_i = \frac{1}{n}(100i - 50) = 100(i - 0.5)/n \qquad \text{(Eq. 9.12)}$$

For instance, considering the above data set and using Equation 9.13 we have

$$p1 = 100(1 - 0.5)/7 = 7.1$$

$$p2 = 100(2 - 0.5)/7 = 21.4$$

$$p3 = 100(3 - 0.5)/7 = 35.7$$

$$p4 = 100(4 - 0.5)/7 = 50$$

$$p5 = 100(5 - 0.5)/7 = 64.29$$

$$p6 = 100(6 - 0.5)/7 = 78.57$$

$$p7 = 100(7 - 0.5)/7 = 92.85$$

This information is summarized in the Table 9.4.

Example 9.11

Assume that you can travel to any continental within one to two hours by year 2100. Considering the inflation and competitions the price variations are listed below.

2800	1600	2500	4500	3500	4500	7400	6400	7570	13500
12300	12500	8500	12000	8500	16300	15400	17400	15300	14500
14900	17500	18400	21300	17500	22000	21500	23300	24500	24000

(a) Using the empirical rule, estimate the interval in which approximately 68% of the data will lie.
(b) Find the 25th, 50th, and 75th percentile for the data above. Interpret these values.

Table 9.4. Sample numbers and percentile

Original	6	3	4	10	7	11	9
x_i	3	4	6	7	9	10	11
I	1	2	3	4	5	6	7
p_i	7.1%	21.4%	35.7%	50%	64.29%	78.57%	92.85%

Solution

To find the required percentile for these data set, first we arrange the 30 values in increasing order as follows:

1600	2500	2800	3500	4500	4500	6400	7400	7570	8500
8500	12000	12300	12500	13500	14500	14900	15300	15400	16300
17400	17500	17500	18400	21300	21500	22000	23300	24000	24500

(a) To find the 68th percentile, first we find i using Equation 9.11:

$$i = \left(\frac{p}{100}\right)n + \frac{1}{2} = \left(\frac{68}{100}\right)30 + \frac{1}{2} = 20.9 \ (round \ 20.9 \ to \ 21) \approx 21$$

Thus, the 68th percentile is given by the value of the 21th sample number in the arranged data set, which is 17,400.

(b) Since there is an even numbers of data set (30 values), the value of the middle term (50th percentile) is determined by the average of 15th and the 16th values as discussed in Section 9.1.3.

$$50\text{th percentile} = \frac{x_{\left(\frac{n}{2}\right)} + x_{\left(\frac{n}{2}+1\right)}}{2} = \frac{x_{(15)} + x_{(16)}}{2} = \frac{13500 + 14500}{2} = 14,000$$

Since there are odd (15) data set values below median and odd (15) dataset values above median, thus the 25th and 75th percentiles are calculated using Equation 9.11

$$i = \left(\frac{p}{100}\right)n + \frac{1}{2} = \left(\frac{25}{100}\right)30 + \frac{1}{2} = 8$$

The 25th percentile is given by the value of the 8th sample number in the arranged data set, which is 7400.

$$i = \left(\frac{p}{100}\right)n + \frac{1}{2} = \left(\frac{75}{100}\right)30 + \frac{1}{2} = 23$$

The 75th percentile is given by the value of the 23th sample number in the arranged data set, which is 17500.

Example 9.12

Biotech Companies involved in R&D for heart attack drugs have got the attentions of many investors. The following companies are promising on a cure for heart attacks are listed with their market profits.

	Company	Market profits (in millions of $)
1	Fizer	8620
2	Ugouron	1110
3	Imclone Cure	26
4	BioPharmaceya	252
5	Sugenia Science	236
6	Maganine	153
7	Unixum	116
8	Ribozymax	47
9	Geneticsall	47

Calculate the mean, median, variance, standard deviation, *Pearsonian Coefficient of Skewness* (SK), and interquartile range for the market profits data.

Solution

To find the mean, median, variance, standard deviation, Pearsonian Coefficient of Skewness (SK), and interquartile range for the market profits data, we will use the equations as discussed before. But first we arrange the data set in increasing order then use the equation as before:

$$47, 47, 116, 153, 236, 252, 264, 1110, 8620$$

$$Mean = \bar{x} = \frac{\sum_{i=1}^{i=n} x_i}{n} = \frac{47+47+116+153+236+255+264+1110+8620}{9} = 1205.33$$

Since there are odd numbers of data set, so the median is

$$i = \left(\frac{p}{100}\right)n + \frac{1}{2} = \left(\frac{50}{100}\right)9 + \frac{1}{2} = 5$$

The 50th percentile is given by the value of the 5th sample number in the arranged dataset, which is 236. Or

$$Md = \left(\frac{n+1}{2}\right) = \left(\frac{9+1}{102}\right) = 5. \text{ Thus, the median value is Md} = 236.$$

To find the variance we need to determine the deviation and squared deviation as in Table 9.5

Table 9.5. Deviation and squared deviation for Pearson Coefficient of Skewness

x	$x_i - \bar{x}$	$(x_i - \bar{x})^2$
47	$47 - 1205.33 = -1158.33$	$(-1158.33)^2 = 1{,}341{,}736$
47	$47 - 1205.33 = -1158.33$	$(-1158.33)^2 = 1{,}341{,}736$
116	$116 - 1205.33 = -1089.33$	$(-1089.33)^2 = 1{,}186{,}647$
153	$153 - 1205.33 = -1052.33$	$(-1052.33)^2 = 1{,}107{,}405$
236	$236 - 1205.33 = -969.33$	$(-969.33)^2 = 939{,}607$
255	$255 - 1205.33 = -950.33$	$(-950.33)^2 = 903{,}133$
264	$264 - 1205.33 = -941.33$	$(-941.33)^2 = 886{,}108$
1110	$1110 - 1205.33 = -95.33$	$(-95.33)^2 = 9{,}088$
8620	$8620 - 1205.33 = 7414.67$	$(7414.67)^2 = 54{,}977{,}282$
$\sum x = 10{,}848$	$\sum(x - \bar{x}) = 0$	$\sum(x_i - \bar{x})^2 = 62{,}692{,}744$

Using the information from Table 9.5 and Equations 9.6 and 9.7, we can calculate the variance and standard deviation as follows:

$$S^2 = \frac{\sum_{i=1}^{n}(x_i - \bar{x})^2}{n-1} = \frac{62{,}698454}{9-1} = 7{,}837{,}307$$

$$s = \sqrt{7{,}837{,}307} = 2799.52$$

$$Pearsonian\ Coefficient\ of\ Skewness = Sk = \frac{3(\bar{x} - Md)}{s} = \frac{3(\frac{\sum_{i=1}^{i=n}x_i}{n} - Md)}{\sqrt{\frac{\sum(x - \bar{x})^2}{n-1}}}$$

$$= \frac{3(1205 - 236)}{2799} = 1.038$$

For accuracy, this problem also was calculated in the computer. Thus, in order to get accurate number one must use computer.

The interquartile range values for can be obtained from above calculation, which is equal to $Q_2 = 236$ and that is also equal to median.

However, since there are four values in the data set below $Q_2 = 236$, and since there are even number below Q_2 therefore we use Equation 9.3 to find Q_1.

$$Q_1 = \frac{x_{\left(\frac{n}{2}\right)} + x_{\left(\frac{n}{2}+1\right)}}{2} = \frac{x_{(2)} + x_{(3)}}{2} = \frac{47 + 116}{2} = 81.5$$

Likewise, there are four values in the data set above $Q_2 = 236$, and since there are even number above Q_2 therefore we use Equation 9.3 once more to find Q_3.

$$Q_3 = \frac{x_{\left(\frac{n}{2}\right)} + x_{\left(\frac{n}{2}+1\right)}}{2} = \frac{x_{(7)} + x_{(8)}}{2} = \frac{264 + 1110}{2} = 687$$

The value of interquartile is given by the difference between the values of the third and the first quartile.

$$\text{IQR} = Q_3 - Q_1 = 687 - 81.5 = 605.5$$

9.1.4.10 The Z-Score (Standard Score)

The standard normal distribution (the z-distribution) is a normal distribution with a mean of zero ($\mu = 0$) and a standard deviation of one ($\sigma = 1$). Normal distributions can be transformed to standard normal distribution (z) by the expression

$$z = \frac{x - \mu}{\sigma} \quad or \; z = \frac{x - \bar{x}}{s} \qquad \text{(Eq. 9.13)}$$

where x is a value of interest from original normal distribution, μ is the population mean of the original normal distribution, $\bar{x}$ is the sample mean, σ is the population standard deviation of the original normal distribution, and s is the sample standard distribution. Since s and σ are positive values, thus any value below the mean will have a negative z-score, equal to mean will have zero z-score, and above the mean will have a positive z-score.

Example 9.13

The cell phone stocks are doing very well. One method of measurement how well they are is to look at the price-to-sale ratio. Since some the companies have not reported their earnings, price-to-earnings ratio doesn't make a sense to evaluate. Thus, price-to- sale ratio is used to evaluate.

x	Cell phone companies	Price-to-sale ratio
1	Super phone	38.0
2	Classic Phone	14.0
3	Future Phone	10.2
4	TV Phone	10.0
5	Internet Phone	7.5
6	Affordable Phone	3.4

What is the z-score for the Super phone? What is the z-score for Classic Phone? What does this explains about the price-to-sale ratio for Super phone?

Table 9.6. Deviation and squared deviation for z-square calculation

x	$x_i - \bar{x}$	$(x_i - \bar{x})^2$
38	$38.0 - 13.85 = 24.15$	$(24.15)^2 = 583.22$
14	$14.0 - 13.85 = 0.15$	$(0.15)^2 = 000.0225$
10.2	$10.2 - 13.85 = -3.65$	$(-3.65)^2 = 13.32$
10	$10.0 - 13.85 = -3.85$	$(-3.85)^2 = 14.82$
7.5	$7.5 - 13.85 = -6.35$	$(-6.35)^2 = 40.32$
3.4	$3.4 - 13.85 = -10.45$	$(-10.45)^2 = 109.20$
$\sum x = 83.1$	$\sum(x - \bar{x}) = 0$	$\sum(x_i - \bar{x})^2 = 760.92$

Solution

To find z-score we need to find the mean and standard deviation:

$$\bar{x} = \sum_{i=1}^{n} x_i / n = \frac{1}{n}\sum_{i=1}^{n} x_i = \frac{x_1 + x_2 + \cdots + x_{n-1} + x_n}{n}$$

$$\bar{x} = \frac{38.0 + 14.0 + 10.2 + 10.0 + 7.5 + 3.4}{6} = 13.85$$

Using the information from Table 9.6 and Equations 9.6 and 9.7, we can calculate the variance and standard deviation as follows:

$$S^2 = \frac{\sum_{i=1}^{n}(x_i - \bar{x})^2}{n-1} = \frac{760.92}{5} = 152.18$$

$$s = \sqrt{64} = 12.34$$

We find the z-score for these two values of Super and classic phone as follows:

For Super phone $\quad x = 38, \; z = \dfrac{x - \bar{x}}{s} = \dfrac{38 - 13.85}{12.34} = 1.96$

For Classic phone $\quad x = 14, \; z = \dfrac{x - \bar{x}}{s} = \dfrac{14 - 13.85}{12.34} = 0.012$

Answer: Super phone is far overvalued

Example 9.14

If the z-score $= -1.45$, the mean is $\bar{x} = 35$, and $x = 17$. What is the value of variance?

Solution

To find variance we rearrange the z-score formula

$$Variance = s^2 = \left(\frac{x-\bar{x}}{z}\right)^2 = \left(\frac{(17-35)}{-1.45}\right)^2 = 154.102$$

9.2 DESCRIPTIVE MEASURES

9.2.1 MEASUREMENT SYSTEM ANALYSIS

Measurement system analysis is an experimental and mathematical method of finding how much the variation within the measurement process contributes to overall process variability. It is a technique that evaluates the reliability and accuracy of a measurement system to reduce the factors that contribute to the variation to the data. The goal is to ratify the measurement system used to obtain the process information is valid. Normally, it compares data set against a standard to obtain facts and knowledge about the component, product, services, features, or transaction.

The following components of variables are required to be investigated and measured before developing process capability decision making from the data, i.e., variable or continuous data and attribute or discrete data. These components are bias (system variation error), stability (accuracy), linearity, discrimination (gage variation error), and precision: repeatability or reproducibility (gage R&R).

9.2.2 ACCURACY/BIAS

Bias is a systematic deviation of a value from a reference point. In other words, it is the difference between the true value and the value obtained from the measurement system.

For best possibility of data accuracy, just record the process measurement as it is without any rounding off the data. Carefully document all the details about the process including the equipment, manpower, process conditions, material, measurement systems, and of course end-to-end timing. Calibrate gages, calipers, micrometers, and any other measurement system prior to data collection. See truth mean versus measure mean value (Figure 9.19).

9.2.3 STABILITY (CONSISTENCY)

The total variation in the measurements obtained with a gage on the same part over a long period of time. The stability of a measurement system is determined and monitored using the statistical control charts for consistency over time. Figure 9.20 illustrates this concept.

9.2.4 LINEARITY

Linearity (see Figure 9.21) is the difference in the bias values through the expected operating range of the gage. Unlike the repeatability and reproducibility that focus on the deviation or

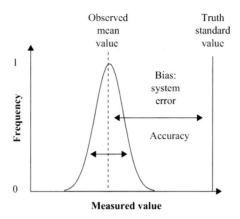

Figure 9.19. Truth mean versus measured mean value.

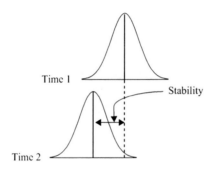

Figure 9.20. Stability distribution cases.

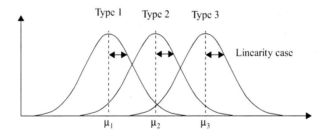

Figure 9.21. Linearity distribution cases.

accuracy of the measuring system, linearity and precision studies quantify the bias. The inspector should think (or doubt) about that whether measuring instrument has the same accuracy for all sizes of object being measured. Could there be linearity error in the equipment? For example, if in an experiment the measurement is off by 5% measuring the length of a 10-feet tree, but is off by 15% when measuring the length of a 15-feet tree, the instrument bias is nonlinear to changes over the dimension of the use. As shown in Figure 9.22 the nonlinear portion could go above or below the linear measurement. Figure 9.23 presents the ideal curve (linear) versus measured curve (nonlinear) illustrating linearity error.

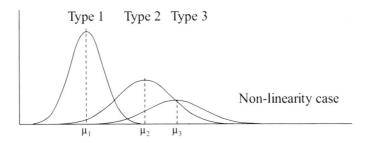

Figure 9.22. Nonlinearity distribution cases.

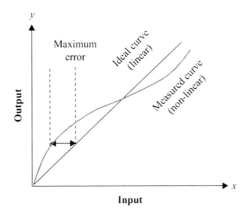

Figure 9.23. Ideal curve (linear) versus measured curve (nonlinear) illustrating linearity error.

9.2.5   GAGE REPEATABILITY AND REPRODUCIBILITY (OR GAGE R&R)

Gage R&R is a study of variability in a product or process. It investigates to determine how much of your monitored process variation is due to measurement system variation. The overall variation is broken down into three categories: part-to-part, repeatability, and reproducibility. The reproducibility component can be further broken down into its operator and operator by part and components.

Repeatability (precision): Someone taking the same measurement (more than once) on the same item with the same instrument (gage) will get the same answer (normally more than 10 parts maximum). This is demonstrated in Figure 9.24. The criteria measurements for repeatability are as follows:

- On the same part
- In the same position on the same part
- Under the same condition (s) of the application
- For a limited period of time

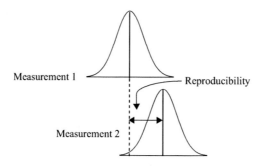

Figure 9.24. Gage R&R distribution cases.

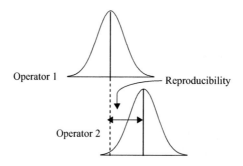

Figure 9.25. Gage R&R distribution cases.

Reproducibility: Different people measuring the same item with the same instrument (gage) will get the same answer. This is shown in Figure 9.25. The criteria measurements for reproducibility are as follows:

- Use the same measuring device
- Apply exact the same techniques of measurement
- Measure the same location of the same part
- Under the same condition(s) of the application
- For a limited period of time.

The gage R&R components are listed in Figure 9.26.

9.2.6 MEASUREMENT SYSTEM COMPONENTS

Measurement system bias recognized through calibration system (accuracy) as

$$\mu_{total} = \mu_{product} + \mu_{gage} \qquad \text{(Eq. 9.14)}$$

and measurement system variability assessed by gage R&R (see Figure 9.27) analysis that is

Total Variance = Product Variance + Measurement Variance (gage)

$$\sigma^2_{total} = \sigma^2_{product} + \sigma^2_{gage} \qquad \text{(Eq. 9.15)}$$

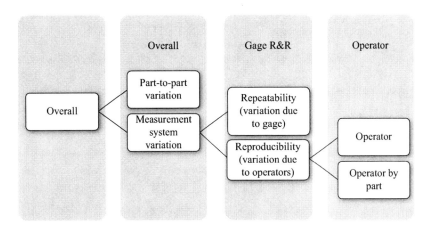

Figure 9.26. Gage R&R components.

Figure 9.27. Gage R&R variances.

where

$$\text{Measurement variance (gage)} = \text{Repeatability} + \text{Reproducibility}$$

or

$$\sigma^2_{gage} = \sigma^2_{repeatability} + \sigma^2_{reproducibility} \qquad \text{(Eq. 9.16)}$$

9.3 PROBABILITY DISTRIBUTIONS AND CONCEPTS

9.3.1 DEFINITION, EXPERIMENT, OUTCOME, AND SAMPLE SPACE

Probability is a measure of the expectation that an event in the future will happen. It can only accept a value between 0 and $1 (0 \le P(E_i) \le 1$ or $0 \le P(A) \le 0)$ and the sum of the probabilities of events (or outcomes) in a sample space always equal to 1.0. The higher the probability of an event, the more possibility of the event will occur. A value close to zero confirms the event is not likely to happen. A value close to one confirms it is likely to happen. There are three definitions of probability: objective (*classical, empirical*) and subjective.

1. *Objective probability:*
   i. *Classical probability*: The classical definition applies when there are *n* equally chance outcomes.

$$P(E_i) = \frac{Number\ of\ events\ E_i\ can\ happen}{Total\ number\ of\ events} \qquad \text{(Eq. 9.17)}$$

ii. *Empirical probability*: The empirical definition applies when the number of times the event (m) happens is divided by the number of observations (n). $P(x)$ equal to probability that event x takes place. For example, the probability of event m is happening in n outcomes (empirical).

$$P(x) = \frac{m}{n} \qquad \text{(Eq. 9.18)}$$

where m is possible of n outcomes.

2. *Subjective probability*: Subjective probability is based on whatever information is available. For example, estimating the probability that the price of electricity will go up more than 10% this summer, estimating the probability of flu rate will top 3% this winter, and estimating the probability that it will start raining within the next couple of hours.

Experiment: An experiment is the measure of some activity or the act of taking some measurement.

Outcome: An outcome (y_{ii}) is the particular result of an experiment. In the case of the rolling die, the possible outcomes are the numbers {1, 2, 3, 4, 5, and 6}.

Event: An event is the collection of one or more outcomes of an experiment. That is, we collect the outcomes 1, 3, and 5. *Elementary event* is an outcome from a sample space with one characteristic for example a red card from a deck of cards. However, *Event* may involve two or more outcomes simultaneously; for example, an ace that is also red from a deck of cards.

Sample Space: A sample space is the set of all possible outcomes (Σ outcomes) from an experiment, e.g., all six faces of a die or all 52 cards of a bridge deck.

Example 9.15

The sample space some experiments are as follows.

1. What is the probability of each outcome when a nickel or dime is tossed?
   The sample space of experiment 1 is: {Head, Tail}
   Probabilities: $P(1) = \frac{1}{2}$, $P(2) = \frac{1}{2}$
2. What is the probability of each outcome when a watermelon is cut in half?
   The sample space of experiment 2 is: {Red, yellow}
   Probabilities: $P(1) = \frac{1}{2}$, $P(2) = \frac{1}{2}$
3. What is the probability of each outcome when a six sided die is rolled?
   The sample space of experiment 3 is: {1, 2, 3, 4, 5, and 6}
   Probabilities: $P(1) = 1/6$, $P(2) = 1/6$, $P(3) = 1/6$, $P(4) = 1/6$, $P(5) = 1/6$, $P(6) = 1/6$

9.3.2 PROBABILITY OF EVENT (EI) AS RELATIVE FREQUENCY

Relative frequency of an event is given in Equation 9.19:

$$Empircial\ (relative\ frequency\ of\ E_i) = \frac{Number\ of\ times\ E_i\ occurs}{N} \qquad \text{(Eq. 9.19)}$$

where E_i = Event of interest, N = Total number of trials

Example 9.16

Coach J.D. throughout his basketball teaching and coaching career has won 650 out of 920 games. What is the probability that Coach J.D. will win in his next basketball game?

Solution

Let N indicates the total number of basketball games played E the number of the games won. As a consequence

$$N = 920 \quad \text{and} \quad E = 650$$

By applying the relative frequency concept, we attain

$$Relative\ frequency = \frac{number\ of\ successful\ trials}{total\ number\ of\ trials}$$

$$P\left(relative\ frequency\ of\ wining\ the\ next\ game\right) = \frac{650}{920} = 0.707$$

Thus, the relative frequencies (approximate probabilities) are

$$P\ (wining\ next\ game) = 0.707$$

Example 9.17

The U.S. Senate consists of 100 senators, 55 are "in favor of" working every other Saturday, 35 are "opposed to" and 10 remained "neutral." What is the probability of that a randomly chosen from the 100 senators is in favor of working every other Saturday?

Solution

Let N indicates the total number of senators in the house of senate and E the number of senates in favor of working every other Saturday. As a consequence

$$N = 100 \quad \text{and} \quad E = 55$$

Table 9.7. Frequency and relative frequency distribution

Outcome	E	P (Relative frequency)
Favor	55	$\dfrac{55}{100} = 0.55$
Oppose	35	$\dfrac{35}{100} = 0.35$
Neutral	10	$\dfrac{10}{100} = 0.10$
	$N = 100$	$\text{Sum} = \sum P(E) = 1.0$

By applying the relative frequency concept, we attain

$$P(randomly\ chosen\ senator\ in\ favor) = \frac{E}{N} = \frac{55}{100} = 0.55$$

The probability shows the actual relative frequency of the outcome. Table 9.7 lists the frequency and relative frequency of the three outcomes for Example 9.17.

Thus, the relative frequencies (approximate probabilities) are

$$P\ (\text{Favor}) = 0.55$$

$$P\ (\text{Oppose}) = 0.35$$

$$P\ (\text{Neutral}) = 0.10$$

Thus, an activity for which the outcome is uncertain is called an experiment and an event consists of one or more possible outcomes of the experiment such as shown in Example 9.18.

Example 9.18

Suppose you toss a coin one time. Then, what will be the probability of event A?

Solution

Tossing a coin one time will give two possible outcomes for event "A" either head (H) or tail (T). This is an equally likely event when all of the outcomes in a sample space have the same probability of occurrence. If a sample space contains n outcomes, all of which are equally likely, then

$$P(E) = \frac{1}{n}$$

For every outcome E_i in S,

$$P(A) = \frac{m}{n} = \frac{1}{2} = 0.50$$

Nickel (x)	Probability $P(x)$
H	½
T	½

Observing one head and one tail,

$$\text{Sample Space} = \text{SS} = \{H, T\}$$

Example 9.19

Suppose you toss two coins (nickel and dime) two times. What will be the probabilities of head (H) and tail (T)?

Solution

Tossing two coins two times will give four possible outcomes.

$$P(A) = \frac{m}{n} = \frac{1}{4} = 0.25$$

n	Nickel (x_1)	Dime (x_2)	Probability $P(x)$
1	H	H	¼
2	H	T	¼
3	T	H	¼
4	T	T	¼

n: possible outcome (m possible of n outcome)

Two outcome for event A (observing one Head and one Tail) $m = 2$, $n = 4$
$P(A) = (m/n) = (2/4) = 0.50$;

$$\text{Sample Space} = \text{SS} = \{(HH), (HT), (TH), (TT)\}$$

9.3.3 MARGINAL AND CONDITIONAL PROBABILITIES

The probability of a single event used to define the contingency table. For example, suppose you took your kids to fish store and they purchased 150 small fishes for your fish tank. Now you put them in a contingency table format such as Table 9.8.

Table 9.8. Contingency table

	American	European	Total
Gold Fish	40	20	60
Dark Fish	60	30	90
Total	100	50	150

Table 9.9. Contingency table

	Ace	Not ace	Total
Red	2	24	26
Black	2	24	26
Total	4	48	52

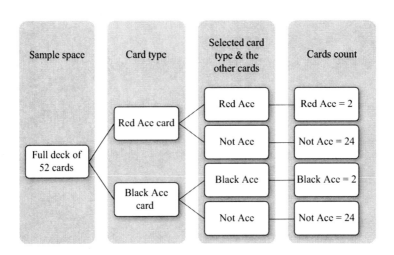

Figure 9.28 Space sample tree diagram.

The four boxes containing 40 (American gold fish), 20 (European gold fish), 60 (America dark fish), and 30 (European dark fish) called cells. The row and column containing the total values are called margins. Marginal probabilities can be determined by taking the certain total number from the total column or row and dividing that number by the grand total for the contingency table.

Now let's create contingency table for red and black ace using 52 cards of a bridge deck.

Table 9.9 represents total number of space samples and Figure 9.28 applies tree diagram to illustrate classification of space sample.

Example 9.20

Refer to data given in the Table 9.8. What is the probability that one of your children takes gold, dark, American, or European fish at random from the fish tank?

Solution

Using the information given in Table 9.8, we see that 60 gold fish out of 150, 90 dark fish out of 150, 100 American out of 150, and 50 European out of 150. Therefore,

$$P(Gold\ fish) = \frac{Total\ number\ of\ gold\ fish}{Total\ number\ of\ gold\ and\ dark\ fish} = \frac{60}{150} = 0.40$$

$$P(Dark\ fish) = \frac{Total\ number\ of\ dark\ fish}{Total\ number\ of\ gold\ and\ dark\ fish} = \frac{90}{150} = 0.60$$

$$P(American\ fish) = \frac{Total\ number\ of\ gold\ fish}{Total\ number\ of\ American\ and\ European\ fish} = \frac{100}{150} = 0.67$$

$$P(European\ fish) = \frac{Total\ number\ of\ gold\ fish}{Total\ number\ of\ American\ and\ European\ fish} = \frac{50}{150} = 0.33$$

9.3.4 THE RULES OF PROBABILITY (UNION OF EVENTS)

As shown in Figure 9.29 the probability of events are between zero and one. Or, some of all the probabilities must be between zero and one.

The union of events is the probability of either event A or event B occurring: $P(A \text{ or } B)$

Let's recall the information from American and European gold/dark fish in the Table 9.9.

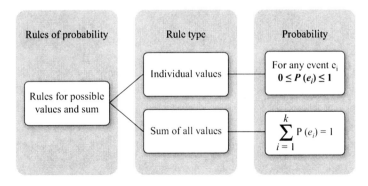

Figure 9.29. Rules of probability.
where k = Number of elementary events in the sample space and
e_i = ith elementary event

Table 9.10. Contingency table

Type	American	European	Total
Gold Fish	40	20	60
Dark Fish	60	30	90
Total	100	50	150

We are interested in selecting a fish with one of the following criteria:

G: Gold fish and or

E: European fish

This scenario is called the union of two events and is written as (G or E) [or (G∪E)]. This union is consists of all fish that are gold or European as

$$P(\text{G or E}) = P(\text{Gold fish}) + P(\text{European fish}) - P(\text{Gold and European})$$

$$= \frac{\text{Total G}}{\text{Grand total}} + \frac{\text{Total E}}{\text{Grand total}} - \frac{P(\text{European gold})}{\text{Grand total}} \qquad \text{(Eq. 9.20)}$$

$$P(G \; or \; E) = P(G) + P(E) - P(G \; and \; E) = \frac{60}{150} + \frac{50}{150} - \frac{20}{150} = \frac{60+50-20}{150} = \frac{90}{150} = 0.60$$

9.3.4.1 *Mutually Exclusive Events*

If E_1 (Event 1) takes place, then E_2 (Event 2) cannot take place. E_1 and E_2 have no common factors. For example, a card cannot be black and red concurrently, when flipping a coin heads and tails are mutually exclusive, walking and running simultaneously mutual exclusive, and when playing card Kings and Aces are mutually exclusive. The probability that A and B will occur together is equal to zero. In other words, it is impossible. Thus, in mathematical notation

$$P(\text{A and B}) = 0$$

The probability of A or B equals to the sum of the probability of A and probability of B

$$P(\text{A and B}) = P(\text{A}) + P(\text{B})$$

On the other hand, what is not mutually exclusive: when playing a card because you can have a king of hearts; thus it is not mutually exclusive. Walking and talking on the phone is not mutually exclusive?

9.3.4.2 *Addition Rule of Probability*

The probability of an event E_i is equal to the sum of the probabilities of the elementary events forming E_i. That is, if the following is established:

$$E_i = \{e_1, e_2, e_3, e_4, e_5\} \text{ sample space for elementary probabilities}$$

Therefore, we have

$$P(E_i) = P(e_i) + P(e_2) + P(e_3) + P(e_4) + P(e_5)$$

Addition Rule 1: When two events, A and B are "mutually" exclusive, the probability that A or B will occur is the sum of the probability of each event.

$$P(A \text{ or } B) = P(A) + P(B)$$

Let's use this rule to compute the probability for Rule 1.

Example 9.21

A pencil box contains 2 red, 6 green, 4, blue, and 8 yellow pencils. If a single pencil is selected at random from the pencil box, what is the probability that it red or blue?

Solution

The probability of red or blue pencil is equal to sum of the probably of red pencil and probability of blue pencil.

$$P(Red\ pencil) = \frac{2}{20} = 0.10$$

$$P(Blue\ pencil) = \frac{4}{20} = 0.20$$

$$P(Red\ or\ Blue) = P(Red) + P(Blue) = 0.10 + 0.20 = 0.30$$

Example 9.22

Suppose you rolled a six-sided die. What is the probability of rolling a 3 or a 6?

Solution

There are two possibilities: (1) the number rolled can be a 3. (2) The number rolled can be a 6. Since they cannot occur at the same time, thus these events are mutually exclusive.

$$P(3) = \frac{1}{6}$$

$$P(6) = \frac{1}{6}$$

$$P(3 \text{ or } 6) = P(3) + P(6) = \frac{1}{6} + \frac{1}{6} = \frac{2}{6} = \frac{1}{3}$$

Addition Rule 2: When two events A and B are "not mutually" exclusive, then P(A or B) is given by the following formula:

$$P(\text{A or B}) = P(\text{A}) + P(\text{B}) - P(\text{A and B})$$

In the rule number 2, P(A and B) refers to the overlap of the two events. Let's apply rule number 2 in Example 9.23.

Example 9.23

Let say in New York City 15% of people watch morning news, 25% of people watch evening news, and 10% of people watch both (morning and evening). Find a person who watches morning, evening, or both?

Solution

Since some of them overlap, thus the two events are not mutually exclusive.

$$P(\text{M or E}) = P(\text{M}) + P(\text{E}) - P(\text{M and E})$$
$$= 0.15 + 0.25 - 0.10$$
$$= 0.50$$

50% of people watch at least morning or evening news.

Example 9.24

In a Lean Six Sigma class of 40 students, 25 are manufacturing engineers and 15 are design engineers. On the mid-term test, 5 manufacturing and 4 design engineers made an "A" grade. If a student is selected at random from the class, what is the probability of selecting a manufacturing engineer or an A student?

Solution

The two events are not mutually exclusive. Thus, we have

$$P(manufacturing \text{ or } design \text{ } engineer)$$
$$= P(manufacturing \text{ } engineer) + P(A) - P(manufacturing \text{ and } A)$$
$$= \frac{25}{40} + \frac{9}{40} - \frac{5}{40} = \frac{29}{40}$$

9.3.4.3 Complementary Events

It is a particular case of mutually exclusive events that take place when these two events consist of all the outcomes in the sample space. These mutually exclusive events are called complementary events and each event considered to complement the other event.

Complementary probability is used to determine the probability of an event occurring by subtracting the probability of the event not occurring from 1.

If $P(A)$ is the probability of event A

$P(\hat{A})$ is the complement of A,

Then, we have $P(\hat{A}) = 1.0 - P(A)$

Example 9.25

Suppose you rolled a six-sided die. What is the probability of rolling a number that is not 5?

$$P(\hat{A}) + P(A) = 1.0$$

$$P(\hat{A}) = 1.0 - P(A)$$

$$P(\text{not } 5) = 1.0 - P(5) = 1.0 - \frac{1}{6} = \frac{5}{6}$$

9.3.5 THE RULES OF PROBABILITY (INTERSECTION OF EVENTS)

9.3.5.1 Independent and Dependent Events

In Section 9.3.3, we studied marginal and conditional probabilities by applying the distribution of the type of the fish and the conclusion regarding the American or European fish.

From the value in Table 9.8, we calculated the forgoing probabilities:

$$P(American) = \frac{Total\ American\ type}{Grand\ Total} = \frac{100}{150} = 0.67$$

$$P(American\ Gold\ fish) = \frac{American\ gold\ fish}{Total\ Gold\ fish} = \frac{40}{60} = 0.67$$

i. *Independent Events*: Occurrence of one does not influence the probability of occurrence of the other.

Now, assume another fish tank with a second contingency table, which also includes fish type. Table 9.11 gives the distribution of fish type and location of fish for 200 of them.

Table 9.11. Contingency table for American and European fish

Type	American (A)	European (E)	Total
Gold Fish (G)	50	35	85
Dark Fish (D)	70	45	115
Total	120	80	200

Computing the probabilities of the same events as above

$$P(American) = \frac{Total\ American\ type}{Grand\ Total} = \frac{50}{200} = 0.67$$

In this case, the type of fish has no impact on the probability that European fish was selected.

Here, we are interested in selecting a fish with one of the following criteria:

G: Gold fish, and or

E: European fish

$$P(G) = P(G \mid E)$$

If the $P(G) = P(G \mid E)$ then event G is said to be independent of event E. That is one doesn't influence the other. Thus event G is independent of event E. From the values of Table 9.5 we can determine the probability of gold fish:

$$P(G) = \frac{Total\ Number\ of\ Gold\ fish}{Total\ number\ of\ Gold\ and\ dark\ fish} = \frac{85}{200} = 0.425$$

$$P(G) = P(G \mid E) = 0.40$$

In addition, events A and B are independent if and only if the following rules (Equation 9.21) applies.

$$P(A \mid B) = P(A) \quad (assuming\ P(B) \neq 0),\ or$$

$$P(B \mid A) = P(B) \quad (assuming\ P(A) \neq 0),\ or$$

$$P(A\ and\ B) = P(A) \cdot P(B)$$

$$e.g.,\ P(A \mid B) = \frac{P(A) \cdot P(B)}{P(B)} = P(A);\ P(B) \neq 0 \qquad \text{(Eq. 9.21)}$$

Furthermore, events *A, B,* and *C* are independent if all the following (9.22) are true:

$$P(A\ and\ B) = P(A) \cdot P(B)$$

$$P(A\ and\ C) = P(A) \cdot P(C)$$

$$P(\text{B and C}) = P(\text{B}) \cdot P(\text{C})$$

$$P(\text{A and B and C}) = P(\text{A}) \cdot P(\text{B}) \cdot P(\text{C}) \qquad \text{(Eq. 9.22)}$$

As mentioned above, events are independent if the occurrence of one event does not affect the occurrence of another like fair coin toss and fair die roll.

E_1 = Heads on one flip of fair coin

E_2 = Heads on second flip of same coin

The results of second flip don't depend on the result of the first flip.

ii. *Dependent Events*: Occurrence of one affects the probability of the other such as a jar of balls contains 20 blue balls and 20 yellow balls.

E_1 = Heavy traffic due to accident on the news

E_2 = Leave early to work

The probability of the second event is affected by the occurrence of the first event.

9.3.5.2 Multiplication Rule: Independent

The special rule of multiplication requires that two events A and B are independent if the occurrence of one event has no effect on the probability of the occurrence of the other event.

This rule is written as (9.23)

$$P(\text{A and B}) = P(\text{A})P(\text{B}), \text{ otherwise } P(\text{A and B}) = P(\text{A})P(\text{B}|\text{A}) \qquad \text{(Eq. 9.23)}$$

The general rule of multiplication is used to find the joint probability that two events will occur. It states that for two events A and B, the joint probability that both events will happen is found by multiplying the probability that event A will happen by the conditional probability of B given that A has already occurred.

For any two events A and B the joint probability, $P(\text{A and B})$ is given by the following formula:

$$P(\text{A and B}) = P(\text{A})P(\text{B}|\text{A}) \text{ or } P(\text{B and A}) = P(\text{B})P(\text{A}|\text{B}) \text{ and that is}$$

$$P(\text{A and B}) = P(\text{A}) \cdot P(\text{B}|\text{A}) = P(\text{B}) \cdot P(\text{A}|\text{B}) \qquad \text{(Eq. 9.24)}$$

As mentioned, if A and B are independent, then $P(\text{B}|\text{A}) = P(\text{B})$, thus the above Equation 9.24 is reduced to Equation 9.25:

$$P(\text{A and B}) = P(\text{A}) \cdot P(\text{B}) \qquad \text{(Eq. 9.25)}$$

where

$P(\text{A})$ = probability that event A occurs

$P(\text{B})$ = probability that event B occurs

$P(\text{A}|\text{B})$ = the conditional probability that event A occurs given that event B has occurred already

$P(\text{B}|\text{A})$ = the conditional probability that event B occurs given that event A has occurred already

Example 9.26

Independent Events: Suppose you rolled a six-sided die. What is the probability of rolling a 3 and then a 6?

Solution

These events are independent. That is rolling a three doesn't change the probability of rolling a six or vice versa.

$$P(A) = \frac{1}{n} \rightarrow P(3) = \frac{1}{6}$$

$$P(B) = \frac{1}{n} \rightarrow P(6) = \frac{1}{6}$$

$$P(A \text{ and } B) = P(A) \bullet P(B)$$

$$P(3 \text{ and } 6) = \frac{1}{6} \bullet \frac{1}{6} = \frac{1}{36}$$

There is a 1 in 36 chance of rolling 3 and then rolling a 6.

Example 9.27

Dependent Events: What is the probability of drawing a king and then drawing a queen from a deck of cards?

Solution

Drawing a king and then drawing a queen from a deck of cards, without replacement of the king back. These events are dependent because drawing a king without replacement of the king back changes the equation and the probability of drawing a queen. Furthermore, without the king in the deck of the cards the probability of drawing a queen changes from 4/52 to 4/51.

$$P(A \text{ and } B) = P(A) \bullet P(B|A)$$

$$P(King \text{ and } Queen) = \frac{4}{52} \bullet \frac{4}{51} = \frac{16}{2652} = 0.006 = 0.60\%$$

Approximately, 0.60% chance of drawing a king, and then drawing a queen without replacement from a deck of cards.

Conditional Probability

When you have been given information about two events and you are asked to determine the probability based on the provided information, then the result is a conditional probability $P(A|B)$. In other words, the conditional probability of an event "A" in relationship to event "B" is the probability that is based on the fact that event "B" has already occurred. The mathematical expression for conditional probability is $P(A|B)$, that is to say the probability of A given B.

If event A and event B are independent then,

$$P(A \mid B) = P(A)$$

$$P(B \mid A) = P(B)$$

$$P(A \text{ and } B) = P(A) \cdot P(B)$$

Conditional probability rule for any two events A and B are given as

$$P(A \mid B) = \frac{P(A \text{ and } B)}{P(B)} = \frac{P(A) \cdot P(B)}{P(B)} = P(A) \quad (P(B) > 0)$$

and

$$P(B \mid A) = \frac{P(B \text{ and } A)}{P(A)} = \frac{P(B) \cdot P(A)}{P(A)} = P(B) \quad (P(A) > 0)$$

(Eq. 9.26)

Let's apply this concept in Example 9.28.

Example 9.28

In an industrial park, 60% of the buildings have central heating system, 30% have solar energy system, and 20% have both systems. What is the probability that a building has solar energy system, given that it has central heating system?

Solution

Table 9.12 shows the distribution of energy system(s) in each building.

Given central heating (CH), we only consider the top row (60% of the buildings). Of these, 20% have central heating (CH). That is 20% of the 60% is about 33.33%

Table 9.12. Contingency table for solar energy and central heating system

	Solar energy (SE)	**No solar energy (NSE)**	**Total**
Central Heating (CH)	0.20	0.40	0.60
No Central Heating (NCH)	0.10	0.30	0.40
Total	0.30	0.70	1.0

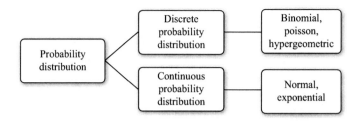

Figure 9.30. Classifications of probability distributions.

$$P(SE|CH) = \frac{P(SE \text{ and } CH)}{P(CH)} = \frac{0.20}{0.60} = 0.3333$$

Probability Distributions

The probability distribution is a fundamental concept in statistics. They are applied in theoretical and practical applications. Some of the practical applications of probability distributions are: to determine the critical area for hypothesis testing, to find the confidence interval for variables, and computer simulation studies particularly random number generator for optimum process improvement.

The classifications of the probability distribution (see Figure 9.30) are discussed in the following sections.

9.4 DISCRETE RANDOM VARIABLES: PROBABILITY DISTRIBUTION

A discrete random variable is a variable that can consider or accept only an integer (whole) number of values. For many possible outcomes such as number of grocery customers per day, number of students in the class, and number of doorbell rings before the door is opened. For absolutely two possible outcomes like gender (male or female), automotive color black or silver, and answer for any question yes or no.

Let x be a discrete random variable that takes the numerical values $x_1, x_2, ..., x_{n-1}, x_n$ with probabilities of $P(x_1), P(x_2), ..., P(x_{n-1}), P(x_n)$ accordingly. A discrete random variable probability distribution consists of the values of the random variable x and their probabilities $P(x)$. Therefore, the probabilities of $P(x)$ are $\sum_{i=1}^{i=n} P(x_i) = 1.0$. Furthermore, if the sum of the probabilities is equal one, then the probability distribution of random variable is called probability of mass function. Let's apply the above probability equation in Examples 9.29 and 9.30.

Example 9.29

A person goes to the farm market. He buys a watermelon and takes it to his family. Let "R" represent Red and "Y" represent yellow. Suppose that both outcomes are equally likely. Then, let "X" equals to the number of "Y."

(a) List all outcomes (Red or Yellow) after cutting the watermelon.
(b) What is $P(x = 0) = ?$
(c) What is $P(x = 1) = ?$
(d) Is the conclusion a probability mass function (PMF)?

Solution

There are two possible outcomes red or yellow and the probabilities are equally likely.

(a) Possible outcomes

Watermelon	$P(x_i)$
R	½
Y	½

Thus, the probability of equally likely is $P(x = R) = P(x = Y) = 0.50$
(b) $P(x = R) = P(x = 0) = P(0) = 0.50$
(c) $P(x = Y) = P(x = 1) = P(1) = 0.50$
(d) To find the probability of mass function sum of the properties must be equal to 1.0.

$$\sum_{i=1}^{n} P(x_i) = P(x_R) + P(x_Y) = P(x = 0) + P(x = 1) = 0.50 + 0.50 = 1.0, \quad \text{since the}$$

sum of the probabilities equal to one thus the function is PMF.

Example 9.30

Is the following function a probability mass function (PMF)?

$$P(X = x) = \frac{5}{6(5 - x)!x!} \quad \text{For } x = 0, 1, 2, 3$$

Solution

Lets the equation above and the given values for $x(s)$ then we have

$$P(X = 0) = \frac{5}{6(5 - 0)!0!} = 0.0069$$

$$P(X = 1) = \frac{5}{6(5 - 1)!1!} = 0.0347$$

$$P(X = 2) = \frac{5}{6(5 - 2)!2!} = 0.0694$$

$$P(X = 3) = \frac{5}{6(5 - 3)!3!} = 0.0694$$

If $PMF = \sum_{x=0}^{x=n} P(X=x) = 1.0$ then the above equation is probability of mass function

$$PMF = \sum_{x=0}^{x=3} P(X=x) = P(0) + P(1) + P(2) + P(3) = 0.0069 + 0.047 + 0.0694 + 0.0694 = 0.1804$$

This is not a probability mass function (PMF) since the sum does not equal 1.00.

9.4.1 BINOMIAL PROBABILITY DISTRIBUTION

It is a discrete probability distribution that has only (absolutely) two possible outcomes, i.e., good or bad, yes or no, on or off, success or fail, and high or low. The probabilities are the same from experiment to experiment. Thus, it is used to achieve probability of x success (or defects) in m independent repetitions (or trials), with probability of success on a single experiment represented by or equal to "p." Then, given binomial probability mass function is defined in the following equation:

$$P(x, p, m) = m![x!(m-x)!]^{-1} p^x (1-p)^{m-x} = {}_m C_x p^x q^{m-x} \qquad \text{(Eq. 9.27)}$$

where

$x = 0, 1, 2, \ldots, n-3, n-2, n-1, n$ (number of occurrences success in m trials)
m = number of repetitions or total number of trials
p = probability of success
$q = 1 - p$ = probability of failure
$m - x$ = number of failure in m trials
!: Called factorial (is the product of multiplying (Equation 9.28) all the numbers beginning from 1 to specified number) for instance,

$$n! = 1 \times 2 \times 3 \times \cdots \times (n-3) \times (n-2) \times (n-1) \times n \qquad \text{(Eq. 9.28)}$$

By definition the factorial of zero is always equal to 1.0.

$$0! = 1.0 \qquad \text{(Eq. 9.29)}$$

Example 9.31

Find 3! and $(8-3)!$.

Solution

The value of 3! And $(8-3)!$ are given by the product of all the integers from 1 to 3 and from 1 to 5. Thus,

$$3! = (1)(2)(3) = 6$$

$$(8-3)! = 5! = (1)(2)(3)(4)(5) = 120$$

Additional properties are

(a) For $p = 0.5$ the binomial distribution shape looks a like normal (bell-shaped).
(b) For $p < 0.5$ the distribution curve skewed right and the skewness increases as p decreases.
(c) For $p > 0.5$ the distribution curve skewed left and the skewness increases as p increases and nears 1.0.

The mean and variance are

$$mean = \mu = mp$$

$$Variance = \sigma^2 = mpq = mp(1-p)$$

$$Standard\ deviation = \sigma = \left(mp(1-p)\right)^{0.50}$$

$$Range = 0 \text{ to } m$$

The shape of the binomial distribution depends on the value of p and m. Figure 9.31 illustrates the value of the $m = 5$ and $p = 0.10$.

Let's apply the above equation to Figure 9.30 when $n = 5$, and $p = 0.10$ we have

$$\mu = mp = (5)(0.10) = 0.50$$

$$\sigma = \sqrt{(5)(0.10)(1.0-0.10)} = 0.6708$$

However, Figure 9.32 shows that as the sample number increases the distribution approaches normal (symmetrical shape) no matter what the p value is.

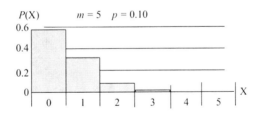

Figure 9.31 Binomial distributions skewed to the left.

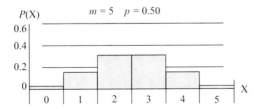

Figure 9.32. Binomial distributions: symmetrical shape.

Similarly, let's apply the above equation to Figure 9.31 when $n = 5$, and $p = 0.50$ we have

$$\mu = mp = (5)(0.50) = 2.50$$

$$\sigma = \sqrt{(5)(0.50)(1.0-0.50)} = 1.118$$

One can use binomial table values in the formulas listed in Table 9.13 to determine the probability values without too much calculation.

The cumulative binomial probability distribution values can be obtained from Table X in the appendix. Table X was developed using equation

$$P(x \leq x_1) = \sum_{k}^{x_1} \binom{n}{k} p^k (1-p)^{n-k}$$

$$P(x \leq x_1) = \sum_{k}^{x_1} \left(\frac{n!}{k!(n-k!)} \right) p^k (1-p)^{n-k}$$

where the probability of x_1 or fewer successes in n trials, or of up to and including x_1 successes in n trials, when each trial has a probability of succeeding p. So for a probability of up to 8 frequency out of 10 trials if $p = 0.95$ we have

$$P(x \leq x_1) = \sum_{k=0}^{x_1} \binom{10}{k} p^k (1-p)^{n-k}$$

Table 9.13. Binomial random variables for a given n samples: to be applied using binomial table

$$P(x \leq k) = \sum_{x=0}^{x=k} P(x) = P(x=0) + P(x=1) + \cdots + P(x=k)$$

Or:

$$P(x \leq k) = 1 - [P(x=k+1) + P(x=k+2) + \cdots + P(x=n)]$$

$$P(x \geq k) = \sum_{x=k}^{x=n} P(x) = P(x=k) + P(x=k+1) + \cdots + P(x=n)$$

Or:

$$P(x \geq k) = 1 - P(x < k) = 1 - P(x \leq k-1) = 1 - [P(x=k-1) + \cdots + P(x=0)]$$

$$P(x \geq k) = P(x \leq n) - P(x \leq k-1); \quad For\ k = n\ P(x \leq k) = P(x \leq n) = 1$$

$$P(x = k) = P(x \leq k) - P(x \leq k-1)$$

$$P(w \leq x \leq z) = P(x=w) + P(x=w+1) + P(x=w+2) + \cdots + P(x=z)$$

$$P\left(x \le x_1\right) = \sum_{k=0}^{8}\left(\frac{10!}{k!\left(10-k!\right)}\right)(0.95)^k \left(1-0.95\right)^{10-k}$$

$$P\left(x \le 8\right) = 0.0861$$

The Excel version is

$$P(x \le x_1) = \text{BINOM.DIST}(\$B12,\$A\$4,AG\$3,\text{TRUE}) = 0.0861$$

where, $\$B12 = x_1 = 8, \$A\$4 = n = 10$, AG$3 = \alpha = 0.95$
    Or,

$$P(x \le 8) = \text{BINOM.DIST}(8,10,0.95,\text{TRUE}) = 0.0861$$

Example 9.32

Binomial Table application: For a binomial random variable with $n = 5$ and $P = 0.30$, compute the following probabilities:

$$P(x = 1),\ P(x \ge 1),\ P(x \le 3),\ P(1 \le x \le 3),\ P(x \le 1)$$

Solution

Using Table 9.12 and binomial table, we can determine the probabilities.

1. From binomial Table at $n = 5$ and $p = 0.30$ the probability of $P(x = 1)$ equal to

$$P(x = 1) = 0.360$$

2. Since some of all the probabilities are equal to 1, if we substrate the probability of zero from 1.0 then we will get the $P(x \ge 1)$. Again, using binomial table one can obtain all the given probabilities:

$$P(x \ge 1) = 1 - P(x = 1 - 1) = 1 - P(x = 0) = 1 - P(0) = 1 - 0.168 = 0.8319$$
$$Or: = P(x = 1) + P(x = 2) + P(x = 3) + P(x = 4) + P(x = 5)$$
$$= 0.360 + 0.309 + 0.132 + 0.028 + 0.002 = 0.8319$$

3. Likewise $P(x \le 3)$:

$$P(x \le 3) = P(x = 0) + P(x = 1) + P(x = 2) + P(x = 3)$$
$$= 0.168 + 0.360 + 0.309 + 0.132 = 0.97$$

Or:

$$P(x \le 3) = 1 - [P(x = 4) + P(x = 5)] = 1 - P(x = 4) - P(x = 5) = 1.0 - 0.028 - 0.002 = 0.97$$

4. Similarly $P(1 \leq x \leq 3)$:

$$P(1 \leq x \leq 3) = P(x = 1) + P(x = 2) + P(x = 3) = 0.360 + 0.309 + 0.132 = 0.801, \text{ Or}$$

$$P(1 \leq x \leq 3) = P(x = 3) + P(x = 4) + P(x = 5) = 0.8012$$

5. Just like in part *b* either we can calculate all the probabilities for $P(x \leq 1)$ or we can subtract all the probabilities of $P(x \geq 1)$ from 1.0. In either case we will end up with the same answer.

$$P(x \leq 1) = P(x = 0) + P(x = 1) = 0.168 + 0.360 = 0.528,$$

Or:

$$P(x \leq 1) = 1 - [P(x = 2) + P(x = 3) + P(x = 4) + P(x = 5)] = 0.528$$

Example 9.33

Suppose in a basketball game a player was fouled out during a 2-point basket. He was asked then to shoot two point baskets. What is the probability of ball going through the basket?

Solution

Now, the probability of the ball going through the basket is listed in Table 9.14. Assuming that if the ball goes through the basket then it is success, otherwise (no basket) it is a failure.

By substituting the value in the Equation 9.30 we have

$$P(x, p, m) = m![x!(m-x)!]^{-1} p^x (1-p)^{m-x} \qquad \text{(Eq. 9.30)}$$

$$P(0, 0.5, 2) = 2![0!(2-0)!]^{-1} 0.5^0 (1-0.5)^{2-0} = 0.25$$

$$P(1, 0.5, 2) = 2![1!(2-1)!]^{-1} 0.5^1 (1-0.5)^{2-1} = 0.50$$

$$P(2, 0.5, 2) = 2![2!(2-2)!]^{-1} 0.5^2 (1-0.5)^{2-2} = 0.25$$

The summary of the probabilities are provided in the Table 9.15.

Table 9.14. The probability of the ball going through the basket

Outcome	First Try	Second try
1	Success	Success
2	Success	Fail
3	Fail	Success
4	Fail	Fail

Table 9.15. Probability of the ball getting 0, 1, and 2 baskets

Number of baskets	Probability
0	0.25
1	0.50
2	0.25

Example 9.34

Binomial: What is the probability that a binomial random variable with $n = 15$ and $p = 0.20$ does not exceed 1.0?

Solution

Using binomial equation (9.30) we have

$$PMF = P(X = x) = P(x) = m![x!(m-x)!]^{-1} p^x (1-p)^{m-x}$$

For $x = 0, 1, 2,\ldots, m$
Given that $m = 15$ and $p = 0.20$
Substituting m and p in the above equation we get

$$P(x = 0) = \{(15!)/[0!(15-0)!]\}^{-1}(0.20)^0 (0.80)^{15-0}$$

$$P(x = 1) = \{(15!)/[1!(15-1)!]\}^{-1}(0.20)^1 (0.80)^{15-1}$$

$P(x \leq 1) = P(x = 0) + P(x = 1)$

$$= \{(15!)/[0!(15-0)!]\}^{-1}(0.20)^0 (0.80)^{15} + \{(15!)/[1!(15-1)!]\}^{-1}(0.20)^1 (0.80)^{14}$$

$$= 0.167$$

Or, one may just look for the values using the binomial table we get

$$P(x \leq 1) = 1 - [P(x = 2) + P(x = 3) + P(x = 4) + \cdots + P(x = 15)] = 0.167$$

9.4.1.1 *Mean and Standard Deviation of Discrete Random Variables*

The mean, variance, and standard deviation of discrete random variable are denoted by μ, σ^2, and σ, respectively for all normal variables. Thus, what is the expected value? The expected value is the outcome that will most likely occur. Here are the mathematical definitions of mean (9.31), variance (9.32), and standard deviation (9.33).

$$Mean\ (\mu) = \sum_{x=0}^{x=n} xP(x) \qquad \text{(Eq. 9.31)}$$

$$variance\ (\sigma^2) = \sum_{x=0}^{x=n} x^2 P(x) - \mu^2 \qquad \text{(Eq. 9.32)}$$

$$Standard\ deviation\ (\sigma) = (Variance)^{0.50} = \left(\sum_{x=0}^{x=n} x^2 P(x) - \mu^2 \right)^{0.50} \qquad \text{(Eq. 9.33)}$$

Examples 9.35 and 9.36 illustrate the calculation of the mean, variance, and standard deviation of a discrete random variable.

Example 9.35

The manager of a biotech company would like to know what the average number of absentees is daily and also what the standard deviation (σ) is of the daily employee absentee rate. Determine these two values from the following probability mass function (*PMF*), which was constructed from historical data of the company.

Number of daily absentee	0	1	2	3	4	5	6
Probability $P(x)$	0.35	0.25	0.11	0.16	0.07	0.4	0.2

Solution

To find the mean number of daily absentee (x), we multiply each value of x by its probability and add these products. This sum gives the probability distribution of x. The product of $xP(x)$ is listed on the third column of Table 9.16.

Table 9.16. Probability distribution of x

x	$P(x)$	$xP(x)$	x^2	$x^2 P(x)$
0	0.35	(0)(0.35) = 0.00	0	0
1	0.25	(1)(0.25) = 0.25	1	0.25
2	0.11	(2)(0.11) = 0.22	4	0.44
3	0.16	(3)(0.16) = 0.48	9	1.44
4	0.07	(4)(0.07) = 0.28	16	1.12
5	0.04	(5)(0.04) = 0.20	25	1.00
6	0.02	(6)(0.02) = 0.12	36	0.72
		$\mu = \Sigma\ xP(x) = 1.55$	$\Sigma x^2 P(x) = 4.97$	

(a) Mean $= \mu = \Sigma\, xP(x) = 1.55$

(b) Variance: $\sigma^2 = \{\Sigma x^2\, P(x)\} - \mu^2 = 4.97 - 2.4025 = 2.5675$

(c) Standard Deviation $= (\sigma^2)^{0.50} = (2.5675)^{0.50} = 1.60$

Example 9.36

The Human Resources (HR) of a phone company is looking at the number of employees claiming finger injury over a period of time. Let's define the random variable x to be number of injuries reported. Based on the history HR came up with probability distribution for x:

x	$P(x)$: Probability	Report
0	0.40	40% of the week no report
1	0.30	30% of the week 1 report
2	0.15	15% of the week 2 reports
3	0.10	10% of the week 3 reports
4	0.05	5% of the week 4 reports
	1.00	

Hint: $\mu = \Sigma\, xP(x);\ \sigma^2 = \{\Sigma x^2\, P(x)\} - \mu^2$

(a) Determine the mean of x?

(b) What is the variance and standard deviation of the random variable concerning the injury report?

Solution

Table 9.17 summarizes the probability distribution of x.

(a) $\mu = \Sigma\, xP(x) = (0)(0.4) + (1)(0.3) + (2)(0.15) + (3)(0.10) + (4)(0.05) = 1.10$ injury report on the average per week.

Table 9.17. Probability distribution of x

x	$P(x)$	$x\, P(x)$	x^2	$x^2P(x)$
0	0.40	0.00	0.00	0.00
1	0.30	0.30	1.00	0.30
2	0.15	0.30	4.00	0.60
3	0.10	0.30	9.00	0.90
4	0.05	0.20	16.00	0.80
	1.00	$\mu = xP(x) = 1.10$		$\Sigma x^2P(x) = 2.60$

(b) $\sigma^2 = \{\Sigma x^2\, P(x)\} - \mu^2 = 2.60 - (1.10)^2 = 1.39$
(c) $\sigma = (1.39)^{0.50} = 1.18$

9.4.2 POISSON PROBABILITY DISTRIBUTION

Poisson probability distribution is a discrete distribution in terms of manufacturing environment. It counts the number of times of a product defect occurrence over the specified volume, time, area, and length. Thus, for x to be a Poisson random variable, it has to be statistically independent that is the occurrence has to happen separately at different times. For example, an occurrence might happen between 1:00 and 4:00 AM and another between 4:00 and 6:00 PM.

1. The occurrence happens randomly in time or space, not in groups.
2. Normally it applies to processes that have probability of defects less than or equal 10%.
3. Mean $= \mu$
4. Variance $= \sigma^2$

The Poisson probability of mass function, or the equation calculating probability for Poisson distribution, is given in Equations 9.34 and 9.35.

$$p(x,\theta) = e^{-\theta}\theta^x (x!)^{-1} \text{ for } x = 1, 2, 3, \ldots, m \qquad \text{(Eq. 9.34)}$$

or

$$p(x) = e^{-\theta}\theta^x (x!)^{-1} \qquad \text{(Eq. 9.35)}$$

where
$p(x)$ = probability of x occurrences
θ = is the expected number of event occurrences over the specified period of time or space interval, and is equal to mp.
m = sample numbers
p = proportion defective (a constant, i.e., 5%)
! = called factorial (is the product of multiplying all the numbers beginning from 1 to specified number) for instance, $3! = (1)(2)(3) = 6$
$e = 2.718281828$

The shape of the Poisson distribution depends on the parameters in number of successes in segment of unit size and the size of the segment of interest, which is equal to variance.

Example 9.37

A quality assurance engineer was assigned to inspect a section (area) of a product. The engineer found that the average defect percentages of the product is about 3% using sample sizes (m) of 150 products. Find the probability of discovering only $x = 4$ defective in the sample products.

Solution

We substitute the values in the Equation 9.35 to the get probability of Poisson

$$p(x) = e^{-\theta} \theta^x (x!)^{-1}$$

$$= e^{-mp} (mp)^x (x!)^{-1}$$

$$= e^{-(150 \times 0.03)} (150 \times 0.03)^4 (4!)^{-1}$$

$$= (0.0111090)(17.086)$$

$$= 0.1898 \text{ or } 18.98\%$$

So there is an 18.98% chance of finding 4 defectives in the 150 sample parts.

Example 9.38

Poisson: Let X is a *Poisson* random variable with a mean of 4.

1. What is the probability that X will be equal to 4?
2. What is the probability that X will be exceed 2?
3. What is the probability that X will be at least equal to 3 but no more than 5?

Solution

Using the Poisson Table, we can calculate the probability.

1. From Poisson table at $\mu = 4$ and $x = 4$ the probability is

$$P(x = 4) = 0.1954$$

2. It is faster to find the probability of $P(x > 2)$ by subtracting the probability of $P(x \leq 2)$ from 1.0 and using the Poisson table similar to part 1. We have

$$P(x > 2) = 1 - P(x \leq 2) = 1 - [P(x = 0) + P(x = 1) + P(x = 2)]$$

$$= 1.0 - (0.0183 + 0.0733 + 0.1465]$$

$$\approx 0.76$$

3. $P(3 \leq x \leq 5) = P(x = 3) + P(x = 4) + P(x = 5)$

$$= 0.1954 + 0.1954 + 0.1563$$

$$= 0.5471$$

9.4.3   THE HYPERGEOMETRIC PROBABILITY DISTRIBUTION

The properties that apply to hypergeometric distribution and make it different than Poisson or binomial are as follows:

1. Discrete processes
2. Small sample size or lots
3. Sampling with no replacement
4. Processes that number of defects are known.

Thus, the probability of mass function (PMF) for hypergeometric distribution for random variables is given in Equation 9.36:

$$p(x) = C_x^k C_{n-x}^{N-k} \left(C_n^N \right)^{-1}$$
(Eq. 9.36)

where

$p(x)$ = probability of discovering x defects
n = sample numbers
N = population size
K = occurrence in the population
$C_x^k = k!(x!(k-x)!)^{-1}$ Combination

$$C_{n-x}^{N-k} = (N-n)!\left((n-x)!\left((N-k)-(n-x)\right)!\right)^{-1}$$

$$\left(C_n^N \right)^{-1} = \left(N!\left(n!(N-n)!\right)^{-1} \right)^{-1}$$

Example 9.39

A young growing company is making products in small lots. An inspector is assigned to do sampling from a particular manufacturing process. So, inspector takes a sample size of $n = 5$ from lot size of $N = 18$ parts, where $k = 4$ occurrence in the population. Find the probability of only $x = 1$ defects in the sample.

Solution

Using Equation 9.36, the probability is

$$p(x) = C_x^k C_{n-x}^{N-k} (C_x^N)^{-1}$$

$$n = 5$$

$$N = 18$$

$$K = 4$$

$$C_x^k = C_1^4 = k!(x!(k-x)!)^{-1} = 4!(1!(4-1)!)^{-1} = 24(6)^{-1} = 4$$

$$C_{n-x}^{N-k} = (N-n)! \{(n-x)![(N-k)-(n-x)]!\}^{-1} =$$

$$C_{5-1}^{18-4} = C_4^{14} = 14!(4!(14-4)!)^{-1} = (8.718E10)(1.10E-8) = 1001$$

$$(C_n^N)^{-1} = [N!(n!(N-n)!)^{-1}]^{-1} =$$

$$(C_5^{18})^{-1} = [18!(5!(18-5)!)^{-1}]^{-1} = 0.000037973 = 3.7973E-5$$

$$p(x) = (4)(1001)(3.7973E-5) = 0.1520 \text{ (or 15.20\%)}$$

Example 9.40

Three Light bulbs were selected from 10 light bulbs. Out of the 10 there were 4 defective. What is the probability that 2 of the 3 selected are defective?

Solution

Similarly we use Equation 9.36

$$p(x) = C_x^k C_{n-x}^{N-k} (C_x^N)^{-1}$$

$$n = 3, \ N = 10, \ K = 4, \ x = 2$$

$$C_x^k = C_1^4 = k!(x!(k-x)!)^{-1} = 4!(2!(4-2)!)^{-1} = 24(4)^{-1} = 6$$

$$C_{n-x}^{N-k} = (N-n)! \{(n-x)![(N-k)-(n-x)]!\}^{-1} =$$

$$C_{3-2}^{10-4} = C_1^6 = 6!(1!(6-1)!)^{-1} = (720)(120)^{-1} = 6$$

$$(C_n^N)^{-1} = (C_3^{10})^{-1} = [10!(3!(10-3)!)^{-1}]^{-1} = (120)^{-1} = 0.00833$$

$$p(x) = (6)(6)(1/120) = 0.30 \text{ (Or 30\%)}$$

9.5   CONTINUOUS RANDOM VARIABLES PROBABILITY DISTRIBUTIONS

By definition a continuous random variable is a variable that considers an infinite number of possible values (can accept any value on a continuum) such as dimension of an object (thickness, height), temperature measurement, and amount of time needed to achieve a Lean Six Sigma project. These can take any values, depending on the measurement capability and accuracy.

9.5.1   NORMAL PROBABILITY DISTRIBUTION

The concept of the normal distribution is the most important continuous distribution in statistics. The normal distributions play a key role in statistical methodology, applications, and are probability curves that have the same symmetric shapes. They are symmetric with data values more focused in the center than in the tails. The term *bell-shaped curve* is often used to describe normal probability distribution. The entire area under the normal distribution curve is equal to one. The height of a normal distribution can be expressed mathematically in two parameters, of mean (μ) and the standard deviation (σ). The mean is a measure of center or location of average and the standard deviation is a measure of spread. The mean can be any value from negative infinity to positive infinity (in between $\pm \infty$) and the standard deviation must be positive. The mean, median, and mode are equal in symmetric normal curve (Figure 9.33). Changing the mean μ shifts the distribution left or right and changing standard deviation σ increases or decreases the spread as shown in Figure 9.34.

By varying the parameters μ and σ, we can obtain different normal distributions.

Figure 9.35 illustrates that the probability is measured by the area under the curve.

The total area under the curve is 1.0, and the curve is symmetric, so half is above the mean, and the other half is below the mean as shown in Figure 9.36 and Equation 9.37.

Thus, mathematically the probability of $f(x)$ (see Equation 9.37) is equal to one. Suppose that x has a continuous distribution, then for any given value of x, the function must meet the following criteria:

$$f(x) \geq 0 \text{ for any event } x \text{ in the domain of } f$$

Since the normal distribution curve (symmetric from the mean) meets the x-axis in the infinity (as shown in Figure 9.37), the area under the curve and above the x-axis is one. This can

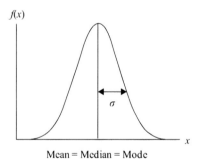

Figure 9.33. Normal distribution curve.

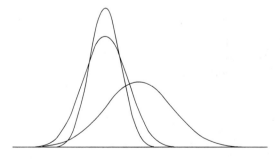

Figure 9.34. Different normal distributions.

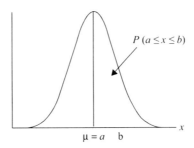

Figure 9.35. Different normal distributions.

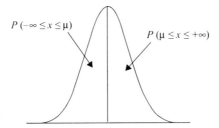

Figure 9.36. Different normal distributions.

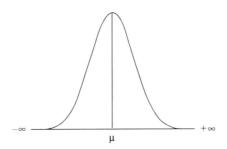

Figure 9.37. Normal distribution curve (symmetric) area equal to one.

be calculated by integrating the probability density on a continuous interval from minus infinity to plus infinity (Equation 9.38).

$$\text{Area under the normal distribution curve} = A(x) = \int_{-\infty}^{+\infty} f(x)\,dx = P(-\infty < x < +\infty) = 1.0$$

(Eq. 9.37)

where the height of a normal curve (the normal density function) for random variable x is defined as

$$f(x) = \frac{1}{\sqrt{2\pi\sigma^2}}\, e^{\frac{-(x-\mu)^2}{2\sigma^2}} \quad \text{For } x \text{ interval } (-\infty \le x \le +\infty)$$

(Eq. 9.38)

where $f(x)$ is the height of a normal distribution curve [$f(x) \geq 0$]; μ is the mean; π is the constant 3.14159; e is the base of natural logarithms, which is equal to 2.718282; σ is the standard deviation of population; and $\dfrac{x-\mu}{\sigma} = z$ (this will be discussed in the z-distribution Section 9.5.1.1). The normal distribution curve for $n\sigma$ (n-sigma) is shown in Figure 9.38.

When $n = 3$ then, for statistical quality control purposes, USL is equal to the mean (μ) plus three times standard deviation ($\mu + 3\sigma$), and LSL is equal to the mean minus three times standard deviation ($\mu - 3\sigma$).

For any upper specification limit (USL) and lower specification limit (LSL) probability density function (Equation 9.39) can be mathematically expressed as

$$A(x) = \int_{x=LSL}^{x=USL} \frac{1}{\sqrt{2\pi\sigma^2}} e^{\frac{-(x-\mu)^2}{2\sigma^2}} dx = \frac{1}{\sqrt{2\pi\sigma^2}} \int_{x=LSL}^{x=USL} e^{\frac{-(x-\mu)^2}{2\sigma^2}} dx$$

$$= \frac{1}{\sqrt{2\pi\sigma^2}} \int_{x=LSL}^{x=USL} e^{\frac{-(\frac{x-\mu}{\sigma})^2}{2}} dx \qquad \text{(Eq. 9.39)}$$

Thus, the probability (by definition probability is the area under the normal distribution curve $A(x)$) will be equal to 0.9973 when the process is centered on the target. This is the probability that 99.73% of data will fall within $\mu \pm 3\sigma$. This is also called 3σ capability, as shown in Equation 9.40.

$$A(x) = \int_{x=\mu-3\sigma}^{x=\mu+3\sigma} \frac{1}{\sqrt{2\pi\sigma^2}} e^{\frac{-(x-\mu)^2}{2\sigma^2}} dx = \frac{1}{\sqrt{2\pi\sigma^2}} \int_{x=\mu-3\sigma}^{x=\mu+3\sigma} e^{\frac{-(x-\mu)^2}{2\sigma^2}} dx = 0.9973 \qquad \text{(Eq. 9.40)}$$

Now, by implementing the random variable z (standard z-transform or standard normal deviation) the new probability density function $f(z)$ is described in the preceding section.

The Empirical Rule – What can we say about the distribution of values around the mean? There are some general rules called *empirical rules*, provides the knowledge about the normal distribution curve, the mean μ and standard deviation σ as shown in Figure 9.39. The rules say the following.

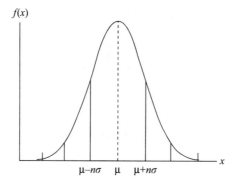

Figure 9.38. Normal probability distribution curve for $n\sigma$ (n-sigma).

1. The interval from $\mu - \sigma$ to $\mu + \sigma$ includes about 68.27% of the values or

$$P[(\mu - \sigma) \leq x \leq (\mu + \sigma)] \approx 0.6827$$

2. The interval from $\mu - 2\sigma$ to $\mu + 2\sigma$ includes about 95.45% of the values

$$P[(\mu - 2\sigma) \leq x \leq (\mu + 2\sigma)] \approx 0.9545$$

3. The interval from $\mu - 3\sigma$ to $\mu + 3\sigma$ includes about 99.73% of the values

$$P[(\mu - 3\sigma) \leq x \leq (\mu + 3\sigma)] \approx 0.9973$$

9.5.1.1 The Standard Normal Distribution (or z-Distribution)

The standard normal distribution (the z-distribution) is a normal distribution (see Equations 9.37 and 9.38 and Figure 9.40) with a mean of zero ($\mu = 0$) and a standard deviation of one ($\sigma = 1$). Normal distributions can be transformed to standard normal distribution (z) by the expression

$$z = \frac{x - \mu}{\sigma}$$

(Eq. 9.41)

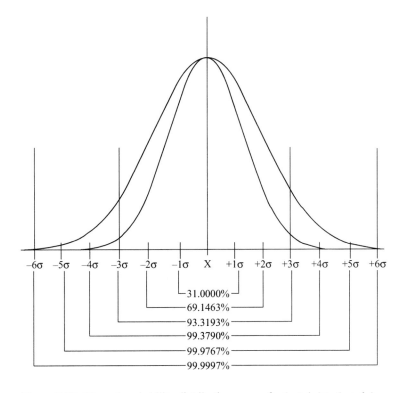

Figure 9.39. Normal probability distribution curves for $(\mu \pm 1\sigma)$ to $(\mu \pm 6\sigma)$.

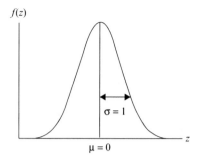

Figure 9.40. Standard normal distribution curve.

where x is a value from original normal distribution, μ is the mean of the original normal distribution, and σ is the standard deviation of the original normal distribution. If we replace x with USL in Equation 9.41 then the area under the normal distribution curve beyond the USL will indicate the nonconformance probability above the specification limit. This is shown in Equation 9.42:

$$z = \frac{USL - \mu}{\sigma}$$

(Eq. 9.42)

As mentioned above, the standard normal distribution also is called the z-distribution. A z-value refers to the number of standard deviations (right or left of the mean) from the mean for a particular score. For instance, if a student scored an 85 (USL = 85) on the final test with a mean of 60 (μ = 60) and a standard deviation of 10 (σ = 10), then using Equation 9.42 they scored 2.5 standard deviations above the mean. So, a z-score of 2.5 means the original score was 2.5 standard deviations above the mean. The z-distribution will only be a normal distribution if the original distribution (x) is normal.

Equation 9.41 will always yield a standard z-transform or standard normal deviation with a mean equal to zero (μ = 0) and a standard deviation of one (σ = 1). Furthermore, the shape of the distribution will not be changed by the conversion. However, if x is not normal, then the standard z-transform will not be normal either.

Now, by substituting the above information and Equation 9.41 in Equation 9.38 will result in Equation 9.43.

$$f(z) = \frac{1}{\sqrt{2\pi}} e^{\frac{-z^2}{2}}, \text{ where } z \text{ equal to } (-\infty < z < +\infty)$$

(Eq. 9.43)

The area within an interval $(-\infty, z)$ is given by Equation 9.44:

$$A(z) = \int_{-\infty}^{z} f(z)dz = \int_{-\infty}^{z} \frac{1}{\sqrt{2\pi}} e^{\frac{-z^2}{2}} dz = \frac{1}{\sqrt{2\pi}} \int_{-\infty}^{z} e^{\frac{-z^2}{2}} dz$$

(Eq. 9.44)

There is no need for integration of Equation 9.44. One can obtain the area of $A(z)$ for various z-numbers from the z-table. (This table is available in the appendix.) The standard normal z-distribution curve for $z = \pm n$ is shown in Figure 9.41.

Example 9.41

What is the nonconformance probability when $z = 1.84$?

Solution

From z-table (see Appendix) at 1.84, the area of standard normal distribution is equal to 0.4671 as shown in Figure 9.42. This represents the area between mean ($z = 0$) and $z = +1.84$ (area at $0 \leq z \leq 1.84$). One should keep in mind that the area in z-table is designed for only one half of the distribution curve.

Considering the two tails probability, the areas of the tails are
Area of right tail ≥ 1.84 equal to 0.0329
Likewise the area of left tail ≤ -1.84 equal to 0.0329
The total area $= 2 \times 0.0329 = 0.0658$
Thus, the probability of defects $= 6.58\%$

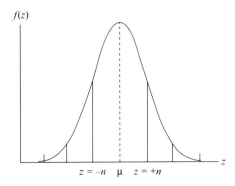

Figure 9.41. Standard normal z-distributions.

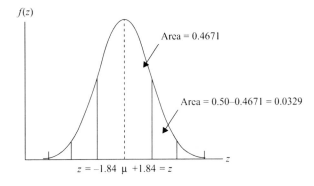

Figure 9.42. Area of z-distribution curve.

The normal probability distribution (for sample numbers $n \geq 30$) is the most common and popular in statistical analysis. However, there are other probability distributions (e.g., t-distribution), for sample numbers less than 30 where normal distribution will not give better results.

Example 9.42

If x is distributed normally with mean of $\mu = 100$ and the standard deviation $s = 50$, then find the following probabilities for the standard normal distribution using z-distribution table. Hint: transform from x to standard normal distribution (z-distribution).

Determine the followings: 1) $P(x > 105)$, 2) $P(x > 95)$, 3) $P(80 < x < 108)$, 4). Find z value at 90 percentile, 5) Find a such that $P(x > a) = 0.5$, 6) $P(65 < x < 88)$.

Solution

By using the normal distribution table (z-table), we use Equation 9.41 to convert the above probabilities into standard normal distribution.

1. Given $P(x > 105)$ at $x = 105$, $\mu = 100$, and $s = 50$ we have

   $z = \dfrac{x - \mu}{\sigma} = \dfrac{105 - 100}{50} = 0.10,$ then from z-table at 0.10 we find the area (probability)

   equal to 0.03983 as shown below. This analysis is expanded and visualized in Figure 9.42.

   Z-Table: See actual Table in appendix

Z	0.0000...0.0100...0.0200.............0.04000
0.0	...↓...
0.1 →	0.03983..
...	....
...	....
1.6	.................................0.44950

   So, $P(0 < Z < 0.10) = 0.03983$

   $$P(x > 105) = P(Z > 0.10)$$
   $$= P(0.10 < Z < -\infty)$$
   $$= P(Z > 0) - P(0 < Z < 0.10)$$
   $$= 0.50 - 0.03983$$
   $$= 0.4602$$

   Figure 9.43 visualizes the above calculation

2. To find z-score for $P(x > 95)$ we use $x = 95$, $\mu = 100$, $s = 50$ in Equation 9.41.

   $z = \dfrac{x - \mu}{\sigma} = \dfrac{95 - 100}{50} = -0.10,$ then from z-table at $z = 0.10$ we find the area (probability) equal to 0.03983. This is also illustrated in Figure 9.44. Thus,

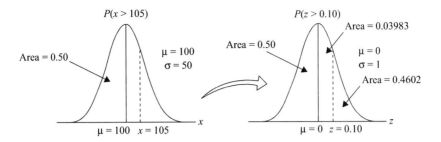

Figure 9.43. Transforming $P(x > 105)$ to $P(z > 0.10) = 0.4602$.

$$P(x > 95) = P\,(Z > 0.10)$$

$$= P(Z > 0) + P(0 < Z < 0.10)$$

$$= P(Z > 0) + P(-0.1 < Z < 0)$$

$$= 0.50 + 0.03983$$

$$= 0.53983$$

Likewise Figure 9.43 visualizes the above computation:

3. We apply the same concept as before to find z-score for $P(80 < x < 108)$ using Equation 9.41.

$$z = \frac{x - \mu}{\sigma}$$

To find the z-scores we equate the values in the probability $P(80 < x < 108)$

$$P\big(80 < x < 108\big) = P\left(\frac{80 - \mu}{\sigma} < \frac{x - \mu}{\sigma} < \frac{108 - \mu}{\sigma}\right)$$

Now, we substitute the values of $x = 80$, $\mu = 100$, $\sigma = 50$ and $x = 108$, $\mu = 100$, $\sigma = 50$

$$P\big(80 < x < 108\big) = P\left(\frac{80 - 100}{50} < \frac{x - \mu}{\sigma} < \frac{108 - 100}{50}\right)$$

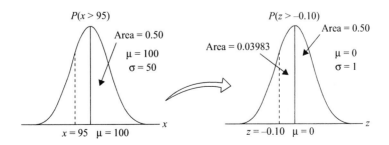

Figure 9.44. Transforming $P(x > 95)$ to $P(z > -0.10) = 0.53983$.

$$P(80 < x < 108) = P(-0.40 < z < 0.16)$$

$$P(-0.40 < z < 0.16) = P(0 < z < 0.16) + P(0 < z < 0.40)$$
$$= P(0 < z < 0.16) + P(-0.40 < z < 0)$$

From z-Table at Z = 0.16 and Z = 0.40 area results in

$$P(0 < Z < 0.16) = 0.06356$$

$$P(0 < Z < 0.40) = 0.15542$$

Thus; $P(0 < Z < 0.16) + P(-0.40 < Z < 0) = 0.06356 + 0.15542 = 0.21$
And similarly Figure 9.45 visualizes the computation:
Therefore, probability of $P(80 < x < 108) = P(-0.40 < z < 0.16) = 0.21$ also shown in Figure 9.44.

4. To find z-value and x-value at 90 percentile we use area of two tailed symmetric (Figure 9.45) normal distribution to find z-value from the z-table. Thus, we have

$$P(z = a) = 0.90$$

Since the normal distribution is symmetric so from the z-table at area equal to $(0.90/2) = 0.45 \approx$ (or 0.4505 value from z-table) the z-value ("a") is equal to $a = 1.645$. This is also shown in Figure 9.46.

$$P(z = 1.645) = 0.90 \approx 0.901 \text{ Two tail}$$

Now we will use Equation 9.41 and information given above to find x as follows:

$$Z = (x - \mu)/\sigma \qquad \text{(Eq. 9.41)}$$

$$1.645 = (x - 100)/50$$

$$(1.645)(50) = x - 100$$

$$x = 182.25$$

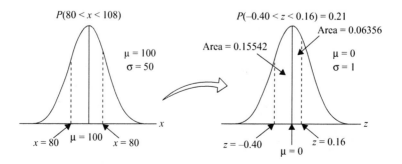

Figure 9.45. Transforming $P(80 < x < 108)$ to $P(-0.40 < z < 0.16) = 0.21$.

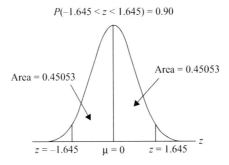

Figure 9.46. z-values at 90 percentile.

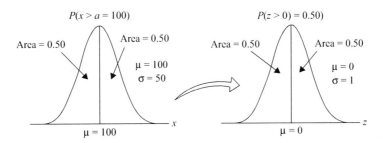

Figure 9.47. Region corresponding to $P(z > 0.50)$.

5. In this problem we are interested to find "a" in $P(x > a) = 0.50$. However, we already know that in normal distribution both sides of symmetric are equal area, which is 0.50 with the mean of $\mu = 0$ and $\sigma = 1$. Thus, "a" in this case is equal to 0 which is the same as mean (Figure 9.47).

6. We want to find $P(65 < x < 88)$, which is calculated as follows:

We use Equation 9.41($z = \dfrac{x - \mu}{\sigma}$), as before to find the z-scores. First we equate the values in the probability $P(65 < x < 88)$.

$$P\left(65 < x < 88\right) = P\left(\frac{65 - \mu}{\sigma} < \frac{x - \mu}{\sigma} < \frac{88 - \mu}{\sigma}\right)$$

Now, we substitute the values of $x = 65$, $\mu = 100$, $\sigma = 50$ and $x = 88$, $\mu = 100$, $\sigma = 50$

$$P\left(65 < x < 88\right) = P\left(\frac{65 - 100}{50} < \frac{x - \mu}{\sigma} < \frac{88 - 100}{50}\right)$$

$$P\left(65 < x < 88\right) = P\left(-0.70 < z < -0.24\right).$$

Using z-table we have

$$P\left(-0.70 < z < -0.24\right) = P\left(0 < z < 0.70\right) - P\left(0 < z < 0.24\right)$$
$$= P\left(-0.70 < z < 0\right) - P\left(-0.24 < z < 0\right) = 0.25804 - 0.0948 = 0.16321$$

This is visualized in Figure 9.48.

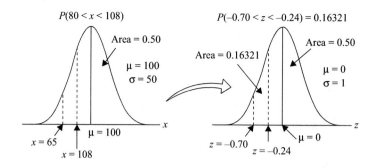

Figure 9.48. Region corresponding to $P(-0.70 < z < -0.24) = 0.16321$.

Example 9.43

A yogurt shop sells freshly made vanilla yogurt. Any unsold yogurt at the end of the day are either discarded or sold at a loss. The request for this vanilla yogurt has lowered a normal distribution with $\mu = 40$ yogurt and $\sigma = 9$ yogurt. How many yogurts should the shop make so that they can meet the demand 95% of the time?

Hint: Area under the normal curve is provided.

Solution

From problem statement we need to find how many yogurts (x_0) should the shop make so that they can meet the demand 95% of the time $P(X \leq x_0) = 0.95$. The computations are as follows.

First we need to find z-value. However, first we need to subtract 95% from 1.

$$1.0 - 0.95 = 0.05$$

Now it is time for Reverse Engineering, using area $= 0.50 - 0.05 = 0.45$
From Z-table at area $= 0.45$ we have $Z = 1.64$ (approximate) as shown below:
Z-Table:

Z 0.0000...0.0100...0.0200..............**0.04000**
0.0 ....**↑**...
0.1 ....|....
....|...
....|...
1.6 ..**↞**..................................**0.44950** ≈ **0.45** = **area**

From Figure 9.49 we know $P(0 < z < z_0) = 0.45$

$$P(0 \leq Z \leq 1.64) = 0.45$$

Thus, the equation of probability $P(X \leq x_0) = 0.95$ become $P(0 \leq Z \leq 1.64) = 0.45$. It also means

$$P(Z \leq 1.64) = 0.45 + 0.50 = 0.95$$

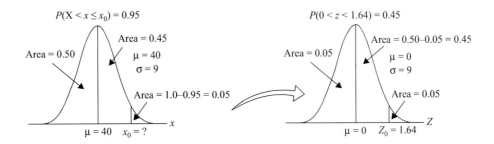

Figure 9.49. Region corresponding to $P(0 < z < 1.64) = 0.45$.

Now to find x_0 we use Equation 9.41

$$Z = (x_0 - \mu)/\sigma \qquad (9.41)$$

Or: rearranging the Equation 9.41 and substituting the values results in

$$x_0 = Z_0\sigma + \mu = (1.64)(9) + 40 = 14.76 + 40 = 54.76 \approx 55.$$

The yogurt shop will need only 55 yogurts to meet the demand at 95% of the time.

Example 9.44

Let Z has a standard normal distribution. Find the following. The answers are already included:

1. $P(Z > 1.60) = 0.0548$
2. $P(Z > 2.65) = 0.004$
3. $P(Z < 1.85) = 0.9678$
4. $P(Z < -1.35) = 0.50 - P(0 \leq Z \leq 1.35) = 0.50 - 0.41149 = 0.08825$
5. $P(Z > -3.15) = 0.50 - P(0 \leq Z \leq 3.15) = 0.50 - 0.49918 = 0.00082$
6. $P(0 < Z < 1.25) = 0.3944$
7. $P(-0.85 < Z < 0) = P(0 \leq Z \leq 0.85) = 0.30234$
8. $P(1.50 < Z < 3.50) = P(0 \leq Z \leq 3.50) - P(0 \leq Z \leq 1.50) = 0.49977 - 0.43319 = 0.0666$
9. $P(-3.18 < Z < -2.75) = P(0 \leq Z \leq 3.18) - P(0 \leq Z \leq 2.75) = 0.49926 - 0.49702 = 0.0022$
10. $P(-1.65 < Z < 3.16) = P(0 \leq Z \leq 1.65) + P(0 \leq Z \leq 3.16) = 00.45053 + 0.49921 = 0.9497$

Example 9.45

Find the value of z_0 for the probability listed below and sketch the appropriate area. $P(0 \leq Z \leq z_0) = 0.40988$, $P(z_0 \leq Z \leq 0) = 0.48500$

Solution

To find Z_0 the calculations are as follows.

1. For probability of $P(0 \leq Z \leq z_0) = 0.40988$ from Z-Table at $P(z) = 0.40988$ (or area = 0.40988) we found z_0 equal to 1.34 ($z_0 = 1.34$). This is illustrated in Figure 9.50.

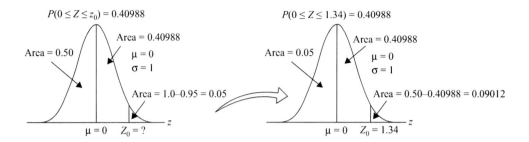

Figure 9.50. Region corresponding to $P(0 < z < 1.34) = 0.40988$.

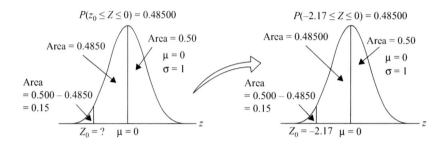

Figure 9.51. Region corresponding to $P(z_0 \leq Z \leq 0) = 0.48500$.

2. For probability of $P(z_0 \leq Z \leq 0) = 0.48500$ similarly from Z-Table at $P(z) = 0.48500$ (or area = 0.48500) we found z_0 equal to 2.17 ($z_0 = 2.17$). This is visualized in Figure 9.51.

Example 9.46

Let X be a random variable with normal distribution having a mean of $\mu = 25$ and standard deviation of $\sigma = 15$. What is the probability that X is less than 8? What is the probability that X is greater than 60?, What is the probability that X is in between 25 and 50?

Solution

The calculations are as follows.

1. To find the probability X is less than 8 ($P(x < 8)$, first we apply the given information in the Equation 9.41 to find the z-value.

$$Z = (x - \mu)/\sigma => Z = (8 - 25)/15 = -1.133$$

$$Z = (x - \mu)/\sigma => Z = (25 - 25)/15 = 0.00$$

In this case we want to find the area between $-\infty$ and -1.133. We need to locate the row corresponding to $z = 1.1$ and the column headed by .033. Finding their intersection from Z-table at $Z = 1.133$ and $Z > 0$ area are as follows:

$$P(Z > 0) = 0.50$$

$$P(0 < Z < 1.13) = 0.37076$$

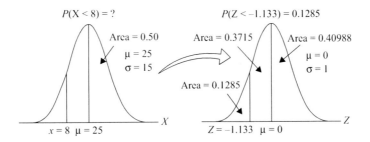

Figure 9.52. Region corresponding to $P(X < 8) = P(Z < -1.133) = 0.1285$.

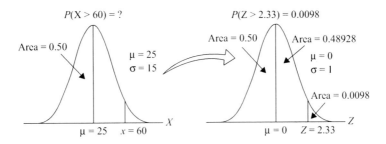

Figure 9.53. Region corresponding to $P(X > 60) = P(Z > 2.33) = 0.0098$.

Thus, $P(X < 8) = P(Z > 0) - P(0 < Z < 1.133) = 0.50 - 0.37076 = 0.1285$
Or, $P(Z < -1.133) = 0.1285$
This result is visualized in Figure 9.52.

2. To find the probability X is more than 60 ($P(x > 60)$, first we apply the given information in the Equation 9.41 to find the z-value.

$$Z = (x - \mu)/\sigma => Z = (60 - 25)/15 = 2.33$$

$$Z = (x - \mu)/\sigma => Z = (25 - 25)/15 = 0.00$$

From Z-table at Z less than 2.33 ($Z < 2.33$) and Z more than zero ($Z > 0$) the area is equal to 0.48928. This is shown in Figure 9.53.
Such that $P(0 < Z < 2.33) = 0.48928$
Thus, $P(X > 60) = P(Z > 0) - P(0 < Z < 2.33) = 0.50 - 0.48928 = 0.0098$
Or, $P(X > 60) = 0.0098$

3. What is the probability that X is in between 25 and 50?
To find the probability X is more than 25 and less than 50 $P(25 < X < 50)$, first we apply the given information in the Equation 9.41 to find the z-values.

$$z = \frac{x - \mu}{\sigma}$$

To find the z-scores we equate the values in the probability of $P(25 < x < 50)$

$$P(25 < x < 50) = P\left(\frac{25 - \mu}{\sigma} < \frac{x - \mu}{\sigma} < \frac{50 - \mu}{\sigma} \right)$$

Now, we substitute the values of $x = 25$, $\mu = 25$, $\sigma = 15$ and $x = 50$, $\mu = 25$, $\sigma = 15$ as shown in Figure 9.54.

$$P(80 < x < 108) = P\left(\frac{25-25}{15} < \frac{x-\mu}{\sigma} < \frac{50-25}{15} \right)$$

$$P(25 < x < 50) = P(0 < z < 1.67)$$

From Z-Table at Z = 1.67 the area is
Thus, $P(0 < Z < 1.67) \approx 0.45254$

9.5.2 T-DISTRIBUTION

This is similar to normal probability distribution ($n \geq 30$) bell-shaped symmetrical curve, with exception that is used for small sample size ($n < 30$), thicker tails, and lesser height in the median frequency. In the course of time, as sample size approaches 30 and higher, the distribution resembles a normal distribution. Other preceding properties differentiate t-distribution from normal distribution.

1. Applies to any sample size $n < 30$, the result is different for different sample sizes n.
2. Degrees of freedom are required.
3. The mean is zero ($\mu = 0$) the same as normal distribution.
4. Variance $= \lambda(\lambda - 2)^{-1}$, variance is larger than one, finally gets closer to one as sample size increases.
5. Range is $(-\infty, +\infty)$.
6. Distribution is symmetrical with respect to mean.
7. Population standard deviation is unknown.

Probability density function (pdf) for t-distribution is defined in Equation 9.45.

$$f(x) = G[0.5(\lambda+1)][(\lambda\pi)^{0.5}G(0.5\lambda)]^{-1}(1+x^2\lambda^{-1})^{-0.5(\lambda+1)} \tag{Eq. 9.45}$$

where G is the Gamma function (the Gamma function extends the factorial function ($n!$)) and λ positive parameter called "degrees of freedom," as λ becomes larger, the t-distribution approaches normal distribution.

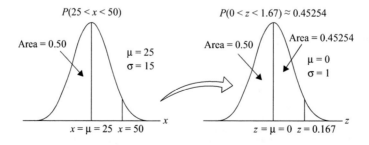

Figure 9.54. Region corresponding to $P(25 > X > 50) = P(0 < Z < 1.67)$ ≈ 0.45254.

Since the *t-distribution* is normally used to establish confidence interval and hypothesis tests, the preceding discussion concentrates on this concept for *t*-distribution testing.

9.5.3 NORMALITY TEST

The two important normality tests Kurtosis and Anderson Darling are described in the following sections.

9.5.3.1 Kurtosis

Kurtosis is a factor that measures how much peak or flat a bell shaped probability distribution curve has. Consider the two probability density functions (PDF) in Figure 9.55. It is difficult to estimate which distribution has a larger standard deviation. In fact, it is impossible to estimate. By a glance one might assume that the distribution on the right has a lower standard deviation due to the higher peak at the target. At the same time, one might think it has a higher standard deviation due to flatter tails.

The various shapes of the two distribution curves represent Kurtosis. Thus, Kurtosis is based on the size of distribution's tails. In Figure 9.54 distribution on the right hand side has larger Kurtosis than the one on the left. Sample Kurtosis can be calculated using Equation 9.46

$$\text{Kurtosis} = \frac{\sum_{i=1}^{n}(x_i - \mu)^4}{n\sigma^4} - 3 \qquad \text{(Eq. 9.46)}$$

where μ is mean and σ standard deviation of x_i

Note that with higher distribution still the area of curve is maintained equal to 1.0.

9.5.3.2 Anderson Darling

A normality test is a normal probability plot that is used to conduct a statistical test, that is, a hypothesis to find out if the distribution curve is normal or not. A best straight line represents

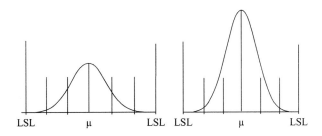

Figure 9.55. These curves represent the conception of Kurtosis. The PDF on right has higher Kurtosis than the PDF on the left. It is more peaked at the center, and it has flatter tails.

a cumulative distribution for a sample/population from which data has been collected. After distribution data have been plotted for normality, one may also test for p-value. If p-value is less than 0.05 (p-value < 0.05), then it is not normal. On the other hand, if p-value ≥ 0.05, then distribution curve is normal. Similarly, if p-value > 0.05, then the hypothesis test fail to reject.

9.5.4 EXPONENTIAL DISTRIBUTION

The exponential distribution is a continuous distribution that is extensively used in reliability engineering and estimation (Section 9.5.5) to describe the lifetime failure of a certain product or component of a machine (or system) in a specified period of time. If the random variable t represents time, α mean time between failure and $\beta = \alpha^{-1}$ failure rate of the system then probability density function of failure rate for exponential distribution can be written as

$$P_F(t) = 1 - e^{-\beta t} \tag{Eq. 9.47}$$

Example 9.47

Suppose a cell phone battery has a usage life defined by exponential distribution with a mean of 25 hours. Determine the probability that a battery won't function before its expected product lifetime of $\alpha = 25$ hours.

Solution

The battery lifecycle (t) has an exponential distribution with $t = 25$ and

$$\beta = \alpha^{-1} = (25)^{-1} = 0.04$$

$$P_F(t) = 1 - e^{-\beta t} = 1 - e^{-\left(25^{-1}\right)(25)} = 0.632 = 63.25\%$$

Therefore, there is 63.20 percent chance that the battery won't function.

9.5.5 RELIABILITY ENGINEERING

Reliability engineering is the application of all variety of techniques and reliability concepts to product lifetime, instrument in testing, and improving of product or services. Therefore, reliability is defined as probability of defect free product, or services for a pre-calculated period of time. Further, successfully lasting functionality life of a product is equal to or more than pre-decided lifecycle. Reliability is quantified by the following three concepts and techniques.

Reliability function: The reliability function $P_R(t)$ (probability of defect-free product or services) is given in Equation 9.48:

$$\log_e P_R(t) = -\frac{t}{\alpha} = -\beta t \tag{Eq. 9.48}$$

or

$$P_R(t) = e^{-\frac{t}{\alpha}} = e^{-\beta t}$$

where $\beta = \dfrac{1}{\alpha}$ and

$P_R(t)$ = The reliability function or probability of defect-free product or services for a time period t or beyond means no failure within the interval time (or no failure will occur until time t expires)

t = Specified period of defect-free services

α = Mean time between failures, when the system can be fixed

β = Constant failure rate of the system (defect rate)

Additionally $\dfrac{1}{\beta}$ is called mean time to failure when the system cannot be fixed.

What Equation 9.47 applies to is that the lifetime of any product under research is presumed to begin at time $t = 0$ and any problems occurring with respect to functionality is irrelevant before the time $t \leq t_{specified}$ or time interval $(0, t)$. We know that reliability is a time-based quality; it is also a function of other factors that impact the time. This is given in Equation 9.49

$$P_R(x) = f(x_1, x_2, ..., x_{n-1}, x_n) \tag{Eq. 9.49}$$

where, $x_1, x_2, ..., x_{n-1}, x_n$ are the variables that affects "time"-based qualities of the system. To achieve product improvement objectives, reliability engineering has to be accurately defined, measured, analyzed, tested, verified, controlled, and sustained in the field or product application.

Example 9.47

Let's say a product has shown 6,000 hours (250 days of manufacturing hours) of mean time between failures. Now, consider a constant defect rate, what is the probability of a defect-free operation over a 24-hour interval?

Solution

Assuming the reliability of a product exponentially decays through period of time t, as cited in Equation 9.47, the equation can be stated in the form of an exponential given in Equation 4.50.

$$P_R(t) = e^{-\frac{t}{\alpha}} = e^{-\beta t} \tag{Eq. 9.50}$$

$$\beta = \frac{1}{6000} = 0.000167 = 1.67 \times 10^{-4} = 1.67\text{E} - 4$$

$$P_R(t) = e^{-\frac{24}{6000}} = e^{-(24)(1.67 \times 10^{-4})} = 0.996008$$

Therefore, the operation is 99.601 percent defect-free over a 24-hour interval.

Probability function: The probability function of failure time is defined as the failure time distribution function $P_F(t)$, which is a supplement of the reliability function $P_R(t)$. In other words, the sum of the both probabilities is equal to 1.0 (100%). That is:

$$P_R(t) + P_F(t) = 1.0 \tag{Eq. 9.51}$$

Equation 9.51 can be expressed as

$$P_R(t) = 1 - P_F(t) \tag{Eq. 9.52}$$

Or

$$P_F(t) = 1 - P_R(t) = 1 - e^{-\beta t}$$

By taking the derivative of Equation 9.51 one can obtain the probability of the density function of failure, as shown in Equation 9.53.

$$P_f(t) = \beta e^{-\beta t} \tag{Eq. 9.53}$$

where, $P_f(t)$ is the failure function, β failure rate, and t is the duration of a product lifetime.

Hazard: The Hazard (or failure) rate function $P_H(t)$, is defined as the limit of "failure rate $(P_H(t))$" as the time interval ($\Delta t \to 0$) nears zero.

$$P_H(t) = \frac{P_f(t)}{P_R(t)} = \frac{P_f(t)}{1 - P_F(t)} = \frac{\beta e^{-\beta t}}{1 - (1 - e^{-\beta t})} = \frac{\beta e^{-\beta t}}{e^{-\beta t}} = \beta \tag{Eq. 9.54}$$

where $P_H(t)$ is the hazard rate function and β is the positive constant failure. Equations 9.53 and 9.54 can also be expressed as in Equation 9.55.

$$P_f(t) = P_R(t) \times P_H(t) = \beta e^{-\beta t} \tag{Eq. 9.55}$$

Example 9.48

What is the reliability of the a light bulb at 3,000 hours, if the target service of the bulb was 2,000 hours for 4,000 bulbs and the total number of failures was 200?

Solution

Using an exponential distribution Equation 9.47 we have

$$\beta = \frac{1}{n\alpha} = \frac{1}{(2000)(4000)} = 0.000000125 = 1.25E - 7$$

$$P_R(t) = e^{-\frac{t}{\alpha}} = e^{-\beta t}, \text{ Where } e = 2.71828183$$

$$P_R(3000) = e^{-(1.25E-7)(3000)} \cong 0.99962 = 99.96\%$$

9.6 INFERENTIAL STATISTICS AND SAMPLING DISTRIBUTION

The goal of inferential statistics is to exercise a sample study from the population and compose inferences about the population. Inferential statistics includes sampling distribution of the mean and central limit theorem, numerical integration of probability distribution and their inverses, confidence intervals, hypothesis testing, analysis of variance, contrast testing, and simple and multiple regressions.

9.6.1 RANDOM SAMPLING AND THE DISTRIBUTION OF THE SAMPLE MEAN

By definition a "sampling distribution is a distribution of statistics." If a population under study is normally distributed, with a mean of μ and a standard deviation of σ, then the sampling distribution of the sample mean $\bar{X}$ is also normally distributed with a mean equal to mean of the population $\mu_{\bar{x}} = \mu$ and a standard deviation equal to standard deviation of the population divided by square root of the sample count is called standard deviation of sampling distribution as shown in Equation 9.56.

$$\sigma_{\bar{x}} = \frac{\sigma}{\sqrt{n}} \qquad \text{(Eq. 9.56)}$$

Sample values should be selected randomly from the population, one at a time, from the population. The sample mean $(\bar{x})$ is used to estimate the population means (μ).

$$Random\ Sample\ Mean = \bar{X} = \frac{\sum_{i=1}^{n} X_i}{n}$$

Thus, the random sampling distribution of the mean is a distribution of the all sample means.

Finite Population correction factor: If the sample is being conducted without the replacement, then standard deviation of sampling distribution need to be modified. Particularly if the sample count is large enough compare to the population size, say about 5 percent of the population. Thus,

$$\text{Mean} = \mu_{\bar{x}} = \mu = \frac{\sum_{i=1}^{n} X_i}{n} = \frac{1}{n} \sum_{i=1}^{n} X_i$$

and

$$\text{Standard deviation} \left(\text{Std. error}\right) = \sigma_{\bar{x}} = \frac{\sigma}{\sqrt{n}} \sqrt{\frac{N-n}{N-1}} = \frac{\sigma}{\sqrt{n}} \left(\frac{N-n}{N-1}\right)^{1/2} \qquad \text{(Eq. 9.57)}$$

where N = Population size, and n = Sample size, and

$$\frac{N-n}{N-1} = \text{Finite Population Correction} \ (fpc) \qquad \text{(Eq. 9.58)}$$

The $fpc = \dfrac{N-n}{N-1}$ this term can be ignored if $\dfrac{n}{N} < 0.05$

And the adjusted z-value for the finite correction is

$$Z = \frac{\bar{X} - \mu}{\dfrac{\sigma}{\sqrt{n}} \sqrt{\dfrac{N-n}{N-1}}}$$ (Eq. 9.59)

Further, to be sure that a simple random sample is obtained from a finite population the measurements should be numbered from 1 to N.

The following examples present the random sampling concepts of the distribution of the sample mean for a small population or sample size.

Example 9.49

Sampling distribution. What are the possible sample combinations of size 2 without replacement from a population of 6, 9, 12, 15, and 18? Then, develop the sample distribution of the sample means and confirm the outcomes.

Solution

We have the population values 6, 9, 12, 15, 18, population size $N = 5$ and sample size $n = 2$. Therefore, the number of possible samples (Table 9.18) that can be collected without replacement is

$$\binom{N}{n} = \binom{5}{2} = 10$$

The sampling distribution of the sample means $(\bar{X})$, and the mean of sample means $\propto_{\bar{x}}$, and standard deviation are calculated using Table 9.19.

Table 9.18. Sample distribution of the sample means

Sample no.	Sample values	Sample mean ($\bar{X}$)	Sample no.	Sample values	Sample mean ($\bar{X}$)
1	6, 9	7.5	6	9, 15	12
2	6, 12	9	7	9, 18	13.5
3	6, 15	10.5	8	12, 15	13.5
4	6, 18	12	9	12, 18	15
5	9, 12	10.5	10	15, 18	16.5

Table 9.19. Sample distribution of the sample means, the mean of sample means, and the standard deviation

$\bar{X}$	$\bar{X}^2$	Frequency	$f(\bar{X})$	$\bar{X}f(\bar{X})$	$\bar{X}^2f(\bar{X})$
7.5	56.25	1	1/10	7.5/10	56.25/10
9.0	81.0	1	1/10	9.0/10	81.0/10
10.5	110.25	2	2/10	21.0/10	220.5/10
12.0	144.0	2	2/10	24.0/10	288.0/10
13.5	182.25	2	2/10	27.0/10	364.5/10
15.0	225.0	1	1/10	15.0/10	225.0/10
16.5	272.25	1	1/10	16.5/10	272.25/10
Total	1071	10	1	120/10	1507.5/10

$$\text{Mean} = \mu_{\bar{x}} = \mu = \frac{\sum_{i=1}^{n} X_i}{n} = \frac{1}{n}\sum_{i=1}^{n} X_i = \sum \bar{X}f(\bar{X}) = \frac{120}{10} = 12$$

$$\sigma_{\bar{X}}^2 = \sum \bar{X}^2 f(\bar{X}) - \left(\sum \bar{X}f(\bar{X})\right)^2 \qquad \text{(Eq. 9.60)}$$

$$= \frac{1507.5}{10} - \left(\frac{120}{10}\right)^2 = 150.75 - 144 = 6.75$$

$$\sigma_{\bar{X}} = \sqrt{\sigma_{\bar{X}}^2} = 2.598$$

Table 9.20 is used to verify the mean and standard deviation of the sampling distribution. Sample mean is

$$\mu = \frac{\sum X}{N} = \frac{60}{5} = 12$$

Similarly their variance

$$\sigma^2 = \frac{\sum X^2}{N} - \left(\frac{\sum X}{N}\right)^2 = \frac{810}{5} - \left(\frac{60}{5}\right)^2 = 162 - 144 = 18$$

Table 9.20. Sum of the samples and their squares

X	6	9	12	15	18	$\sum X = 60$
X^2	36	81	144	225	325	$\sum X^2 = 810$

Confirming the outcomes,

$$\mu_{\bar{x}} = \mu = 12$$

$$\sigma_{\bar{X}}^2 = \frac{\sigma^2}{n}\left(\frac{N-n}{N-1}\right) = \frac{18}{2}\left(\frac{5-2}{5-1}\right) = 6.75$$

$$\sigma_{\bar{X}} = \sqrt{\sigma_{\bar{X}}^2} = \frac{\sigma}{\sqrt{n}}\sqrt{\frac{N-n}{N-1}} = \sqrt{6.75} = 2.598$$

Example 9.50

Sampling distribution: A random sample combinations of size three collected without replacement from a population that includes four values 7, 8, 8, 9. First determine sample mean $\bar{X}$ for each sample and develop sampling distribution of $\bar{X}$. Then, compute the mean and standard deviation of this sampling distribution and compare the outcomes with population parameters.

Solution

We have the population values 7, 8, 8, 9, population size N = 4 and sample size n = 3. Therefore, the number of possible samples (Table 9.21) that can be collected without replacement is

$$\binom{N}{n} = \binom{4}{3} = 4$$

The sampling distribution of the sample means$\left(\bar{X}\right)$, its mean $\mu_{\bar{x}}$, and standard deviation are calculated using Table 9.22.

$$\text{Mean} = \mu_{\bar{x}} = \mu = \frac{\sum_{i=1}^{n} X_i}{n} = \frac{1}{n}\sum_{i=1}^{n} X_i = \sum \bar{X}f\left(\bar{X}\right) = \frac{96}{12} = 8$$

$$\sigma_{\bar{X}}^2 = \sum \bar{X}^2 f\left(\bar{X}\right) - \left(\sum \bar{X}f\left(\bar{X}\right)\right)^2 = \frac{2306}{36} - \left(\frac{96}{12}\right)^2 = 64.06 - 64.0 = 0.06$$

Table 9.21. Possible samples obtained from population

Sample No.	Sample Values	Sample Mean ($\bar{X}$)
1	7, 8, 8	23/3
2	7, 8, 9	24/3
3	7, 8, 9	24/3
4	8, 8, 9	25/3

Table 9.22. Sampling distribution calculation

$\bar{X}$	$\bar{X}^2$	*Frequency*	$f(\bar{X})$	$\bar{X}f(\bar{X})$	$\bar{X}^2 f(\bar{X})$
23/3	529/9	1	1/4	23/12	529/36
24/3	576/9	2	2/4	48/12	1152/36
25/3	625/9	1	1/4	25/12	625/36
Total	1730/9	4	1	96/12	2306/36

$$\sigma_{\bar{X}} = \sqrt{\sigma_{\bar{X}}^2} = \sqrt{0.06} = 0.24$$

To verify the mean and standard deviation of the sampling distribution we have Table 9.23. Sample distribution mean is

$$\mu = \frac{\sum X}{N} = \frac{32}{4} = 8$$

And the variance of sample distribution

$$\sigma^2 = \frac{\sum X^2}{N} - \left(\frac{\sum X}{N}\right)^2 = \frac{258}{4} - \left(\frac{32}{4}\right)^2 = 64.50 - 64.0 = 0.50$$

$$\sigma = \sqrt{\sigma^2} = \sqrt{0.50} = 0.71$$

Confirming the outcome,

$$\mu_{\bar{X}} = \mu = 8$$

$$\sigma_{\bar{X}}^2 = \frac{\sigma^2}{n}\left(\frac{N-n}{N-1}\right) = \frac{0.50}{3}\left(\frac{4-3}{4-1}\right) = \frac{0.50}{3}\left(\frac{1}{3}\right) = \frac{1}{18} = 0.06$$

$$\sigma_{\bar{X}} = \sqrt{\sigma_{\bar{X}}^2} = \frac{\sigma}{\sqrt{n}}\sqrt{\frac{N-n}{N-1}} = \sqrt{0.06} = 0.24$$

Example 9.51

Statistical Inferences and Sampling: A team of Quality Control Engineers is assigned to similar projects. It is expected that the time it takes an engineer to analyze his/her project is normally distributed with a mean of 30 minutes and a standard deviation of 4 minutes.

Table 9.23. Sum of the samples and sum of their squares

X	7	8	8	9	$\sum X = 32$
X^2	49	64	64	81	$\sum X^2 = 258$

1. Determine the probability that an engineer will analyze the data in less than 27 minutes.
2. In randomly selected sample of six engineers, find the probability that the mean time it takes them to analyze the given data in less than 27 minutes.

Solution

We are interested in finding the probability of $P(x < 27)$ and $P(\bar{x} < 27)$ with the given information on the sampling $\mu = 30$, $\sigma = 4$, $n = 4$, $\bar{x} = 27$. Due to the fact that Z-table are based on the positive z-values and we know that normal distribution curves are symmetric, thus we will find the area using $P(x > 27)$ and then apply the result to the left side of the normal distribution. Once again we will utilize Equation 9.41 to calculate the z-value. Again we apply Equation 9.41 in computing z-value or from the Z-table:

$$Z = \frac{\bar{x} - \mu}{\sigma}$$

1. $P(x > 27) = P\left(\frac{\bar{x} - \mu}{\sigma} > \frac{27 - 30}{4}\right) = P(Z > 0.75) = P(Z < -0.75)$

Or, from Z-Table at $Z = 0.75$ the area is equal to 0.2737. Thus we have

$$P(Z < -0.75) = P(Z > 0.75) = 0.50 - P(0 < Z < 0.75) = 0.50 - 0.2737 = 0.2266$$

2. Similarly, we apply the same concept to $P(\bar{x} < 27)$. However, here the standard deviation for the average mean $\bar{x}$ is equal to $\sigma / \sqrt{n}$

$$Z = \frac{\bar{x} - \mu}{\sigma_{\bar{x}}} = \frac{\bar{x} - \mu}{\dfrac{\sigma}{\sqrt{n}}}$$

$$P(\bar{x} < 27) = P\left(\frac{\bar{x} - \mu}{\sigma / \sqrt{n}} < \frac{27 - 30}{4 / \sqrt{6}}\right) = P(z < -1.837)$$

Or, from Z-table at $Z = 1.837$ the area is equal to 0.4671. Thus we have

$$P(Z < -1.837) = P(Z > 1.837) = 0.50 - P(0 < Z < 1.837) = 0.50 - 0.4671 = 0.0329$$

Now, the next examples present the random sampling concepts of the distribution of the sample mean for a large population or sample size.

Example 9.52

Statistical Inferences and Sampling: A bio-tech company employs 253 chemists who have a mean of 6.2 years laboratory experience with standard deviation of 2.10 years.

1. If a sample of 35 chemists were chosen randomly "Without Replacement," then find probability that mean number of work experience would be more than 6.8 years.
2. If the same number of chemists (35) were chosen randomly "With Replacement," then find probability that mean number of work experience would be more than 6.8 years.

Solution

From problem statement we have
Population: $N = 253$, $\mu = 6.20$, $\sigma = 2.10$, $n = 35$

1. *Sampling Without Replacement:* The goal is to find the probability of $P(\bar{x} > 6.80)$ using the following equations at given conditions.

$$\text{Mean} = \mu_{\bar{x}} = \mu = \frac{\sum_{i=1}^{n} X_i}{n} = \frac{1}{n}\sum_{i=1}^{n} X_i$$

$$\text{Standard deviation}\left(\text{Std. error}\right) = \sigma_{\bar{x}} = \frac{\sigma}{\sqrt{n}}\sqrt{\frac{N-n}{N-1}} = \frac{\sigma}{\sqrt{n}}\left(\frac{N-n}{N-1}\right)^{1/2}$$

$$\frac{N-n}{N-1} = \text{Finite Population Correction } (fpc)$$

$$\text{The } fpc = \frac{N-n}{N-1} \text{ this term can be ignored if } \frac{n}{N} < 0.05$$

In this case $\dfrac{n}{N} = \dfrac{35}{253} = 0.135 > 0.05$, thus

$$\sigma_{\bar{x}} = \frac{2.1}{\sqrt{35}}\sqrt{\frac{253-35}{253-1}} = \frac{2.1}{\sqrt{35}}\sqrt{\frac{218}{252}} = 0.3302$$

Now we have obtained all the information to use in finding the probability of $P(\bar{x} > 6.80)$. By substituting in the Equation 9.41 ($Z = \dfrac{\bar{x} - \mu}{\sigma_{\bar{x}}}$)

$$P(\bar{x} > 6.80) = P\left(\frac{\bar{x} - 6.20}{0.3302} > \frac{6.80 - 6.20}{0.3302}\right) = P(Z > 1.82)$$

From *Z*-Table we obtain the area correspond with $Z = 1.82$. That is $(\text{Area})_{(z=1.82)} = 0.4656$

$$P(Z > 1.82) = 0.50 - 0.4656 = 0.0344$$

Figure 9.56 illustrate this result.

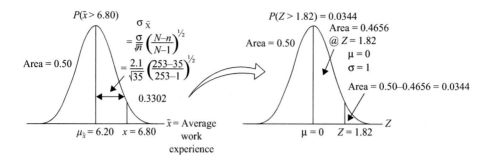

Figure 9.56. Region corresponding to $P(\bar{x} > 6.80) = P(z > 1.82) = 0.0344$.

2. *Sampling With Replacement*: The goal is to find the probability of $P(\bar{x} > 6.80)$

$$\text{Mean} = \mu_{\bar{x}} = \mu = \frac{\sum_{i=1}^{n} X_i}{n}$$

$$\text{Standard deviation}\left(\text{Std. error}\right) = \sigma_{\bar{x}} = \frac{\sigma}{\sqrt{n}} = \frac{2.1}{\sqrt{35}} = 0.355$$

Now we have obtained all the information to use in finding the probability of $P(\bar{x} > 6.80)$. By substituting in the Equation 9.41 ($Z = \dfrac{\bar{x} - \mu}{\sigma_{\bar{x}}}$)

$$P(\bar{x} > 6.80) = P\left(\frac{\bar{x} - 6.20}{0.3550} > \frac{6.80 - 6.20}{0.3550}\right) = P(z > 1.69)$$

From Z-Table we obtain the area correspond with $Z = 1.69$. That is $(\text{Area})_{(z = 1.69)} = 0.4545$

$$P(Z > 1.82) = 0.50 - 0.4545 = 0.0455$$

Figure 9.57 illustrate this result.

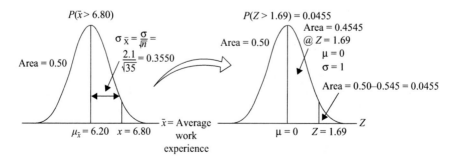

Figure 9.57. Region corresponding to $P(25 > X > 50) = P(0 < Z < 1.67) \approx 0.45254$.

Example 9.53

Statistical Inferences and Sampling: Happy Pizza Company catering division reported that in 2009 nationwide 5153 orders were placed online. The mean of order size was $213 with a standard deviation of $50. What is the probability that 500 samples selected online from 5153 have a mean order size between $210 and $215? (1) Assume without replacement concept. (2) Apply replacement concept.

Solution

1. *Sampling without Replacement*: From problem statement we have $N = 5153$, $\mu = 213$, $\sigma = 50$, $n = 500$. We are interested in finding the probability of $P(210 < \bar{x} < 215)$. To find the probability we use the Equation 9.41 ($z = \dfrac{\bar{x} - \mu}{\sigma_{\bar{x}}}$) and $\sigma_{\bar{x}} = \dfrac{\sigma}{\sqrt{n}}$.

$$P(210 < \bar{x} < 215) = P((210 - \mu) < (\bar{x} - \mu) < (215 - \mu))$$

$$= P\left(\frac{210 - \mu}{\sigma_{\bar{x}}} < \frac{\bar{x} - \mu}{\sigma_{\bar{x}}} < \frac{215 - \mu}{\sigma_{\bar{x}}}\right)$$

$$= P\left(\frac{210 - \mu}{\dfrac{\sigma}{\sqrt{n}}} < \frac{\bar{x} - \mu}{\dfrac{\sigma}{\sqrt{n}}} < \frac{215 - \mu}{\dfrac{\sigma}{\sqrt{n}}}\right)$$

$$= P\left(\frac{210 - 213}{\dfrac{50}{\sqrt{500}}} < \frac{\bar{x} - \mu}{\dfrac{\sigma}{\sqrt{n}}} < \frac{215 - 213}{\dfrac{50}{\sqrt{500}}}\right)$$

$$= P(-1.342 < Z < 0.894)$$

$$= 0.4099 + 0.3133 = 0.7232$$

This can be visualized in Figure 9.58.

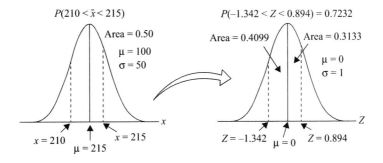

Figure 9.58. Regions corresponding to $P(210 < \bar{x} < 215) = P(-1.342 < Z < 0.894) = 0.7232$.

2. *Sampling With Replacement*: From problem statement we have $N = 5153$, $\mu = 213$, $\sigma = 50$, $n = 500$. We are interested in finding the probability of $P(210 < \bar{x} < 215)$. To find the probability we use again the Equation 9.41 $(Z = \dfrac{\bar{x} - \mu}{\sigma_{\bar{x}}})$ and $\sigma_{\bar{x}} = \dfrac{\sigma}{\sqrt{n}}$

$$P(210 < \bar{x} < 215) = P\left((210 - \mu) < (\bar{x} - \mu) < (215 - \mu)\right)$$

$$= P\left(\frac{210 - \mu}{\sigma_{\bar{x}}} < \frac{\bar{x} - \mu}{\sigma_{\bar{x}}} < \frac{215 - \mu}{\sigma_{\bar{x}}}\right)$$

$$= P\left(\frac{210 - \mu}{\dfrac{\sigma}{\sqrt{n}}\sqrt{\dfrac{N-n}{N-1}}} < \frac{\bar{x} - \mu}{\dfrac{\sigma}{\sqrt{n}}\sqrt{\dfrac{N-n}{N-1}}} < \frac{215 - \mu}{\dfrac{\sigma}{\sqrt{n}}\sqrt{\dfrac{N-n}{N-1}}}\right)$$

$$= P\left(\frac{210 - 213}{\dfrac{50}{\sqrt{500}}\sqrt{\dfrac{5153 - 500}{5153 - 1}}} < \frac{\bar{x} - \mu}{\dfrac{\sigma}{\sqrt{n}}\sqrt{\dfrac{N-n}{N-1}}} < \frac{215 - 213}{\dfrac{50}{\sqrt{500}}\sqrt{\dfrac{5153 - 500}{5153 - 1}}}\right)$$

$$= P\left(\frac{-3}{0.2125} < Z < \frac{2}{0.2125}\right)$$

$$= P(-1.4118 < Z < 0.9412)$$

$$\approx 0.42073 + 0.32639 = 0.7471$$

Thus,

$$P(210 < \bar{x} < 215) = P(-1.4118 < Z < 0.9412) \cong 74.71\%$$

9.6.2 CENTRAL LIMIT THEOREM (CLT)

By definition, for random samples of n measurements drawn from the population with a mean of μ and standard deviation of σ, at any shape of the population's distribution curve, with the given sample size (large enough), the distribution of the sample mean $(\bar{X})$ will be almost normal with a mean of equal to the same as population mean $(\mu_{\bar{x}} = \mu)$ and the standard deviation equal to the standard deviation of the population divided by the square root of sample count $(\sigma_{\bar{x}} = \dfrac{\sigma}{\sqrt{n}})$. This is shown in Figure 9.59.

In other terms, the central limit theorem (CLT) basically describes that as the sample size (n) increases and becomes large, the following criteria takes place:

1. The sampling distribution of the mean becomes approximately normal regardless of the distribution of the original variables.
2. The distribution of the sample mean $(\bar{X})$ is centered at the population mean, μ, of the original variable. Furthermore, the standard deviation of the sampling distribution of the mean nears $(\sigma_{\bar{x}} = \dfrac{\sigma}{\sqrt{n}})$ Equation 9.57.

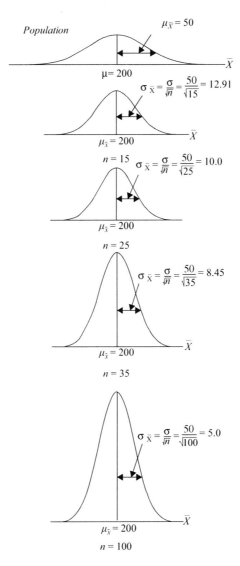

Figure 9.59. CLT: Corresponding to sampling distribution as n increases $(\sigma_{\bar{x}} = \dfrac{\sigma}{\sqrt{n}})$.

Equation 9.57 indicates that as the sample size increases, the standard deviation decreases. Thus, sampling mean becomes normal distribution as n increases, and variability decreases.

Example 9.54

Future-Tec has concluded that production assembly time for component F1 normally distributed with mean of $\mu = 20$ minutes and standard deviation of $\sigma = 3$ minutes.

Find the following:

1. What is the probability that a person on the production line takes longer than 22 minutes to produce one component F1?

2. What is probability that the average production time for 15 people exceeds 22 minutes? $P(x\square > 22) = ?$

3. What is probability that the average production time for 15 people is between 19 and 21 minutes? $P(19 < x\square < 21) = ?$

Solution

We are interested in finding the probabilities of the following:

1. $P(x > 22) =$?

$$P(x > 22) = P\left(\frac{x - \mu}{\sigma} > \frac{22 - 20}{3}\right) = P(Z > 0.67)$$

From the Z-Table at $Z = 0.67$ the area is equal to 0.2486. Thus, we have

$P(Z > 22) = 0.50 - 0.2486 = 0.2514$ or 25.14%. This is shown in Figure 9.60.

2. Now we will find probability of $P(\bar{x} > 22)$ When $n = 15$.

$$\text{Mean} = \mu_{\bar{x}} = \mu = \frac{\sum_{i=1}^{n} X_i}{n} = 20 \text{ minutes}$$

$$\text{Standard deviation (Std. error)} = \sigma_{\bar{x}} = \frac{\sigma}{\sqrt{n}} = \frac{3}{\sqrt{15}} = 0.77$$

Now we have obtained all the information to use in finding the probability of $P(\bar{x} > 22)$. By substituting in Equation 9.41 ($Z = \frac{\bar{x} - \mu}{\sigma_{\bar{x}}}$)

$$P(\bar{x} > 22) = P\left(\frac{\bar{x} - 22}{0.77} > \frac{22 - 20}{0.77}\right) = P(z > 2.60)$$

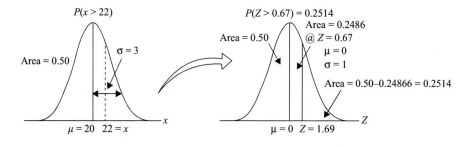

Figure 9.60. CLT-Corresponding to sampling distribution as n increases ($\sigma_{\bar{x}} = \frac{\sigma}{\sqrt{n}}$).

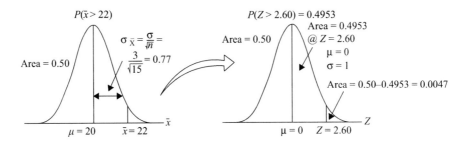

Figure 9.61. CLT corresponding to probability of $P(Z > 2.60) = 0.4953$.

From Z-table we obtain the area correspond with Z = 2.60. That is $(\text{Area})_{(z=2.60)} = 0.4953$

$$P(Z > 2.60) = 0.50 - 0.4953 = 0.0047$$

Thus, average production time for 15 people will be less than 1% probability (Figure 9.61).
3. Now we will find probability of $P(19 < \bar{x} < 21)$ when $n = 15$.

$$\text{Mean} = \mu_{\bar{x}} = \mu = \frac{\sum_{i=1}^{n} X_i}{n} = 20 \text{ minutes}$$

$$\text{Standard deviation} (\text{Std. error}) = \sigma_{\bar{x}} = \frac{\sigma}{\sqrt{n}} = \frac{3}{\sqrt{15}} = 0.77$$

Now we have obtained all the information to use in finding the probability of $P(19 < \bar{x} < 21)$. By substituting in the Equation 9.41 $(Z = \frac{\bar{x} - \mu}{\sigma_{\bar{x}}})$

$$P(19 < \bar{x} < 21) = P\left(\frac{19 - 20}{0.77} < \frac{\bar{x} - 20}{0.77} < \frac{21 - 20}{0.77}\right) = P(-1.30 < Z < 1.30)$$

From Z-table we obtain the area correspond with Z = 1.30. That is $(\text{Area})_{(z=1.30)} = 0.4032$

$$P(-1.30 < Z < 1.30) = 0.4032 + 0.4032 = 0.8064$$

Thus, average production time for 15 people will be less than 81% probability (Figure 9.62).

9.6.3 CONFIDENCE INTERVAL FOR THE MEAN (μ) OF NORMAL POPULATION (σ KNOWN)

The goal of taking a random sample from a population and estimating a statistic, like the mean $(\bar{X})$ from the measurement, is to estimate the mean of the population. How well the sample

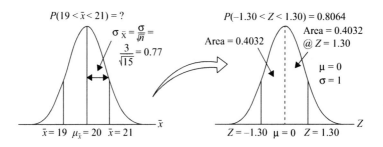

Figure 9.62. CLT corresponding to probability of $P(-1.30 < Z < 1.30) = 0.8064$.

statistic estimates the vital population value is a continuous concern. A confidence interval states this concern because it supports a range of values that is likely to include the population parameter of interest.

Confidence level: Confidence intervals are built at a *confidence level*, such as 90%, 95%, or 99%, chosen by the analyst. What this means that if the same population is sampled on different time and interval estimates are made on each time, the outcome intervals would support the true population parameter in about 90%, 95%, or 99%, of the cases. Thus, a confidence level affirms at a "$1-\alpha$" level can be assumed as the inverse of risk factor, α.

Not let's consider normal population distribution with known σ and we want to estimate the unknown μ. We know that $\bar{X}$ is normally distributed with μ and standard deviation $\dfrac{\sigma}{\sqrt{n}}$.

From Equation 9.41 we know the Z equal to

$$Z = \frac{\bar{X} - \mu}{\sigma_{\bar{X}}} = \frac{\bar{X} - \mu}{\sigma / \sqrt{n}}$$

The probability for the Z value is 0.95 for two tails with significance level of 5%. Thus, we have

$$P(-Z_{\frac{\alpha}{2}} \le Z \le Z_{\frac{\alpha}{2}}) = 0.95; \Leftarrow \text{the value of } Z \text{ equal to } 1.96 \text{ for 95\% Confidence}$$

$$P(-Z_{\frac{0.05}{2}=0.025} \le Z \le Z_{\frac{0.05}{2}=0.025}) = 0.95; \quad \alpha = 1.0 - 0.95 = 0.05$$

Because the area under the normal distribution curve between -1.96 and 1.96 is 0.95, we have

$$P(-1.96 \le Z \le 1.96) = 0.95$$

Substitute Z by its value and solve for μ

$$P(-1.96 \le \frac{\bar{X} - \mu}{\sigma / \sqrt{n}} \le 1.96) = 0.95$$

$$P\left(-1.96(\sigma / \sqrt{n}) \le \bar{X} - \mu \le 1.96(\sigma / \sqrt{n})\right) = 0.95$$

Add $-\bar{X}$ to the equation

$$P\left(-\bar{X} - 1.96(\sigma / \sqrt{n}) \le -\mu \le -\bar{X} + 1.96(\sigma / \sqrt{n})\right) = 0.95$$

Simple algebra tells us that

$$P\left(\bar{X} - 1.96(\sigma / \sqrt{n}) \le \mu \le \bar{X} + 1.96(\sigma / \sqrt{n})\right) = 0.95$$

95% Confidence interval formula is

$$\bar{X} - 1.96(\sigma / \sqrt{n}) \ \ to \ \ \bar{X} + 1.96(\sigma / \sqrt{n}) \qquad \text{(Eq. 9.61)}$$

And a more concise expression is

$$\bar{X} \pm 1.96(\sigma / \sqrt{n})$$

Or for a particular sample mean $(\bar{X})$, $(1 - \alpha) \cdot 100\%$ confidence interval.

$$\bar{X} - Z_{\alpha/2}(\sigma / \sqrt{n}) \ \ to \ \ \bar{X} + Z_{\alpha/2}(\sigma / \sqrt{n}) \qquad \text{(Eq. 9.62)}$$

Let Z_α represent value of Z so that the area to the right of Z equal to α. Thus, from Z-table one can conclude

$$Z_\alpha = Z_{0.025} = 1.96$$

$$Z_\alpha = Z_{0.05} = 1.645$$

$$Z_\alpha = Z_{0.10} = 1.280$$

This is visualized in Figure 9.63. So, the higher the confidence level, the wider the interval level. Therefore, the confidence interval applies for $n > 30$. This is based on central limit theorem (CTL).

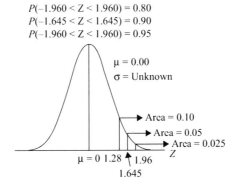

Figure 9.63. Confidence interval for different significance level.

Example 9.55

Confidence Intervals for the Mean of Normal population (σ Known): A reliability engineer is concerned about the tensile strength of a plastic material being molded in an environmental condition of an organic solvent. A sample of 25 parts were randomly collected and measured for tensile strength. The analysis indicates that tensile strength (in psi) is normally distributed with a standard deviation of 2. Determine a 95% confidence interval for the mean tensile strength of parts. The recorded data are as follows:

23, 27, 23, 22, 24, 22, 21, 24, 23, 23, 25, 22, 24, 21, 26, 19, 21, 20, 24, 20, 26, 25, 21, 23, 21

Solution

Let's consider α = 0.95, and from problem statement we have σ = 2, n = 25
Estimate of μ sample:

Normal random variable = $\bar{X} = \dfrac{\sum_{i=1}^{n} X_i}{n} = \dfrac{23+27+\cdots+21}{25} = 22.8$. Tensile strength (psi)

Thus, we use the following equations from the previous sections: Mean= $\mu_{\bar{x}} = \mu$, $\sigma_{\bar{x}} = \dfrac{\sigma}{\sqrt{n}}$,

Standard normal variable $Z = \dfrac{\bar{x}-\mu}{\sigma_{\bar{x}}} = \dfrac{\bar{x}-\mu}{\dfrac{\sigma}{\sqrt{n}}}$

Now, we are interested to find the 95% confidence interval:

$$P(-Z_{\frac{\alpha}{2}} \le Z \le Z_{\frac{\alpha}{2}}) = 0.95; \text{ where the value of } Z \text{ equal to } Z = \dfrac{\bar{X}-\mu}{\sigma_{\bar{X}}} = \dfrac{\bar{X}-\mu}{\sigma/\sqrt{n}}$$

From Z-table at 95% the Z-value equal to $Z = 1.96$ as shown in Figure 9.64. The probability of 95% confidence interval equal to

$$P(-1.96 \le Z \le 1.96) = 0.95$$

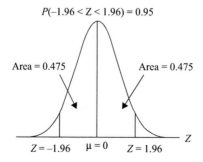

Figure 9.64. 95% Confidence interval with σ-known.

Then we substitute Z by its value and solve for μ:

$$P(-1.96 \le \frac{\bar{X}-\mu}{\sigma/\sqrt{n}} \le 1.96) = 0.95$$

$$P[-1.96(\sigma/\sqrt{n}) \le \bar{X}-\mu \le 1.96(\sigma/\sqrt{n})] = 0.95, \text{ then, Add}\left(-\bar{X}\right) \text{ to the equation}$$

$$P\left(-\bar{X}-1.96(\sigma/\sqrt{n}) \le -\mu \le -\bar{X}+1.96(\sigma/\sqrt{n})\right) = 0.95$$

$$P\left(\bar{X}-1.96(\sigma/\sqrt{n}) \le \mu \le \bar{X}+1.96(\sigma/\sqrt{n})\right) = 0.95$$

The 95% confidence interval formula equal to

$$\bar{X}-1.96(\sigma/\sqrt{n}) \text{ to } \bar{X}+1.96(\sigma/\sqrt{n})$$

$$22.80-1.96(\frac{2}{\sqrt{25}}) \text{ to } 22.80+1.96(\frac{2}{\sqrt{25}}); \text{ or } 21.95 \text{ to } 23.65$$

In conclusion we are 95% confident that the average "μ" production time to produce component tensile strength falls in between 21.95 and 23.65 psi.

9.6.4 CONFIDENCE INTERVAL FOR THE MEAN (μ) OF NORMAL POPULATION (σ UNKNOWN)

If the population mean μ is not known then it can be estimated by $\bar{X}$, and if the population standard deviation σ is unknown, the logic is to replace it by its estimate standard deviation s. That is

$$\text{replace } \frac{\bar{X}-\mu}{\sigma/\sqrt{n}} \text{ by } \frac{\bar{X}-\mu}{s/\sqrt{n}}; \text{ note that } Z = \frac{\bar{X}-\mu}{\sigma/\sqrt{n}}$$

$$\text{or, replace } (\bar{X}-\mu)n^{0.50}\sigma^{-1} \text{ by } (\bar{X}-\mu)n^{0.50}s^{-1}$$

$$\frac{\bar{X}-\mu}{s/\sqrt{n}} \text{ no longer standard normal variable.}$$

Now, $(\bar{X}-\mu)/(\frac{s}{\sqrt{n}})$ follows the t-distribution also called student's t-distribution

Similar to Z-distribution the t-distribution is symmetric with respect to mean, however wider tails compare to standard normal distribution. But, its shape is based on the sample size "n". Consequently, one must take the sample size "n" into account. This is possible by degrees of freedom $= df = n - 1$. Note that as the sample size increase and gets closer to 30 the curve become normal as shown in Figure 9.65.

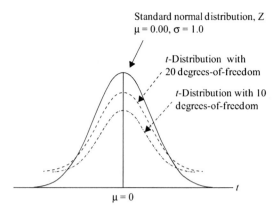

Figure 9.65. Confidence interval for *t*-distribution
(small sample size).

Confidence interval for *t*-distribution (small samples) can be stated as follows:

$$\bar{X} \mp t_{\alpha/2,n-1}\left(\frac{s}{\sqrt{n}}\right)$$ (Eq. 9.61)

or,

$$\bar{X} - t_{\alpha/2,n-1}\left(\frac{s}{\sqrt{n}}\right) \quad \text{to} \quad \bar{X} + t_{\alpha/2,n-1}\left(\frac{s}{\sqrt{n}}\right)$$

where, *t*-values $t_{\alpha/2,n-1}$ can be obtained from *t*-distribution table in the appendix.

Example 9.56

Confidence Intervals for the Mean of Normal Population (σ Unknown): A financial consultant reported that return on cetain mutual funds are normally distributed. A sample of 23 mutual funds were chosen and their annual growths for the past year were recorded as follows:

8.0, 12.0, 11.5, 16.0, 25.0, 18.0, 14.5, 10.9, 23.0, 19.0, 18.0, 32.0, 9.6, 11.0, 15.1, 18.0, 11.8, 3.2, 8.4, 16.9, 22.2, 21.0, 12.0

Determine (1) 90% and (2) 95% confidence interval for the *mean* of annual return on investment of mutual funds.

Solution

Calculations for 90% and 95% confidence interval are as follows.

1. To find the 90% (α =1.0 − 0.90 = 0.10) confidence interval we use equations from the previous section in finding point estimate for sample mean and standard deviation.

$$\bar{X} = \frac{\sum_{i=1}^{n} X_i}{n} = \frac{\sum_{i=1}^{23} X_i}{23} = \frac{8.0 + 12.0 + \ldots + 21.0 + 12.0}{23} = 15.523$$

$$s = \sqrt{\frac{\sum (X - \bar{X})^2}{n-1}} = \sqrt{\frac{(8.0 - 15.523)^2 + (12.0 - 15.523)^2 + \cdots + (12.0 - 15.523)^2}{23 - 1}} = 6.455$$

$\bar{X} - t_{\frac{\alpha}{2}, df=n-1}\left(\frac{s}{\sqrt{n}}\right)$ to $\bar{X} + t_{\frac{\alpha}{2}, df=n-1}\left(\frac{s}{\sqrt{n}}\right)$; the value of $t_{\frac{\alpha}{2}, df=n-1}$ can be obtained from t-distribution table in the appendix.

$$= 15.523 - t_{0.05,22}\left(\frac{6.455}{\sqrt{23}}\right) \quad \text{to} \quad 15.523 + t_{0.05,22}\left(\frac{6.455}{\sqrt{23}}\right)$$

$$= 15.523 - 1.717\left(\frac{6.455}{\sqrt{23}}\right) \quad \text{to} \quad 15.523 + 1.717\left(\frac{6.455}{\sqrt{23}}\right);$$

$$= 13.215 \text{ to } 17.837$$

$$\text{Note: Maximum error} = E = 1.717\left(\frac{6.455}{\sqrt{23}}\right) = 2.311$$

Conclusion: the financial consultant is 90% confident that mutual funds' return on investment is between 13.215 and 17.837 annual percent.

2. The 90% ($\alpha = 1.0 - 0.95 = 0.05$) confidence interval (CI) for mutual funds return on investment is $\bar{X} - t_{\frac{\alpha}{2}, df=n-1}\left(\frac{s}{\sqrt{n}}\right)$ to $\bar{X} + t_{\frac{\alpha}{2}, df=n-1}\left(\frac{s}{\sqrt{n}}\right)$; the value of $t_{\frac{\alpha}{2}, df=n-1}$ can be obtained from t-distribution table

$$= 15.523 - t_{0.025,22}\left(\frac{6.455}{\sqrt{23}}\right) \quad \text{to} \quad 15.523 + t_{0.025,22}\left(\frac{6.455}{\sqrt{23}}\right)$$

$$= 15.523 - 2.074\left(\frac{6.455}{\sqrt{23}}\right) \quad \text{to} \quad 15.523 + 2.074\left(\frac{6.455}{\sqrt{23}}\right);$$

$$= 12.735 \text{ to } 18.317$$

$$\text{Note: Maximum error} = E = 2.074\left(\frac{6.455}{\sqrt{23}}\right) = 2.79$$

In conclusion the financial consultant at 95% confident that mutual funds return on investment is between 12.735 and 18.317 annual percent.

9.6.5 SELECTING THE NECESSARY SAMPLE SIZE

Sampling approaches are random, stratified random, systematic, and subgroup sampling. They are defined as:

    i. *Random*–each unit has the same chance of being selected.
    ii. *Stratified random*–randomly sample a proportionate number from each group.
    iii. *Systematic*–sample every *n*th one (for instance, every 6th, or 12th, etc.).
    iv. *Subgroup*–sample *n* units every 4th time (for instance, 4 units every hour) then calculate the mean for each subgroup.
    v. *Continuous data*–based on continuum of numbers, for example, currency, time, weight, and height Sample size can be calculated using Equation 9.62.

$$n = \left(\frac{zS}{E} \right)^2 \qquad \text{(Eq. 9.62)}$$

where *n* is the sample number, *z* the standard normal distribution (*Z* score), and sample standard deviation, *E* is the measurement error used to determine the essential sample size to provide the specified level of accuracy and specified prior to sampling.

    vi. *Discrete* (*attribute/discontinue*) *data*: This is based on the counts or classes, for example, flipping a coin to yield a head or a tail. Sample size depends on the level of accuracy for the given application and can be calculated using Equation 9.63:

$$n = \frac{(z^2)pq}{E^2} \qquad \text{(Eq. 9.63)}$$

where $q = p - 1$, *p* is the estimated proportion of correct observation, and *q* is the estimated proportion of defects (scraps). Equation 9.63 becomes Equation 9.64.

$$n = \frac{(z^2)p(p-1)}{E^2} \qquad \text{(Eq. 9.64)}$$

Or, the necessary sample size when the standard deviation σ is known.

$$n = \left(\frac{Z_{\sigma/2} \cdot \sigma}{E} \right)^2 \qquad \text{(Eq. 9.65)}$$

$$E = Z_{\sigma/2 \frac{\sigma}{\sqrt{n}}} \qquad \text{(Eq. 9.66)}$$

Example 9.57

A local hospital survey shows that customer's satisfaction rates are below national average. A random sample of 15 from a normally distributed population is as follows:

$$68, 52, 73, 38, 26, 48, 50, 58, 54, 42, 47, 50, 40, 45, 54$$

1. What is the minimum sample number for the sample mean to be within 3 units (E = 3) of the population mean with 95% confidence? Apply sample standard deviation as an estimate for the population standard deviation.
2. Repeat part (a) by applying High (H) and Low (L) values of the surveys to estimate the population standard deviation.

Solution

1. The 95% (α =1.0 – 0.95 = 0.05) confidence interval are as follows:
   The "point estimate" of μ is $\overline{X}$

$$\overline{X} = \frac{\sum_{i=1}^{n} X_i}{n} = \frac{\sum_{i=1}^{15} X_i}{15} = \frac{68 + 52 + \cdots + 45 + 54}{15} = 49.67$$

The "point estimate" of σ is s

$$s = \sqrt{\frac{\sum (X - \overline{X})^2}{n-1}} = \sqrt{\frac{(X_1 - \overline{X})^2 + \cdots + (X_{n-1} - \overline{X})^2 + (X_n - \overline{X})^2}{n-1}}$$

$$= \sqrt{\frac{(68 - 49.67)^2 + (52 - 49.67)^2 + \cdots + (45 - 49.67)^2 + (54 - 49.67)^2}{15 - 1}} = 11.57$$

Thus, $\overline{X} = 49.67$, $s = 11.57$, E = 3, $Z_{\frac{\alpha}{2} = \frac{0.05}{2} = 0.025} = 1.96$

$$n = \left(\frac{Z_{\frac{\alpha}{2}} \bullet s}{E} \right)^2 = \left(\frac{(1.96)(11.57)}{3} \right)^2 = 57.14 \cong 58 \text{ round to the next integer}$$

2. To obtain a rough approximate of σ use the following formula: The expected range = $H - L$

$$\sigma \cong \frac{H - L}{4} = \frac{73 - 26}{4} = 11.75$$

$$n = \left(\frac{Z_{\frac{\alpha}{2}} \bullet \sigma}{E} \right)^2 = \left(\frac{(1.96)(11.75)}{3} \right)^2 = 58.93 \cong 59$$

Consequently, they would require 59 sample sizes to make a statement regarding customer satisfaction.

Example 9.58

SI incorporated receiving inspection manager is interested in the average number of *x*-inch valve parts from off-shore that can be inspected by his employees.

1. Assuming that the number of parts that are inspected each hour by an employee follow a normal distribution (concentrated at μ),
2. The manager wants to estimate μ with 90% confidence.
3. The estimate must be within one unit
4. The manager estimates that high is 45 and low is 25 parts based on engineering recommendation 50 parts were requested from off-shore and just received them.

Find: How large a sample will be necessary?

Solution

To obtain a rough approximate of σ use the following formula: The expected range $= H - L$.

$$\sigma \cong \frac{H-L}{4} = \frac{45-25}{4} = 5 \text{ parts}$$

$$Z_{\frac{\alpha}{2}} = Z_{\frac{0.10}{2}} = Z_{0.05} = 1.645 \text{ (From } Z\text{-Table)}$$

$$\sigma = 5, \quad E = 1$$

$$n = \left(\frac{Z_{\alpha/2} \cdot \sigma}{E} \right)^2 = \left(\frac{(1.645)(5)}{1} \right)^2 = 67.7$$

9.7   HYPOTHESIS TESTING, INFERENCES PROCEDURES, AND PROPORTIONS TESTING

As we discussed before, statistical inferences are involved with decisions making about a population based on the information contained in a random sample from the same population. The hypothesis testing is a method also used in inferential statistical analysis and related to confidence intervals. The hypothesis tests state the uncertainty of the sample estimate. Instead of offering an interval, a hypothesis test tries to discredit a specific claim about a population parameter based on the sample data. It consists of hypothesis testing for the mean (μ) and variances (σ^2) of the population; inferences procedure for two populations; estimating and testing for population proportion (p). For instance, the hypothesis tests might be one of the following:

*   the population mean (μ) is $\mu = 12$
*   the population standard deviation (σ) is $\sigma = 7$
*   the two population means are equal ($\mu_1 = \mu_2$)
*   the two or more population standard deviation are equal ($\sigma_1 = \sigma_2 = \cdots = \sigma_{n-1} = \sigma_n$)

To reject a hypothesis means that it is false. On the other hand, to accept a hypothesis doesn't mean that it is true, we just don't have proof to be certain of alternatively. Therefore, hypothesis tests are normally expressed in both circumstances that is questioning (null hypothesis) and a case that is believed (alternative hypothesis). It covers both small and large samples hypothesis testing.

By definition if a quality control inspector claims that the mean of a population is equal to $\mu = \alpha$, then, the fact is that either $H_0: \mu = \alpha$, or $H_a: \mu \neq \alpha$. So, one needs to test and determine to reject $H_0: \mu = \alpha$ or fail to reject H_0, where H_0 is called null hypothesis which the investigator wishes to discredit that $(H_0: \mu = \alpha)$, and H_a is called alternative hypothesis, which the investigator wishes to support that $(H_a: \mu \neq \alpha)$. The following phases are the techniques for hypothesis testing.

1. State the problem conditions and focus on the test, for instance:
   (a) Null hypothesis $\quad\quad H_0: \mu = \mu_0$
   (b) Alternative hypothesis $\quad H_a: \mu \neq \mu_0$
   Note: The conditions for samples with z-test and population mean are as follows:
   (a) The normal population or large samples $n \geq 30$ and σ are known.
   (b) If the tests are two tails then alternative H_a has a sign of $(H_a: \mu \neq \mu_0)$.
   (c) If it is left tail H_a has a sign of less than $(H_a: \mu < \mu_0)$ and
   (d) If it is right tail then H_a has a sign of greater than $(H_a: \mu > \mu_0)$.
2. Define the critical values and test statistic, i.e., for testing large samples $n \geq 30$, use Z-Table (or t) and define the rejection region.

$$Z = \frac{(\overline{X} - \mu_o)\sqrt{n}}{\sigma} \approx \frac{(\overline{X} - \mu_o)\sqrt{n}}{s}$$

3. Calculate the value of the test statistic.
4. Determine whether to reject H_0 or fail to reject H_0.
5. Report a conclusion in terms of the original problem.

The preceding discussion is detailed in Example 9.59.

Example 9.59

Recall the light bulb reliability concept from Example 9.48 with number of samples 50 and average lifecycle of 400 hours with the standard deviation of 40 hours. What conclusion one could make using a significance level of $\alpha = 0.10$?

Solution

Use the preceding steps:

1. Define and formulate a hypothesis test condition:
   Null hypothesis $\quad\quad H_0: \mu = 485$
   Alternate hypothesis $\quad H_a: \mu \neq 485$

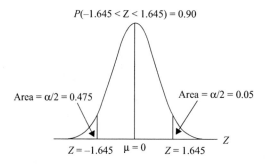

Figure 9.66. Normal distribution curve for Example 9.59.

2. State Z-test statistic

$$Z = \frac{(\bar{X} - \mu_o)\sqrt{n}}{\sigma} \approx \frac{(\bar{X} - \mu_o)\sqrt{n}}{s}$$

$$= \frac{(\bar{X} - 495)\sqrt{150}}{\sigma} \approx \frac{(\bar{X} - 495)\sqrt{150}}{s}$$

3. Determine the rejection region. Reject if $Z > 1.645$ or $Z < -1.645$. This is visualized in Figure 9.66.
4. Find the z-value.

$$Z = \frac{(\bar{X} - \mu_o)\sqrt{n}}{s} = \frac{(500 - 485)\sqrt{50}}{40} = 2.65$$

Since calculated z-value 2.65 > 1.645, then the conclusion is to reject H_0. In Figure 9.61, z-test falls in the rejection area.

5. Make a conclusion: according to the sample data, there is enough evidence to conclude that the average lifecycle of light bulbs is not 485 hours.

The null hypothesis, H_0, is generally the hypothesis or claim being tested and must contain the equality (or inequality) sign; $=, \geq, \leq$. The alternative hypothesis, H_α, represents the population values other than those contained in the null hypothesis. In other words, a statement in contradiction to the null hypothesis and must contain signs of $\neq, >, <$.

Example 9.60

Milk packaging company- The containers of milk are filled with 128 oz. (1-gal) or more ounces of milk on the average (H_0: $\mu \geq 128$ oz.) or the containers are filled with less than 128 ounces on average (H_α: $\mu < 32$ oz.)

1. Formulate a hypothesis test condition:
   Null hypothesis H_0: $\mu \geq 128$ oz.
   Alternate hypothesis H_α: $\mu < 128$ oz.

2. Support or Refute the null hypothesis. There are two possible errors.

   Type I error: The investigator may decide that the process is filling containers with an average less than 128 ounces, when the average fill is 128 or more ounces. In this case reject a true null hypothesis H_0: $\mu \geq 128$ oz. Or, reject when in fact it is true. This is called a "*type I statistical error*".

   Type II error: The investigator may decide that the process is filling containers with an average of 128 or more ounces when, in fact it is not. In this case accept null hypothesis when false H_a: $\mu < 128$ oz. Or, fail to reject when in fact it is not true. This is called "type II statistical error."

3. State of the nature is listed in the Table 9.24.

4. Decision making rule: The goal of hypothesis testing is to use information obtained from sample to decide whether to accept or reject the null hypothesis about the population. Thus,

   If $\bar{X} \geq k$, mean that the null hypothesis is correct and accept H_0 or fail to reject H_0

   If $\bar{X} < k$, means that the null hypothesis is false and reject H_0.

5. **Hypothesis Testing: 5 Step Procedure**
   1. Set up the null (H_0) and alternative hypothesis (H_a)
   2. Define the "test statistic"

$$Z = \frac{(\bar{X} - \mu_o)\sqrt{n}}{\sigma} \approx \frac{(\bar{X} - \mu_o)\sqrt{n}}{s}$$

   3. Define the rejection region (Figure 9.67): α, $\alpha/2$, ...

      If, $Z^* > Z_\alpha$, reject H_0, where Z^* is the calculated Z-value and Z_α is the Z-value at "α" from Z-Table.

$$\text{Reject } H_0 \text{ if } -Z_{\alpha=-k} = |\frac{(\bar{X} - \mu_o)\sqrt{n}}{\sigma}| > Z_{\alpha=k}$$

Table 9.24. Hypothesis testing outcome possibilities

Outcome	Actual Situation	
	H_0: $\mu \geq 128$ oz. H_0_ True	H_a: $\mu < 128$ oz. H_0_ False
Reject null hypothesis (H_0) Samples fails test. H_a: $\mu < 128$ oz.	"*Type I error*" (α) 128 oz milk containers not distributed	"*No error*" ($1 - \beta$) Less than 128 oz milk containers distributed *Correct decision*
Accept (H_0)/ or fail to reject null hypothesis (H_0) Samples passes test H_0: $\mu \geq 128$ oz.	"*No error*" ($1 - \alpha$) 128 oz milk containers not Distributed *Correct decision*	"*Type II error*" (β) Less than 128 oz of milk containers distributed

where α = probability of rejecting (H_0) when (H_0) is true = $P(type\ I\ error)$
β = probability of failing to reject (H_0) when (H_0) is false = $P(type\ II\ error)$

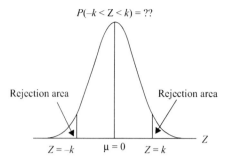

Figure 9.67. Normal distribution with hypothesis rejection area.

R the absolute value term is equal to $-k > \dfrac{\left(\bar{X} - \mu_o\right)\sqrt{n}}{\sigma} > k$ and the Z value can be calculated using equation

$$Z = (\bar{X} - \mu)\sqrt{n}\sigma^{-1}$$

4. Calculate the "test statistic."
5. Give a conclusion in terms of the problem.

9.7.1 HYPOTHESIS TESTING FOR THE MEAN (μ) AND VARIANCE (σ^2) OF THE POPULATION

Table 9.25 states the formulas and testing conditions for large sample on a normal population mean. That is for two tailed and one-tailed tests. These equations and conditions are applied in Examples 9.61 through 9.63 for testing mean and variances of the population. The population may be normal or not. It doesn't matter for large samples as stated in the central limit theorem (CLT) or normally $n > 30$. Once the α-value has been chosen the steps follows as in the Examples 9.59 and 9.60.

Example 9.61

Hypothesis Testing on the Mean (μ) of Population (Large Sample): The manager of Happy Pizza thinks that average time customers wait before being served is 10 minutes. To test his thought, the manager selects 50 customers and finds their average wait time 11.9 minutes with a standard deviation of 1.4 minutes.

1. Determine 95% confidence interval for the average waiting time of a customer.
2. Find if the data shows that mean waiting time is different than 10 minutes, using α = 5% significance level?

Table 9.25. Formulas for testing conditions: large sample tests on a normal population mean

Two-Tailed Test

$H_0: \mu \geq \mu_o$

$H_a: \mu < \mu_o$

Reject H_0 if $|Z_{\text{Calculated}} = Z^*| > Z_{\alpha/2}$

$$Z = \frac{\bar{X} - \mu_0}{\sigma / \sqrt{n}} \approx \frac{\bar{X} - \mu_0}{\dfrac{s}{\sqrt{n}}}; \mu_o = \mu \text{ specified in } H_0; \bar{x} = \frac{\sum x}{n}; s = \sqrt{\frac{\sum(x_i - \bar{x})^2}{n-1}}$$

Or,

$$Z = \frac{(\bar{x} - \mu_o)\sqrt{n}}{\sigma} \approx \frac{(\bar{x} - \mu_o)\sqrt{n}}{s} = \frac{\left(\dfrac{\sum x}{n} - \mu_o\right)\sqrt{n}}{\sqrt{\dfrac{\sum(x_i - \bar{x})^2}{n-1}}} = \frac{\left(\sum x - n\mu_o\right)\sqrt{n(n-1)}}{n\sqrt{\sum(x_i - \bar{x})^2}}$$

One-Tailed Test

$H_0: \mu \leq \mu_o$	$H_0: \mu \geq \mu_o$
$H_a: \mu > \mu_o$	$H_a: \mu < \mu_o$
Reject H_0 if $Z^* > Z_\alpha$	Reject H_0 if $Z^* < -Z_\alpha$
($Z_\alpha = 1.645$ for $\alpha = 0.05$)	($-Z_\alpha = -1.645$ for $\alpha = 0.05$)

Solution

The calculations are as follows.

1. From problem statement we have

    $\bar{X} = 11.90$, $s = 1.4$, at 95% confidence interval $Z_{\alpha = 0.05} = 1.96$, $n = 50$

STEP 1

$H_0: \mu = 10$ versus $H_a: \mu \neq 10$

STEP 2

Standardize $\bar{X}$, we have

$$Z = \frac{(\bar{X} - \mu_0)\sqrt{n}}{\sigma_x} = \frac{(\bar{X} - \mu_0)\sqrt{n}}{s}$$

Confidence for the mean of population is

$$P\left(-Z_{\frac{\alpha}{2}} \leq Z \leq Z_{\frac{\alpha}{2}}\right) = 0.95 \text{ where } Z = \frac{\bar{X} - \mu}{\sigma_X} = \frac{\bar{X} - \mu}{\sigma / \sqrt{n}}$$

From Z-table at 95% the Z-value equal to $Z = 1.96$ as shown in Figure 9.67. The probability of 95% confidence interval equal to

$$P(-1.96 \leq Z \leq 1.96) = 0.95$$

where

$$Z = \frac{\bar{X} - \mu_0}{\sigma_x / \sqrt{n}} = \frac{\bar{X} - \mu_0}{s / \sqrt{n}}; \mu_0 = \mu \text{ specified in } H_0$$

Or,

$$Z = \frac{(\bar{X} - \mu_0)\sqrt{n}}{\sigma_x} = \frac{(\bar{X} - \mu_0)\sqrt{n}}{s} = \frac{(11.90 - 10)\sqrt{50}}{1.40} = 9.60$$

Now, we substitute Z by its value and solve for μ:

$$P\left(-1.96 \leq \frac{\bar{X} - \mu}{\sigma / \sqrt{n}} \leq 1.96\right) = 0.95$$

$\bar{X} \pm k\left(\dfrac{\sigma}{\sqrt{n}}\right)$; the value of $\dfrac{\sigma}{\sqrt{n}}$ can be estimated $by \dfrac{s}{\sqrt{n}}$ due to large sample size

$\bar{X} \pm k\left(\dfrac{s}{\sqrt{n}}\right)$;

$$\bar{X} - k\left(\frac{s}{\sqrt{n}}\right) to \bar{X} + k\left(\frac{s}{\sqrt{n}}\right)$$

$$\bar{X} - 1.96\left(\frac{\sigma}{\sqrt{n}}\right) to \bar{X} + 1.96\left(\frac{\sigma}{\sqrt{n}}\right)$$

$$11.9 - 1.96\left(\frac{1.4}{\sqrt{50}}\right) to 11.9 + 1.96\left(\frac{1.4}{\sqrt{50}}\right)$$

$$11.90 \pm 0.388$$

$$11.512 \text{ to } 12.288$$

2. Since 10 doesn't lie in interval, thus reject H_0

STEP 3

The rejection area is

$$Reject \ H_0 \ if \ \left| \frac{(\bar{X} - 5.9)\sqrt{n}}{\sigma} \right| > Z_{\frac{\alpha}{2}} = k = 1.960$$

$$-1.96 > \frac{(\bar{X} - 10)\sqrt{n}}{\sigma}; \ or; \ \frac{(\bar{X} - 10)\sqrt{n}}{\sigma} > 1.96$$

$$Z = \frac{(\bar{X} - 10)\sqrt{n}}{\sigma} \approx \frac{(\bar{X} - 10)\sqrt{n}}{s} = \frac{(11.90 - 10)\sqrt{50}}{1.40} = 9.60$$

STEP 4

Therefore, reject H_0 if $|Z| > k$; or $Z^* > Z_{\alpha/2} = k$ or $Z < Z_{\alpha/2} = -k$. Value of k obtained from Z-table at $\alpha = 0.05$.

STEP 5

Since $Z^* = Z_{Calculated} = 9.60 > 1.96$ so reject H_0. The average population means $\mu \neq 10$. This is visualized in Figure 9.68.

Example 9.62

Hypothesis testing on the Mean (μ) of Population (Large Sample): The life of a battery in a disposable cell phone is estimated to be 425 hours ($\bar{X} = 425$) from a random sample number of 45 cell phones. The life of a cell phone battery is considered to be normally distributed with a population of standard deviation of 25 hours.

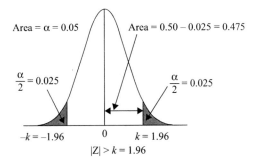

Figure 9.68. Confidence interval.

1. Determine the 95% confidence interval for the mean life of the disposable cell phone battery.
2. What is the actual mean life of a cell phone battery if different than 650? Using a 10% significance level (90% confidence).

Solution

The calculations are as follows.

1. From problem statement we have the following given:

$$\mu = 425, \sigma = 25, \text{ at 95\% confidence interval } Z_{\alpha = 0.05} = 1.96, n = 50$$

STEP 1

$H_0: \mu = 425$ hours $H_a: \mu \neq 425$ hours

STEP 2

Standardize $\bar{X}$, we have

$$Z = \frac{(\bar{X} - \mu_0)\sqrt{n}}{\sigma_x} = \frac{(\bar{X} - \mu_0)\sqrt{n}}{s}$$

Confidence interval for the population is

$$P\left(-Z_{\frac{\alpha}{2}} \leq Z \leq Z_{\frac{\alpha}{2}}\right) = 0.95 \text{ where } Z = \frac{\bar{X} - \mu}{\sigma_X} = \frac{\bar{X} - \mu}{\sigma / \sqrt{n}}$$

$$P(-1.96 \leq Z \leq 1.96) = 0.95$$

where

$$Z = \frac{\bar{X} - \mu_0}{\sigma_x / \sqrt{n}} = \frac{\bar{X} - \mu_0}{s / \sqrt{n}}; \mu_0 = \mu \text{ specified in } H_0$$

Or,

$$Z = \frac{(\bar{X} - \mu_0)\sqrt{n}}{\sigma_x} = \frac{(\bar{X} - \mu_0)\sqrt{n}}{s} = \frac{(425 - 650)\sqrt{45}}{25} = -60.37$$

From Z-table at 95% the Z-value equal to $Z = 1.96$ as shown in Figure 9.68. The probability of 95% confidence interval equal to

$$P(-1.96 \leq Z \leq 1.96) = 0.95$$

Now, we substitute Z by its value and solve for μ:

$$P\left(-1.96 \leq \frac{\bar{X} - \mu}{\sigma / \sqrt{n}} \leq 1.96\right) = 0.95$$

Or,

$$\bar{X} \pm k\left(\frac{\sigma}{\sqrt{n}}\right); \text{ the value of } \frac{\sigma}{\sqrt{n}} \text{ can be estimated } by \frac{s}{\sqrt{n}} \text{ due to large sample size}$$

$$\bar{X} \pm k\left(\frac{s}{\sqrt{n}}\right);$$

$$\bar{X} - k\left(\frac{s}{\sqrt{n}}\right) \text{ to } \bar{X} + k\left(\frac{s}{\sqrt{n}}\right)$$

$$\bar{X} - 1.96\left(\frac{\sigma}{\sqrt{n}}\right) \text{ to } \bar{X} + 1.96\left(\frac{\sigma}{\sqrt{n}}\right)$$

Substituting values and simplifying we have

$$425 - 1.96\left(\frac{25}{\sqrt{45}}\right) \text{ to } 425 + 1.96\left(\frac{25}{\sqrt{45}}\right)$$

$$425 \pm 7.304$$

$$417.69 \text{ to } 432.30$$

2. Since 650 doesn't lie in interval of 95%, thus reject H_0. Now, we calculate with $\alpha = 0.10$,

$$H_0 : \mu = 650, H_a : \mu \neq 65$$

$$Z = \frac{(\bar{X} - \mu_o)\sqrt{n}}{\sigma} = \frac{(425 - 650)\sqrt{45}}{25} = -60.30$$

$$\text{Reject } H_0 \text{ if } |Z^*| = \left|\frac{(\bar{X} - \mu_o)\sqrt{n}}{\sigma} = \frac{(425 - 650)\sqrt{45}}{25}\right| > Z_{\frac{\alpha}{2}} = k = 1.645$$

Or, reject H_0 if $(-k = -1.645) > Z > (+k = 1.645)$
Reject H_0; since $(Z^* = -60.30) < (Z = -1.645)$, the average population mean $\mu \neq 650$
This is visualized in Figure 9.69.

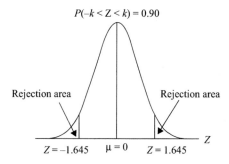

Figure 9.69. Confidence interval.

Example 9.63

Hypothesis Testing on the Mean (μ) of Population (Large Sample): A random sample of 55 was collected. *Test* (H_a) that the mean of population is less than 120 by applying the significance level of $\alpha = 0.05$. The following statistical values were calculated from the collected samples: $\Sigma x = 6325$, $\sum_{i=1}^{55}\left(X_i - \bar{X}\right)^2 = 2125$.

Solution

Given: $\mu = \mu_o = 120$, at $\alpha = 0.05$, interval $Z = 1.645$, $n = 55$, $\sum_{i=1}^{55}\left(X_i - \bar{X}\right)^2 = 2125$, $\Sigma x = 6325$
One-tailed test:

STEP 1

The hypothesis test is

H_0: $\mu \geq \mu_0 = 120$
H_α: $\mu < \mu_0 = 120$

STEP 2

The test statistic is

$$Z^* = \frac{\bar{x} - \mu_o}{\sigma/\sqrt{n}} = \frac{\bar{x} - \mu_o}{s/\sqrt{n}} = \frac{(\bar{x} - \mu_o)\sqrt{n}}{s}$$

STEP 3

Reject H_0 if $Z^* < -Z_\alpha$ from Z-table $-Z_\alpha = -1.645$ for significance level of $\alpha = 0.05$. This is shown in Figure 9.69.

STEP 4

Calculate the conditions

$$\bar{x} = \frac{\sum x}{n} = \frac{6325}{55} = 55$$

$$s = \sqrt{\frac{\sum(x_i - \bar{x})^2}{n-1}} = \left(\frac{2125}{55-1}\right)^{\frac{1}{2}} = 6.27$$

$$Z^* = \frac{\bar{x} - \mu_o}{\sigma / \sqrt{n}} = \frac{\bar{x} - \mu_o}{s / \sqrt{n}} = \frac{(\bar{x} - \mu_o)\sqrt{n}}{s} = \frac{(115 - 120)\sqrt{55}}{6.27} = -5.91; \ \mu_o = \mu \text{ specified in } H_o$$

STEP 5

Reject H_0 if $Z^* = -5.91 < -Z_\alpha = -Z_{\alpha = 0.05} = -1.645,$ Thus, reject H_0 as shown in Figure 9.70.

Reject H_0; $-5.91 < -1.645$, the population mean $\mu < 120$

9.7.2 p-VALUE APPLICATION

The *p-value* is the value of significance level "α" at which the testing procedure for hypothesis changes the outcome based on the provided data. It is the biggest value of significance level (α) that you will fail to reject null hypothesis H_0 or the *p-value* is the smallest value of α for which the H_0 could be rejected. The smaller the *p-value*, is the stronger the facts versus H_0.

The *p-value* can be computed by taking the place of area of "α" with the area of calculated Z-value (Z^*) of the "test statistic" step. Table 9.26 refers to testing conditions of *p-value*.

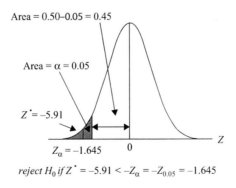

reject H_0 if $Z^* = -5.91 < -Z_\alpha = -Z_{0.05} = -1.645$

Figure 9.70. Rejection area for example 9.63.

Table 9.26. Testing conditions for *p-value*

For:	*p*-value equal to	Reason "when using *a*"
H_a: $\mu \neq \mu_0$	2(area outside of $Z_{\text{Computed}} = Z^*$)	the value of *a* represents a two-tailed area.
H_a: $\mu > \mu_0$	area to the right of $Z_{\text{Computed}} = Z^*$	the value of *a* represents a right tailed area.
H_a: $\mu < \mu_0$	area to the left of $Z_{\text{Computed}} = Z^*$	the value of *a* represents a left-tailed area.

Interpreting the *p*-value: Classical approach

Reject H_0	If *p-value* < α
Fail to reject H_0	If *p-value* $\geq \alpha$

General rule of thumb

Reject H_0	if *p-value* is small ($p < 0.01$)
Fail to reject H_0	if *p-value* is large ($p > 0.1$)
Consequently, the data are "inconclusive"	if $0.01 \leq$ *p-value* ≤ 0.10

Example 9.64

Hypothesis testing using *p-value* approach A protein bar company claims that anyone who uses its newly developed B1-protein bar will give them energy for $\mu = 5$ hours period. A random sample of $n = 50$ people were tested. The result showed that effective time is 4.65 hours with a sample standard deviation of $s = 0.75$ hours. Use the *p*-value criteria to find if there is enough evidence to support the hypothesis that mean value of protein bar effect is 5 hours.

Solution

This is large sample to test on normal population mean. The application is for two-tailed test.

1. Test condition
   - H_0: $\quad \mu = \mu_0 = 5$
   - H_a: $\quad \mu \neq \mu_0 = 5$
2. Rejection condition
   - Reject H_0 if *p-value* < α
   - Fail to reject H_0 if *p-value* $\geq \alpha$
   - General rule of thumb:
   - Reject H_0 if *p-value* < 0.01
   - Fail to reject H_0 if *p-value* ≥ 0.10

3. Now, calculate Z^*

$$Z^* = \frac{\bar{x} - \mu_o}{\sigma / \sqrt{n}} = \frac{\bar{x} - \mu_o}{s / \sqrt{n}} = \frac{(\bar{x} - \mu_o)\sqrt{n}}{s}$$

where $\mu_o = \mu$ specified in H_0

$$\bar{x} = \frac{\sum x}{n}, \text{ and } s = \sqrt{\frac{\sum (x_i - \bar{x})^2}{n-1}}$$

or

$$Z^* = \frac{(\bar{x} - \mu_o)\sqrt{n}}{\sigma} \approx \frac{(\bar{x} - \mu_o)\sqrt{n}}{s} = \frac{\left(\dfrac{\sum x}{n} - \mu_o\right)\sqrt{n}}{\sqrt{\dfrac{\sum (x_i - \bar{x})^2}{n-1}}} = \frac{\left(\sum x - n\mu_o\right)\sqrt{n(n-1)}}{n\sqrt{\sum (x_i - \bar{x})^2}}$$

$$Z^* = \frac{\bar{x} - \mu_o}{s / \sqrt{n}} = \frac{(4.65 - 5)\sqrt{50}}{0.75} = -3.30$$

4. From Z-table at $Z = 3.30$, the area i equal to 0.49952

   The p-$Value$ = 2 (area outside of Z^*) = 2(area at $Z < -3.30$) = 2(0.50 – 0.49952) = 0.00096

   Thus, reject H_0 because the area 0.00096 at computer Z^*-value is less than predetermine $\alpha = 0.01$ area:

   Reject H_0; $0.00096 < 0.01$, the population mean $\mu \neq 5$

9.7.3 HYPOTHESIS TESTING USING p-VALUE APPROACH (USING EQUAL MEAN)

The general rule of thumb that is used in describing p-$value$ hypothesis testing on the mean based on the typical α-values of 0.01 and 0.10 are as

Reject H_0 if the p-$value$ is smaller than 0.01 ($p < 0.01$)
Fail to reject H_0 if the p-$value$ is larger than 0.1 ($p > 0.1$)

Therefore, if $0.01 \leq p$-$value \leq 0.10$, the sample data is considered to be inconclusive (no conclusion). Furthermore, if p-$value = 0.0001$ then we strongly reject H_0 because $p = 0.0001$ is extremely small compare to any average α-value. So, this supports the H_α. On the other hand, if p-$value = 0.65$, without doubt we would fail to reject H_0. So, this also means fail to support H_α. In other words, the outcome supports H_0 and rejects H_α. Because $p = 0.65$ is large compared to average α-value. The following example will expand on this concept.

Example 9.65

Hypothesis testing using *p-value* approach (using equal mean) The University of North Ice-Green Land claims that students who live off campus pay $350 monthly utility bill. A random sample of 250 students surveyed shows that students who live off campus on average pay $356 utility bill with standard deviation of s = $55 a month.

1. Use the *p*-value criteria to find if there is enough evidence to support the hypothesis that mean value of off campus students utility bill is ***about*** $350 per month.
2. Use the *p*-value criteria to find if there is enough evidence to support the hypothesis that mean value of off campus student's utility bill is ***more than*** $350 per month.
3. Find that *null hypothesis H_0* will be rejected if α = 0.01? α = 0.05? α = 0.10?

Solution

This is a large sample test using a *p*-value application at n = 250, s = 55, μ = 350, and $\bar{x}$ = 356

STEP 1

One-tailed test due to inequality condition
$$H_0: \ \mu \le \mu_0 = 350$$
$$H_\alpha: \ \mu > \mu_0 = 350$$

STEP 2

Reject H_0 if *p-value* < α
Fail to reject H_0 if *p-value* $\ge$ α
Again the general rule of thumb for hypothesis testing using *p*-value is
Reject H_0 if *p-value* < 0.01
Fail to reject H_0 if *p-value* $\ge$ 0.10

STEP 3

1. Now, we can calculate the Z^*-value using the same equation as we used in previous example:

$$Z^* = \frac{(\bar{x}-\mu_o)\sqrt{n}}{\sigma} \approx \frac{(\bar{x}-\mu_o)\sqrt{n}}{s} = \frac{\left(\frac{\sum x}{n}-\mu_o\right)\sqrt{n}}{\sqrt{\frac{\sum(x_i-\bar{x})^2}{n-1}}} = \frac{\left(\sum x-n\mu_o\right)\sqrt{n(n-1)}}{n\sqrt{\sum(x_i-\bar{x})^2}}$$

$$Z^* = \frac{\bar{x}-\mu_o}{\sigma/\sqrt{n}} = \frac{\bar{x}-\mu_o}{s/\sqrt{n}} = \frac{356-350}{55/\sqrt{50}} = 1.72$$

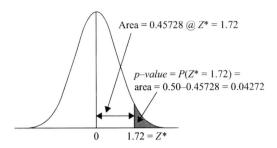

Figure 9.71. Rejection area and *p-value*.

STEP 4

From the Z-table in the appendix at $Z = 1.72$, the area equal to 0.45728
    The *p-value* = area to the right of $Z^* = 0.50 - 0.45728 = 0.04272$ (Figure 9.70).

2. Interpreting the *p*-value: classical approach
    Reject H_0 if *p-value* < α
    Fail to reject H_0 if *p-value* $\geq \alpha$
        Thus, the *p-value* = 0.04272; for $\alpha = 0.01$, fail to reject H_0 and for $\alpha = 0.05$ reject H_0, for $\alpha = 0.1$ reject H_0. This is visualized in Figure 9.71.

Example 9.66

The Owner of Los Angeles Wellness Center club believes that the recent program changes have greatly improved the club's activity for members.

1. They now stay longer at the club per visit than previous club called Fitness Center.
2. Study shows that previous mean (population) time was 36 minutes with population Standard deviation = 11 min.
3. A simple random sample of $n = 200$ visits is selected, the current sample mean is 36.80 minutes.
4. Test the owner's claim and justify for the changes of project. Use $\alpha = 0.05$.

Solution

From problem statement we have $\overline{X} = 36.80$ minutes, $\sigma = 11$ minutes, $\mu = 36$ minutes, $\alpha = 0.05$

STEP 1

Hypothesis test:
    H_0: $\mu \geq \mu_0 = 36$ minutes versus H_α: $\mu < \mu_0 = 36$ minutes (Owner's claim)

STEP 2

The test statistic is

$$Z^* = \frac{(\bar{x} - \mu_o)\sqrt{n}}{\sigma} \approx \frac{(\bar{x} - \mu_o)\sqrt{n}}{s} \approx \frac{(36.80 - 36.0)\sqrt{200}}{11} = 1.0286 = 1.03$$

STEP 3

The rejection area is defined as
Fail to reject H_0 if $p\text{-value} = P(Z \geq Z^* = 1.03) = 0.1515 \geq \alpha = 0.05$
Or, reject H_0 if $p\text{-value} = P(Z < Z^* = 1.03) = 0.1515 < \alpha = 0.05$
This is illustrated in Figure 9.72.

STEP 4

We compute the value of the test statistic. The *p-value* is the probability of a Z-value from standard normal distribution (Z-table) being at least as large as 1.03.

Because the *p-value* = 0.1515 > α = 0.05, so we don't reject the H_0.

STEP 5

The difference between sample mean and hypothesized population mean is not large enough to attribute to anything but sampling error.

9.7.4 HYPOTHESIS TESTING ON THE MEAN (μ) OF A NORMAL POPULATION FOR SMALL SAMPLE

The method to apply the hypothesis testing for small samples when the standard deviation is not known, we change from Z-standard normal distribution to *t*-distribution. The testing conditions and test statistics are listed in Table 9.27.

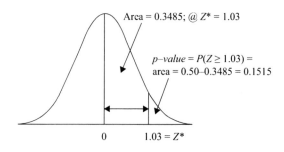

Figure 9.72. Rejection area of *p-value* (Z-is obtained from Z-table).

Table 9.27. Testing conditions: small sample tests on a normal population mean

Two-Tailed Test:

H_0: $\mu = \mu_0$

H_α: $\mu \neq \mu_0$

Reject H_0 if $|t_{\text{Calculated}} = t^*| > t_{\alpha/2,\, df = n-1}$

$$t^* = \frac{\bar{X} - \mu_0}{\sigma / \sqrt{n}} \approx \frac{\bar{X} - \mu_0}{\dfrac{s}{\sqrt{n}}}; \mu_0 = \mu \text{ specified in } H_0;\ \bar{x} = \frac{\sum x}{n};\ s = \sqrt{\frac{\sum (x_i - \bar{x})^2}{n-1}}$$

Or,

$$t^* = \frac{(\bar{x} - \mu_o)\sqrt{n}}{\sigma} \approx \frac{(\bar{x} - \mu_o)\sqrt{n}}{s} = \frac{\left(\dfrac{\sum x}{n} - \mu_o\right)\sqrt{n}}{\sqrt{\dfrac{\sum(x_i - \bar{x})^2}{n-1}}} = \frac{\left(\sum x - n\mu_o\right)\sqrt{n(n-1)}}{n\sqrt{\sum(x_i - \bar{x})^2}}$$

One-Tailed Test:

H_0: $\mu \leq \mu_0$ H_0: $\mu \geq \mu_0$

H_α: $\mu > \mu_0$ H_α: $\mu < \mu_0$

Reject H_0 if $t^* > t_{\alpha,\, df = n-1}$ Reject H_0 if $t^* < -t_{\alpha,\, df = n-1}$

Further, since the data size is less than $n = 30$, it is essential to check the data for normality and that the t-test would be valid unless it is stated in the problem statement. One method of quick checking for normality is to do stem-and-leaf. Or, a better way to test for normally as discussed earlier in this chapter is Anderson-Darling normality test.

Example 9.67

Hypothesis testing on the mean (μ) of a normal population: Small sample two-tailed test
The following random sample data were obtained from a normally distributed population. Test the mean of the population is different than $\mu = 20$. Use $\alpha = 0.05$ for samples 18, 14, 15, 20, 12, 23, 18, 19, 21, 16

Solution

This is a small-sample test on normal population mean with $\alpha = 0.05$, and $n = 10$

STEP 1

Two-tailed test

     H_0: $\mu = \mu_0 = 20$

     H_α: $\mu \neq \mu_0 = 20$

STEP 2

For this problem the "test statistic" is

$$t = \frac{\overline{X} - \mu_0}{\sigma/\sqrt{n}} \approx \frac{\overline{X} - \mu_0}{\frac{s}{\sqrt{n}}}$$

STEP 3

Reject H_0 if $|t_{Calculated} = t^*| > t_{\alpha/2, df = n-1} = t_{0.05/2, df = 10-1} = t_{0.025, 9}$

STEP 4

Calculate the conditions
The mean:

$$\overline{X} = \frac{\sum_{i=1}^{i=n=10} x}{n} = \frac{18 + 14 + \cdots + 21 + 16}{10} = \frac{176}{10} = 17.60$$

The standard deviation,

$$s = \sqrt{\frac{\sum (x_i - \overline{x})^2}{n-1}} = 3.37$$

And $t_{calculated} = t^*$ value

$$t^* = \frac{\overline{X} - \mu_o}{\sigma/\sqrt{n}} \approx \frac{X - \mu_o}{s/\sqrt{n}}$$

Or $t^* = \dfrac{(\overline{x} - \mu_o)\sqrt{n}}{\sigma} \approx \dfrac{(\overline{x} - \mu_o)\sqrt{n}}{s} = \dfrac{\left(\dfrac{\sum x}{n} - \mu_o\right)\sqrt{n}}{\sqrt{\dfrac{\sum (x_i - \overline{x})^2}{n-1}}} = \dfrac{(\sum x - n\mu_o)\sqrt{n(n-1)}}{n\sqrt{\sum (x_i - \overline{x})^2}}$

Thus,

$$t^* = \frac{(\sum x - n\mu_o)\sqrt{n(n-1)}}{n\sqrt{\sum (x_i - \overline{x})^2}} = \frac{(176 - 10(20))\sqrt{10(10-1)}}{10\sqrt{102.41}} = \frac{-227.684}{101.20} = -2.249$$

STEP 5

Reject H_α if $|t_{Calculated} = t^*| > t_{\alpha/2, df = n-1} = t_{0.05/2, df = 10-1} = t_{0.025, 9} = 2.262$

Since $|t^*| = |-2.249| = 2.249 > t_{0.025, 9} = 2.262$, this means we fail to reject.

Fail to reject H_0; $2.25 < 2.262$; the average population mean $\mu = 20$

Example 9.68

Small sample one-tailed test: A pharmaceutical company's customer service call center where customers can call regarding their product questions. The following data was obtained about the time that customer service spends with customer over the phone.

1. Previous studies indicate that the distribution of time required to each call is normally distributed, with a mean of 500 sec.
2. Customer service manager has collected a random sample of 16 calls.
3. They wish to find if the mean call time is now fewer than 500 sec after a training program given to customer service.

What conclusion would you reach using significance level (risk factor) of $a = 0.01$?

Solution

We will compute the mean and standard deviation before the step 1.

$$n = 16, \bar{X} = \frac{\sum_{i=1}^{i=n} x}{n} = 470 \text{ sec}, s = \sqrt{\frac{\sum(x_i - \bar{x})^2}{n-1}} = 45 \text{ sec}$$

STEP 1

Hypothesis test

H_0: $\mu \geq \mu_0 = 500$ versus H_α: $\mu < \mu_0 = 500$

STEP 2

Test statistics

$$t^* = \frac{(\bar{x} - \mu_o)\sqrt{n}}{\sigma} \approx \frac{(\bar{x} - \mu_o)\sqrt{n}}{s} = \frac{(470 - 500)\sqrt{16}}{45} = -2.67$$

STEP 3

Rejection area defined as

Reject H_0 if $t^* < -t_{\alpha, df = n-1}$ or, reject H_0 if $t^* < -t_{a = 0.01, 15} = -2.602$
t_α is obtained from table in the appendix and t^*-computed as in step 4.

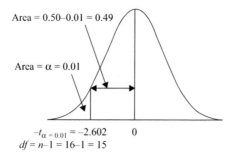

Figure 9.73. Rejection region.

STEP 4

Test statistic, also see step 2.

$$t^* = \frac{(\bar{x} - \mu_o)\sqrt{n}}{s} = \frac{(470 - 500)\sqrt{16}}{45} = -2.67$$

All the above values is visualized in Figure 9.73.

STEP 5

At $\alpha = 0.01$ there is enough prove to support customer service claim of mean time reduced below 500 seconds.

9.7.5   INFERENCE PROCEDURES FOR TWO POPULATIONS: APPLYING THE CONCEPTS

Figure 9.74 shows the normal distribution of population 1 with mean μ_1 standard deviation σ_1, and the population 2 with mean μ_2 standard deviation σ_2. Both distributions use large samples: that is, $n_1 > 30$ and $n_2 > 30$. Due to comparison of two populations we have two of everything. Here the random sample n_1 could be independent from or dependent to n_2 and vice versa. Further, $\bar{X}_1$ is point estimate of μ_1 and $\bar{X}_2$ is the point estimate of μ_2 with s_1 and s_2 standard deviation of samples 1 and 2, respectively.

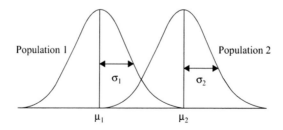

Figure 9.74. Two populations with their means and standard deviation.

Independent samples: The event of a process in the first sample has an impact in the values of the second sample outcome. $\mu_2 = f(\mu_1)$. In other words, the outcome of second sample is a function of first process sample variations.

Dependent samples: In the dependent samples or paired samples the event of a process in the first sample has no impact in the values of the other samples outcome. $\mu_2 = \mu_1 \dots$

Table 9.28 states the conditions for the two large population means as well as confidence interval when the standard deviations (σ_1, σ_2) are known and unknown case.

Table 9.28. Testing conditions for two populations mean (μ_1 and μ_2): large sample

Two-Tailed Test

H_0: $\mu = \mu_1$

H_a: $\mu \neq \mu_2$

Reject H_0 if $|Z_{\text{Calculated}} = Z^*| > Z_{\alpha/2}$

$$Z = \frac{\bar{X}_1 - \bar{X}_2}{\sqrt{\dfrac{\sigma_1^2}{n_1} + \dfrac{\sigma_2^2}{n_2}}} = \frac{\bar{X}_1 - \bar{X}_2}{\left(\dfrac{\sigma_1^2}{n_1} + \dfrac{\sigma_2^2}{n_2}\right)^{1/2}} \approx \frac{\bar{X}_1 - \bar{X}_2}{\left(\dfrac{s_1^2}{n_1} + \dfrac{s_2^2}{n_2}\right)^{1/2}}$$

where,

$$\bar{x} = \frac{\sum x}{n}; \; s = \sqrt{\frac{\sum (x_i - \bar{x})^2}{n-1}}$$

One-Tailed Test

H_0: $\mu_1 \leq \mu_2$ H_0: $\mu_1 \geq \mu_2$

H_a: $\mu_1 > \mu_2$ H_a: $\mu_1 < \mu_2$

Reject H_0 if $Z^* > Z_\alpha$ Reject H_0 if $Z^* < -Z_\alpha$

($Z_\alpha = 1.645$ for $\alpha = 0.05$)   ($-Z_\alpha = -1.645$ for $\alpha = 0.05$)

Confidence interval when σ_1 and σ_2 are known:

$$(\bar{X}_1 - \bar{X}_2) \pm Z_{\alpha/2} \sqrt{\frac{\sigma_1^2}{n_1} + \frac{\sigma_2^2}{n_2}}$$

Confidence interval when σ_1 and σ_2 are unknown:

$$(\bar{X}_1 - \bar{X}_2) \pm Z_{\alpha/2} \sqrt{\frac{s_1^2}{n_1} + \frac{s_2^2}{n_2}}$$

Sample sizes for n_1 and n_2:

$$n_1 = \frac{Z_{\alpha/2}^2 s_1 (s_1 + s_2)}{E^2} \quad \text{and} \quad n_2 = \frac{Z_{\alpha/2}^2 s_2 (s_1 + s_2)}{E^2}$$

Notice that confidence interval in Table 9.28 is very similar to confidence interval for a single population mean applied to a large sample size. Such that

$$\text{Point estimate} \pm Z_{\alpha/2} \cdot (\text{standard deviation of point estimate})$$

In addition, Table 9.28 includes the formulas for estimating the minimum total sample sizes with sample errors (E_1 and E_2). See the application of these equations in the following examples.

Example 9.69

Comparing two means ($\mu_1 - \mu_2$) using Two Large, Independent Samples: Using the Method of Constructing Confidence Interval A leading agriculture company would like to test (investigate) hypothesis that *rock-n-roll* music increases the productivity of strawberry collection (H_a: $\mu_1 > \mu_2$). Samples of two different groups of 50 strawberry collectors were randomly selected from across the company. The identical area was given to each group. The rating from 1 to 100 is recorded for each individual. The first group was allowed to listen to music and the second group without music. The result showed that group who listened to *rock-n-roll* music had a mean rating of 75 ($\mu_1 = 75$) with a sample standard deviation of $\sigma_1 = 5.5$. The second group had a mean rating of 71 ($\mu_2 = 71$) with a sample standard deviation of $\sigma_2 = 4.8$.

1. Use a significance level of 10% ($\alpha = 0.10$) to test the company's theory (note: $Z_{a = 0.10} = 1.28$ from Z-table).
2. Find a 90% confidence interval for the difference in the mean ($\mu_1 - \mu_2$) rating of the two groups.
3. Describe if the same strawberry collectors were used in each group. Would the assumption for the validity of the test change?
4. Find out the sample sizes with maximum error, $E = 1.5$ when $a = 0.10$.

Solution

Large sample tests on normal two populations mean ($\mu_1 - \mu_2$).

1. From problem statement, we have $\bar{X}_1 = 75, \bar{X}_2 = 71, n_1 = n_2 = 50, \sigma_1 = 5.5, \sigma_2 = 4.8$

STEP 1

One-tailed test:

H_0: $\mu_1 \le \mu_2$ rearrange H_0: $\mu_1 - \mu_2 \le 0$
H_a: $\mu_1 > \mu_2$ rearrange H_a: $\mu_1 - \mu_2 > 0$

STEP 2

Reject H_0 if $Z^* > Z_{\alpha = 0.10}$ for one tail use full alpha α

STEP 3

Calculate the value of Z^*.

$$Z^* = \frac{\bar{X}_1 - \bar{X}_2}{\sqrt{\dfrac{\sigma_1^2}{n_1} + \dfrac{\sigma_2^2}{n_2}}} = \frac{\bar{X}_1 - \bar{X}_2}{\left(\dfrac{\sigma_1^2}{n_1} + \dfrac{\sigma_2^2}{n_2}\right)^{1/2}} = \frac{\bar{X}_1 - \bar{X}_2}{\left(\dfrac{(5.5)^2}{50} + \dfrac{(4.8)^2}{50}\right)^{1/2}} = 3.8746$$

One-tailed test: H_0: $\mu_1 \leq \mu_2$ and H_a: $\mu_1 > \mu_2$

STEP 4

From Z-table at α (area) = 0.10 the value of Z equal to 1.28. Therefore, since $Z_{\text{Computed}} = Z^* = 3.8746 > Z_{\alpha = 0.10} = 1.28$, thus we reject null hypothesis H_0. This is featured in Figure 9.75.

2. 90% confidence interval for the mean difference $\mu_1 - \mu_2$

From the Table 9.28, we have the confidence interval model:

$$(\bar{X}_1 - \bar{X}_2) \pm Z_{\alpha/2} \sqrt{\frac{s_1^2}{n_1} + \frac{s_2^2}{n_2}}$$

$$(75 - 71) \pm (Z_{\frac{0.10}{2} = 0.05} = 1.645)\left(\frac{(5.5)^2}{50} + \frac{(4.8)^2}{50}\right)^{1/2}$$

$$4 \pm (1.645)(1.032)$$

This is reduced to

$$(4 - 1.698) \text{ to } (4 + 1.698)$$

$$4 \pm 1.698$$

Thus, 90% confidence interval falls in 2.302 to 5.658.

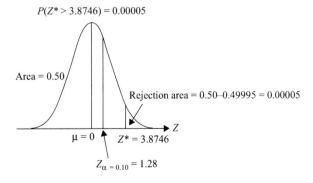

$P(Z^* > 3.8746) = 0.00005$

Area = 0.50

Rejection area = 0.50–0.49995 = 0.00005

$\mu = 0$ $Z^* = 3.8746$

$Z_{\alpha = 0.10} = 1.28$

Figure 9.75. Rejection region.

Table 9.29 *P-value*

For:	$p = p$-value	Reason "when using α"
$H_a: \mu_1 \neq \mu_2$	2(area outside of $Z_{Computed} = Z^*$)	the value of *a* represent *s* a two tailed area.
$H_a: \mu_1 > \mu_2$	area to the right of $Z_{Computed} = Z^*$	the value of *a* represent *s* a right tailed area.
$H_a: \mu_1 < \mu_2$	area to the left of $Z_{Computed} = Z^*$	the value of *a* represent *s* a left tailed area.

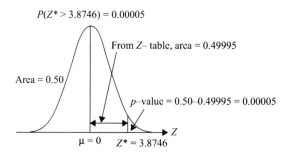

Figure 9.76. Rejection area of *p-value*.

3. The procedure in part (2) is valid only for independent *t* samples. If the same strawberry collectors were used in each group, then the samples were not independent *t*.

4. $n_1 = \dfrac{Z_{a/2}^2 s_1 (s_1 + s_2)}{E^2} = \dfrac{(1.645)^2 (5.5)(5.5 + 4.8)}{(1.5)^2} = 68.1 \approx 69$

   $n_2 = \dfrac{Z_{a/2}^2 s_2 (s_1 + s_2)}{E^2} = \dfrac{(1.645)^2 (4.8)(5.5 + 4.8)}{(1.5)^2} = 59.5 \approx 60$

Using the *p-value* for the data in Example 9.69, what would be the outcome by applying the classical method with α = 0.10? For this case, the *p-value* will be the area under the computed test statistic Z curve to the right, which we reject H_0 in the tail for this case of the computed test statistic ($Z^* = 3.8746$), especially in the cases illustrated in Table 9.29.

So, the rejection area of *p*-value for the above Example 9.69 can be pictured in Figure 9.76.

9.7.6 COMPARING TWO NORMAL POPULATION MEANS ($\mu_1 - \mu_2$) USING TWO SMALL, INDEPENDENT SAMPLES: *APPLY THE MECHANICS*

When using small samples ($n_1 < 30$ and or $n_2 < 30$) from two populations, Central Limit Theorem (CLT requires: $n_1 > 30$ and or $n_2 > 30$) no longer exists. In the previous sections, we discussed the conversion from large sample to small sample both being normal populations, the hypothesis testing procedures and confidence interval continues entirely the same, only with one difference that we use *t*-distribution in place of Z-distribution to determine the test statistics. Thus, we will apply the same method for small samples from two populations. Table 9.30 consists of all

Table 9.30. Testing conditions for two populations mean (μ_1 and μ_2): small sample

Two-Tailed Test:

H_0: $\mu_1 - \mu_2 = D_0$

H_α: $\mu_1 - \mu_2 \neq D_0$

($D_0 = 0$ for H_0: $\mu_1 = \mu_2$)

Reject H_0 if $|t_{\text{Calculated}} = t^*| > t_{\alpha/2,\,df}$; where $df = n_1 + n_2 - 2$

Case # 1: $\sigma_1 = \sigma_2$

$$t = \left(\left(\bar{x}_1 - \bar{x}_2 \right) - D_0 \right) \left(S_c \left(n_1^{-1} + n_2^{-1} \right)^{1/2} \right)^{-1}$$

where n_1 = sample 1 count, n_2 = sample 2 count, S_c^2 = combined sample variance, merely a weight average of S_1^2 and S_2^2 defined by

$$s_c^2 = \left(\left(n_1 - 1 \right) s_1^2 + (n_2 - 1) s_2^2 \right) \left(n_1 + n_2 - 2 \right)^{-1}$$

And combined standard deviation is

$$S_C = \left(\left(\left(n_1 - 1 \right) s_1^2 + (n_2 - 1) s_2^2 \right) \left(n_1 + n_2 - 2 \right)^{-1} \right)^{0.50}$$

Degrees of freedom when $\sigma_1 = \sigma_2$

$df = n_1 + n_2 - 2$

$$S_1 = \sqrt{\frac{\sum (x_i - \bar{x})^2}{n-1}}, \; S_1 = \sqrt{\frac{\sum (x_i - \bar{x})^2}{n-1}}$$

Case # 2: when population standard are not equal $\sigma_1 \neq \sigma_2$

$$t = \left(\left(\bar{x}_1 - \bar{x}_2 \right) - D_0 \right) \left(\left(s_1^2 n_1^{-1} + s_2^2 n_2^{-1} \right)^{1/2} \right)^{-1}$$

$$S_C = \left(\left(\left(n_1 - 1 \right) s_1^2 + (n_2 - 1) s_2^2 \right) \left(n_1 + n_2 - 2 \right)^{-1} \right)^{0.50}$$

The degrees of freedom when $\sigma_1 \neq \sigma_2$:

$$df = \frac{\left(s_1^2 n_1^{-1} + s_2^2 n_2^{-1} \right)^2}{\dfrac{\left(s_1^2 n_1^{-1} \right)^2}{n_1 - 1} + \dfrac{\left(s_2^2 n_2^{-1} \right)^2}{n_2 - 1}} = \left(s_1^2 n_1^{-1} + s_2^2 n_2^{-1} \right)^2 \left(\frac{\left(s_1^2 n_1^{-1} \right)^2}{n_1 - 1} + \frac{\left(s_2^2 n_2^{-1} \right)^2}{n_2 - 1} \right)^{-1}$$

Sample standard deviation:

$$s_1 = \sqrt{\frac{\sum (x_i - \bar{x})^2}{n-1}}, \; s_2 = \sqrt{\frac{\sum (x_i - \bar{x})^2}{n-1}}$$

(Continued)

Table 9.30. (*Continued*)

One-Tailed Test:

$H_0: \mu_1 - \mu_2 \leq D_0$ $H_0: \mu_1 - \mu_2 \geq D_0$

$H_a: \mu_1 - \mu_2 > D_0$ $H_a: \mu_1 - \mu_2 < D_0$

$(D_0 = 0 \text{ for } H_0: \mu_1 \leq \mu_2)$ $(D_0 = 0 \text{ for } H_0: \mu_1 \geq \mu_2)$

Reject H_0 if $t^* > t_{\alpha/2, df}$ Reject H_0 if $t^* < -t_{\alpha/2, df}$

Confidence Interval when $\mu_1 - \mu_2$ (small sample)

$$(\bar{X}_1 - \bar{X}_2) \pm t_{\alpha/2, df} \sqrt{\frac{s_1^2}{n_1} + \frac{s_2^2}{n_2}}$$

where, if *df* is not an integer number then round this value to the nearest integer value. Further, *df* value depends on the both cases when $\sigma_1 = \sigma_2$ and $\sigma_1 \neq \sigma_2$.

---

conditions for case # 1 when assuming both standard deviations are equal $\sigma_1 = \sigma_2$ and case # 2 when not both standard deviations are equal $\sigma_1 \neq \sigma_2$.

The following examples given apply all the concepts cited in Table 9.29.

Example 9.70

The data below concluded from independent random samples from *approximate* normal populations.

Sample 1: 23, 25, 23, 22, 24, 19 $n_1 = 6$
Sample 2: 25, 27, 24, 28, 26 $n_2 = 5$

1. Conduct a hypothesis test to find if the mean of second population is greater than mean of first population ($H_0: \mu_1 < \mu_2$). Assume that the population standard deviations are equal $(\sigma_1 - \sigma_2)$. Use a 0.05 significance level ($\alpha = 0.05$).
2. Carry out the statistical test assuming that population variances aren't equal $(\sigma_1^2 \neq \sigma_2^2)$.

Solution

Two small samples:

1. Given $n_1 = 5$, $n_2 = 6$, assume $\sigma_1^2 = \sigma_2^2$

STEP 1

Define the hypothesis:
$H_0: \mu_1 \geq \mu_2$ rearrange $H_0: \mu_1 - \mu_2 \geq 0$
$H_a: \mu_1 < \mu_2$ rearrange $H_a: \mu_1 - \mu_2 < 0$

STEP 2

Reject H_0 if $t^* < -t_{\alpha,df}$

STEP 3

Assume $\sigma_1^2 = \sigma_2^2$. Consequently the combined variance is calculated by the following sample means and standard deviations.

Sample means:

$$\bar{X}_1 = \frac{\sum_{i=1}^{i=n=6} x}{n} = \frac{23+25+23+22+24+19}{6} = \frac{133}{6} = 22.67$$

$$\bar{X}_2 = \frac{\sum_{i=1}^{i=n=5} x}{n} = \frac{25+27+24+28+26}{5} = \frac{130}{5} = 26$$

Sample standard deviations:

$$s_1 = \sqrt{\frac{\sum(x_i - \bar{x})^2}{n-1}} = \sqrt{\frac{(23-22.67)^2 + \cdots + (19-22.67)^2}{6-1}} = \sqrt{\frac{26.83}{5}} = 2.07$$

$$s_2 = \sqrt{\frac{\sum(x_i - \bar{x})^2}{n-1}} = \sqrt{\frac{(25-26)^2 + \cdots + (26-26)^2}{5-1}} = \sqrt{\frac{10}{4}} = 1.58$$

Combined sample variance and degree of freedom are

$$S_C^2 = ((n_1-1)s_1^2 + (n_2-1)s_2^2)(n_1+n_2-2)^{-1}; \quad \text{where } df = n_1 + n_2 - 2$$

$$S_C = \sqrt{\frac{(n_1-1)s_1^2 + (n_2-1)s_2^2}{n_1+n_2-2}} = \left(\left((n_1-1)s_1^2 + (n_2-1)s_2^2\right)(n_1+n_2-2)^{-1}\right)^{0.50}$$

$$S_C = \left(\left((6-1)(2.07)^2 + (5-1)(1.58)^2\right)(6+5-2)^{-1}\right)^{0.50} = 1.87$$

$$t = \frac{(\bar{x}_1 - \bar{x}_2) - D_0}{S_C\sqrt{\frac{1}{n_1} + \frac{1}{n_2}}} = \left((\bar{x}_1 - \bar{x}_2) - D_0\right)\left(S_C(n_1^{-1} + n_2^{-1})^{0.5}\right)^{-1} \text{ where assuming that } \sigma_1^2 = \sigma_2^2$$

Substituting the values in the above equation we get

$$t = \left((22.67 - 26) - 0\right)\left(1.87(6^{-1} + 5^{-1})^{0.5}\right)^{-1} = -2.94; \; t = -2.94$$

STEP 4

Thus, $t_{\text{Calculated}} = t^* = -2.94 < -t_{\alpha,df=n1+n2-2} = t_{\alpha=0.05, \, df=9} = -1.833$
So, reject H_0, because $t^* = -2.94 < t_{\alpha, \, df} = t_{\alpha=0.05, \, df=9} = -1.833$

2. Assume $\sigma_1^2 \neq \sigma_2^2$

STEP 1

Test condition

H_0: $\mu_1 \geq \mu_2$ or rearrange to H_0: $\mu_1 - \mu_2 \geq 0$

H_α: $\mu_1 < \mu_2$ or rearrange to H_α: $\mu_1 - \mu_2 < 0$

STEP 2

Reject H_0 if $t^* < -t_{\alpha,df}$

STEP 3

Assume $\sigma_1^2 \neq \sigma_2^2$. Consequently the combined variance given by
Sample means:

$$\bar{X}_1 = \frac{\sum_{i=1}^{i=n=6} x}{n} = \frac{23 + 25 + 23 + 22 + 24 + 19}{6} = \frac{133}{6} = 22.67$$

$$\bar{X}_2 = \frac{\sum_{i=1}^{i=n=5} x}{n} = \frac{25 + 27 + 24 + 28 + 26}{5} = \frac{130}{5} = 26$$

Sample standard deviations:

$$s_1 = \sqrt{\frac{\sum (x_i - \bar{x})^2}{n-1}} = \sqrt{\frac{(23 - 22.67)^2 + \cdots + (19 - 22.67)^2}{6-1}} = \sqrt{\frac{26.83}{5}} = 2.07$$

$$s_2 = \sqrt{\frac{\sum (x_i - \bar{x})^2}{n-1}} = \sqrt{\frac{(25 - 26)^2 + \cdots + (26 - 26)^2}{5-1}} = \sqrt{\frac{10}{4}} = 1.58$$

The combined variance:

$$s_c^2 = \left((n_1 - 1)s_1^2 + (n_2 - 1)s_2^2 \right)\left(n_1 + n_2 - 2 \right)^{-1};$$

Where the degrees of freedom is

$$df = \frac{\left(s_1^2 n_1^{-1} + s_2^2 n_2^{-1} \right)^2}{\dfrac{\left(s_1^2 n_1^{-1} \right)^2}{n_1 - 1} + \dfrac{\left(s_2^2 n_2^{-1} \right)^2}{n_2 - 1}}$$

$$S_C = \sqrt{\frac{(n_1 - 1)s_1^2 + (n_2 - 1)s_2^2}{n_1 + n_2 - 2}} = \left(\left((n_1 - 1)s_1^2 + (n_2 - 1)s_2^2 \right)\left(n_1 + n_2 - 2 \right)^{-1} \right)^{0.50}$$

$$S_C = \left(\left((6-1)(2.07)^2 + (5-1)(1.58)^2 \right)(6+5-2)^{-1} \right)^{0.50} = 1.87$$

The calculated t-distribution is

$$t = \frac{(\bar{x}_1 - \bar{x}_2) - D_0}{\sqrt{\dfrac{s_1^2}{n_1} + \dfrac{s_2^2}{n_2}}} = \left((\bar{x}_1 - \bar{x}_2) - D_0 \right) \left((s_1^2 n_1^{-1} + s_2^2 n_2^{-1})^{0.5} \right)^{-1}$$

Where *not* assuming that $\sigma_1 = \sigma_2$

$$t = \left((22.67 - 26) - 0 \right) \left(\left((2.07)^2 (6)^{-1} + (1.58)^2 (5)^{-1} \right)^{0.50} \right)^{-1} = -3.02 \; ; t_{Calculated} = -3.02$$

$$df = \frac{\left(s_1^2 n_1^{-1} + s_2^2 n_2^{-1} \right)^2}{\dfrac{\left(s_1^2 n_1^{-1} \right)^2}{n_1 - 1} + \dfrac{\left(s_2^2 n_2^{-1} \right)^2}{n_2 - 1}}$$

$$df = \frac{\left((2.07)^2 (6)^{-1} + (1.58)^2 (5)^{-1} \right)^2}{\dfrac{\left((2.07)^2 (6)^{-1} \right)^2}{6-1} + \dfrac{\left((1.58)^2 (5)^{-1} \right)^2}{5-1}} = 8.96 \approx 9.0$$

STEP 4

Thus, $t_{Calculated} = t^* = -3.02 < -t_{\alpha, \, df} = t_{\alpha = 0.05, \, df \;\; 9} = -1.833$
So, reject H_0, because $t^* = -3.02 < -t_{\alpha, \, df} = t_{\alpha = 0.05, \, df \;\; 9} = -1.833$

Example 9.71

A reliability engineer was requested to evaluate the tensile strength of two different thermoplastics material called homopolymer acetal and copolymer. The related data for both tensile strength of acetals are as follows: Homopolymer acetal $\bar{x} = 118$, $s = 17$, $n = 9$. Copolymer acetal $\bar{x} = 143$, $s = 24$, $n = 16$. The tensile strength for both acetals are normally distrbuted.

(a) Use the *p-value* to find if data supportsthat the mean of tensile strength for both acetals are different. Use significance level of 2% ($a = 0.02$). Do not assume population variances are equal ($\sigma_1^2 \neq \sigma_2^2$).
(b) Construct 99% confidece interval for $\mu_1 - \mu_2$. Do not assume population variances are equal ($\sigma_1^2 \neq \sigma_2^2$).
(c) Perform hypothesis test using combined estimate of standard deviation and compare the two results.

Solution

Two small samples

(a) Assume $\sigma_1^2 \neq \sigma_2^2$

Given: $n_1 = 9$, $n_2 = 16$, $\overline{X}_1 = 118$, $\overline{X}_2 = 143$, $s_1 = 17$, $s_2 = 24$, $\alpha = 0.02$, $\dfrac{\alpha}{2} = .01$

STEP 1

Define the hypothesis
$H_0: \mu_1 - \mu_2 = 0$ rearrange to $H_0: \mu_1 = \mu_2$
$H_a: \mu_1 - \mu_2 \neq 0$ rearrange to $H_a: \mu_1 \neq \mu_2$

STEP 2

Define the test statistics. The proper test for this problem is

$$t = \frac{(\overline{x}_1 - \overline{x}_2) - D_0}{\sqrt{\dfrac{s_1^2}{n_1} + \dfrac{s_2^2}{n_2}}} = \left((\overline{x}_1 - \overline{x}_2) - D_0\right)\left((s_1^2 n_1^{-1} + s_2^2 n_2^{-1})^{0.5}\right)$$

STEP 3

Define the rejection area
Reject H_0, if $|t^*| > t_{\alpha/2, df}$
Reject H_0, if p-value $< \alpha$
Fail to reject H_0, if p-value $\geq \alpha$, state inconclusive if $0.01 < p$-value < 0.10
General rule of thumb: reject H_0, if p-value > 0.01

STEP 4

Calculate the value of the test statistic and carry out the test. The compounded value of t is

$$\overline{X}_1 = \frac{\sum_{i=1}^{i=n} x}{n} = 118, \quad \overline{X}_2 = \frac{\sum_{i=1}^{i=n} x}{n} = 143$$

$$s_1 = \sqrt{\frac{\sum (x_i - \overline{x})^2}{n-1}} = 17, \quad s_2 = \sqrt{\frac{\sum (x_i - \overline{x})^2}{n-1}} = 24$$

$$s_c^2 = \left((n_1 - 1)s_1^2 + (n_2 - 1)s_2^2\right)\left(n_1 + n_2 - 2\right)^{-1}$$

$$t = \left((\bar{x}_1 - \bar{x}_2) - D_0 \right) \left(\left(s_1^2 n_1^{-1} + s_2^2 n_2^{-1} \right)^{1/2} \right)^{-1} ; \text{ where not assuming } \sigma_1 = \sigma_2$$

$$t = \left((118 - 143) - 0 \right) \left(\left((17)^2 (9)^{-1} + (24)^2 (16)^{-1} \right)^{0.50} \right)^{-1} = -3.03; \ |t| = |-3.03| = 3.03$$

$$df = \frac{\left(s_1^2 n_1^{-1} + s_2^2 n_2^{-1} \right)^2}{\dfrac{\left(s_1^2 n_1^{-1} \right)^2}{n_1 - 1} + \dfrac{\left(s_2^2 n_2^{-1} \right)^2}{n_2 - 1}}$$

$$df = \frac{\left((17)^2 (9)^{-1} + (24)^2 (16)^{-1} \right)^2}{\dfrac{\left((17)^2 (9)^{-1} \right)^2}{9 - 1} + \dfrac{\left((24)^2 (16)^{-1} \right)^2}{16 - 1}} = \frac{4439.12}{215.29} = 21.55 \approx 22$$

Thus, $|t_{Calculated}| = |t^*| = |-3.03| = 3.03 > t_{\alpha/2, \, df = n1 + n2 - 2} = t_{\alpha/2 = 0.01, \, df = 22} = 2.508$
p-value = 0.0062 from Excel computer output.
p-value is less than 2(0.01) = 0.02, where p-value is inconclusive if it is more than 0.01 and less than 0.10.

STEP 5

Reject the null hypothesis H_0, because $|t^*| > t_{\alpha/2, \, df}$

$$t_{\alpha/2, 22} = -3.03, \quad \frac{a}{2} = .0031, \quad p = a = .0062$$

$$reject \ H_0 \ 0.0062 < 0.02$$

(b) 99% confidence interval at $\alpha = 0.01$, $\alpha/2 = 0.005$, and assume $\sigma_1^2 \neq \sigma_2^2$

$$(\bar{x}_1 + \bar{x}_2) \pm t_{\frac{\alpha}{2}, df} \left(s_1^2 n_1^{-1} + s_2^2 n_2^{-1} \right)^{0.50}$$

Substituting the values

$$(118 + 143) \pm t_{\frac{\alpha}{2} = 0.005, df = 22} \left((17)^2 (9)^{-1} + (24)^2 (16)^{-1} \right)^{0.50}$$

$$-25 \pm (2.819)(8.253)$$

$$-25 \pm 23.27$$

And finally

$$-48.27 \text{ to } -1.73$$

Thus, interval lower limit = −48.26
And interval higher limit = −1.7370

(c) Hypothesis testing. From problem statement we have the following
(d)

$$n_1 = 9, \quad n_2 = 16, \quad \bar{X}_1 = 118, \quad \bar{X}_2 = 143, \quad s_1 = 17, \quad s_2 = 24, \quad \alpha = 0.02, \quad \frac{\alpha}{2} = 0.01$$

STEP 1

Define the hypothesis
$H_0: \mu_1 - \mu_2 = 0$ *rearrange to* $H_0: \mu_1 = \mu_2$
$H_\alpha: \mu_1 - \mu_2 \neq 0$ *rearrange to* $H_\alpha: \mu_1 \neq \mu_2$

STEP 2

Define the test statistics.

$$t = \left(\left(\bar{x}_1 - \bar{x}_2 \right) - D_0 \right) \left(S_C \left(n_1^{-1} + n_2^{-1} \right)^{1/2} \right)^{-1}$$

STEP 3

The rejection area
Reject null hypothesis H_0, if $|t^*| > t_{\alpha/2, \, df}$

STEP 4

Now, we can calculate the value of the test statistic and carry out the test. The compounded values of t are as follows:

$$\bar{X}_1 = \frac{\sum_{i=1}^{i=n} x_i}{n} = 118, \quad \bar{X}_2 = \frac{\sum_{i=1}^{i=n} x_i}{n} = 143$$

$$s_1 = \sqrt{\frac{\sum (x_i - \bar{x})^2}{n-1}} = 17, \quad s_2 = \sqrt{\frac{\sum (x_i - \bar{x})^2}{n-1}} = 24$$

$$s_c^2 = \left((n_1 - 1)s_1^2 + (n_2 - 1)s_2^2 \right)(n_1 + n_2 - 2)^{-1}; \text{ Where } df = n_1 + n_2 - 2$$

$$S_C = \left(\left((n_1 - 1)s_1^2 + (n_2 - 1)s_2^2 \right)(n_1 + n_2 - 2)^{-1} \right)^{0.50}$$

$$S_C = \left(\left((9-1)(17)^2 + (16-1)(17)^2 \right)(9 + 16 - 2)^{-1} \right)^{0.50} = 21.821$$

$$t = \left(\left(\bar{x}_1 - \bar{x}_2 \right) - D_0 \right) \left(S_C \left(n_1^{-1} + n_2^{-1} \right)^{1/2} \right)^{-1}; \text{ Where assuming } \sigma_1 = \sigma_2$$

$$t = \left((118-143)-0\right)\left(21.821\left(9^{-1}+16^{-1}\right)^{1/2}\right)^{-1} = -2.75;\ |t^*| = |-2.75| = 2.75$$

$$df = n_1 + n_2 - 2 = 9 + 16 - 2 = 23$$

STEP 5

In conclusion, $|t_{\text{Calculated}}| = |t^*| = |-2.75| = 2.75 > t_{\alpha\ 2,df\ \ n1\ \ n2-2} = t_{\alpha/2 = 0.01,\ df\ \ 23} = 2.50$
So *reject* null hypothesis H_0, because $|t^*| = 2.75 > t_{\alpha/2,df} = 2.50$

9.7.7 COMPARING THE VARIANCE OF TWO NORMAL POPULATIONS $(\sigma_1^2 - \sigma_2^2)$ USING INDEPENDENT SAMPLES-F TEST (SMALL SAMPLE SIZE): APPLY THE MECHANICS

Just like in the previous section the focus here will be on the two independent samples from normal population. In estimating the standard deviation of sample one versus sample two, the means of both sample will not be much of a concern, whether they are equal or not. In previous sections when examining the difference of means we used point estimators $\overline{X}_1 - \overline{X}_2$. If $\overline{X}_1 - \overline{X}_2$ was large, we rejected the null hypothesis H_0: $\mu_1 = \mu_2$. When examining variances, we use the ratio of the two sample variances, s_1^2 and s_2^2, in testing a hypothesis test and creating a confidence intervals. Thus, we have

$$F = \frac{s_1^2}{s_2^2}$$

Due to sensitivity of this equation to larger sample sizes, it is advised to test the shape of sample data when using *F*-statistic. *F*-test conditions are listed in the Table 9.31.

Example 9.72

The following data were collected from normally distributed populations.

Sample 1: 61, 82, 74, 22, 16, 25, 36
Sample 2: 62, 55, 84, 46, 45, 65, 58, 70

(a) Find out if the data support enough evidence that standard deviation of second population is less that standard deviation of first population. Use $\alpha = 0.05$ significance level.
(b) What is the *p-value* in part (a).

Solution

Hypothesis testing with significance level $\alpha = 0.05$, $\alpha/2 = 0.005$, assume $\sigma_1^2 \neq \sigma_2^2$
Given: $n_1 = 7$, $n_2 = 8$

Table 9.31. *F*-test (testing conditions): comparing the variance of two normal populations (σ_1^2 and σ_2^2; figure 9.77) using independent samples: small sample

Two-Tailed Test:

H_0: $\sigma_1 - \sigma_2 = 0$ rearrange to H_0: $\sigma_1 = \sigma_2$

H_α: $\sigma_1 - \sigma_2 \neq 0$ rearrange to H_α: $\sigma_1 \neq \sigma_2$

$$F = \frac{s_1^2}{s_2^2} = \frac{\dfrac{\sum (x_i - \bar{x})^2}{n_1 - 1}}{\dfrac{\sum (x_i - \bar{x})^2}{n_2 - 1}}$$

Reject H_0 if $F_{Calculated} = F^* > F_{\alpha/2, df_1 = n_1 - 1, df_2 = n_2 - 1}$ *right tail*

Reject H_0 if $F_{Calculated} = F^* < F_{1-\frac{\alpha}{2}, df_1 = n_1 - 1, df_2 = n_2 - 1}$ *left tail*

One-Tailed Test:

H_0: $\sigma_1 \leq \sigma_2$ H_0: $\sigma_1 \geq \sigma_2$

H_α: $\sigma_1 > \sigma_2$ H_α: $\sigma_1 < \sigma_2$

$$F = \frac{s_1^2}{s_2^2} = \frac{\dfrac{\sum (x_i - \bar{x})^2}{n_1 - 1}}{\dfrac{\sum (x_i - \bar{x})^2}{n_2 - 1}} \qquad F = \frac{s_1^2}{s_2^2} = \frac{\dfrac{\sum (x_i - \bar{x})^2}{n_1 - 1}}{\dfrac{\sum (x_i - \bar{x})^2}{n_2 - 1}}$$

Reject H_0 if $F_{Calculated} = F^* > F_{\alpha, df_1, df_2}$ Reject H_0 if $F_{Calculated} = F^* < F_{1-\alpha, df_1, df_2}$

Where $df_1 = n_1 - 1$ and $df_2 = n_2 - 1$ Where $df_1 = n_1 - 1$ and $df_2 = n_2 - 1$

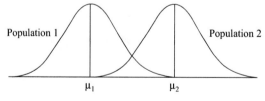

Figure 9.77. Two populations and their means.

STEP 1

Define the hypothesis

H_0: $\sigma_1 \leq \sigma_2$ *rearrange* H_0: $\sigma_1 - \sigma_2 \leq 0$

H_α: $\sigma_1 > \sigma_2$ *rearrange* H_α: $\sigma_1 - \sigma_2 > 0$

STEP 2

Define the test statistics. The proper test for this problem is

$$F = \frac{s_1^2}{s_2^2} = \frac{\dfrac{\sum(x_i - \bar{x})^2}{n_1 - 1}}{\dfrac{\sum(x_i - \bar{x})^2}{n_2 - 1}}$$

STEP 3

Define the rejection area.

$$\text{Reject } H_0 \text{ if } |F^*| > F_{\alpha, \, df_1 = n_1 - 1, \, df_2 = n_2 - 1}$$

$$df_1 = n_1 - 1 = 7 - 1 = 6 \text{ and } df_2 = n_2 - 1 = 8 - 1 = 7$$

STEP 4

Say $\sigma_1^2 \neq \sigma_2^2$

$$\bar{X}_1 = \frac{\sum_{i=1}^{i=n} x_i}{n} = 45.14, \quad \bar{X}_2 = \frac{\sum_{i=1}^{i=n} x_i}{n} = 60.625$$

$$s_1 = \sqrt{\frac{\sum(x_i - \bar{x})^2}{n_1 - 1}} = \sqrt{\frac{4316.854}{7 - 1}} = 26.82, \quad s_2 = \sqrt{\frac{\sum(x_i - \bar{x})^2}{n_2 - 1}} = \sqrt{\frac{1151.875}{8 - 1}} = 12.112$$

(a) Hypothesis for F-testing
(b) One-tailed test: $H_0: \sigma_1 \leq \sigma_2, \quad H_a: \sigma_1 > \sigma_2$

$$F = \frac{s_1^2}{s_2^2} = \frac{\dfrac{\sum(x_i - \bar{x})^2}{n_1 - 1}}{\dfrac{\sum(x_i - \bar{x})^2}{n_2 - 1}} = \frac{\dfrac{4316.854}{7 - 1}}{\dfrac{1151.875}{8 - 1}} = \frac{719.476}{164.554} = 4.372$$

STEP 5

In conclusion: reject H_0 if $F_{Calculated} = F^* = 4.372 > F_{\pm, df_1 = n_1 - 1, df_2 = n_2 - 1} = F_{0.05, 6, 7} = 3.866$

Thus, we reject H_0 since $F^* = 4.372 > F_{0.05, 6, 7} = 3.866$

where $df_1 = n_1 - 1$ and $df_2 = n_2 - 1$

Table 9.32. Excel output for *f*-test two-sample variable 1 and variable 2 at *a* = 0.05

Source	Variable 1	Variable 2
Mean	45.14286	60.625
Observations	7	8
df	6	7
F	4.3722	
$P(F \leq f)$ one-tail	0.037181	*p-value* = 0.037181
F Critical one-tail	3.8659	

(c) *P*-Value: $0.01 < p\text{-}value < 0.05$

The Excel output is presented in the Table 9.32 and the *p-value* = 3.866.

9.7.8   ESTIMATION AND TESTING FOR POPULATION PROPORTIONS

Sample proportion is defined as a statistic parameter that estimates the proportion of a population and consists some property:

$$Point\ estimate\ of\ population\ proportion\ mean = \frac{number\ of\ successes\ in\ the\ sample}{Total\ number\ of\ sample\ size}$$

$$\hat{p} = \frac{x}{n} = xn^{-1} \qquad \text{(Eq. 9.67)}$$

where $\hat{p}$ is point estimate of *p* population proportion *mean*, *x* the number of success of sample components, and *n* is the total sample size.

Just like in the previous cases, the population means μ and standard deviation σ, we used point estimate $\bar{X}$ and "*s*," respectively. Here once again we will use $\hat{p}$ to estimate the population proportion *p*. One must note that the actual values of μ, σ, and *p* will remain unknown unless complete investigation is carried out to obtain the 100% actual outcome.

A population proportion test is a binomial case, which indicates each component of the group is certain attribute, either a success or a failure. Thus, *p* is defined to be a success proportion of the population.

9.7.9   CONFIDENCE INTERVAL FOR A POPULATION PROPORTION:
LARGE SAMPLE

As previously discussed the large sample sizes (i.e., *n* > 30) follow the CLT. Consequently, we can use the CLT concept to $\hat{p}$ and confirm that point estimate is normal random variable for the cases where CLT applies. Further it supports the testing qualification by applying the terms *np* > 5 and $n(1 - p) > 5$ for large sample sizes of $\hat{p}$ distribution. We have

$$\sigma_p = \left(\frac{p(1-p)}{n}\right)^{1/2} \qquad \text{(Eq. 9.68)}$$

Standardizing the outcome we get

$$Z = (\hat{p} - p)\left(\frac{p(1-p)}{n}\right)^{1/2} \qquad \text{(Eq. 9.69)}$$

The confidence interval for population proportion follows the same standard as normal distribution

$$(Point\ Estimate) \pm Z_{\alpha/2} \cdot (Standard\ Deviation\ of\ Point\ Estimator) \qquad \text{(Eq. 9.70)}$$

The standard deviation of point estimate can be expressed as

$$s_{\hat{p}} = \left(\frac{\hat{p}(1-\hat{p})}{n-1}\right)^{1/2} = \left((\hat{p}(1-\hat{p}))(n-1)^{-1}\right)^{1/2} \qquad \text{(Eq. 9.71)}$$

By applying Equation 9.71 in the Equation 9.70, a $(1 - \alpha) \cdot 100\%$ confidence interval for population proportion p with large sample ($n > 30$), $n\hat{p} > 5$ and $n(1 - \hat{p}) > 5$ is

$$\hat{p} \pm Z_{\alpha/2}\left((\hat{p}(1-\hat{p}))(n-1)^{-1}\right)^{1/2} \qquad \text{(Eq. 9.72)}$$

Example 9.73

Suppose a manufacturing company is under investigation. The result of investigation and audit indicates that 14 errors in 250 shipments have occurred this year. What is 90% confidence interval for proportion of all shipments in the current year that includes shipment errors?

Solution

Using Z-Table, $Z_{\alpha/2} = Z_{0.05} = 1.645$. In addition, $\hat{p} = \frac{14}{250} = 0.056$. Hence, the 90% confidence interval for p is

$$\hat{p} \pm Z_{\alpha/2}\left((\hat{p}(1-\hat{p}))(n-1)^{-1}\right)^{1/2}$$

$$0.056 \pm 1.645\left((0.056(1-0.056))(250-1)^{-1}\right)^{1/2}$$

$$0.056 \pm 0.015$$

$$0.041 \text{ to } 0.071$$

Based on the result of sample, there is a 90% confidence that the percentage of shipments containing errors is between 4.1% and 7.1%.

Suppose that in Example 9.73 you want your point estimate, $\hat{p}$, to be within some number around actual value proportion, p. In Example 9.73 the maximum error, E, was 0.015, that is 1.5%. Normally the larger the sample numbers the lower the maximum errors. What if customer's specification requires that we estimate the parameter p no more than within 1% with 90% confidence?

$$E = 0.01 = 1.645 \left(\frac{\hat{p}(1-\hat{p})}{n-1} \right)^{1/2} \qquad \text{(Eq. 9.73)}$$

Thus,

$$0.01 = 1.645 \left(\frac{0.015(1-0.015)}{250-1} \right)^{1/2}$$

Taking the square of both sides of equation leads to

$$n = \frac{(1.645)^2 (0.015)(1-0.015)}{(0.01)^2} + 1 = 400.82 \approx 401$$

By solving Equation 9.73 for n, provides the essential sample size to estimate p with maximum error E and confidence level $(1 - \alpha) \cdot 100\%$:

$$n = \frac{Z_{\alpha/2}^2 \hat{p}(1-\hat{p})}{E^2} + 1 \qquad \text{(Eq. 9.74)}$$

9.7.10 HYPOTHESIS TESTING FOR A POPULATION PROPORTION

How can one statistically reject or accept statement that about 75% of universities will raise their tuition in the next couple of years. Maybe someone can guess at the value of 75% or can research prove to reject or just leave it alone due to insufficient facts conclude that this percentage is less than 75%? We will conduct a hypothesis test as before we did for standard deviation and sample mean.

Small sample testing: Since confidence interval used to carry out hypothesis testing, likewise t-distribution and Z-distribution table, we will use confidence interval for population proportion table to conduct hypothesis testing. Further Table 9.33 will be used to set the conditions.

The values Confidence Interval (C.I.) for population proportion, small samples in Table XI was developed using the following equations in the Excel Microsoft Office application.

$$P_L = \frac{1}{1 + \left(\frac{n-x+1}{x} \right) F_{((1-\alpha/2); 2(n-x)+2; 2x)}}$$

Table 9.33. Hypothesis testing (small samples: $5 \leq n \leq 20$)

Two-Tailed Test: $H_0: p = p_0$ versus $H_a: p \neq p_0$

1. Acquire $(1 - \alpha) \cdot 100\%$ confidence interval from Confidence Interval for Population Proportion-Table, that is (P_L, P_U), using value of x (the number of success)
2. Reject H_0 if p_0 doesn't fall in between (P_L and P_U)
3. Accept H_0 or fail to reject H_0 if p_0 meets the condition as $P_L \leq p_0 \leq P_U$

One-Tailed Test: $H_0: p \leq p_0$ versus $H_a: p > p_0$

1. Acquire $(1 - \alpha) \cdot 100\%$ confidence interval from Confidence Interval for Population Proportion-Table that is (P_L, P_U), using value of x (the number of success)
2. Reject H_0 if $p_0 < p_L$
3. Fail to reject H_0 if $p_0 \geq p_L$

One-Tailed Test: $H_0: p \geq p_0$ versus $H_a: p < p_0$

1. Acquire $(1 - \alpha) \cdot 100\%$ confidence interval from Confidence Interval for Population Proportion-Table in the appendix that is (P_L, P_U), using value of x (the number of success).
2. Reject H_0 if $p_0 > p_L$
3. Fail to reject H_0 if $p_0 \leq p_L$

In one-tailed test, we multiply α by 2 when determining the confidence interval for p from table. For instance, in the case of $\alpha = 0.02$, $2\alpha = 0.10$, and this would give us 90% confidence interval from the table. Notice that some particular binomial tables only can be used for $\alpha = 0.025$ or $\alpha = 0.05$ for one-tailed tests.

$$P_U = \frac{1}{1 + \left(\dfrac{n - x}{x + 1}\right) F_{\left(\frac{\alpha}{2}; 2(n-x)+2; 2x+2\right)}}$$

where n = sample number, x is the sample frequency, and α is confidence level (i.e., 90%, 95%, and 99%). The Excel version is

$P_L = 1/(1 + (\$B\$1 - \$C4 + 1)/\$C4 * F.INV(1 - \$E\$1/2,2 * (\$B\$1 - \$C4) + 2,2 * \$C4)) = 0.012$

where Cell B1 = n, C4 = x, E1= α,
    For example: B1 = n = 20, C4 = x = 2, E1= α = 0.05
    Likewise for P_U

$P_U = 1/(1 + (\$B\$1 - \$C4)/(\$C4 + 1) * F.INV(\$E\$1/2,2 * (\$B\$1 - \$C4),2 * \$C4 + 2)) = 0.317$

where Cell B1 = n, C4 = x, E1= α,
    For example: B1 = n = 20, C4 = x = 2, E1= α = 0.05

Example 9.74

Small samples: Suppose that a company is interested in purchasing chocolates as a Christmas present for their staff if the bakery claims that the proportion, p, of all the chocolates will be tasty more than 0.80 (80%). Does the data support this claim with $\alpha = 0.025$?

Solution

The claim under investigation falls into the alternative hypothesis H_α. The hypothesis testing conditions are

$$\text{One-Tailed Test: } H_0: p \le p_0 = 0.80 \text{ versus } H_\alpha: p > p_0 = 0.80$$

We noticed that in the samples of 18 chocolate boxes distribution, 16 people liked it (success). Because $\alpha = 0.025$, we multiply this by 2 and get $2\alpha = 0.05$ and refer to table in the appendix for 95% ($\alpha = 0.05$) confidence interval for p when $n = 18$, $x = 16$. This is

$$P_L = 0.653 \text{ to } P_U = 0.986 \text{ or } (P_L, P_U) = (0.653 \text{ to } 0.986)$$

If $p_0 = 0.80$ fell to the left of P_L, then we would reject H_0, but now $p_0 = 0.80$ is greater than $P_L = 0.653$, thus we fail to reject H_0.

Large sample testing: As before we carried out five steps procedure for large samples ($n > 30$), here as well the same concepts will be applied in testing null hypothesis H_0 against alternate hypothesis H_α. The proper test statistic equation (Table 9.34) for sample np_0 and $n(1 - p_0)$ both being greater than 5. The reject area can be determined by computing test statistic "Z" in the two and one tailed test and the boundary value of p in the null hypothesis H_0.

Table 9.34. Hypothesis testing (large sample; np_0 and $n(1 - p_0)$ both > 5)

Two-Tailed Test: $H_0: p = p_0$ versus $H_\alpha: p = p_0$

Reject H_0 if $|Z| > Z_{\alpha/2}$

Test statistic

$$Z = \frac{\hat{p} - p}{\sqrt{\dfrac{p(1-p)}{n-1}}} \quad \text{or} \quad Z = \frac{\hat{p} - p_0}{\sqrt{\dfrac{p_0(1-p_0)}{n-1}}}; \quad p = mean; \; \hat{p} = \frac{x}{n}$$

One-Tailed Test

$H_0: p \le p_0$	$H_0: p \ge p_0$
$H_\alpha: p > p_0$	$H_\alpha: p < p_0$
Reject H_0 if $Z_{\text{Calculated}} = Z^* > Z_\alpha$	Reject H_0 if $Z_{\text{Calculated}} = Z^* < -Z_\alpha$

Example 9.75

Estimation and Confidence Interval (CI) for Population Proportion: An independent survey was carried out in the West Tech show. The survey showed that 40% of the male who signed up for show also brought a guest with them. On the other hand, 30% of the female who signed up for the show brought a guest.

1. Develop a 90% confidence interval for the percent of male that bring a guest having a sample size of 8.
2. Develop a 90% confidence interval for the percent of female that bring a guest having a sample size of 8.
3. Construct 90% confidence interval for part (1) with a sample size of 85. Compare to C.I. with the one obtained in (1).
4. Construct 90% confidence interval for part (2) with a sample size of 85. Compare to C.I. with the one obtained in (2).
5. Find the sample size for part (1) within 2% of maximum error E.

Solution

Calculations are as follows:

1. 90% confidence interval at $n = 8$, $\hat{p} = 40\% = 0.40$. For percent of male,

$$\hat{p} = \frac{x}{n} = \frac{Sample\,observation\,having\,specified\,attribute}{Sample\,Size}$$

Or, $x = n\hat{p} = (8)(0.35) = 2.80$

$$\hat{p} \pm Z_{\frac{\alpha}{2}}\sqrt{\frac{\hat{p}(1-\hat{p})}{n-1}} = 0.40 \pm 1.645\sqrt{\frac{0.40(1-0.40)}{8-1}} = 0.095\ to\ 0.705$$

Confidence Interval (C.I.) – width = 0.705 – 0.095 = 0.61

2. For percent of female

$$\hat{p} \pm Z_{\frac{\alpha}{2}}\sqrt{\frac{\hat{p}(1-\hat{p})}{n-1}} = 0.30 \pm 1.645\sqrt{\frac{0.30(1-0.30)}{8-1}} = 0.015\ to\ 0.585$$

CI – width = 0.585 – 0.015 = 0.57

3. Apply sample numbers $n = 85$ in part (1) and compare the CI (Figure 9.73).

$$\hat{p} \pm Z_{\frac{\alpha}{2}}\sqrt{\frac{\hat{p}(1-\hat{p})}{n-1}} = 0.40 \pm 1.645\sqrt{\frac{0.40(1-0.40)}{85-1}} = 0.312\ to\ 0.488$$

CI – width = 0.488 – 0.312 = 0.176

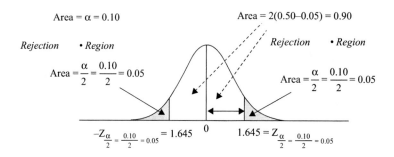

Figure 9.78. Rejection area for population proportion.

4. Apply sample numbers $n = 85$ in part (2) and compare the CI (Figure 9.78).

$$\hat{p} \pm Z_{\frac{\alpha}{2}}\sqrt{\frac{\hat{p}(1-\hat{p})}{n-1}} = 0.30 \pm 1.645\sqrt{\frac{0.30(1-0.30)}{85-1}} = 0.218 \; to \; 0.382$$

$$CI - width = 0.382 - 0.218 = 0.164$$

5. $\quad n = \dfrac{Z_{a/2}^2 \hat{p}(1-\hat{p})}{E^2} + 1 = \dfrac{(1.645)^2(0.40)(1-0.40)}{(0.02)^2} + 1 = 1624.62 \approx 1625$

Example 9.76

Hypothesis Testing for a Population Proportion

In each case as listed below, test the hypothesis $H_0: p = p_0$ against $H_a: p \neq p_0$ for random sample of size n.

(a) $\hat{p} = 0.35$, $\quad p_0 = 0.20$, $\quad \alpha = 0.05$, $\quad n = 10$
(b) $\hat{p} = 0.45$, $\quad p_0 = 0.25$, $\quad \alpha = 0.10$, $\quad n = 20$
(c) $\hat{p} = 0.95$, $\quad p_0 = 0.80$, $\quad \alpha = 0.01$, $\quad n = 50$
(d) $\hat{p} = 0.06$, $\quad p_0 = 0.15$, $\quad \alpha = 0.05$, $\quad n = 100$
(e) $\hat{p} = 0.80$, $\quad p_0 = 0.85$, $\quad \alpha = 0.10$, $\quad n = 300$

Solution

We will solve the problem based on the following three steps:

$\quad H_0: p = p_0 \quad$ versus $\quad H_a: p \neq p_0$

STEP 1

Find Confidence interval C.I. (P_L, P_U) from table in the appendix

STEP 2

Reject H_0 if p_0 does not lie between P_L, and P_U

STEP 3

Fail to reject H_0 if $P_L \leq p_0 \leq P_U$

(a) Given from problem statement $\hat{p} = 0.35$, $p_0 = 0.20$, $\alpha = 0.05$, $n = 10$, 95% confidence interval.

$n = 10$ falls in the testing condition of "$5 \leq$ Small sample ≤ 20." From Equation 9.67 we have

$$\hat{p} = \frac{x}{n}; x = \hat{p}n = (0.35)(10) = 3.50 \approx 4.0$$

Thus, confidence interval for population proportion, small sample from table at

$x = 4$, $\alpha = 0.05$, $n = 10$, equal to $P_L = 0.122$, $P_U = 0.738$; $(P_L, P_U) = (0.122, 0.738)$

$p_0 = 0.20$ lie in between; $(P_L, P_U) = (0.122, 0.738)$

$$\hat{p} \pm Z_{\frac{\alpha}{2}} \sqrt{\frac{\hat{p}(1-\hat{p})}{n-1}} = 0.35 \pm 1.96 \sqrt{\frac{0.35(1-0.35)}{10-1}} = 0.03838 \; to \; 0.6616$$

Fail to reject H_0

$$CI - Width = 0.6616 - 0.03838 = 0.6232$$

(b) $\hat{p} = 0.45$, $p_0 = 0.25$, $\alpha = 0.10$, $n = 20$, 90% confidence interval
$n = 20$ falls in the testing condition: $5 \leq$ Small sample ≤ 20

$$\hat{p} = \frac{x}{n}; x = \hat{p}n = (0.45)(20) = 9.0$$

Thus, confidence interval for population proportion, small sample from table at

$x = 9$, $\alpha = 0.10$, $n = 20$, equal to $P_L = 0.259$, $P_U = 0.653$; $(P_L, P_U) = (0.259, 0.653)$

$p_0 = 0.25$ lie in between; $(P_L, P_U) = (0.259, 0.653)$

$$\hat{p} \pm Z_{\frac{\alpha}{2}} \sqrt{\frac{\hat{p}(1-\hat{p})}{n-1}} = 0.45 \pm 1.645 \sqrt{\frac{0.45(1-0.45)}{20-1}} = 0.02623 \; to \; 0.6377$$

Fail to reject H_0

$$CI - Width = 0.6377 - 0.2623 = 0.3755$$

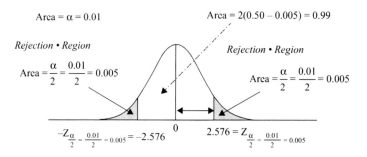

Figure 9.79. Rejection area for population proportion.

(c) $\hat{p} = 0.95$, $p_0 = 0.80$, $\alpha = 0.01$, $n = 50$, 99% confidence interval.

$n = 50$ falls in the testing condition: Large sample; np_0 and $n(1 - p_0)$ both > 5

$np_0 = (50)(0.80) = 40$ and $n(1 - p) = 50 (1 - 0.80) = 10$ both > 5

Two-tailed test:

$H_0: p = p_0$ versus $H_a: p \neq p_0$

Reject H_0 if $|Z^*| > Z_{\alpha/2}$

$$Z = \frac{\hat{p} - p_0}{\sqrt{\dfrac{p_0(1 - p_0)}{n - 1}}} = \frac{0.95 - 0.80}{\sqrt{\dfrac{0.80(1 - 0.80)}{50 - 1}}} = \frac{0.15}{0.05714} = 2.625$$

Since $|Z_{\text{Calculated}}| = |Z^*| = |2.625| = 2.625 \gg Z_{\alpha/2 = 0.01/2 = 0.005} = 2.576$, so reject H_0 (Figure 9.79).

(d) $\hat{p} = 0.06$, $p_0 = 0.15$, $\alpha = 0.05$, $n = 100$, 95% confidence interval

$n = 100$ falls in the testing condition: Large sample; np_0 and $n(1 - p_0)$ both > 5

$np_0 = (100)(0.15) = 15$ and $n(1 - p_0) = 100(1 - 0.15) = 75$ both > 5

Two-tailed test:

$H_0: p = p_0$ versus $H_a: p \neq p_0$

Reject H_0 if $|Z^*| > Z_{\alpha/2}$

$$Z = \frac{\hat{p} - p_0}{\sqrt{\dfrac{p_0(1 - p_0)}{n - 1}}} = \frac{0.06 - 0.15}{\sqrt{\dfrac{0.15(1 - 0.15)}{100 - 1}}} = \frac{-0.09}{0.03589} = -2.514$$

Since $|Z_{\text{Calculated}}| = |Z^*| = |-2.514| = 2.514 \gg Z_{\alpha/2 = 0.01/2 = 0.005} = 1.960$, so reject H_0 (Figure 9.80)

(e) $\qquad\qquad \hat{p} = 0.80$, $p_0 = 0.85$, $\alpha = 0.10$, $n = 300$

$$Z = \frac{\hat{p} - p_0}{\sqrt{\dfrac{p_0(1 - p_0)}{n - 1}}} = \frac{0.80 - 0.85}{\sqrt{\dfrac{0.85(1 - 0.85)}{300 - 1}}} = -2.42$$

Since $|Z_{\text{Calculated}}| = |Z^*| = |-2.42| = 2.42 \gg Z_{\alpha/2 = 0.1/2 = 0.05} = 1.645$, so reject H_0 (Figure 9.81).

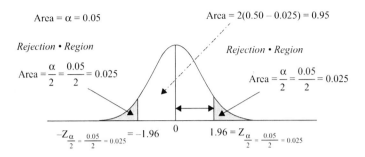

Figure 9.80. Rejection area for population proportion.

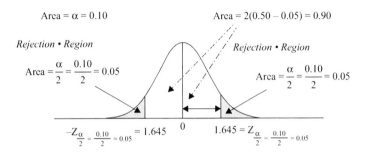

Figure 9.81. Rejection area for population proportion.

9.7.11 COMPARING POPULATION PROPORTION: TWO LARGE INDEPENDENT SAMPLES

Let's take a look at a few examples as

Is the university test scores in the Unites States are higher than China?
Are there more athletic people in the California than New York?

These questions entail two proportions from two different populations. Again as we covered before in comparing two population mean and standard deviation, here as well we will do the same or what is identical to single population proportion with only one difference that there two of everything. In this section, we assume all the independent populations in our analysis. Thus, the point estimate for p_1 (population 1) and p_2 (population 2) are as follows:

$$\hat{p}_1 = \frac{x_1}{n_1} \ and \ \hat{p}_2 = \frac{x_2}{n_2}$$

Confidence interval for $p_1 - p_2$: For the two populations the confidence interval interest will be the difference of $p_1 - p_2$. So for large independent samples (where np_0 and $n(1 - p_0)$ becomes $n_1 \hat{p}_1, n_1(1 - \hat{p}_1)$ and $n_2 \hat{p}_2, n_2(1 - \hat{p}_2)$ are each larger than 5) the estimators will follow $\hat{p}_1 - \hat{p}_2$ nearly a normal random variable with

$$\text{Mean} = p_1 - p_2 \qquad \text{(Eq. 9.75)}$$

and the standard deviation equal to

$$\sigma_{p_1-p_2} = \left(p_1\left(1-p_1\right)n_1^{-1} + p_2\left(1-p_2\right)n_2^{-1} \right)^{1/2} \qquad \text{(Eq. 9.76)}$$

And the standard deviation of estimate $\hat{p}_1 - \hat{p}_2$ is

$$s_{\hat{p}_1-\hat{p}_2} = \left(\hat{p}_1\left(1-\hat{p}_1\right)\left(n_1-1\right)^{-1} + \hat{p}_2\left(1-\hat{p}_2\right)\left(n_2-1\right)^{-1} \right)^{1/2} \qquad \text{(Eq. 9.77)}$$

The outcome of confidence interval for $p_1 - p_2$ by applying large independent samples ($n_1\hat{p}_1$, $n_1\left(1-\hat{p}_1\right)$, and $n_2\hat{p}_2$, $n_2\left(1-\hat{p}_2\right)$ each larger than 5)

$$(\hat{p}_1 - \hat{p}_2) \pm Z_{\frac{\alpha}{2}} \left(\hat{p}_1\left(1-\hat{p}_1\right)\left(n_1-1\right)^{-1} + \hat{p}_2\left(1-\hat{p}_2\right)\left(n_2-1\right)^{-1} \right)^{1/2} \qquad \text{(Eq. 9.78)}$$

Example 9.77

A well-known university hospital admission reported that a sample 96 boys and 123 girls between ages 19 and 22 years admitted for flu cases in one week. After treatments 18 of the boys and 60 of the girls stayed overnight for observation. We assume that the girls form a random sample from a population of similar girls and the same for boys. Develop a 99 percent (99%) confidence interval between the two populations.

Solution

Given $n_1 = 123$ and $n_2 = 96$

$$\hat{p}_1 = \frac{x_1}{n_1} = \frac{60}{123} = 0.4878 \text{ and } \hat{p}_2 = \frac{x_2}{n_2} = \frac{18}{96} = 0.1875$$

$$(\hat{p}_1 - \hat{p}_2) \pm Z_{\frac{\alpha}{2}} \left(\hat{p}_1\left(1-\hat{p}_1\right)\left(n_1-1\right)^{-1} + \hat{p}_2\left(1-\hat{p}_2\right)\left(n_2-1\right)^{-1} \right)^{1/2}$$

Or,

$$(\hat{p}_1 - \hat{p}_2) \pm Z_{\frac{\alpha}{2}} \sqrt{\frac{\hat{p}_1\left(1-\hat{p}_1\right)}{\left(n_1-1\right)} + \frac{\hat{p}_2\left(1-\hat{p}_2\right)}{\left(n_2-1\right)}}$$

$$(0.4878 - 0.1875) \pm 2.575 \sqrt{\frac{0.4878\left(1-0.4878\right)}{\left(123-1\right)} + \frac{0.1875\left(1-0.1875\right)}{\left(96-1\right)}}$$

$$0.3003 \pm 2.757\ (0.0604)$$

$$0.3003 \pm 0.1666$$

$$0.1337 \text{ to } 0.4334$$

The difference between two populations, $p_1 - p_2$ is 0.3003. We are 99% confident that the true value of the difference between the two population proportions lies between 0.1337 and 0.4334.

Sample sizes: Two population proportion sample sizes can be determined by applying the following relations:

$$n_1 = \frac{Z_{\alpha/2}^2 (a+b)}{E^2} + 1 \qquad \text{(Eq. 9.79)}$$

$$n_2 = \frac{Z_{\alpha/2}^2 (c+b)}{E^2} + 1 \qquad \text{(Eq. 9.80)}$$

where

$$E = Z_{\frac{\alpha}{2}} \sqrt{\frac{\hat{p}_1 (1-\hat{p}_1)}{(n_1 - 1)} + \frac{\hat{p}_2 (1-\hat{p}_2)}{(n_2 - 1)}} \qquad \text{(Eq. 9.81)}$$

$$a = \hat{p}_1 (1 - \hat{p}_1) \qquad \text{(Eq. 9.82)}$$

$$b = \sqrt{\hat{p}_1 \hat{p}_2 (1 - \hat{p}_1)(1 - \hat{p}_2)} \qquad \text{(Eq. 9.83)}$$

$$c = \hat{p}_2 (1 - \hat{p}_2) \qquad \text{(Eq. 9.84)}$$

Hypothesis testing for p_1 and p_2: We will use the same format of comparing two normal population means (μ_1 and μ_2) using two large independent samples with the *test statistic p_1 versus p_2* (large independent samples $n_1 \hat{p}_1$, $n_1 (1 - \hat{p}_1)$, and $n_2 \hat{p}_2$, $n_2 (1 - \hat{p}_2)$ each larger than 5). Table 9.35 illustrates all the testing conditions and the following examples demonstrate the two large independent samples for population proportions.

Example 9.78

Comparing Two Populations Proportion (Large, Independent Samples)
A quality engineer wants to find if there is a difference in the number of nonconforming soda bottle caps produced by two different injection molding machines. A bottle cap is nonconforming if the cap is not soda gas (CO_2) tight. To research this, the engineer samples 400 caps from each molding machine. The results showed that one process produced 10 nonconforming and the other produced 15 nonconforming.

(a) Based on these data, can the quality engineer conclude that there is a significance difference in the nonconforming bottle cap produced by injection molding processes? Use $a = 0.10$

(b) Find *p-value* to back up your result. Use $\alpha = 0.10$

Solution

(a) $H_0: p_1 = p_2$ versus $\quad H_a: p_1 \neq p_2$

Table 9.35. Formula for hypothesis testing condition: population proportion large independent: samples $(\hat{p}, \bar{p}, p_0)$

Two Tailed Test: $H_0 : p_1 = p_2$ *versus* $H_a : p_1 \neq p_2$

One-Tailed Test: $H_0 : p_1 \leq p_2$ *versus* $H_a : p_1 > p_2$

$\qquad\qquad\quad H_0 : p_1 \geq p_2$ *versus* $H_a : p_1 < p_2$

Test statistic:

$n_1\hat{p}_1, n_1(1-\hat{p}_1)$, and $n_2\hat{p}_2, n_2(1-\hat{p}_2)$ each larger than 5

$$Z = \frac{\hat{p}_1 - \hat{p}_2}{\sqrt{\bar{p}(1-\bar{p})\left(\frac{1}{n_1}+\frac{1}{n_2}\right)}} = (\hat{p}_1 - \hat{p}_2)\left(\bar{p}(1-\bar{p})\left(\frac{1}{n_1}+\frac{1}{n_2}\right)\right)^{-1/2}$$

If $n_1 = n_2$ then

$$Z = (\hat{p}_1 - \hat{p}_2)\left(2\bar{p}(1-\bar{p})n_1^{-1}\right)^{-1/2}$$

$$\hat{p}_1 = \frac{x_1}{n_1} \qquad \hat{p}_2 = \frac{x_2}{n_2} \qquad \bar{p} = \frac{x_1+x_2}{n_1+n_2}$$

Rejection area:

** For Two-Tailed Test **: $H_0 : p_1 = p_2$ *versus* $H_a : p_1 \neq p_2$ *reject* H_0 *if* $|Z| > Z_{\alpha/2}$

** For One-Tailed Test **: $H_0 : p_1 \leq p_2$ *versus* $H_a : p_1 > p_2$ *reject* H_0 *if* $|Z| > Z_\alpha$

$\qquad\qquad\qquad\qquad\quad H_0 : p_1 \geq p_2$ *versus* $H_a : p_1 < p_2$ *reject* H_0 *if* $|Z| < -Z_\alpha$

$x_1 = 10$, $x_2 = 15$, $\alpha = 0.10$, $n = 400$, 90% confidence interval

$n = 400$ falls in the testing condition: Large sample; np_0 and $n(1-p)$ both > 5

STEP 1

Define the hypothesis: Two-tailed test

$\qquad H_0: p_1 = p_2 \qquad$ versus $\qquad H_a: p_1 \neq p_2$

STEP 2

Define the test statistics.

$$Z = \frac{\hat{p}_1 - \hat{p}_2}{\sqrt{\frac{\bar{p}(1-\bar{p})}{n_1}+\frac{\bar{p}(1-\bar{p})}{n_2}}}$$

STEP 3

Define the rejection area:
   Reject H_0 if $|Z^*| > Z_{\alpha/2}$

STEP 4

Calculate the value of the test statistic and carry out the test.

$$\hat{p}_1 = \frac{x_1}{n_1} = \frac{10}{400} = 0.025$$

$$\hat{p}_2 = \frac{x_2}{n_2} = \frac{15}{400} = 0.0375$$

Since comparing the two populations the both percentile should apply: so

np_0 and $n(1-p)$ both should be more than 5 (>5)

$(400)(0.025) = 10$ and $400(1-0.025) = 390$ both > 5

$(400)(0.0375) = 15$ and $400(1-0.0375) = 385$ both > 5

$$\bar{p} = \frac{x_1+x_2}{n_1+n_2} = \frac{10+15}{400+400} = \frac{25}{800} = 0.03125 \ or \ (3.125\%)$$

$$Z = \frac{\hat{p}_1 - \hat{p}_2}{\sqrt{\dfrac{\bar{p}(1-\bar{p})}{n_1} + \dfrac{\bar{p}(1-\bar{p})}{n_2}}} = \frac{\hat{p}_1 - \hat{p}_2}{\sqrt{\bar{p}(1-\bar{p})\left(\dfrac{1}{n_1} + \dfrac{1}{n_2}\right)}} = \left(\hat{p}_1 - \hat{p}_2\right)\left(\bar{p}(1-\bar{p})\left(\dfrac{1}{n_1} + \dfrac{1}{n_2}\right)\right)^{-1/2}$$

Thus, we have

$$Z = \frac{\hat{p}_1 - \hat{p}_2}{\sqrt{\bar{p}(1-\bar{p})\left(\dfrac{1}{n_1} + \dfrac{1}{n_2}\right)}} = \frac{0.025 - 0.0375}{\sqrt{0.03125(1-0.03125)\left(\dfrac{1}{400} + \dfrac{1}{400}\right)}} = -1.02$$

STEP 5

Conclusion

   Since $|Z_{\text{Calculated}}| = |Z^*| = |-1.02| = 1.02 > Z_{\alpha/2 = 0.1/2 = 0.05} = 1.645$, Fail to reject H_0

(b) *p-value* = 2(area outside of Z computed = Z^*), we will use $\alpha = 0.10$.
   Thus, at $Z^* = Z$ computed = 1.02 the area = 0.3461
   p-value = 2(area outside of Z computed = Z^*) = 2(0.50 − 0.3461) = 0.3078

Example 9.79

An agriculture magazine believed that the proportion of families with more than four cows in farm city 1 was greater than the probability of families with more than four cows in farm city 2. The magazine collected random samples of size $n_1 = 190$ and $n_2 = 165$ from city 1 and city 2, respectively. The number of families with more than four cows was $x_1 = 84$ from city 1 and $x_2 = 71$ from city 2.

(a) Using these data, can one conclude that the proportion of families with more than four cows is higher in city1 than in city 2? Apply a significance level of $a = 0.01$.

(b) Determine the *p-value*. Find if one, can reject null hypothesis using *p-value* at $\alpha = 0.05$?

Solution

(a) Given: $n_1 = 190$, $x_1 = 84$, and $n_2 = 165$, $x_2 = 71$

STEP 1

Define the hypothesis: One-tailed test

$$H_0: p_1 \leq p_2 \qquad \text{versus} \qquad H_a: p_1 > p_2$$

STEP 2

The "test statistics" is

$$Z = Z^* = \frac{\hat{p}_1 - \hat{p}_2}{\sqrt{\dfrac{\overline{p}(1-\overline{p})}{n_1} + \dfrac{\overline{p}(1-\overline{p})}{n_2}}} = \frac{\hat{p}_1 - \hat{p}_2}{\sqrt{\overline{p}(1-\overline{p})\left(\dfrac{1}{n_1} + \dfrac{1}{n_2}\right)}} = (\hat{p}_1 - \hat{p}_2)\left(\overline{p}(1-\overline{p})\left(\dfrac{1}{n_1} + \dfrac{1}{n_2}\right)\right)^{-1/2}$$

STEP 3

The "rejection" area is
From Z-table at $\alpha = 0.01$, The $Z_{\alpha = 0.10} = 2.326$
Thus, we reject H_0 if $Z^* > Z_{\alpha = 0.10} = 2.326$

STEP 4

Now, we can calculate the value of the "test statistic" and carry out the test.
Since, $n_1 = 190$, $x_1 = 84$, and $n_2 = 165$, $x_2 = 71$

$$\hat{p}_1 = \frac{x_1}{n_1} = \frac{84}{190} = 0.442$$

$$\hat{p}_2 = \frac{x_2}{n_2} = \frac{71}{165} = 0.4303$$

$$\overline{p} = \frac{x_1 + x_2}{n_1 + n_2} = \frac{84 + 71}{190 + 165} = \frac{155}{355} = 0.4366 \ or \ (43.66\%)$$

$$Z = \frac{\hat{p}_1 - \hat{p}_2}{\sqrt{\overline{p}(1-\overline{p})\left(\frac{1}{n_1} + \frac{1}{n_2}\right)}} = \frac{0.442 - 0.4303}{\sqrt{0.4366(1-0.4366)\left(\frac{1}{190} + \frac{1}{165}\right)}} = 0.2217$$

STEP 5

Since, $Z_{calculated} = Z^* = 0.2217 > Z_{\alpha = 0.10} = 2.326$, thus, the conclusion is we fail to reject H_0.

(b) *p-value* = area outside of Z computed = Z^*, use $\alpha = 0.05$ at $Z^* = 0.2217$ the area = 0.0909
p-value = $(0.50 – 0.0909) = 0.4091$, since *p-value* = $0.4091 > \alpha = 0.05$ fail to reject H_0

9.8 ADVANCED ANALYSIS OF VARIANCE (ANOVA)

Analysis of variance (ANOVA) is a methodology for hypothesis testing about the mean. It is a frequently used science of statistical inference technique for analyzing the output of design of experiment output data or transactional data in the form of mathematics. ANOVA is useful in comparing two, three, or multiple means. So, the hypothesis testing becomes

H_0: $\mu_1 = \mu_2 = \mu_3 = \cdots = \mu_n$
H_a: not all the means (μ's) are equal.

9.8.1 ONE-WAY ANALYSIS OF VARIANCE

The one-way (also called the one-factor) ANOVA simply describes that a hypothesis test can be applied in determining if the mean of one or more populations is different or does not depend on one independent factor. The requirement steps are as follows:

1. Samples are randomly and independently drawn from each population.
2. Assumptions: Samples from each population follow normal distribution with the same standard distribution (σ) and common variance (σ^2).
3. H_0: $\mu_1 = \mu_2 = \mu_3 = \mu_4 = \cdots = \mu_{m-2} = \mu_{m-1} = \mu_m$
   H_a : Not all means are equal. This is always a right tail test.
   Note: H_a defines that at least two of the means (μ's) are different or unequal.
       Examples: Accident rates for highway 405, 502, and 207. Expected mileage for four different brands of tires.
4. Determine the difference among the means of two, three, or more populations μ_1, μ_2, $\mu_3, \cdots \mu_m$

Table 9.36. One-factor ANOVA

Source of inconsistency (variation)	Degrees of freedom (*df*)	Sum of squares (*SS*)	Mean squares (*MS*)	Statistical test (*F*)
Factor	$k-1$	SS (*factor*)	$MS(factor) = \dfrac{SS(factor)}{k-1}$	$F = \dfrac{MS(factor)}{MS(error)}$
Error	$n-k$	SS(*error*)	$MS(error) = \dfrac{SS(error)}{n-k}$	
Total	$n-1$	SS(total)		

5. Apply α-*value* (significance level).
6. Establish One-factor ANOVA Table 9.36.

Let's consider the following case throughout our discussion on "Hypothesis of One-Way ANOVA."

Suppose Grocery Store "Xeon" has branches in the north, south, east, and west coast of United States. Recently the VP of sales questioned whether the company's mean price (per pound) on medium apple differed by region. Survey showed the following data:

	East	West	North	South	
1	$4.10	$6.90	$4.6	$12.50	
2	5.90	9.10	11.40	7.50	
3	10.45	13.00	6.15	6.25	
4	11.55	7.90	7.85	8.75	
5	5.25	9.10	4.30	11.15	
6	7.75	13.40	8.70	10.25	
7	4.78	7.60	10.20	6.40	
8	6.22	5.00	10.80	9.20	
					Grand Mean ($\bar{\bar{X}}$)
Mean ($\bar{X}$)	$7.00	$9.00	$8.00	$9.00	$8.25
Variance (σ^2)	$7.341	$8.423	$7.632	$5.016	

Construct ANOVA Table.

Solution

The detailed calculations are as follows.

The ANOVA assumptions

$$H_0: \mu_1 = \mu_2 = \mu_3 = \mu_4 = \cdots = \mu_{m-2} = \mu_{m-1} = \mu_m$$

H_a: At least two population means are different. This is always a right tail test.

m = Number of populations

The above test is visualized in Figures 9.82 through 9.84.

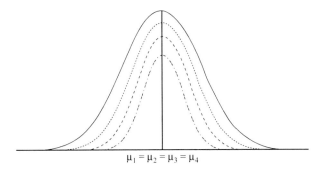

$$\mu_1 = \mu_2 = \mu_3 = \mu_4$$

Figure 9.82. All the means are the same.

Or,

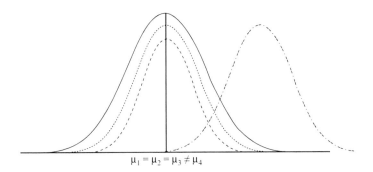

$$\mu_1 = \mu_2 = \mu_3 \neq \mu_4$$

Figure 9.83. At least one mean is different.

Or,

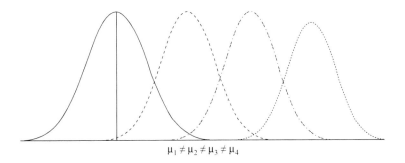

$$\mu_1 \neq \mu_2 \neq \mu_3 \neq \mu_4$$

Figure 9.84. All the means are different.

Partitioning the Variations The variation can be split into two parts:
First Sum of Squares of Total (SST)

$$SST = SSB + SSW \qquad \text{(Eq. 9.83)}$$

where SST (Total Sum of Squares (or total variation)), the variation of the individual data values across the various factor levels, SSB (Sum of Squares Between (or *factors*)), variation between the factor (X_i) sample means (μ), and SSW (Sum of Squares Within (or *error*)), variation that exists within a particular factor (X_i) level.

Thus, total sum of the squares can be calculated by

$$SST = \sum_{i=1}^{k}\sum_{j=1}^{n_i}\left(x_{ij} - \overline{\overline{x}}\right)^2 = \left(x_{11} - \overline{\overline{x}}\right)^2 + \left(x_{12} + \overline{\overline{x}}\right)^2 + \cdots + \left(x_{kn_k} + \overline{\overline{x}}\right)^2 \qquad \text{(Eq. 9.84)}$$

where SST is total sum of squares, k is number of populations (level of treatments), n_i sample size from population i, x_{ij} is j^{th} measurement from population i, and $\overline{\overline{x}}$ is grand mean (mean of all data values) mean of all group means, and n_k = Number of observations in the kth sample.

In the case of "Xeon" we have

$$SST = \sum_{i=1}^{k}\sum_{j=1}^{n_i}\left(x_{ij} - \overline{\overline{x}}\right)^2$$

$$= \left(x_{11} - \overline{\overline{x}}\right)^2 + \left(x_{12} + \overline{\overline{x}}\right)^2 + \cdots + \left(x_{21} + \overline{\overline{x}}\right)^2 + \left(x_{22} + \overline{\overline{x}}\right)^2 + \cdots + \left(x_{31} + \overline{\overline{x}}\right)^2 + \left(x_{32} + \overline{\overline{x}}\right)^2$$

$$+ \cdots + \left(x_{41} + \overline{\overline{x}}\right)^2 + \left(x_{42} + \overline{\overline{x}}\right)^2 + \cdots + \left(x_{48} + \overline{\overline{x}}\right)^2$$

$$= (4.11 - 8.25)^2 + (5.9 - 8.25)^2 + \cdots + (6.9 - 8.25)^2 + (9.10 - 8.25)^2 + \cdots + (4.60 - 8.25)^2$$
$$+ (11.40 - 8.25)^2 + \cdots (12.50 - 8.25)^2 + (7.50 - 8.25)^2 + \cdots + (9.20 - 8.25)^2$$

$$= 220.88$$

Or, it can be calculated using the following equation

$$SS\left(total\right) = \sum_{i=1}^{n} x^2 - \frac{T^2}{n} \qquad \text{(Eq. 9.85)}$$

$$\sum_{i=1}^{k}\sum_{j=1}^{n_i} x_{ij}^2 = x_{11}^2 + x_{12}^2 + \cdots + x_{21}^2 + x_{22}^2 + \cdots + x_{31}^2 + x_{32}^2 + \cdots + x_{41}^2 + x_{42}^2 + \cdots + x_{48}^2$$

$$= 4.10^2 + 5.90^2 + \cdots + 6.90^2 + 9.10^2 + \cdots + 4.60^2 + 11.40^2 + \cdots + 12.50^2 + 7.50^2 + \cdots$$
$$+ 9.20^2$$

$$= 2398.88$$

$$T = T_1 + T_2 + \cdots + T_{m-1} + T_m$$

$$T = T_1 + T_2 + T_3 + T_4 = T_{\text{East}} + T_{\text{West}} + T_{\text{North}} + T_{\text{South}}$$

$$= 56 + 72 + 64 + 72$$

$$= 264$$

$$n = n_1 + n_2 + \cdots + n_{m-1} + n_m$$

$$n = n_1 + n_2 + n_3 + n_4 = 8 + 8 + 8 + 8 = 32$$

Thus, SS (total) = $2398.88 - ((264)^2 / 32) = 2398.88 - (69696/32) = 2398.88 - 2178 =$ 220.88

This can be visualized in Figure 9.85.

Sum of Squares Between **(factors) – Group Variation:**

$$SST = \textbf{SSB} + SSW$$

$$SSB = \sum_{i=1}^{k} n_i \left(\bar{X}_i - \bar{\bar{X}} \right)^2 \qquad \text{(Eq. 9.86)}$$

where SSB is sum of squares between (or factors), k is the number of populations, n_i is sample size from population i, $\bar{X}_i$ sample mean from population i, and $\bar{\bar{X}}$ is grand mean, the mean of all the data values in the investigation.

The mean square between = SSB/degrees of freedom

$$MSB = \frac{SSB}{df} = \frac{SSB}{k-1} \qquad \text{(Eq. 9.87)}$$

As shown in Figure 9.86.

Thus, sum of squares between group variations can be calculated by

$$SSB = \sum_{i=1}^{k} n_i \left(\bar{X}_i - \bar{\bar{X}} \right)^2 = n_1 \left(\bar{X}_1 - \bar{\bar{X}} \right)^2 + n_2 \left(\bar{X}_2 - \bar{\bar{X}} \right)^2$$

$$+ \cdots + n_{k-2} \left(\bar{X}_{k-2} - \bar{\bar{X}} \right)^2 + n_{k-1} \left(\bar{X}_{k-1} - \bar{\bar{X}} \right)^2 + n_k \left(\bar{X}_k - \bar{\bar{X}} \right)^2$$

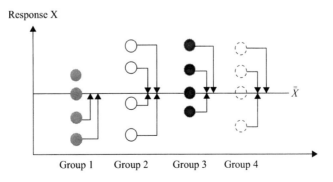

Figure 9.85. Representation of total variations.

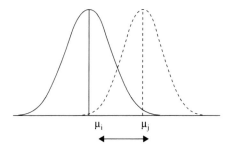

Figure 9.86. Variation due to differences among the groups.

where

$$\bar{\bar{X}} = \frac{\sum_{i=1}^{k} \bar{x}_i}{k} = \frac{\bar{x}_1 + \bar{x}_2 + \bar{x}_3 + \cdots + \bar{x}_{k-1} + \bar{x}_k}{k}$$

In the case of "Xeon" we have

$$\bar{\bar{X}} = \frac{\sum_{i=1}^{k} \bar{x}_i}{k} = \frac{\bar{x}_1 + \bar{x}_2 + \bar{x}_3 + \bar{x}_4}{k} = \frac{7.00 + 9.00 + 8.00 + 9.00}{4} = 8.25$$

$$SSB = \sum_{i=1}^{k} n_i \left(\bar{X}_i - \bar{\bar{X}} \right)^2 = n_1 \left(\bar{X}_1 - \bar{\bar{X}} \right)^2 + n_2 \left(\bar{X}_2 - \bar{\bar{X}} \right)^2 + \cdots + n_3 \left(\bar{X}_3 - \bar{\bar{X}} \right)^2 + n_4 \left(\bar{X}_4 - \bar{\bar{X}} \right)^2$$

SSB (or SS_{factor}) = $8(7 - 8.25)^2 + 8(9 - 8.25)^2 + 8(8 - 8.25)^2 + 8(9 - 8.25)^2 = 22.00$

This is illustrated in Figure 9.87.

Or, it can be calculated using the following equation:

$$SS\left(factor \right) = \left(\frac{T_1^2}{n_1} + \frac{T_2^2}{n_2} + \frac{T_3^2}{n_3} + \cdots + \frac{T_{m-2}^2}{n_{m-2}} + \frac{T_{m-1}^2}{n_{m-1}} + \frac{T_m^2}{n_m} \right) - \frac{T^2}{n} \qquad \text{(Eq. 9.88)}$$

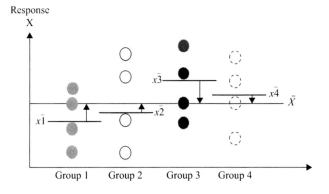

Figure 9.87. Between group variation.

$$T = T_1 + T_2 + \cdots + T_{m-1} + T_m$$

$$T = T_1 + T_2 + T_3 + T_4 = T_{East} + T_{West} + T_{North} + T_{South}$$

$$= 56 + 72 + 64 + 72$$

$$= 264$$

$$T^2 = 264 = 69,696$$

$$n = n_1 + n_2 + \cdots + n_{m-1} + n_m$$

$$n = n_1 + n_2 + n_3 + n_4 = 8 + 8 + 8 + 8 = 32$$

$$SS\left(factor\right) = \left(\frac{T_1^2}{n_1} + \frac{T_2^2}{n_2} + \frac{T_3^2}{n_3} + \frac{T_4^2}{n_4}\right) - \frac{T^2}{n} = \left(\frac{56^2}{8} + \frac{72^2}{8} + \frac{64^2}{8} + \frac{72^2}{8}\right) - \frac{69,696}{32}$$

$$= 2200 - 2178 = 22.00$$

Sum of Squares Within (errors)–Group Variation:

$$SST = SSB + SSW$$

$$SSW = \sum_{i=1}^{k} \sum_{j=1}^{n_i} \left(x_{ij} - \bar{x}_i\right)^2 = SST - SSB(or\ SS_{factor}) \qquad \text{(Eq. 9.89)}$$

where SSW is the sum of squares within (or errors), k is the number of populations, n_i is sample size from population i, $\bar{x}_i$ sample mean from population i, and x_{ij} is j^{th} measurement from population i.

The mean square within = SSB/degrees of freedom

$$MSW = \frac{SSW}{df} = \frac{Mean\ Square\ Within\ (or\ SS_{error})}{N - K} \qquad \text{(Eq. 9.90)}$$

Variation within each group pictured in Figure 9.88.

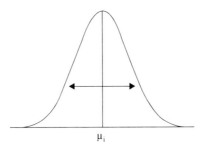

Figure 9.88. Summing the variation within each group and then adding over all groups.

Thus, sum of squares within the group's variations can be calculated by

$$SSW = \sum_{i=1}^{k} \sum_{j=1}^{n_j} \left(x_{ij} - \overline{x}_i\right)^2 = \left(x_{11} - \overline{x}_1\right)^2 + \left(x_{12} - \overline{x}_1\right)^2 + \cdots + \left(x_{21} - \overline{x}_2\right)^2 + \left(x_{22} - \overline{x}_i\right)^2$$
$$+ \cdots + \left(x_{31} - \overline{x}_3\right)^2 + \left(x_{32} - \overline{x}_3\right)^2 + \cdots + \left(x_{41} - \overline{x}_4\right)^2 + \left(x_{42} - \overline{x}_4\right)^2 + \cdots + \left(x_{48} - \overline{x}_4\right)^2$$

Note: $k = 4$; $n = 8$; $i = 1, 2, 3, ..., 8$; $j = 1, 2, 3, ..., 8$

$$SSW = \sum_{i=1}^{k} \sum_{j=1}^{n_j} \left(x_{ij} - \overline{x}_i\right)^2 = \left(4.1 - 7.0\right)^2 + \left(5.9 - 7.0\right)^2 + \cdots + \left(6.90 - 9.0\right)^2 + \left(9.10 - 9.0\right)^2 + \cdots$$
$$+ \left(4.60 - 8.0\right)^2 + \left(11.40 - 8.0\right)^2 + \cdots + \left(12.50 - 9.0\right)^2 + \left(7.50 - 9.0\right)^2 + \cdots + \left(9.20 - 9.0\right)^2$$
$$= 198.88$$

Or simply substrate the sum of the squares between the groups from sum of the squares total we get

$$SSW = SST - SSB(or\ SS_{factor}) = 220.88 - 22.00 = 198.88$$

Or, it can be calculated using the following equation:

$$SSW\left(or\ SS_{error}\right) = SS\left(total\right) - SSB\left(or\ SS_{factor}\right)$$
$$= \left(\sum_{i=1}^{n} x^2 - \frac{T^2}{n}\right) - \left(\left(\frac{T_1^2}{n_1} + \frac{T_2^2}{n_2} + \frac{T_3^2}{n_3} + \cdots + \frac{T_{m-2}^2}{n_{m-2}} + \frac{T_{m-1}^2}{n_{m-1}} + \frac{T_m^2}{n_m}\right) - \frac{T^2}{n}\right)$$
$$= \sum_{i=1}^{n} x^2 - \left(\frac{T_1^2}{n_1} + \frac{T_2^2}{n_2} + \frac{T_3^2}{n_3} + \cdots + \frac{T_{m-2}^2}{n_{m-2}} + \frac{T_{m-1}^2}{n_{m-1}} + \frac{T_m^2}{n_m}\right)$$

Figure 9.89 illustrates the within the group variations.

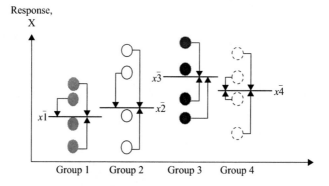

Figure 9.89. Within group variation.

One-Factor ANOVA: F-*Test Statistics*

Let's continue *F*-test with "Xeon" case

1. Set the hypothesis conditions
   Overall:
   $H_0: \mu_1 = \mu_2 = \mu_3 = \mu_4 = \cdots = \mu_{k-2} = \mu_{k-1} = \mu_k$
   H_α: Not all means are equal. This is always a right tail test.
   Note: H_0 defines that at least two of the means (μ's) are different or unequal.
   Apply this concept for "Xeon"
   $H_0: \mu_1 = \mu_2 = \mu_3 = \mu_4$
   H_α: At least two populations mean are different.
2. Calculate the "test statistic"

$$F = \frac{MSB}{MSW} = \frac{\dfrac{SSB}{df_1}}{\dfrac{SSW}{df_2}} = \frac{\dfrac{SSB}{k-1}}{\dfrac{SSW}{n-k}} = \frac{\dfrac{22.00}{4-1}}{\dfrac{198.88}{32-4}} = \frac{7.3333}{7.10286} = 1.03244 \qquad \text{(Eq. 9.91)}$$

where MSB is mean square between sample variances, MSW is mean square within sample variances, *k* is number of populations, *N* is sum of sample sizes from all the population, and the *F*-statistic is the ratio of the between estimate of variance and the within estimate of variance with the following criteria.
   - The *F*-ratio always must be positive and the outcome of ratio will be close than one if the null hypothesis is true and ratio will be larger than one if the null hypothesis is false.

$$H_0: \mu_1 = \mu_2 = \mu_3 = \mu_4 = \cdots = \mu_{k-2} = \mu_{k-1} = \mu_k$$

   - The term degree of freedom $df_1 = k - 1$, normally will be small.
   - The term degree of freedom $df_1 = N - k$, normally will be large.
3. Rejection area
   If $F > F_{\alpha = 0.05}$ reject H_0 otherwise do not reject H_0.
   $df_1 = k - 1 = 4 - 1 = 3$
   $df_2 = N - k = 32 - 4 = 28$
4. Decision making
   Because $F = 1.03244 < F_{\alpha = 0.05} = 2.95$, thus do not reject it (Figure 9.90).
   Reject H_0 if $F_{calculated} > F_{\alpha, k-1, n-k}$
   where $F_{\alpha, k-1, n-k}$ is obtained from *F*-Table in appendix.
   k = Number of levels or treatment
   n = Total number of observations
5. Conclusion. There is evidence that all mean are equal.

Construction of One-Way ANOVA Table: Insert all the information obtained in the table format as in Table 9.37

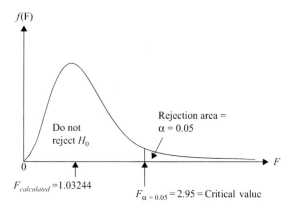

Figure 9.90. *F*-statistic rejection region.

Table 9.37. One-factor ANOVA table

Source of inconsistency (variation)	Degrees of freedom (*df*)	Sum of squares (*SS*)	Mean squares (*MS*)	Statistical test (*F-ratio*)
Between Samples (Factor)	$k - 1 = 4 - 1$ $= 3$	*SSB* (SS *Factor*) $= 22.00$	MSB = SSB/k − 1 $= 22.300/(4 - 1)$ $= 7.333$	F = MSB/MSW $= 7.3333/7.10286$ $= 1.03244$
Within Salmples (Error)	$n - k$ $= (8)(4) - 4$ $= 28$	SSW (or SS *Error*) $= 198.88$	MSW = SSW/n − k $= 198.88/(32 - 4)$ $= 7.10286$	
Total	$n - 1 = 32 - 1$ $= 23$	SS(total) $= 220.88$		

where degrees of freedoms are

$$df(\,factor\,) = k - 1$$

$$df(error) = n - k$$

$$df(total) = df(\,factor\,) + df(error) = k - 1 + (n - k) = k - 1 + n - k = n - 1$$

Example 9.80

One Factor ANOVA Comparing More than Two Means: A manufacturer of tire company wanted to test whether there is a difference in the lifetime of four different designs for icy environment condition. The quality/reliability control engineer used four different automotives for each design in the pilot study. The following data (in months) were obtained:

Car 1	Car 2	Car 3	Car 4
42	33	36	34
36	38	31	28
49	47	25	37
41	43	29	28
46	42	31	29
50	40	30	25

(a) Construct an One-Way ANOVA table
(b) Apply 95% confidence interval.

Solution

First we calculate the total, average, and variances of each pilot.

$$SST = SSW \text{ (or, } SS_{error}) + SSB \text{ (or } SS_{factor})$$

1. All four populations are obtained independently and randomly.
2. Hypothesis testing:
   $H_0: \mu_1 = \mu_2 = \mu_3 = \mu_4$
   H_a: At least two population means are different. This is always a right tail test.
3. $\alpha = 0.05$
4. Calculations are as follows. First construct Table 9.38 and compute the mean and variances

Table 9.38 The average and variance values of the cars

	Car #1	Car #2	Car #3	Car #4
1	42	33	36	34
2	36	38	31	28
3	49	47	25	37
4	41	43	29	28
5	46	42	31	29
6	50	40	30	25
Total	264	243	182	181
Average	$\bar{x}_1 = 44$	$\bar{x}_2 = 40.50$	$\bar{x}_3 = 30.33$	$\bar{x}_4 = 30.17$
Variance (s^2)	28.40	22.70	12.67	19.77

$n = 24$

Reading per treatment = $n_1 = n_2 = n_3 = n_4 = 6$

Number of levels (number of populations) = $m = 4$

Now, we need to calculate the mean between the groups, sum of the squares of total, sum of square in between the group, and sum of the squares within the groups.

$$\bar{\bar{x}} = \frac{\sum_{i=1}^{k} \bar{x}_i}{k} = \frac{\bar{x}_1 + \bar{x}_2 + \bar{x}_3 + \bar{x}_4}{4} = \frac{44 + 40.50 + 30.33 + 30.17}{4} = 36.25$$

Method #1:

$$SST = \sum_{i=1}^{k} \sum_{j=1}^{n_i} (x_{ij} - \bar{\bar{x}})^2 = (x_{11} - \bar{\bar{x}})^2 + (x_{12} - \bar{\bar{x}})^2 + \cdots + (x_{21} - \bar{\bar{x}})^2 + (x_{22} - \bar{\bar{x}})^2 + \cdots$$

$$+ (x_{31} - \bar{\bar{x}})^2 + (x_{32} - \bar{\bar{x}})^2 + \cdots + (x_{41} - \bar{\bar{x}})^2 + (x_{42} - \bar{\bar{x}})^2 + \cdots + (x_{46} - \bar{\bar{x}})^2$$

$$= (42 - 36.25)^2 + (36 - 36.25)^2 + \cdots + (33 - 36.25)^2 + (38 - 36.25)^2 + \cdots$$

$$+ (36 - 36.25)^2 + (31 - 36.25)^2 + \cdots + (34 - 36.25)^2 + (28 - 36.25)^2 + \cdots$$

$$+ (25 - 36.25)^2$$

$$= 1318.50$$

$$SSB \, (or \, SS_{factor}) = n_1 (\bar{x}_1 - \bar{\bar{x}})^2 + n_2 (\bar{x}_2 - \bar{\bar{x}})^2 + n_3 (\bar{x}_3 - \bar{\bar{x}})^2 + n_4 (\bar{x}_4 - \bar{\bar{x}})^2$$

$$= 6(44 - 36.25)^2 + 6(40.50 - 36.25)^2 + 6(30.33 - 36.25)^2 + 6(30.14 - 36.25)^2$$

$$= 900.83$$

$$SSW \, (or; SS_{error}) = \sum_{i=1}^{k} \sum_{j=1}^{n_i} (x_{ij} - \bar{x}_i)^2 = SST - SSB (or \, SS_{factor})$$

Note: $k = 4; \quad n = 24; \quad i = 1, 2, \ldots 6; \quad j = 1, 2, \ldots, 6$

So,

$$SSW = (x_{11} - \bar{x}_1)^2 + (x_{12} - \bar{x}_2)^2 + \ldots + (x_{kn_k} - \bar{x}_k)^2$$

$$SSW \, (or; SS_{error}) = SST - SSB (or \, SS_{factor}) = 1318.50 - 900.83 = 417.67$$

Method #2:

$$SS(total) = SS\,(factor) + SS\,(error)$$

$$SST = \sum_{i=1}^{k}\sum_{j=1}^{n_i} x_{ij}^2 - \frac{T^2}{n} = \begin{pmatrix} x_{11}^2 + x_{12}^2 + \cdots \\ + x_{21}^2 + x_{22}^2 + \cdots \\ + x_{31}^2 + x_{32}^2 + \cdots \\ + x_{41}^2 + x_{42}^2 + \cdots + x_{46}^2 \end{pmatrix} - \frac{\left(T_1 + T_2 + T_3 + T_4\right)^2}{n_1 + n_2 + n_3 + n_4}$$

$$= (42^2 + 36^2 + \cdots + 29^2 + 25^2) - \frac{(264 + 243 + 182 + 181)^2}{6 + 6 + 6 + 6}$$

$$= 1318.50$$

$$SS(factor) = \left(\frac{T_1^2}{n_1} + \frac{T_2^2}{n_2} + \frac{T_3^2}{n_3} + \cdots + \frac{T_{m-2}^2}{n_{m-2}} + \frac{T_{m-1}^2}{n_{m-1}} + \frac{T_m^2}{n_m}\right) - \frac{T^2}{n}$$

$$= \left(\frac{T_1^2}{n_1} + \frac{T_2^2}{n_2} + \frac{T_3^2}{n_3} + \frac{T_4^2}{n_4}\right) - \frac{T_1 + T_2 + T_3 + T_4}{n_1 + n_2 + n_3 + n_4}$$

$$= \left(\frac{264^2}{6} + \frac{243^2}{6} + \frac{182^2}{6} + \frac{181^2}{6}\right) - \frac{264 + 243 + 182 + 181}{6 + 6 + 6 + 6}$$

$$= 900.83$$

SS $(error)$ = SS$(total)$ − SS $(factor)$

$$= \sum x^2 - \left(\frac{T_1^2}{n_1} + \frac{T_2^2}{n_2} + \frac{T_3^2}{n_3} + \ldots + \frac{T_{m-2}^2}{n_{m-2}} + \frac{T_{m-1}^2}{n_{m-1}} + \frac{T_m^2}{n_m}\right)$$

$$= \left(\sum x^2 - \frac{T^2}{n}\right) - \left((\frac{T_1^2}{n_1} + \frac{T_2^2}{n_2} + \frac{T_3^2}{n_3} + \frac{T_4^2}{n_4}) - \frac{T^2}{n}\right)$$

$$= \sum x^2 - \frac{T^2}{n} - \left(\frac{T_1^2}{n_1} + \frac{T_2^2}{n_2} + \frac{T_3^2}{n_3} + \frac{T_4^2}{n_4}\right) + \frac{T^2}{n}$$

$$= \sum x^2 - \left(\frac{T_1^2}{n_1} + \frac{T_2^2}{n_2} + \frac{T_3^2}{n_3} + \frac{T_4^2}{n_4}\right) = 1318.50 - \left(\frac{264^2}{6} + \frac{243^2}{6} + \frac{182^2}{6} + \frac{181^2}{6}\right)$$

$$= 1318.50 - 900.83$$

$$= 417.67$$

Degrees of freedom (df): since $k = 4$ and $n = 24$ so the (df)s are

$$df(factor) = k - 1 = 4 - 1 = 3$$

$$df(error) = n - k = 24 - 4 = 20$$

$$df(total) = df(factor) + df(error) = k - 1 + (n - k) = k - 1 + n - k = n - 1 = 24 - 1 = 23$$

$$MS(factor) = \frac{SS(factor)}{k-1} = \frac{900.83}{3} = 300.277$$

$$MS(error) = \frac{SS(error)}{n-k} = \frac{417.67}{20} = 20.883$$

$$F = \frac{MS(factor)}{MS(error)} = \frac{300.277}{20.883} = 14.379$$

Now, form ANOVA Table 9.39 and fill in the table with the information obtained above.

5. Make a conclusion

Rejection condition:

Reject H_0 if $F_{Calculated} > F_{\alpha, k-1, n-k}$

where $F_{\alpha, k-1, n-k}$ is obtained from table in appendix.

$F_{Calculated} = 14.379$

$F_{\alpha, df1, df2} = F_{\alpha, k-1, n-k} = F_{0.05, 4-1, 24-4} = F_{0.05, 3, 20} = 3.10$ from F-Table in the appendix. This is a right tail test. Thus with above values that because, $F_{Calculated} = 14.379 > F_{\alpha, k-1, n-k} = 3.10$ null hypothesis (H_0: $\mu_1 = \mu_2 = \mu_3 = \mu_4$) is rejected.

Therefore, the conclusion is that at significance level $\alpha = 0.05$, data shows average life-times for the four different designs are not the same.

95% Confidence Interval for μ_1: from problem statement and table the point estimate of μ_1 for sample 1 is $\bar{x}_1 = 44$.

$$S_C = \left(MS(error)\right)^{0.50} = \left(\frac{SS(error)}{n-k}\right)^{0.50} = (20.883)^{0.50} = 4.570 \quad \text{(Eq. 9.92)}$$

Table 9.39. One-factor ANOVA table

Source of inconsistency (variation)	Degrees of freedom (*df*)	Sum of squares (*SS*)	Mean squares (*MS*)	Statistical test (*F*)
Factor	$k - 1 = 4 - 1 = 3$	SS (*factor*) = 900.83	MS(*factor*) = SS(*factor*)/k – 1 = 300.277	F = MS(*factor*)/ MS(*error*) 14.379
Error	$n - k = 24 - 4 = 20$	SS(*error*) = 417.67	MS(*error*) = SS(error)/n – k = 20.883	
Total	$n - 1 = 24 - 1 = 23$	SS(*total*) = 1318.50		

$$\bar{X}_i \pm t_{\alpha/2, n-k} S_C \left(\frac{1}{n_i}\right)^{0.50} \tag{Eq. 9.93}$$

$$\bar{X}_1 \pm (t_{0.05/2, 24-4} = t_{0.025, 20})(S_C) \left(\frac{1}{n_i}\right)^{0.50}$$

$$44 \pm (2.086)(4.570)(1/6)^{0.50}$$

$$44 \pm 3.8918$$

$$40.1082 \text{ to } 47.8918$$

We are 95% confident that the average lifetime of the design 1 for Car #1 is between 40.10 to 47.89 months.

95% *Confidence Interval for the difference of* $\mu_1 - \mu_2$: The point estimate of μ_1 is $\bar{x}_1 = 44$ and point estimate of μ_1 is $\bar{x}_3 = 30.33$:

$$\left(\bar{X}_i - \bar{X}_j\right) \pm t_{\alpha/2, n-k} S_C \left(\frac{1}{n_i} + \frac{1}{n_j}\right)^{0.50} \tag{Eq. 9.94}$$

$$= \left((\bar{X}_1 = 44) - (\bar{X}_3 = 30.33)\right) \pm (t_{0.05/2, 24-4} = t_{0.025, 20} = 2.086)(S_C = 5.64) \left(\frac{1}{n_1 = 6} + \frac{1}{n_3 = 6}\right)^{0.50}$$

$$= (44 - 30.33) \pm (2.086)(5.64) \left(\frac{1}{6} + \frac{1}{6}\right)^{0.50}$$

$$= 13.67 - (2.086)(5.64)(0.57735) \text{ to } 13.67 + (2.086)(5.64)(0.57735)$$

$$= 13.67 - 6.79255 \text{ to } 13.67 + 6.79255$$

$$= 6.877 \text{ to } 20.46$$

Based on the 95% confidence interval outcome, average lifetime for the Car #1 is between 6.877 to 20.46 months higher than the average life time for Car #3. The computer output for ANOVA is shown in Tables 9.40 and 9.41.

Table 9.40. Computer output for single-factor ANOVA: summary

Groups	Count	Sum	Average	Variance
CAR 1	6	264	44	28.4
CAR 2	6	243	40.5	22.7
CAR 3	6	182	30.3333	12.6667
CAR 4	6	181	30.1667	19.7667

334 • MASTERING LEAN SIX SIGMA

Table 9.41. ANOVA summary

Source of variation	SS	df	MS	F	p-value	F critical
Between Groups	900.833	3	300.278	14.3788	3.2E-05	3.09839
Within Groups	417.667	20	20.8833			
Total	1318.5	23				

Example 9.81

One Factor ANOVA Comparing More than Two Means: The employees at a controller assembly plant would like to earn more breaks to increase their productivity without reducing actual assembly time. The VP of operation claims that increasing the number of 15-minute breaks will affect productivity efficiency. The current breaks are three times in an 8-hour shift. The VP of operation requests the manager in charge to run a test using four groups of five people. The tests decided to be 3, 4, 5, and 6 breaks. The number of controllers assembled each day is recorded for five days.

(a) Test the VP of operation's claim using ANOVA procedure with a 10% significance factor ($\alpha = 0.10$).
(b) Find 95% confidence interval for μ_1 and $\mu_1 - \mu_2$.
(c) Find p-value, by applying a 5% significance factor ($\alpha = 0.05$).

3 breaks	4 breaks	5 breaks	6 breaks k = 4 groups b = 6 blocks/rows
420	430	410	400
390	415	400	380
405	400	415	390
410	385	420	380
415	420	310	390
380	390	380	375

Solution

First we calculate the total, average, and variances of each pilot.

$n = bk$, where b: rows or blocks, k: groups

SST = SSW (or, SS_{error}) + SSB (or SS_{factor})

1. All four populations are obtained independently and randomly.
2. Hypothesis testing condition:

H_0: $\mu_1 = \mu_2 = \mu_3 = \mu_4$

H_a: At least two population means are different. This is always a right tail test.

Table 9.42. The average and variance values of the breaks

	3 Breaks	4 Breaks	5 Breaks	6 Breaks
1	420	430	410	400
2	390	415	400	380
3	405	400	415	390
4	410	385	420	380
5	415	420	310	390
6	380	390	380	375
Total ($\sum$)	$\sum = 2420$	$\sum = 2440$	$\sum = 2335$	$\sum = 2315$
Average ($\bar{x}$)	$\bar{x}_1 = 403.33$	$\bar{x}_2 = 406.67$	$\bar{x}_3 = 3389.17$	$\bar{x}_4 = 385.83$
Variance (s^2)	$s_1^2 = 236.67$	$s_2^2 = 316.67$	$s_3^2 = 1704.17$	$s_4^2 = 84.17$

3. $\alpha = 0.05$
4. Calculations for the average and variances of the breaks are illustrated in the Table 9.42:
   $n = 24$
   Reading per treatment $= n_1 = n_2 = n_3 = n_4 = 6$
   Number of levels (number of populations) $= m = 4$

   Now, we need to calculate the mean between the groups, sum of the squares of total, sum of square in between the group, and sum of the squares within the groups. These can be calculated by two methods as follows:

$$\bar{\bar{x}} = \frac{\sum_{i=1}^{k} \bar{x}_i}{k} = \frac{\bar{x}_1 + \bar{x}_2 + \bar{x}_3 + \bar{x}_4}{k} = \frac{403.33 + 406.67 + 389.167 + 385.83}{4} = \frac{1585}{4} = 396.25$$

Method #1:

$$SST = \sum_{i=1}^{k} \sum_{j=1}^{n_i} (x_{ij} - \bar{\bar{x}})^2 = (x_{11} - \bar{\bar{x}})^2 + (x_{12} - \bar{\bar{x}})^2 + \cdots + (x_{21} - \bar{\bar{x}})^2 + (x_{22} - \bar{\bar{x}})^2 + \cdots$$
$$+ (x_{31} - \bar{\bar{x}})^2 + (x_{32} - \bar{\bar{x}})^2 + \cdots + (x_{41} - \bar{\bar{x}})^2 + (x_{42} - \bar{\bar{x}})^2 + \cdots + (x_{46} - \bar{\bar{x}})^2$$

$$= (420 - 396.25)^2 + (390 - 396.25)^2 + \cdots + (430 - 396.25)^2 + (415 - 396.25)^2 + \cdots$$
$$+ (410 - 396.25)^2 + (400 - 396.25)^2 + \cdots + (400 - 396.25)^2 + (380 - 396.25)^2 + \cdots$$
$$+ (375 - 396.25)^2 = 13612.50$$

$$SSB\,(or\,SS_{factor}) = n_1(\bar{x}_1 - \bar{\bar{x}})^2 + n_2(\bar{x}_2 - \bar{\bar{x}})^2 + n_3(\bar{x}_3 - \bar{\bar{x}})^2 + n_4(\bar{x}_{k4} - \bar{\bar{x}})^2$$

$$= 6(403.33 - 396.25)^2 + 6(406.67 - 396.25)^2 + 6(389.83 - 396.25)^2$$
$$+ 6(385.83 - 396.25)^2 = 1904.167$$

$$SSW(or; SS_{error}) = \sum_{i=1}^{k} \sum_{j=1}^{n_j} (x_{ij} - \bar{x}_i)^2 = SST - SSB(or\ SS_{factor})$$

Note: $k = 4$, $n = 24$; $i = 1, 2, \dots, 6$; $j = 1, 2, \dots, 6$

$$SSW = (x_{11} - \bar{x}_1)^2 + (x_{12} - \bar{x}_2)^2 + \cdots + (x_{kn_k} - \bar{x}_k)^2$$

$$SSW(or; SS_{error}) = SST - SSB(or\ SS_{factor}) = 13612.50 - 1904.167 = 11,708.33$$

Method #2:

$$SST = SS\ (factor) + SS\ (error)$$

$$SST = \sum_{i=1}^{k} \sum_{j=1}^{n_j} x_{ij}^2 - \frac{T^2}{n} = (x_{11}^2 + x_{12}^2 + \cdots + x_{21}^2 + x_{22}^2 + \cdots + x_{31}^2 + x_{32}^2 + \cdots + x_{41}^2 + x_{42}^2 + \cdots + x_{46}^2)$$

$$- \frac{(T_1 + T_2 + T_3 + T_4)^2}{n_1 + n_2 + n_3 + n_4}$$

$$= (420^2 + 390^2 + \cdots + 390^2 + 375^2) - \frac{(2420 + 2440 + 2335 + 2315)^2}{6 + 6 + 6 + 6}$$

$$= 13612.50$$

$$SS(factor) = \left(\frac{T_1^2}{n_1} + \frac{T_2^2}{n_2} + \frac{T_3^2}{n_3} + \cdots + \frac{T_{m-2}^2}{n_{m-2}} + \frac{T_{m-1}^2}{n_{m-1}} + \frac{T_m^2}{n_m} \right) - \frac{T^2}{n}$$

This the case of this problem we have

$$SS(factor) = \left(\frac{T_1^2}{n_1} + \frac{T_2^2}{n_2} + \frac{T_3^2}{n_3} + \frac{T_4^2}{n_4} \right) - \frac{(T_1 + T_2 + T_3 + T_4)^2}{n_1 + n_2 + n_3 + n_4}$$

$$= \left(\frac{2420^2}{6} + \frac{2440^2}{6} + \frac{2335^2}{6} + \frac{2315^2}{6} \right) - \frac{(2420 + 2440 + 2335 + 2315)^2}{6 + 6 + 6 + 6} = 1904.167$$

$$SS\ (error) = SS(total) - SS(factor)$$

$$SS\ (error) = \sum x^2 - \left(\frac{T_1^2}{n_1} + \frac{T_2^2}{n_2} + \frac{T_3^2}{n_3} + \cdots + \frac{T_{m-2}^2}{n_{m-2}} + \frac{T_{m-1}^2}{n_{m-1}} + \frac{T_m^2}{n_m} \right)$$

$$= \left(\sum x^2 - \frac{T^2}{n} \right) - \left((\frac{T_1^2}{n_1} + \frac{T_2^2}{n_2} + \frac{T_3^2}{n_3} + \frac{T_4^2}{n_4}) - \frac{T^2}{n} \right)$$

$$= \sum x^2 - \frac{T^2}{n} - \left(\frac{T_1^2}{n_1} + \frac{T_2^2}{n_2} + \frac{T_3^2}{n_3} + \frac{T_4^2}{n_4} \right) + \frac{T^2}{n}$$

$$= \sum_{i=1}^{k} \sum_{j=1}^{n_i} x_{ij}^2 - \left(\frac{T_1^2}{n_1} + \frac{T_2^2}{n_2} + \frac{T_3^2}{n_3} + \frac{T_4^2}{n_4} \right)$$

$$= 3,781,950 - \left(\frac{2420^2}{6} + \frac{2440^2}{6} + \frac{2335^2}{6} + \frac{2315^2}{6} \right)$$

$$= 3,781,950 - 3,770,242$$

$$= 11,708.33$$

Degrees of freedom (df): since $k = 4$ and $n = 24$ so the (df)s are

$$df\,(factor) = k - 1 = 4 - 1 = 3$$

$$df\,(error) = n - k = 24 - 4 = 20$$

$$df\,(total) = df\,(factor) + df\,(error) = k - 1 + (n - k) = k - 1 + n - k = n - 1 = 24 - 1 = 23$$

$$MS(factor) = \frac{SS(factor)}{k-1} = \frac{1904.167}{3} = 634.72$$

$$MS(error) = \frac{SS(error)}{n-k} = \frac{11,708.33}{20} = 585.417$$

$$F = \frac{MS(factor)}{MS(error)} = \frac{634.72}{585.417} = 1.084 = F_{Calculated}$$

The One-Factor ANOVA values are shown in Table 9.43

Table 9.43. One-Factor ANOVA table

Source of inconsistency (variation)	Degrees of freedom (df)	Sum of squares (SS)	Mean squares (MS)	Statistical test (F)
SSB (or SSFactor)	$df_1 = k - 1$ $= 4 - 1$ $= 3$	SS (factor) $= 1940.167$	MS (factor) $= SS(factor)/k - 1$ $= 634.72$	$F = MS(factor)/$ $MS(error)$ $= 1.084$
SSW (or SSError)	$df_1 = n - k$ $= 24 - 4$ $= 20$	SS (error) $= 11,708.33$	MS (error) $= SS(error)/n - k$ $= 585.417$	
Total	$n - 1 = 24 - 1$ $= 23$	SS (total) $= 13,612.50$		

5. Make a conclusion

$$\text{Reject } H_0 \text{ if } F_{Calculated} > F_{\alpha, k-1, n-k}$$

where $F_{\alpha, k-1, n-k}$ is obtained from F-table in appendix
$F_{Calculated} = 1.084$
$F_{\alpha, df1, df2} = F_{\alpha, k-1, n-k} = F_{0.10, 4-1, 24-4} = F_{0.10, 3, 20} = 2.380$ this comes from F-table in the appendix, in addition this is a right tail test.

$$\text{Reject } H_0 \text{ if } F_{Calculated} > F_{\alpha, k-1, n-k}$$

Because $F_{Calculated} = 1.084 < F_{\alpha, k-1, n-k} = 2.380$ (this value is from F-table)
Failed to reject null hypothesis (H_0: $\mu_1 = \mu_2 = \mu_3 = \mu_4$).

Therefore, the conclusion is that at significance level $\alpha = 0.10$, data shows average lifetimes for the four different designs are the same.

95% *Confidence Interval for* μ_1: From problem statement and table, the point estimate of μ_1 for sample 1 is $\bar{x}_1 = 403.33$.

$$S_C = \left(MS(error)\right)^{0.50} = \left(\frac{SS(error)}{n-k}\right)^{0.50} = \left(\frac{11,708.33}{24-4}\right)^{0.50} = (585.417)^{0.50} = 24.195$$

$$\bar{X}_i \pm t_{\alpha/2, n-k} S_C \left(\frac{1}{n_i}\right)^{0.50}$$

$$\bar{X}_1 \pm (t_{0.05/2,\ 24-4} = t_{0.025,\ 20})(S_C)\left(\frac{1}{n_i}\right)^{0.50}$$

$$403.33 \pm (2.086)(24.195)(1/6)^{0.50}$$

$$403.33 \pm (2.086)(24.195)(0.40825)$$

$$403.33 \pm 20.604$$

$$382.725 \text{ to } 423.935$$

In conclusion, we are 95% confident that the average productivity for 3 breaks is between 382.725 to 423.935 assemblies.

95% *Confidence Interval for the difference of* $\mu_1 - \mu_2$: The point estimate of μ_1 is $\bar{x}_1 = 403.33$ and point estimate of μ_1 is $\bar{x}_3 = 389.167$.

$$\left(\bar{X}_i - \bar{X}_j\right) \pm t_{\alpha/2,\ n-k} S_C \left(\frac{1}{n_i} + \frac{1}{n_j}\right)^{0.50}$$

$$= \left((\bar{X}_1 = 403.33) - (\bar{X}_3 = 389.167)\right)$$

$$\pm (t_{0.05/2,\ 24-4} = t_{0.025,\ 20} = 2.086)(S_C = 24.195)\left(\frac{1}{n_1 = 6} + \frac{1}{n_3 = 6}\right)^{0.50}$$

Table 9.44. Computer output for single-factor ANOVA summary

Groups	Count	Sum	Average	Variance
3 Breaks 1	6	2420	403.3333	236.6667
4 Breaks 2	6	2440	406.6667	316.6667
5 Breaks 3	6	2335	389.1667	1704.167
6 Breaks 4	6	2315	385.8333	84.16667

Table 9.45. One-way ANOVA: 3 breaks, 4 breaks, 5 breaks, 6 breaks

Source of variation	SS	df	MS	$F_{Calculated}$	p-value	F-critical
Between Groups	1904.167	3	634.7222	1.084223	0.378462	3.098391
Within Groups	11708.33	20	585.4167			
Total	13612.5	23				

$$= (403.33 - 389.167) \pm (2.086)(24.195)\left(\frac{1}{6}+\frac{1}{6}\right)^{0.50}$$

$$= 14.163 \pm (2.086)(24.195)(0.57735)$$

$$= 14.163 \pm 29.139$$

$$-14.976 \text{ to } 43.302$$

Based on the 95% confidence interval outcome, average assembly for 3 breaks is between −14.976 to 43.302 higher than the assembly for 5 breaks. (The negative sign is due because the problem is just asking the difference of two mean.)

Tables 9.44 and 9.45 summarize ANOVA table for the above information.

9.8.2 RANDOMIZED BLOCK DESIGN AND ANALYSIS OF VARIANCE

In one-factor (or one-way) ANOVA, we obtained samples randomly from each k populations (levels) describing a single factor. The variable that is being controlled is referred to as the independent variable and the variable that is being measured is called dependent variable.

Rather than acquiring independent samples from k populations, the randomized block design data are arranged into uniform and subgroups structures called *blocks*, such that the variability within the blocks are less than the variability between the blocks. When more than one factor is included in the design of experiment that influences the outcome of dependent variables it is called blocking. Let's consider the following course of action.

Recently, the vice president (VP) of sales and marketing for a high technology company at Silicon Valley conducted a survey regarding their solar system product. The survey was carried out in the east, west, and south of the United States. The goal was requested to find out if the service quality is the same in East, West, and Southern area. VP has developed a test in which he will ask 14 customers point of view in each region. The test rated from 1 to 1000 points (1000 being high). A survey shows as follows:

Customer	East	West	South	Block	Means
1	830	647	630	702.33	
2	652	840	786	789.67	
3	652	747	730	709.67	
4	885	639	617	713.67	
5	814	943	632	796.33	
6	733	916	410	686.33	
7	770	923	727	806.67	
8	829	903	726	819.33	
9	847	760	648	751.67	
10	878	856	668	800.67	
11	728	878	670	758.67	
12	693	990	825	836.00	
13	807	871	564	747.33	
14	901	980	719	866.67	
Treatment Mean	793.57	849.50	668.00	770.36 = Grand Mean	

Develop randomized block ANOVA Table.

Solution

Testing Condition

H_0: $\mu_1 = \mu_2 = \mu_3 = \mu_4$

H_a: At least two populations have different means.

Total variations can be split into three parts.

$$\textbf{SST} = \text{SSB (or SS}_{factor}) + \text{SSBL} + \text{SSW (or, SS}_{error})$$

where SSBL sum of squares between blocks. Let's recall the SST equation

$$SST = \sum_{i=1}^{k}\sum_{j=1}^{n_i}\left(x_{ij} - \overline{\overline{x}}\right)^2 = \left(x_{11} - \overline{\overline{x}}\right)^2 + \left(x_{12} + \overline{\overline{x}}\right)^2 + \cdots + \left(x_{kn_k} + \overline{\overline{x}}\right)^2$$

where SST is total sum of squares, k is number of populations (level of treatments), n_i sample size from population i, x_{ij} is j^{th} measurement from population I, and $\overline{\overline{x}}$ is grand mean (mean of all data values) mean of all group means, and n_k = Number of observations in the kth sample.

In the case of "Xeon" we have

$$SST = \sum_{i=1}^{k}\sum_{j=1}^{n_i}\left(x_{ij} - \overline{\overline{x}}\right)^2$$

$$= \left(x_{11} - \overline{\overline{x}}\right)^2 + \left(x_{12} + \overline{\overline{x}}\right)^2 + \cdots + \left(x_{21} + \overline{\overline{x}}\right)^2 + \left(x_{22} + \overline{\overline{x}}\right)^2 + \cdots + \left(x_{31} + \overline{\overline{x}}\right)^2$$

$$+ \left(x_{32} + \overline{\overline{x}}\right)^2 + \cdots + \left(x_{3:14} + \overline{\overline{x}}\right)^2$$

$$SST = (830 - 770.36)^2 + (652 - 770.36)^2 + \cdots + (647 - 770.36)^2 + (840 - 770.36)^2$$
$$+ \cdots + (630 - 770.36)^2 + (786 - 770.36)^2 + \cdots + (719 + 770.36)^2 = 614,641.60$$

Sum of the squares between variations: The sum of squares between group variations can be calculated by

$$SST = \textbf{SSB (or SS}_{\textbf{factor}}\textbf{)} + SSBL + SSW \text{ (or, } SS_{\text{error}})$$

$$SSB = \sum_{i=1}^{k} n_i \left(\overline{X}_i - \overline{\overline{X}}\right)^2$$

$$= n_1\left(\overline{X}_1 - \overline{\overline{X}}\right)^2 + n_2\left(\overline{X}_2 - \overline{\overline{X}}\right)^2 + \cdots + n_{k-2}\left(\overline{X}_{k-2} - \overline{\overline{X}}\right)^2 + n_{k-1}\left(\overline{X}_{k-1} - \overline{\overline{X}}\right)^2$$

$$+ n_k\left(\overline{X}_k - \overline{\overline{X}}\right)^2$$

where

$$\overline{\overline{X}} = \frac{\sum_{i=1}^{k}\overline{x}_i}{k} = \frac{\overline{x}_1 + \overline{x}_2 + \overline{x}_3 + \cdots + \overline{x}_{k-1} + \overline{x}_k}{k}$$

In the case of "high tech company" we have

$$\overline{\overline{X}} = \frac{\sum_{i=1}^{k}\overline{x}_i}{k} = \frac{\overline{x}_1 + \overline{x}_2 + \overline{x}_3}{k} = \frac{793.57 + 849.50 + 668.00}{3} = 770.30$$

$$SSB = \sum_{i=1}^{k} n_i \left(\overline{X}_i - \overline{\overline{X}}\right)^2 = n_1\left(\overline{X}_1 - \overline{\overline{X}}\right)^2 + n_2\left(\overline{X}_2 - \overline{\overline{X}}\right)^2 + n_3\left(\overline{X}_3 - \overline{\overline{X}}\right)^2$$

$$SSB \text{ (or } SS_{factor}) = 14(793.57 - 770.30)^2 + 14(849.50 - 770.30)^2 + 14(668.00 - 770.30)^2$$
$$= 241,912.70$$

This is visualized in Figure 9.91.

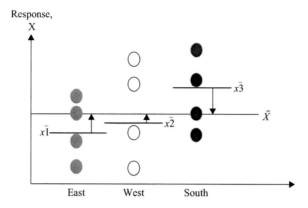

Figure 9.91. Between-group variation.

Sum of the squares for Blocking: Thus, sum of squares for blocking can be computed by

$$SST = SSB \text{ (or } SS_{factor}) + \mathbf{SSBL} + SSW \text{ (or, } SS_{error}) \qquad \text{(Eq. 9.95)}$$

$$SSBL = \sum_{j=1}^{b} k\left(\bar{x}_j - \bar{\bar{x}}\right)^2 \qquad \text{(Eq. 9.96)}$$

where
   k : number of levels for this factor
   b: number of blocks
   $\bar{x}_j$: Sample mean from the jth block
   $\bar{\bar{x}}$: Grand mean (the mean of all the mean or mean of all the data values) as used in SSB
section.
   Recall $\bar{\bar{x}}$:

$$\bar{\bar{X}} = \frac{\sum_{i=1}^{k} \bar{x}_i}{k} = \frac{\bar{x}_1 + \bar{x}_2 + \bar{x}_3}{k} = \frac{793.57 + 849.50 + 668.00}{3} = 770.30$$

$$SSBL = \sum_{j=1}^{b} k\left(\bar{x}_j - \bar{\bar{x}}\right)^2 = k\left(\bar{x}_1 - \bar{\bar{x}}\right)^2 + k\left(\bar{x}_2 - \bar{\bar{x}}\right)^2 + \cdots + k\left(\bar{x}_{14} - \bar{\bar{x}}\right)^2$$

$$SSBL = 3(702.33 - 770.36)^2 + 14(789.67 - 770.36)^2 + \cdots + 14(866.67 - 770.36)^2 = 116,605$$

Sum of the squares within the groups: Thus, sum of squares within the groups can be
computed by subtracting SSB and SSBL from SST

$$SSW \text{ (or, } SS_{error}) = SST - (SSB \text{ (or } SS_{factor}) + SSBL)$$

Substituting information from the calculation above

$$\mathbf{SSW} \text{ (or, } SS_{error}) = 614,641.60 - (241,912.70 + 116,605) = 256,123.90$$

Mean Squares – The following relations are used in calculating the mean squares:

$$\text{Mean square blocking} = MSBL = \frac{SSBL}{b-1} \qquad \text{(Eq. 9.97)}$$

$$\text{Mean square between} = MSB = \frac{SSB}{k-1} \qquad \text{(Eq. 9.98)}$$

$$\text{Mean square within} = MSW = \frac{SSW}{(k-1)(b-1)} \qquad \text{(Eq. 9.99)}$$

Tables 9.46 and 9.47 summarize the ANOVA calculations.

Table 9.46. One-factor ANOVA table

Source of inconsistency (variation)	Degrees of freedom (*df*)	Sum of squares (*SS*)	Mean squares (*MS*)	Statistical test (*F*)
Between Blocks	$b-1$	$SSBL = SST - (SSB + SSW)$	MSBL = SSBL/$(b-1)$	F-ratio = SSBL/MSW
SSB (or SS Factor)	$df_1 = k-1$	SSB (SS Factor)	MSB = SSB/$k-1$	$F = MSB/MSW$
SSW (or SS Error)	$df_2 = (k-1)(b-1)$	SSW (or SS Error)	MSW = SSW/$(k-1)(b-1)$	
Total	$N-1$	SS(*total*) = SSBL + SSB + SSW		

Table 9.47. One-factor ANOVA table

Source of inconsistency (variation)	Degrees of freedom (*df*)	Sum of squares (*SS*)	Mean squares (*MS*)	Statistical test (*F*)
Between Blocks	$b-1$ = 14 – 1 = 13	$SSBL$ = $SST - (SSB + SSW)$ = 116,605.00	MSBL = SSBL/$(b-1)$ = 8,969.60	F-ratio = MSBL/MSW = 0.9106
SSB (or SS Factor)	$df_1 = k-1$ = 3 – 1 = 2	SSB (SS Factor) = 241,912.70	MSB = SSB/$k-1$ = 120,956.40	$F = MSB/MSW$ = 12.279
SSW (or SS Error)	df_2 = $(k-1)(b-1)$ = $(3-1)(14-1)$ = 26	SSW (or SS Error) = 256,132.90	MSW = SSW/$(k-1)(b-1)$ = 9,850.90	
Total	$N-1 = 42-1$ = 41	SS(*total*) = SSBL + SSB + SSW = 614,641.60		

where k is number of population, df the degrees of freedom, N the sum of the sample sizes from all populations.

One Factor ANOVA Blocking Test: F-*Test Statistics*

Let's continue F-test with "high tech" case

1. Set the hypothesis conditions
   Overall:
   H_0: $\mu_1 = \mu_2 = \mu_3 = \mu_4 = \cdots = \mu_{k-2} = \mu_{k-1} = \mu_k$
   H_a: Not all the block means are equal.
   Note: H_0 defines that at least two of the means (μ's) are different or unequal.
   For "High Tech"
   H_0: $\mu_1 = \mu_2 = \mu_3$
   H_a: At least two populations have different means.
2. Calculate the "test statistic"

$$F = \frac{MSBL}{MSW} = \frac{\dfrac{SSBL}{df_1}}{\dfrac{SSW}{df_2}} = \frac{\dfrac{SSBL}{b-1}}{\dfrac{SSW}{(k-1)(b-1)}} = \frac{\dfrac{116,605}{14-1}}{\dfrac{256,123.90}{(3-1)(14-1)}} = \frac{8,969.60}{9,850.90} = 0.91057 \quad \text{(Eq. 9.99)}$$

$$F = \frac{MSB}{MSW} = \frac{\dfrac{SSB}{df_1}}{\dfrac{SSW}{df_2}} = \frac{\dfrac{SSB}{k-1}}{\dfrac{SSW}{(k-1)(b-1)}} = \frac{\dfrac{116,605}{3-1}}{\dfrac{256,123.90}{(3-1)(14-1)}} = \frac{120,956.40}{9,850.90} = 12.279$$

3. Rejection area
   If $F > F_{\alpha = 0.05}$ reject H_0 otherwise do not reject H_0.
   $df_1 = b - 1 =$
   $df_2 = (k-1)(b-1)$
4. Decision making
   H_0: $\mu_1 = \mu_2 = \mu_3$
   H_a: Not all μ_i are the same (blocking is effective).
   Because $F_{Calculated} = 0.9105 < F_{\alpha = 0.05} = 2.119$ at $df_1 = b - 1 = 14 - 1 = 13$ and $df_2 = (k-1)(b-1) = (3-1)(14-1) = 26$. Thus, do not reject the null hypothesis
   H_0: $\mu_1 = \mu_2 = \mu_3$
   H_a: At least two populations have different means.
   Because $F_{Calculated} = 12.279 > F_{\alpha = 0.05} = 2.119$ at $df_1 = k - 1 = 3 - 1 = 2$ and $df_2 = (k-1)$ $(b-1) = (3-1)(14-1) = 26$, thus reject the null hypothesis.

Example 9.82

Randomized block design: The Alpine Medical Insurance Company is reviewing the primary care claims (in dollars) submitted by 10 different families. The survey was conducted in the three different cities. The data randomly collected and recorded by year total listed below.

Can the insurance company conclude that there is a difference in the three population mean by applying 5% significance level.

Family	Dallas	Los Angeles	New York	Total	$k = 3$ number of factor levels
1	720	930	880	2530	$b = 10$ number of blocks or rows
2	890	1200	1150	3240	
3	705	750	900	2355	
4	910	850	660	2420	
5	1250	2000	1500	4750	
6	980	1800	1100	3880	
7	990	1650	850	3490	
8	900	980	1500	3380	
9	1200	1350	750	3300	
10	1650	1100	1000	3750	
Total (T)	$T_1 = 10,195$	$T_2 = 12,610$	$T_3 = 10,290$		
Average ($\bar{x}$)	1,019.5	1,261	1029		
Variance (s^2)	79,702.5	182,676.7	83,232.22		
Standard Deviation	282.32	427.41	288.50		

Solution

Testing Condition

H_0: $\mu_1 = \mu_2 = \mu_3 = \mu_4$

H_a: At least two populations have different means

Total variations can be split into three parts.

i. **Sum of the squares of total**

$$SST = SSB \ (\text{or } SS_{factor}) + SSBL + SSW \ (\text{or, } SS_{error})$$

$$\bar{\bar{X}} = \frac{\sum_{i=1}^{k} \bar{x}_i}{k} = \frac{\bar{x}_1 + \bar{x}_2 + \bar{x}_3}{k} = \frac{1,019.50 + 1,261 + 1029}{3} = 1103.17$$

$$SST = \sum_{i=1}^{k} \sum_{j=1}^{n_i} \left(x_{ij} - \bar{\bar{x}} \right)^2$$

$$= \left(x_{11} - \bar{\bar{x}} \right)^2 + \left(x_{12} + \bar{\bar{x}} \right)^2 + \cdots + \left(x_{21} + \bar{\bar{x}} \right)^2 + \left(x_{22} + \bar{\bar{x}} \right)^2 + \cdots + \left(x_{31} + \bar{\bar{x}} \right)^2$$

$$+ \left(x_{32} + \bar{\bar{x}} \right)^2 + \cdots + \left(x_{3:10} + \bar{\bar{x}} \right)^2$$

$$SST = (720 - 1103.17)^2 + (890 - 1103.17)^2 + \cdots + (930 - 1103.17)^2 + (1200 - 1103.17)^2$$
$$+ \cdots + (880 - 1103.17)^2 + (1150 - 1103.17)^2 + \cdots + (1000 - 1103.17)^2$$
$$= 3,48,624.17$$

ii. ***Sum of the squares between variations***: The sum of squares between group variations can be calculated by

$$SST = \textbf{SSB (or SS}_{\textbf{factor}}\textbf{)} + SSBL + SSW \text{ (or, } SS_{error})$$

$$SSB = \sum_{i=1}^{k} n_i \left(\bar{X}_i - \bar{\bar{X}} \right)^2 = n_1 \left(\bar{X}_1 - \bar{\bar{X}} \right)^2 + n_2 \left(\bar{X}_2 - \bar{\bar{X}} \right)^2 + n_3 \left(\bar{X}_3 - \bar{\bar{X}} \right)^2$$

$$SSB \text{ (or SS}_{factor}) = 10(1,019.50 - 1103.17)^2 + 10(1,261 - 1103.17)^2 + 10(1029 -$$
$$1103.17)^2 = 374,121.67$$

This is visualized in Figure 9.92.

iii. ***Sum of the squares for Blocking***: Thus, sum of squares for blocking can be computed by

$$SST = SSB \text{ (or SS}_{factor}) + \textbf{SSBL} + SSW \text{ (or, } SS_{error})$$

$$SSBL = \sum_{j=1}^{b} k \left(\bar{x}_j - \bar{\bar{x}} \right)^2 = k \left(\bar{x}_1 - \bar{\bar{x}} \right)^2 + k \left(\bar{x}_2 - \bar{\bar{x}} \right)^2 + \cdots + k \left(\bar{x}_{10} - \bar{\bar{x}} \right)^2$$

$$SSBL = 3(2530 - 1103.17)^2 + 3(3240 - 1103.17)^2 + \cdots + 3(3750 - 1103.17)^2 =$$
1,648,974.17

iv. ***Sum of the squares within the groups***: Thus, sum of squares within the groups can be computed by subtracting SSB and SSBL from SST:

$$\textbf{SSW} \text{ (or, } SS_{error}) = SST - (SSB \text{ (or SS}_{factor}) + SSBL)$$

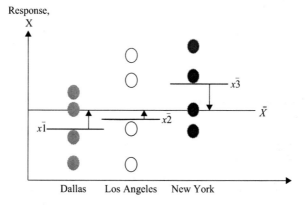

Figure 9.92. Between group variation.

Table 9.48. One-factor ANOVA table

Source of inconsistency (variation)	Degrees of freedom (*df*)	Sum of squares (*SS*)	Mean squares (*MS*)	Statistical test (*F*)
Between Blocks	$b-1$ $= 10-1$ $= 9$	$SSBL$ $= SST - (SSB + SSW)$ $= 1,648,974.20$	$MSBL$ $= SSBL/(b-1)$ $= 183,219.40$	F-ratio $= MSBL/$ MSW $= 2.2565$
SSB (or SS Factor)	$df_1 = k-1$ $= 3-1$ $= 2$	SSB *(SS Factor)* $= 374,121.67$	MSB $= SSB/k-1$ $= 187,060.80$	$F = MSB/$ MSW $= 2.3038$
SSW (or SS Error)	$df_2 = (k-1)(b-1)$ $= (3-1)(10-1)$ $= 18$	SSW (or SS Error) $= 1,461,528.30$	MSW $= SSW/(k-1)$ $(b-1)$ $= 81,196.02$	
Total	$N-1 = 30-1 = 29$	SS *(total)* $= SSBL + SSB + SSW$ $= 3,484,624.20$		

Substituting information from the calculation above,

SSW (or, SS_{error}) = 3,484,624.20 − (374,121.67 + 1,648,974.17) = 1,461,528.30

Construct ANOVA Table 9.48 using information obtained from problem solution.

One-Factor ANOVA Blocking Test: F-*Test Statistics*

Let's continue *F*-test with "high tech" case

5. Set the hypothesis conditions

H_0: $\mu_1 = \mu_2 = \mu_3$

H_a: At least two populations have different means.

6. Calculate the "test statistic"

$$F = \frac{MSBL}{MSW} = \frac{\dfrac{SSBL}{df_1}}{\dfrac{SSW}{df_2}} = \frac{\dfrac{SSBL}{b-1}}{\dfrac{SSW}{(k-1)(b-1)}} = \frac{\dfrac{1,648,974.20}{10-1}}{\dfrac{1,461,528.30}{(3-1)(10-1)}} = 2.2565$$

$$F = \frac{MSB}{MSW} = \frac{\dfrac{SSB}{df_1}}{\dfrac{SSW}{df_2}} = \frac{\dfrac{SSB}{k-1}}{\dfrac{SSW}{(k-1)(b-1)}} = \frac{\dfrac{374,121.67}{3-1}}{\dfrac{1,461,528.30}{(3-1)(10-1)}} = 2.3038$$

7. Rejection area

If $F > F_{\alpha = 0.05}$ reject H_0 otherwise do not reject H_0.

$df_1 = b - 1$

$df_2 = (k - 1)(b - 1)$

8. Decision making

H_0: $\mu_1 = \mu_2 = \mu_3$

H_a: Not all μ_i are the same (blocking is effective).

Because $F_{Calculated} = 2.2565 > F_{\alpha = 0.05} = 2.4563$ at $df_1 = b - 1 = 10 - 1 = 9$ and $df_2 = (k - 1)$ $(b - 1) = (3 - 1)(10 - 1) = 18$. Thus, reject the null hypothesis

H_0: $\mu_1 = \mu_2 = \mu_3$

H_a: At least two populations have different means.

Because $F_{Calculated} = 2.3038 < F_{\alpha = 0.05} = 3.555$ at $df_1 = k - 1 = 3 - 1 = 2$ and $df_2 = (k - 1)$ $(b - 1) = (3 - 1)(10 - 1) = 18$; thus do not reject the null hypothesis.

9.8.3 TWO-WAY ANALYSIS OF VARIANCE

As described earlier, one-way ANOVA is used in analyzing experiment with one independent factor. In one factor, analysis is based on only one single factor with different levels without any interaction. The two-way (also called two factor) ANOVA can be applied in two factor factorial design of experiment (DOE) or designs with more than one factor (Figure 9.93). It examines the effect of two or more independent factors (x's) of interest on the dependent variables (y's) (e.g., effect of pressure (x_A) and temperature (x_B) on the number of defects detected). Further the interaction between the different levels of these two factors ($x_A x_B$) (e.g., does the effect of one particular amount of pressure depend on what temperature is set)?

The goal is to find out if the individual factors (say A or B) have a significant or any effect on the output of experiment. Consequently, there are three sets (A, B, and AB) of hypothesis test, which can be applied to two-factor factorial design. The big difference between the one and two factor ANOVA procedure is in constructing the ANOVA Table. Here the assumptions are based on the populations are normally distributed, populations have equal variances, and independent random samples are drawn. Let's apply this concept in the hypothesis of two-way ANOVA in the following survey.

The HR VP for the Future-Tech would like to examine the effect of gender and employee classification on the yearly Health Care claims for single people. Classification varies from

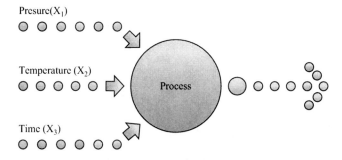

Figure 9.93. General model of a transformation process.

form 1 through 4. Using two-way factorial design the study will show the effect of factor A (gender) and factor B (employee classification) and their interaction.

Survey shows the following data:

Gender	Employee classification (Factor B)					
(Factor A)	Form 1	Form 2	Form 3	Form 4	Total	Average
Male	191,224,200	133,181,101	260,328,352	300,280,240	2790	232.50
Female	236,188,271	270,305,290	150,210,145	150,110,80	2405	200.42

Solution

First we create a Table 9.49 with total and average of each category as follows.

Two-way ANOVA Sources of Variations: Here the source of variations are coming from the two factor of interest are A and B.

a = number of "Factor A" levels = 2
b = number of "Factor B" levels = 4
r = number of replications in each category = 3
n = total number of observations = abr = (2)(4)(3) = 24

The sum of the square always adds up:

$$SS_T = SS_A + SS_B + SS_{AB} + SS_E$$

Total = factor A + factor B + interaction + errors

Table 9.49. Effect of gender and employee classification on the dental claim

Gender	Employee classification (Factor B)					
(Factor A)	Form 1	Form 2	Form 3	Form 4	Total	Average
Male	191,224,200	133,181,101	260,328,352	300,280,240	$T_1 = 2790$,	$\bar{x}_{i=1} = 232.50$
	($R_{11} = 615$)	($R_{12} = 415$)	($R_{13} = 940$)	($R_{14} = 820$)		
	$\bar{x}_{A1B1} : 205$	$\bar{x}_{A1B2} : 138.33$	$\bar{x}_{A1B3} : 313.33$	$\bar{x}_{A1B4} : 273.33$		
Female	236,188,271	270,305,290	150,210,145	150,110,80	$T_2 = 2405$,	$\bar{x}_{i=2} = 200.42$
	($R_{21} = 695$)	($R_{22} = 865$)	($R_{23} = 505$)	($R_{ab} = R_{24} = 340$)		
	$\bar{x}_{A2B1} : 231.67$	$\bar{x}_{A2B2} : 288.33$	$\bar{x}_{A2B3} : 168.33$	$\bar{x}_{A2B4} : 113.33$		
Total:	$S_1 = 1,310$	$S_2 = 1,280$	$S_3 = 1,445$	$S_b = S_4 = 1,160$	**5,195**	
Average:	$\bar{x}_{j=1} = 218.33$	$\bar{x}_{j=2} = 213.33$	$\bar{x}_{j=3} = 240.83$	$\bar{x}_{j=4} = 193.33$	$\bar{\bar{x}} = 216.46$	

where

SS_T = Total variation of all the factors, interactions, and errors

SS_A = Variation due to factor A

SS_B = Variation due to factor B

SS_{AB} = Variation due to interaction between factor A and factor B

SS_E = Inherent variation (error)

Two-Factor ANOVA Equations: The sum of the squares relations for two factors and more are as follows:

Total sum of the squares:

$$SS_T = \sum_{i=1}^{a}\sum_{j=1}^{b}\sum_{k=1}^{r}\left(x_{ijk} - \overline{\overline{x}}\right)^2 \qquad \text{(Eq. 9.100A)}$$

Comparably, we can use alternate format to calculate SS_T:

$$SS_T = \sum x^2 - \frac{T^2}{n} = \sum_{i=1}^{k}\sum_{j=1}^{n_i} x_{ij}^2 - \frac{T^2}{abr}; \text{ where } \sum x^2 \text{ is sum of the squares of all the } (x)s \qquad \text{(Eq. 9.100B)}$$

Sum of the squares of factor A:

$$SS_A = br\sum_{i=1}^{a}\left(\overline{x}_i - \overline{\overline{x}}\right)^2 = br\left(\left(\overline{x}_1 - \overline{\overline{x}}\right)^2 + \left(\overline{x}_2 - \overline{\overline{x}}\right)^2 + \cdots + \left(\overline{x}_{i=a} - \overline{\overline{x}}\right)^2\right) \qquad \text{(Eq. 9.101A)}$$

or, using second technique will give us the same conclusion

$$SS_A = \frac{1}{br}\left(T_1^2 + T_2^2 + \cdots + T_a^2\right) - \frac{T^2}{abr}, \text{ where } T = T_1 + T_2 + \cdots + T_a \qquad \text{(Eq. 9.101B)}$$

Sum of the squares of factor B:

$$SS_B = ar\sum_{j=1}^{b}\left(\overline{x}_j - \overline{\overline{x}}\right)^2 = ar\left(\left(\overline{x}_1 - \overline{\overline{x}}\right)^2 + \left(\overline{x}_2 - \overline{\overline{x}}\right)^2 + \cdots + \left(\overline{x}_{j=b} - \overline{\overline{x}}\right)^2\right) \qquad \text{(Eq. 9.102A)}$$

Likewise we apply the second technique:

$$SS_B = \frac{1}{br}\left(S_1^2 + S_2^2 + \cdots + S_b^2\right) - \frac{T^2}{abr} \qquad \text{(Eq. 9.102B)}$$

Sum of squares of interaction between A and B:

$$SS_{AB} = r\sum_{i=1}^{a}\sum_{j=1}^{b}\left(\overline{x}_{ij} - \overline{x}_i - \overline{x}_j + \overline{\overline{x}}\right)^2 \qquad \text{(Eq. 9.103A)}$$

Similarly, we can use alternate format to calculate SS_{AB}:

$$SS_{AB} = \frac{1}{r}\left(\sum R^2\right) - SS_A - SS_B - \frac{T^2}{abr}; \text{ where } \sum R^2 = \sum_{i=1}^{a}\sum_{j=1}^{b}R_{ij}^2 \qquad \text{(Eq. 9.103B)}$$

$\sum R^2$ = Sum of squares of all replicate totals:
Sum of the squares of errors:

$$SS_E = \sum_{i=1}^{a}\sum_{j=1}^{b}\sum_{k=1}^{r}\left(x_{ijk} - \bar{x}_{ij}\right)^2 \qquad \text{(Eq. 9.104)}$$

where the mean and grand mean of the factors in the populations are calculated as follows:
The mean of each level of factor A:

$$\bar{x}_{i(A)} = \frac{\sum_{j=1}^{b}\sum_{k=1}^{r}x_{ijk}}{br} \qquad \text{(Eq. 9.105)}$$

The mean of each level of factor B:

$$\bar{x}_{j(B)} = \frac{\sum_{i=1}^{a}\sum_{k=1}^{r}x_{ijk}}{ar} \qquad \text{(Eq. 9.106)}$$

The mean of each category (subdivision):

$$\bar{x}_{ij(AB)}\sum_{k=1}^{r}\frac{x_{ijk}}{r} \qquad \text{(Eq. 9.107)}$$

The grand mean:

$$\bar{\bar{x}} = \frac{\sum_{i=1}^{a}\sum_{j=1}^{b}\sum_{k=1}^{r}x_{ijk}}{abr} \qquad \text{(Eq. 9.108)}$$

a = number of "Factor A" levels
b = number of "Factor B" levels
r = total number of replication in each category (subdivision)
The sum of the square calculations for Future Tech is as follows:

$$\bar{x}_{i(A)} = \frac{\sum_{j=1}^{b=4}\sum_{k=1}^{r=3}x_{ijk}}{br} = \frac{x_{i=1,j=1,k=1} + x_{i=1,j=1,k=2} + x_{i=1,j=1,k=3} + \cdots + x_{i=1,j=4,k=1} + x_{i=1,j=4,k=2} + x_{i=1,j=4,k=3}}{br}$$

$$\bar{x}_{1(A)} = \frac{\sum_{j=1}^{4}\sum_{k=1}^{3}x_{ijk}}{br} = \frac{x_{1,1,1} + x_{1,1,2} + x_{1,1,3} + x_{1,2,1} + x_{1,2,2} + x_{1,2,3} + \cdots + x_{1,4,1} + x_{1,4,2} + x_{1,4,3}}{br}$$

$$\overline{x}_{1(A)} = \frac{\sum_{j=1}^{4}\sum_{k=1}^{3}x_{ijk}}{br} = \frac{191+224+200+133+181+101+260+328+352+300+280+240}{(4)(3)}$$

$$= \frac{2790}{12} = 232.50$$

$$\overline{x}_{2(A)} = \frac{\sum_{j=1}^{4}\sum_{k=1}^{3}x_{ijk}}{br} = \frac{236+188+271+270+305+290+150+210+145+150+110+80}{(4)(3)}$$

$$= \frac{2405}{12} = 200.42$$

$$\overline{\overline{x}} = \frac{\sum_{i=1}^{a}\sum_{j=1}^{b}\sum_{k=1}^{r}x_{ijk}}{abr}$$

$$\overline{x}_{j(B)} = \frac{\sum_{i=1}^{a=2}\sum_{k=1}^{r=3}x_{ijk}}{ar} = \frac{x_{j=1,i=1,k=1}+x_{j=1,i=1,k=2}+x_{j=1,i=1,k=3}+x_{j=1,i=2,k=1}+x_{j=1,i=2,k=2}+x_{j=1,i=2,k=3}}{ar}$$

$$\overline{x}_{1(B)} = \frac{\sum_{i=1}^{a=2}\sum_{k=1}^{r=3}x_{ijk}}{ar} = \frac{191+224+200+236+188+271}{(2)(3)} = \frac{1310}{6} = 218.33$$

$$\overline{x}_{2(B)} = \frac{\sum_{i=1}^{a=2}\sum_{k=1}^{r=3}x_{ijk}}{ar} = \frac{133+181+101+270+305+290}{(2)(3)} = \frac{1280}{6} = 213.33$$

$$\overline{x}_{3(B)} = \frac{\sum_{i=1}^{a=2}\sum_{k=1}^{r=3}x_{ijk}}{ar} = \frac{260+328+352+150+210+145}{(2)(3)} = \frac{1445}{6} = 240.83$$

$$\overline{x}_{4(B)} = \frac{\sum_{i=1}^{a=2}\sum_{k=1}^{r=3}x_{ijk}}{ar} = \frac{300+280+240+150+110+80}{(2)(3)} = \frac{1160}{6} = 193.33$$

$$\overline{\overline{x}} = \frac{\sum_{i=1}^{a=2}\sum_{j=1}^{b=4}\sum_{k=1}^{r=3}x_{ijk}}{abr} = \frac{1310+1280+1445+1160}{(2)(4)(3)} = \frac{5195}{24} = 216.46$$

Sum of the squares of factor A:

$$SS_A = br\sum_{i=1}^{a}\left(\overline{x}_i - \overline{\overline{x}}\right)^2 = br\left(\left(\overline{x}_{1(A)} - \overline{\overline{x}}\right)^2 + \left(\overline{x}_{2(A)} - \overline{\overline{x}}\right)^2\right)$$

$$SS_A = (4)(3)\sum_{i=1}^{2}\left(\overline{x}_i - \overline{\overline{x}}\right)^2 = (4)(3)\left((232.50 - 216.46)^2 + (200.47 - 216.46)^2\right) = 6176.04$$

We can achieve the same conclusion using second technique for this calculation:

$$SS_A = \frac{1}{br}\left(T_1^2 + T_2^2 + \cdots + T_a^2\right) - \frac{T^2}{abr} = \frac{1}{(4)(3)}\left(2790^2 + 2405^2\right) - \frac{5195^2}{(2)(4)(3)} = 6176.04$$

Sum of the squares of factor B:

$$SS_B = ar\sum_{j=1}^{b}\left(\overline{x}_j - \overline{\overline{x}}\right)^2 = ar\left(\left(\overline{x}_1 - \overline{\overline{x}}\right)^2 + \left(\overline{x}_2 - \overline{\overline{x}}\right)^2 + \ldots + \left(\overline{x}_{j=b} - \overline{\overline{x}}\right)^2\right)$$

$$SS_B = (2)(3)\sum_{j=1}^{4}\left(\overline{x}_j - \overline{\overline{x}}\right)^2$$

$$= (2)(3)\left(\left(218.33 - 216.46\right)^2 + \left(213.33 - 216.46\right)^2 + \left(240.83 - 216.46\right)^2 + \left(193.33 - 216.46\right)^2\right)$$

$$= 6853.12$$

Or, using second techniques we have

$$SS_B = \frac{1}{br}\left(S_1^2 + S_2^2 + \cdots + S_b^2\right) - \frac{T^2}{abr}$$

$$= \frac{1}{(4)(3)}\left(1310^2 + 1280^2 + 1445^2 + 1160^2\right) - \frac{5195^2}{(2)(4)(3)} = 6853.12$$

Sum of the squares of interactions between factor A and B:

$$\overline{x}_{i=1,j=1(AB)} = \sum_{k=1}^{r}\frac{x_{ijk}}{r} = \frac{x_{i=1,j=1,k=1} + x_{i=1,j=1,k=2} + x_{i=1,j=1,k=3}}{r} = \frac{191 + 224 + 200}{3} = \frac{615}{3} = 205$$

$$\overline{x}_{i=1,j=2(AB)} = \sum_{k=1}^{r}\frac{x_{ijk}}{r} = \frac{x_{i=1,j=2,k=1} + x_{i=1,j=2,k=2} + x_{i=1,j=2,k=3}}{r} = \frac{133 + 181 + 101}{3} = \frac{415}{3} = 313.33$$

$$SS_{AB} = r\sum_{i=1}^{a}\sum_{j=1}^{b}\left(\overline{x}_{ij} - \overline{x}_i - \overline{x}_j + \overline{\overline{x}}\right)^2$$

$$= r\left(\left(\overline{x}_{i=1,j=1} - \overline{x}_{i=1} - \overline{x}_{j=1} + \overline{\overline{x}}\right)^2 + \left(\overline{x}_{i=1,j=2} - \overline{x}_{i=1} - \overline{x}_{j=2} + \overline{\overline{x}}\right)^2 + \cdots + \left(\overline{x}_{i=2,j=4} - \overline{x}_{i=2} - \overline{x}_{j=4} + \overline{\overline{x}}\right)^2\right)$$

$$SS_{AB} = 3\left(\left(205 - 232.50 - 218.33 + 216.46\right)^2 + \left(138.33 - 232.50 - 213.33 + 216.46\right)^2 \right.$$

$$\left. + \cdots + \left(113.33 - 200.42 - 193.33 + 216.46\right)^2\right)$$

$$= 98,578.13$$

Similarly, we can use alternate format to calculate SS_{AB}:

$$SS_{AB} = \frac{1}{r}\left(\sum_{i=1}^{a}\sum_{j=1}^{b}R_{ij}^2\right) - SS_A - SS_B - \frac{T^2}{abr}$$

$$SS_{AB} = \frac{1}{3}\left(615^2 + 415^2 + 940^2 + 820^2 + 695^2 + 865^2 + 505^2 + 340^2\right)$$
$$- 6176.04 - 6853.12 - \frac{5195^2}{(2)(4)(3)}$$
$$= 98,578.13$$

Total sum of the squares:

$$SS_T = \sum_{i=1}^{a=2}\sum_{j=1}^{b=4}\sum_{k=1}^{r=3}\left(x_{ijk} - \overline{\overline{x}}\right)^2 = \left(x_{1,1,1} - \overline{\overline{x}}\right)^2 + \left(x_{1,1,2} - \overline{\overline{x}}\right)^2 + \left(x_{1,1,3} - \overline{\overline{x}}\right)^2$$
$$+ \cdots + \left(x_{2,4,1} - \overline{\overline{x}}\right)^2 + \left(x_{2,4,2} - \overline{\overline{x}}\right)^2 + \left(x_{2,4,3} - \overline{\overline{x}}\right)^2$$

$$SS_T = \sum_{i=1}^{a=2}\sum_{j=1}^{b=4}\sum_{k=1}^{r=3}\left(x_{ijk} - \overline{\overline{x}}\right)^2 = (191 - 216.46)^2 + (224 - 216.46)^2 + (200 - 216.46)^2$$
$$+ \cdots + (150 - 216.46)^2 + (110 - 216.46)^2 + (80 - 216.46)^2$$
$$= 131,173.96$$

Comparably, we can use alternate format to calculate SS_T:

$$SS_T = \sum x^2 - \frac{T^2}{abr}, \text{ where } \sum x^2 \text{ is sum of the squares of all the } (x)s$$

$$SS_T = \left(191^2 + 224^2 + 200^2 + \cdots + 150^2 + 110^2 + 80^2\right) - \frac{5195^2}{(2)(4)(3)} = 131,173.96$$

Consequently,

$$SS_E = \sum_{i=1}^{a=2}\sum_{j=1}^{b=4}\sum_{k=1}^{r=3}\left(x_{ijk} - \overline{x}_{ij}\right)^2$$

Or,

$$SSE = SST - SSA - SSB - SSAB$$
$$= 131,173.96 - 6,176.04 - 6,853.12 - 98,578.13 = 19,566.67$$

Degrees of freedom (df):

The df for individual factor is one less than the number of levels and df for interaction A and B is the product of df of factor A and B. If n is equal to number of readings and a, b, are levels for factors A and B, respectively, then df for single observation are as follows:

$$df \text{ (factor A)} = a - 1$$

$$df \text{ (factor B)} = b - 1$$

$$df \text{ (Interaction A} \bullet \text{B)} = (a - 1)(b - 1) = ab - b - a + 1$$

$$df(\text{total}) = n - 1 = df(\text{factor A}) + df(\text{factor B}) + df(\text{Interaction A} \cdot \text{B})]$$

$$= (a - 1) + (b - 1) + (ab - b - a + 1) = ab - 1$$

For replicated observation design the (df)s are different for "total" and "error"

$$df(\text{total}) = n - 1 = abr - 1$$

$$df(\text{error}) = n - 1 = df(\text{total}) - [df(\text{factor A}) + df(\text{factor B}) + df(\text{Interaction A} \cdot \text{B})]$$

$$= (abr - 1) - [(a - 1) + (b - 1) + (ab - b - a + 1)]$$

$$= ab(r - 1)$$

Degrees of freedom always adds up to

$$df(\text{total}) = n - 1 = (n - ab) + (a - 1) + (b - 1) + (a - 1)(b - 1)$$

Hypothesis Testing Conditions

1. **Factor A:**
   $H_{0,A}$: Factor A is not significant $(\mu_1 = \mu_2 = \mu_3)$
   $H_{\alpha,A}$: Factor A is significant (not all means are equal)

   Reject $H_{0,A}$ if $F_{A(\text{Calculated})} > F_{\alpha, df1, df2}$ with significance level, α, value.

   where F_A is calculated and $F_{\alpha, df1, df2}$ is obtained from F-Table in the appendix, and *degrees freedom* is
   $df_{1(\text{or, A})} = a - 1$ degrees of freedom for factor A
   $df_{2(\text{or, E})} = ab(r - 1)$ degrees of freedom for error
   Further, the mean square factor for A can be computed using

   $$MS_A = \frac{SSA}{df_{1(A)}} = \frac{br \sum_{i=1}^{a} \left(\bar{x}_i - \bar{\bar{x}} \right)^2}{a - 1} \qquad \text{(Eq. 9.109)}$$

   And test statistic

   $$F_A = \frac{MS_A}{MS_E} \qquad \text{(Eq. 9.110)}$$

2. **Factor B:**
   $H_{0,B}$: Factor B is not significant $(\mu_1 = \mu_2 = \mu_3)$
   $H_{\alpha,B}$: Factor B is significant (not all means are equal)

   Reject $H_{0,B}$ if $F_{B(\text{Calculated})} > F_{\alpha, df1, df2}$ with significance level, α, value

   where F_B is calculated and $F_{\alpha, df1, df2}$ is obtained from F-Table in the appendix, and *degrees freedom* is

$df_{1(B)} = b - 1$ degrees of freedom for factor B

$df_{2(E)} = ab(r - 1)$ degrees of freedom for error

In addition, the mean square factor for B can be computed using

$$MS_B = \frac{SS_B}{df_{1(B)}} = \frac{ar\sum_{j=1}^{b}\left(\bar{x}_j - \bar{\bar{x}}\right)^2}{b-1}$$ (Eq. 9.111)

And test statistic

$$F_B = \frac{MS_B}{MS_E}$$ (Eq. 9.112)

3. Interaction AB:

$H_{0,AB}$: No significant interaction between A and B

$H_{\alpha,AB}$: Significant interaction between A and B

Reject $H_{0,AB}$ if $F_{AB(Calculated)} > F_{\alpha,df1,df2}$ with significance level, α, value

where F_{AB} is calculated and $F_{\alpha,df1,df2}$ is obtained from F-Table in the appendix, and *degrees freedom* is

$df_{1(AB)} = (a - 1)(b - 1)$ degrees of freedom for interaction A and B

$df_{2(E)} = ab(r - 1)$ degrees of freedom for error

In addition, the mean square factor for AB can be computer using

$$MS_{AB} = \frac{SS_B}{df_{1(AB)}} = \frac{r\sum_{i=1}^{a}\sum_{j=1}^{b}\left(\bar{x}_{ij} - \bar{x}_i - \bar{x}_j + \bar{\bar{x}}\right)^2}{(a-1)(b-1)}$$ (Eq. 9.113)

And test statistic

$$F_{AB} = \frac{MS_{AB}}{MS_E}$$ (Eq. 9.114)

4. Errors: The mean square errors can be computer using

$$MS_E = \frac{SS_E}{df_{2(E)}} = \frac{\sum_{i=1}^{a}\sum_{j=1}^{b}\sum_{k=1}^{r}\left(x_{ijk} - \bar{x}_{ij}\right)^2}{ab(r-1)}$$ (Eq. 9.113)

where $df_{2(E)} = ab(r - 1)$

And test statistic

$$F_{AB} = \frac{MS_{AB}}{MS_E}$$ (Eq. 9.114)

Note that the denominator of the *F*-test is always the same but the numerator is different.

Table 9.50. Two-factor ANOVA table with interaction (consists of five sources of variation)

Source of inconsistency (variation)	Degrees of freedom (df)	Sum of squares (SS)	Mean squares (MS)	Statistical test (F)
Factor A	$df_1 = a - 1$	SS_A	$MS_A = SS_A/(a-1)$	$F_A\text{-ratio}$ $= MS_A/MS_E$
Factor B	$df_2 = b - 1$	SS_B	$MS_B = SS_B/(b-1)$	$F_B = MS_B/MS_E$
Interaction AB	$df_3 = (a-1)(b-1)$	SS_{AB}	$MS_{AB} = SS_{AB}/((a-1)(b-1))$	$F_{AB} = MS_{AB}/MS_E$
Error	$df_E = n - ab$ $= ab(r-1)$	SS_E	$MS_E = SS_E/(ab(r-1))$	
Total	$df_T = n - 1$ $= abr - 1$	SS_T		

For testing conditions of two-way ANOVA also refer to Table 9.50.
The summary of two-factor ANOVA is listed in the Table 9.51.

Table 9.51. Two-factor ANOVA table with interaction (consists of five sources of variation)

Source of inconsistency (variation)	Degrees of freedom (df)	Sum of squares (SS)	Mean squares (MS)	Statistical test (F)
Factor A (Employee Gender)	$df_1 = a - 1$ $= 2 - 1$ $= 1$	$SS_A = 6{,}176.04$	$MS_A = SS_A/(a-1)$ $= 6{,}176.04$	$F_A\text{-ratio}$ $= MS_A/MS_E$ $= 5.05$
Factor B (Employee Classification)	$df_2 = b - 1$ $= 4 - 1$ $= 3$	$SS_B = 6{,}853.12$	$MS_B = SS_B/(b-1)$ $= 2{,}284.37$	$F_B = MS_B/MS_E$ $= 1.87$
Interaction AB	$df_3 = (a-1)(b-1)$ $= (1)(3) = 3$	$SS_{AB} = 98{,}578.13$	$MS_{AB} = SS_{AB}/((a-1)(b-1))$ $= 32{,}859.38$	$F_{AB} = MS_{AB}/MS_E$ $= 26.87$
Error	$df_E = n - ab$ $= ab(r-1)$ $= (2)(4)(3-1)$ $= 16$	$SS_E = 19{,}566.67$	$MS_E = SS_E/(ab(r-1))$ $= 1{,}222.92$	
Total	$df_T = n - 1$ $= abr - 1$ $= 24 - 1 = 23$	$SS_T = 131{,}173.96$		

Example 9.83

The Two-Way Factorial Design (ANOVA) The molding process engineer would like to investigate the severity of defects (cracks) found during injection molding processes in three different shifts. To determine if there was a difference in three different shifts, an experiment was conducted with one factor having three levels: first shift, second shift, and third shift. The second factor was optimized processes, one level for each of the proposed processes. Data from the experiment, with three replicates per treatment, show the average number of nondefective parts produced per hour over a month period for each shift. Determine if there is any difference in the outcome of three proposed procedures using 10% significant level $\alpha = 0.10$.

Proposed procedure table

Background (Factor A)	Process (Factor B)		
	1	2	3
First Shift	27, 29, 30	26, 27, 24	26, 28, 27
Second Shift	23, 24, 25	19, 18, 17	23, 24, 21
Third Shift	22, 25, 20	19, 15, 14	22, 21, 24

Solution

Let A means background and B mean process, then

Guideline table

Background (Factor A)	Process (Factor B)			Total
	1	2	3	
First Shift	$27 + 29 + 30 = 86$ $= R_{11}$	$26 + 27 + 24 = 77$ $= R_{12}$	$26 + 28 + 27 = 81$ $= R_{13}$	$T_1 = 244$
Second Shift	$23 + 24 + 25 = 72$ $= R_{21}$	$19 + 18 + 17 = 54$ $= R_{22}$	$23 + 24 + 21 = 68$ $= R_{23}$	$T_2 = 194$
Third Shift	$22 + 25 + 20 = 67$ $= R_{31}$	$19 + 15 + 14 = 48$ $= R_{32}$	$22 + 21 + 24 = 67$ $= R_{33}$	$T_3 = 182$
Total	$S_1 = 225$	$S_2 = 179$	$S_3 = 216$	$T = 620$

ANOVA: Apply two factorial design (using stacked data design)
Sum of the square (SS) are:
Factor A (background):

$$SS_A = \frac{1}{br}\left(T_1^2 + T_2^2 + \cdots + T_a^2\right) - \frac{\left(T_1 + T_2 + \cdots + T_a\right)^2}{abr} = \frac{1}{br}\left(T_1^2 + T_2^2 + T_3^2\right) - \frac{\left(T_1 + T_2 + T_3\right)^2}{n}$$

$$SS_A = \frac{1}{(3)(3)}\left(244^2 + 194^2 + 182^2\right) - \frac{\left(244 + 194 + 182\right)^2}{(3)(3)(3)} = \frac{1}{9}(130,296) - \frac{384,400}{27} = 240.296$$

Where, T = total of all $n = abr$ reading that is $T = T_1 + T_2 + T_3 + \cdots + T_a$

Factor B (background):

$$SS_B = \frac{1}{br}\left(T_1^2 + T_2^2 + \cdots + T_a^2\right) - \frac{\left(T_1 + T_2 + \cdots + T_a\right)^2}{abr} = \frac{1}{br}\left(T_1^2 + T_2^2 + T_3^2\right) - \frac{\left(T_1 + T_2 + T_3\right)^2}{n}$$

$$SS_A = \frac{1}{(3)(3)}\left(244^2 + 194^2 + 182^2\right) - \frac{\left(244 + 194 + 182\right)^2}{(3)(3)(3)} = \frac{1}{9}(130,296) - \frac{384,400}{27} = 240.296$$

where T = total of all $n = abr$ reading that is $T = T_1 + T_2 + T_3 + \cdots + T_a$

Interaction AB:

$$SS_{AB} = SS_{Interaction\,AB} = \frac{1}{r}\left(\sum_{i=1}^{a}\sum_{j=1}^{b}R_{ij}^2\right) - SS_A - SS_B - \frac{T^2}{abr}; \text{ where } a = 3,\ b = 3,\ r = 3$$

$$\sum_{i=1}^{a}\sum_{j=1}^{b}R_{ij}^2 = R_{11}^2 + R_{12}^2 + \cdots + R_{1b}^2 + R_{21}^2 + R_{22}^2 + \cdots + R_{2b}^2 + R_{31}^2 + R_{32}^2 + \cdots + R_{3b}^2 + \cdots + R_{a1}^2 + R_{a2}^2 + \cdots + R_{ab}^2$$

$$SS_{AB} = \frac{1}{r}\left(\sum_{i=1}^{a=3}\sum_{j=1}^{b=3}R_{ij}^2 = R_{11}^2 + R_{12}^2 + R_{13}^2 + R_{21}^2 + R_{22}^2 + R_{23}^2 + R_{31}^2 + R_{32}^2 + R_{33}^2\right)$$

$$-\left(SS_A + SS_B + \frac{\left(T_1 + T_2 + T_3\right)^2}{abr}\right)$$

$$SS_{AB} = \frac{1}{3}\left(86^2 + 77^2 + 81^2 + 72^2 + 54^2 + 68^2 + 67^2 + 48^2 + 67^2\right)$$

$$-\left(240.296 + 132.074 + + \frac{\left(244 + 194 + 182\right)^2}{(3)(3)(3)}\right)$$

$$= \frac{1}{3}(43892) - (14609.41) = 21.260$$

Sum of the squares of total replicates are

$$SS_T = \sum x^2 - \frac{T^2}{n} = \sum_{i=1}^{k}\sum_{j=1}^{n_i}x_{ij}^2 - \frac{T^2}{abr}$$

$$SS_T = \left(x_{11}^2 + x_{12}^2 + x_{13}^2 + \cdots + x_{19}^2 + x_{21}^2 + x_{22}^2 + x_{23}^2 + \cdots + x_{29}^2 + x_{31}^2 + x_{32}^2 + x_{33}^2 + \cdots + x_{39}^2\right)$$

$$-\frac{\left(T_1 + T_2 + T_3\right)^2}{abr}$$

$$SS_T = \left(27^2 + 29^2 + 30^2 + 26^2 + 27^2 + 24^2 + 26^2 + 28^2 + 27^2 + 23^2 + 24^2 + 25^2 + 19^2 + 18^2 + 17^2 \right.$$
$$\left. + 23^2 + 24^2 + 21^2 + 22^2 + 25^2 + 20^2 + 19^2 + 15^2 + 14^2 + 22^2 + 21^2 + 24^2 \right)$$
$$- \frac{(244 + 194 + 182)^2}{(3)(3)(3)}$$

$$= 14682 - 14237$$

$$= 444.963$$

$$SS_E = SS_T - SS_A - SS_B - SS_{AB} = 444.963 - (240.296 + 132.074 + 21.260) = 51.333$$

$$MS_A = \frac{SS_A}{a-1} = \frac{240.296}{3-1} = 120.148$$

$$MS_B = \frac{SS_B}{b-1} = \frac{132.074}{3-1} = 66.037$$

$$MS_{AB} = \frac{SS_{AB}}{(a-1)(b-1)} = \frac{21.260}{(3-1)(3-1)} = 5.315$$

$$MS_E = \frac{SS_E}{ab(r-1)} = \frac{51.333}{(3)(3)(3-1)} = \frac{51.333}{18} = 2.852$$

$$F_A = \frac{MS_A}{MS_E} = \frac{120.148}{2.852} = 42.128$$

$$F_B = \frac{MS_B}{MS_E} = \frac{66.037}{2.852} = 23.155$$

$$F_{AB} = \frac{MS_{AB}}{MS_E} = \frac{5.315}{2.852} = 1.864$$

And the two-way ANOVA Table is constructed in Table 9.52.

9.9   LINEAR REGRESSION ANALYSIS

Regression analysis is the process of modeling the estimate of relationship between independent variables (x)s and one or more dependent variable (y)s. The process of explanatory or predictor variable is called simple regression. More than one explanatory variable is referred to multiple regressions.

9.9.1   SCATTER PLOTS AND CORRELATION ANALYSIS

A scatter plot (or scatter diagram) describes linear relationship or association between two dependent and independent variables indicating that a linear regression expression might be applicable. Correlation analysis is used to measure the strength of the linear relationship

Table 9.52. Two-factor ANOVA table with interaction (consists of five sources of variation)

Source of inconsistency (variation)	Degrees of freedom (df)	Sum of squares (SS)	Mean squares (MS)	Statistical test (F)
Factor A (Employee Gender)	$df_1 = a - 1$ $= 3 - 1$ $= 2$	$SS_A = 240.296$	$MS_A = SS_A/(a - 1)$ $= 120.148$	$F_A\text{-}ratio$ $= MS_A/MS_E$ $= 42.128$
Factor B (Employee Classification)	$df_2 = b - 1$ $= 3 - 1$ $= 2$	$SS_B = 132.074$	$MS_B = SS_B/(b - 1)$ $= 66.037$	$F_B = MS_B/MS_E$ $= 23.155$
Interaction AB	$df_3 = (a - 1)$ $(b - 1)$ $= (2)(2) = 4$	$SS_{AB} = 21.260$	$MS_{AB} = SS_{AB}/((a - 1)$ $(b - 1))$ $= 5.315$	$F_{AB} = MS_{AB}/MS_E$ $= 1.864$
Error	$df_E = n - ab$ $= ab(r - 1)$ $= (3)(3)(3 - 1)$ $= 18$	$SS_E = 51.333$	$MS_E = SS_E/(ab(r - 1))$ $= 2.852$	
Total	$df_T = n - 1$ $= abr - 1$ $= 27 - 1 = 26$	$SS_T = 444.963$		

between two variables. A scatter diagram is a plot of the values of response (Y) against the corresponding values of input (X). It is only concerned with strength of the relationship. Let's recall some of the concepts from 8.1.5 for

1. Strong positive linear slope in x–y coordinate means "perfect positive correlation," as shown in Figure 9.94.

$$Slope = \frac{\Delta y}{\Delta x} = +1.0$$

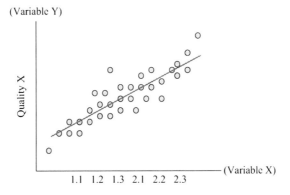

Figure 9.94. Strong positive linear relationship ($r = +1$).

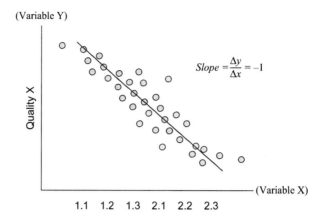

Figure 9.95. Strong negative linear relationship ($r = -1$).

2. Strong negative linear slope in x–y coordinate means "perfect negative correlation," as shown in Figure 9.95.

$$Slope = \frac{\Delta y}{\Delta x} = -1.0.$$

3. Scatter (non-linear) data all over the x–y coordinate (Figure 9.96) means "no correlation." e.g., $Slope = \frac{\Delta y}{\Delta x} = 0.0.$

The population correlation coefficient ρ (*rho*) measures the strength of the relationship between the dependent (y) and independent (x) variables. The sample correlation coefficient "r" is an estimate of the population (ρ) and is used to measure the strength of the linear relationship in the sample drawings. The closer the value of "r" gets to +1 the stronger the positive linear relationship between the two variables. Likewise the closer the value of correlation "r" gets to −1 the stronger the negative linear relationship is between the two variables. Further, as the r nears the zero (0), the weaker the linear relationship exists.

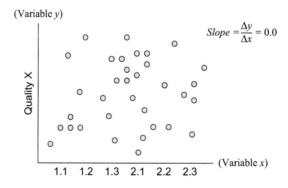

Figure 9.96. Strong negative linear relationship ($r = 0$).

Correlation coefficient calculation:

$$r = \frac{\sum(x-\bar{x})(y-\bar{y})}{\left(\left(\sum(x-\bar{x})^2\right)\left(\sum(y-\bar{y})^2\right)\right)^{1/2}}$$

Or the mathematical algebraic equivalent,

$$r = \frac{n\sum xy - \sum x \sum y}{\left(\left(n\left(\sum x^2\right)-\left(\sum x\right)^2\right)\left(n\left(\sum y^2\right)-\left(\sum y\right)^2\right)\right)^{1/2}} \quad \text{(Eq. 9.115)}$$

where r is unit free sample correlation coefficient, n = sample size, x = independent variable value, y = Dependent variable value.

Suppose that a farmer grows pumpkin and would like to find out the relationship of the pumpkin's high versus the diameter. The farmer obtains the following information from Table 9.53.

$$r = \frac{n\sum xy - \sum x \sum y}{\left(\left(n\left(\sum x^2\right)-\left(\sum x\right)^2\right)\left(n\left(\sum y^2\right)-\left(\sum y\right)^2\right)\right)^{1/2}}$$

$$r = \frac{8(3,142)-(73)(321)}{\left(\left(8(713)-(73)^2\right)\left(8(14,111)-(321)^2\right)\right)^{1/2}} = 0.886$$

The correlation $r = 0.886$ is relatively strong positive linear relationship between the x and y.

Now, we will perform statistical hypothesis testing to find out if there is a linear relationship between pumpkin diameter and its height using $\alpha = 0.05$ significance level.

Table 9.53. Pumpkin growth and their dimension relations

Pumpkin diameter x	Pumpkin height y	xy	x^2	y^2
7	21	147	49	441
7	27	189	49	729
6	33	198	36	1089
8	35	280	64	1225
9	49	441	81	2401
11	45	495	121	2025
12	51	612	144	2601
13	60	780	169	3600
$\sum \bar{x} = 73$	$\sum y = 321$	$\sum xy = 3,142$	$\sum x^2 = 713$	$\sum y^2 = 14,111$

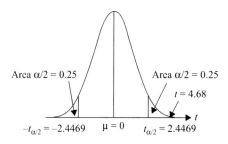

Figure 9.97. Rejection area.

1. Hypothesis testing
   H_0: $\rho = 0$ there is no correlation
   H_α: $\rho \neq 0$ there is correlation
2. Significance test statistic for correlation and calculation:

$$t = r\left(\frac{1-r^2}{df}\right)^{-1/2}, \text{ where } df = n-2$$

$$t = r\left(\frac{1-r^2}{df}\right)^{-1/2} = 0.886\left(\frac{1-0.886^2}{8-2}\right)^{-1/2} = 4.68$$

3. Rejection area
   Reject H_0 if $t_{calculated} > t_{\alpha = 0.05, df = n-2 = 6}$
4. Conclusion
   Reject H_0 because there is enough prove of relationship between two variables at risk factor $\alpha = 0.05$. This is shown in Figure 9.97.

9.9.2 SIMPLE LINEAR REGRESSION MODEL AND ANALYSIS

Scientists always research to discover how one or more variables depend on other one or more independent variables. For instance, is rain water related to air temperature drop? If so, what is the relationship?

Regression analysis is used to predict the value of a dependent variables (y's) based on the one or more independent variable (x's. It describes the effects of the independent variables on the outcome of dependent variables. Thus, changes in the result of "y" caused by the changes of the "x" and relationship between the "x" and "y" referred to a simple linear function ($y = f(x)$) with only one independent variable.

9.9.3 LINEAR REGRESSION MODEL

Population regression model: In regression model settings we have a dependent variable y (also known as response variable) and an independent variable x (further known as repressor or predictor variable). The measured data is collected in pairs (x_1, y_1), (x_2, y_2), ..., (x_{n-1}, y_{n-1}),

(x_n, y_n). The set-up is that the mean of response y depends on the value of X. Thus, the Equation 9.116 mathematically models this concept to as a population linear regression model:

$$y = \beta_0 + \beta_1 X + \epsilon \qquad \text{(Eq. 9.116)}$$

where $y = \beta_0 + \beta_1 X + \epsilon$ is equation of line with β_0 (y-intercept) and β_1 slope.

y: Dependent variable, or response

X: Independent variable

β_0: "y" intercept of population

β_1: Slope coefficient of population

ϵ : Random error term (or residual). Error values are statistically independent and probability distribution of this term is normally distributed and has constant variance for any given value of X.

$\beta_0 + \beta_1 X$: Linear term

Estimated regression model: The sample regression line provides an estimate of the population regression line. This is envisioned in Figure 9.98.

$$\hat{y}_i = b_0 + b_1 x \qquad \text{(Eq. 9.117)}$$

where

$\hat{y}_i$: Predicted (or estimated) y value

b_0: "y" intercept of estimate

b_1: Slope coefficient of estimate

In the estimate model the individual random terms e_i has a mean of zero.

9.9.4 LEAST SQUARE CRITERIA

The "y" intercept (b_0) and slope of estimate (b_1) are developed by computing the values of b_0 and b_1 that minimizes the sum of the squared errors (residuals is the difference between the actual (measured) values in the experiment and the theoretical or true value):

$$\sum e^2 = \sum_{i=1}^{n} (y_i - \hat{y}_i)^2 = \sum (actual - estimate)^2 = \sum_{i=1}^{n} (y_i - (b_0 + b_1 x_i))^2 \qquad \text{(Eq. 9.118)}$$

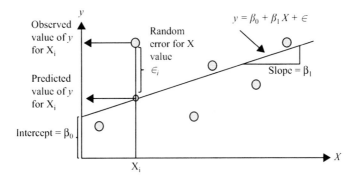

Figure 9.98. Population linear regression.

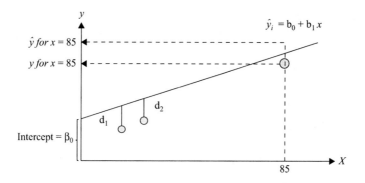

Figure 9.99. Population linear regression least square line.

Least square lin: Figure 9.99 illustrates the concept of population linear regression least square line.

Least square equations: The formulas for least square estimator of the slope b_1 is

$$b_1 = \frac{\sum_{i=1}^{n}(x_i - \bar{x})(y_i - \bar{y})}{\sum_{i=1}^{n}(x_i - \bar{x})^2}$$

The algebraic equivalent expression is

$$b_1 = \frac{Sum\ of\ cross\ product\ for\ xy}{Sum\ of\ squares\ of\ x} = \frac{SCP_{xy}}{SS_x} = \frac{\sum_{i=1}^{n}(x_i - \bar{x})(y_i - \bar{y})}{\sum_{i=1}^{n}(x_i - \bar{x})^2} = \frac{\sum_{i=1}^{n}x_i y_i - \dfrac{\sum_{i=1}^{n}x_i \sum_{i=1}^{n}y_i}{n}}{\sum_{i=1}^{n}x_i^2 - \dfrac{\left(\sum_{i=1}^{n}x_i\right)^2}{n}}$$

(Eq. 9.119)

and least square estimator of the intercept b_0 is

$$b_0 = \bar{y} - b_1\bar{x} = \frac{y}{n} - b_1\left(\frac{\sum_{i=1}^{n}x_i}{n}\right)$$

(Eq. 9.120)

Example 9.84

Bivariate Sample data and Correlation Coefficient: The owner of the Future Tech would like to predict the number of the component FRX sold when they advertised at the low prices. There are no restrictions on the quantity; due to the fact that the company make component FRX as needed. Past history indicates the following records in Table 9.54.

(a) Find the least squares line for X and Y.
(b) The manager thinks that there is a strong relationship between Y and X^2. Find the prediction equation for Y using X^2.

Table 9.54. Component FRX price history

Number of FRX sold (y)	14	16	17	17	13	12	16	17	15	4	13
Price of FRX (x)	2.30	2.10	1.80	1.89	2.50	2.80	1.99	1.90	2.25	2.39	2.70

Solution

i	y	x	$y-\bar{y}$	$(y-\bar{y})^2$	$x-\bar{x}$	$(x-\bar{x})^2$	xy	$Z=x^2$	y^2	$(x^2)y$	Z^2
1	14	2.30	−0.909	0.826	0.062	0.004	32.20	5.29	196	74.06	27.98
2	16	2.10	1.091	1.190	−0.138	0.019	33.60	4.41	256	70.56	19.45
3	17	1.80	2.091	4.371	−0.438	0.192	30.60	3.24	289	55.08	10.50
4	17	1.89	2.091	4.371	−0.348	0.121	32.13	3.57	289	60.69	12.74
5	13	2.50	−1.909	3.644	0.262	0.069	32.50	6.25	169	81.25	39.06
6	12	2.80	−2.909	8.462	0.562	0.317	33.60	7.87	144	94.44	61.94
7	16	1.99	1.091	1.190	−0.248	0.062	31.84	3.96	256	63.36	15.68
8	17	1.90	2.091	4.371	−0.338	0.114	32.30	3.61	289	61.37	13.03
9	15	2.25	0.091	0.008	0.012	0.0001	33.75	5.06	225	75.9	25.60
10	14	2.39	−0.909	0.826	0.152	0.023	33.46	5.71	196	79.94	32.60
11	13	2.70	−1.909	3.644	0.462	0.213	35.10	7.29	169	94.77	53.14
$\Sigma=$	164	24.64	0.001	32.90	0.002	1.134	361.08	56.24	2478	811.42	311.74

(a) The sum of the component calculations are

$$\Sigma x = 2.3 + 2.10 + \cdots + 2.39 + 2.70 = 24.64$$

$$\Sigma y = 14 + 16 + \cdots + 14 + 13 = 164$$

$$\Sigma xy = (14)(2.30) + (16)(2.10) + \cdots + (14)(2.39) + (13)(2.70) = 361.08$$

$$\Sigma x^2 = 2.30^2 + 2.10^2 + 1.80^2 + \cdots + 2.25^2 + 2.39^2 + 2.70^2 = 56.24$$

$$\Sigma y^2 = 14^2 + 16^2 + 17^2 + \cdots + 15^2 + 14^2 + 13^2 = 2478$$

$$\bar{x} = \Sigma x/n = 24.64/11 = 2.24$$

$$\bar{y} = \Sigma y/n = 164/11 = 14.91$$

$$\sum_{i=1}^{n}(x_i - \bar{x})(y_i - \bar{y}) \neq \left(\sum_{i=1}^{n}(x_i - \bar{x})\right)\left(\sum_{i=n}^{n}(y_i - \bar{y})\right)$$

$$b_1 = \frac{Sum\ of\ cross\ product\ for\ xy}{Sum\ of\ squares\ of\ x} = \frac{SCP_{xy}}{SS_x} = \frac{\sum_{i=1}^{n}(x_i - \bar{x})(y_i - \bar{y})}{\sum_{i=1}^{n}(x_i - \bar{x})^2} = \frac{\sum_{i=1}^{n}x_i y_i - \dfrac{\sum_{i=1}^{n}x_i \sum_{i=1}^{n}y_i}{n}}{\sum_{i=1}^{n}x_i^2 - \dfrac{\left(\sum_{i=1}^{n}x_i\right)^2}{n}}$$

$$SCP_{xy} = \sum\nolimits_{i=1}^{n}(x_i - \bar{x})(y_i - \bar{y}) = -5.9818$$

$$SS_x = \sum\nolimits_{i=1}^{n}(x_i - \bar{x})^2 = 1.1328$$

$$SS_y = \sum\nolimits_{i=1}^{n}(y_i - \bar{y})^2 = 32.909$$

$$b_1 = slope = \frac{SCP_{xy}}{SS_x} = \frac{-5.9818}{1.1328} = -5.2805$$

$$b_0 = intercept = \bar{y} - b_1\bar{x} = \frac{y}{n} - b_1\left(\frac{\sum_{i=1}^{n}x_i}{n}\right) = 14.9091 - (-5.2808)(2.2808) = 26.$$

The regression equation is:

$$\hat{y} = b_0 + b_1 x = 26.7279 - 5.2805x$$

$$r = \frac{\sum_{i=1}^{n}(x_i - \bar{x})(y_i - \bar{y})}{\left(\left(\sum_{i=1}^{n}(x_i - \bar{x})^2\right)\left(\sum_{i=1}^{n}(y_i - \bar{y})^2\right)\right)^{1/2}} = \frac{SCP_{xy}}{\left((SS_x)(SS_y)\right)^{1/2}} = \frac{-5.9818}{\left((1.1328)(32.909)\right)^{1/2}} = 0.9797$$

(b) Let $X^2 = Z$, $n = 11$, then the procedure is the same as part (a)

$$\Sigma Z = \Sigma X^2 = 56.2368$$

$$\Sigma y = 164$$

$$\Sigma ZY = \Sigma(X^2)Y = 811.164$$

$$\bar{Z} = 5.1124$$

$$\bar{y} = 14.909$$

$$\Sigma Z^2 = 311.74$$

$$\Sigma Y^2 = 2478$$

$$b_1 = \frac{Sum\ of\ cross\ product\ for\ xy}{Sum\ of\ squares\ of\ x} = \frac{SCP_{zy}}{SS_z} = \frac{\sum_{i=1}^{n}(z_i - \bar{z})(y_i - \bar{y})}{\sum_{i=1}^{n}(z_i - \bar{z})^2} = \frac{\sum_{i=1}^{n}z_i y_i - \dfrac{\sum_{i=1}^{n}z_i \sum_{i=1}^{n}y_i}{n}}{\sum_{i=1}^{n}z_i^2 - \dfrac{\left(\sum_{i=1}^{n}z_i\right)^2}{n}}$$

$$SCP_{zy} = \sum(z - \bar{z})(y - \bar{y}) = -27.2754$$

$$SS_z = \sum(z - \bar{z})^2 = 23.8263$$

$$SS_y = \sum_{i=1}^{n}(y_i - \bar{y})^2 = 32.909$$

$$b_1 = slope = \frac{SCP_{zy}}{SS_z} = \frac{-27.2754}{23.8263} = -5.2805$$

$$b_0 = intercept = \bar{y} - b_1\bar{z} = \frac{y}{n} - b_1\left(\frac{\sum_{i=1}^{n}z_i}{n}\right) = 14.909 - (-1.1448)(5.1124) = 20.7618$$

$$\hat{y} = b_0 + b_1 z = 20.7618 - 1.1448z$$

Or, by replacing x in place of z we get

$$\hat{y} = 20.7618 - 1.1448x^2$$

The FRX computer analysis is shown in Table 9.55.
Regression Analysis: (Y) Sold FRX versus Price (x)
The regression equation is

$$Y = b_0 + b_1 x$$

Or, $Y = y$-Intercept + Slope of line x

$$(Y) \, Sold \, FRX = 26.7 - 5.28 \, Price \, (x)$$

$$S = 0.383076 \quad R\text{-}Sq = 96.0\% \quad R\text{-}Sq \, (adj) = 95.5\%$$

ANOVA

Source	df	SS	MS	F	Significance F
Regression	1	31.5884	31.588	215.256	1.36702E-07
Residual	9	1.32072	0.1467		
Total	10	32.9090			

Table 9.55. Component FRX analysis computer summary output

Regression Statistics	
Multiple R	0.979728217
R Square	0.959867379
Adjusted R Square	0.955408199
Standard Error	0.383076445
Observations	11

	Coefficients	Std. Error	*t*-Stat	*P-value*	Lower 95%	Upper 95%
Intercept	26.73	0.8138	32.84	1.2E-10	24.887	28.569
X Variable 1	−5.281	0.35992	−14.67	1.4E-07	−6.095	−4.467

Mathematical properties of least squares regression coefficients: The least square regression line follows specific mathematical (algebraically) properties such as the following:

1. The sum of the residuals from the least squares regression line is equal to zero $\sum(y-\hat{y})=0$.
2. The least square regression line always passes through the point mean of dependent variable y and the mean independent variable x.

$$\bar{y} = b_0 + b_1\bar{x}$$

3. The mean of estimated or predicted (fitted) value of dependent variable y is equal to the mean of the y values.
4. The least square coefficients are unbiased estimates of population β_0 and β_1.
5. The sum of the squared residuals (errors $\sum(y-\hat{y})^2$) is a minimum.

Explained and unexplained variation (Figure 9.100): The total variation is consisting of the sum of "explained variation" (SS_R) and the "unexplained variation" (SS_E).

$$SS_T = SS_Y = SS_E + SS_R$$

where
SS_T = Total sum of the squares = $\sum_{i=1}^{n}(y_i-\bar{y})^2$ measures the variation of the y_i values around their mean y.
SS_E = Sum of the squares of errors (unexplained variation) = $\sum_{i=1}^{n}(y_i-\hat{y}_i)^2$ attributable to elements (factors) other than x and y relationship.
SS_R = Sum of the squares of regression (explained variation) = $\sum_{i=1}^{n}(\hat{y}_i-\bar{y})^2$ attributable to dependent y and independent variables x relationship. Thus,

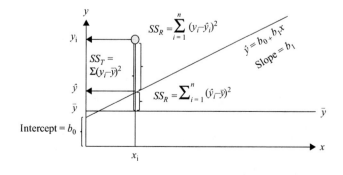

Figure 9.100. Explained and unexplained variation.

$$SS_Y = \sum_{i=1}^{n}(y_i - \bar{y})^2 = \sum_{i=1}^{n}(y_i - \hat{y}_i)^2 + \sum_{i=1}^{n}(\hat{y}_i - \bar{y})^2$$

where

y = Actual values of dependent variables

$\bar{y}$ = Mean values of the dependent variable

$\hat{y}$ = Predicted (estimated) values of y for the given x values

Coefficient of determination, R^2: Known as R^2, it is expressed as the prediction of the future outcome. The higher the R^2, the better the variance that dependent variable is explained by the independent variable. For instance, in Example 9.84 and Table 9.54, $R^2 = 0.96$. This means that the variation in the regression is 96% explained by the independent variable. That is a good regression. However *if* in an ANOVA Table, $R^2 = 0.50$. This means that the variation in the regression is 50% explained by the independent variable. That is not a good regression. Thus,

$$R^2 = \frac{SS_R}{SS_T} = \frac{Sum\ of\ the\ squares\ explained\ by\ regression}{total\ sum\ of\ squares}; where\ 0 \le R^2 \le 1 \quad (\text{Eq. 9.121})$$

In the single independent variable cases, the coefficient of determination is

$$R^2 = r^2$$

where R^2 = Coefficient of determination, r = simple correlation coefficient

When $R^2 = 1$, it indicates perfect x–y linear relationship.

For $R^2 = +1$, 100% of the variation in y is explained by variation in x.

When $0 < R^2 < 1.0$ weaker linear relationship exist between x and y.

And when $R^2 = 0$ no linear relationship exist between x and y. The value of y does not depend on x.

In Example 9.84 and Table 9.54, R^2 is equal to

$$R^2 = \frac{SS_R}{SS_T} = \frac{31.5884}{32.9090} = 0.95987$$

Thus, 95.99% of the variation in number of FRX sold is explained by price of FRX.

Standard error of estimate: The standard deviation of observation about the regression line is estimated by

$$S_\varepsilon = \left(\frac{SS_E}{n-2}\right)^{1/2} \quad (\text{Eq. 9.122})$$

where SS_E = Sum of squares error, n = sample size, and k = number of independent variables (x)s in the model.

Standard deviation of the regression slope: The standard error of the regression slope coefficient (b_1) is estimated by

$$S_{b_1} = S_\varepsilon \left(\sum_{i=1}^{n} (x_i - \bar{x})^2 \right)^{-1/2} = S_\varepsilon \left(\sum_{i=1}^{n} x_i^2 - \frac{\left(\sum_{i=1}^{n} x_i \right)^2}{n} \right)^{-1/2} \qquad \text{(Eq. 9.123)}$$

where

S_{b_1} = Estimate of the standard error of the least squares slope

$S_\varepsilon = \left(\dfrac{SS_E}{n-2} \right)^{1/2}$ = Sample standard error of the estimate

Referring to Example 9.84 and Table 9.54, the standard errors are

$$S_\varepsilon = \left(\frac{SS_E}{n-2} \right)^{1/2} = \left(\frac{1.32072}{11-2} \right)^{1/2} = 0.383076$$

$$S_{b_1} = S_\varepsilon \left(\sum_{i=1}^{n} (x_i - \bar{x})^2 \right)^{-1/2} = (0.383076)(1.134)^{-1/2} = 0.35992$$

Example 9.85

The Simple Linear Regression Model: Future-Tech survey reported that the number of total annual warranty returned from F_1 parts in the west coast (in hundreds) and the number of the engineering staff working on the project are as follows:

i	F_1 Parts (X) in hundreds	Engineering Staff (Y)
1	2.1	15
2	3.8	18
3	4.1	18
4	10.0	59
5	3.2	14
6	3.9	18
7	3.1	24

$n = 7$

1. Find the least square line.
2. Identify the value of the slope and the intercept for the simple linear regression model.
3. Estimate the variance of the error for the model.
4. Find the residual $(y_i - \hat{y}_i)$ for all the y values.

Solution

F_1 parts in hundreds (x)	Engineering staff (y)	xy	x^2	y^2	$x - \bar{x}$	$(x - \bar{x})^2$	$y - \bar{y}$	$(y - \bar{y})^2$
2.1	15	31.50	4.41	225	−2.214	4.903	−8.714	75.939
3.8	18	68.40	14.44	324	−0.514	0.264	−5.714	32.653
4.1	18	73.80	16.81	324	−0.214	0.046	−5.714	32.653
10.0	59	590.00	100.00	3481	5.686	32.327	35.286	1,245.082
3.2	14	44.80	10.24	196	−1.114	1.242	−9.714	94.367
3.9	18	70.20	15.21	324	−0.414	0.172	−5.714	32.653
3.1	24	74.40	9.61	576	−1.214	1.474	0.286	00.082
Σ = 30.20	166	953.10	170.72	5,450	0.00	40.429	0.000	1,513.429

1.
$$\Sigma x = 30.20$$

$$\Sigma y = 166$$

$$\Sigma xy = 953.10$$

$$\bar{x} = 4.314$$

$$\bar{y} = 23.714$$

$$\Sigma x^2 = 170.72$$

$$\Sigma y^2 = 5,450$$

$$n = 7$$

$$b_1 = \frac{\text{Sum of cross product for } xy}{\text{Sum of squares of } x} = \frac{SCP_{xy}}{SS_x} = \frac{\sum_{i=1}^{n}(x_i - \bar{x})(y_i - \bar{y})}{\sum_{i=1}^{n}(x_i - \bar{x})^2} = \frac{\sum_{i=1}^{n} x_i y_i - \dfrac{\sum_{i=1}^{n} x_i \sum_{i=1}^{n} y_i}{n}}{\sum_{i=1}^{n} x_i^2 - \dfrac{\left(\sum_{i=1}^{n} x_i\right)^2}{n}}$$

$$SCP_{xy} = \sum_{i=1}^{n}(x_i - \bar{x})(y_i - \bar{y}) = 236.929$$

$$SS_x = \sum_{i=1}^{n}(x_i - \bar{x})^2 = 40.429$$

$$SS_y = \sum_{i=1}^{n}(y_i - \bar{y})^2 = 1513.429$$

$$b_1 = slope = \frac{SCP_{xy}}{SS_x} = \frac{236.929}{40.429} = 5.860$$

$$b_0 = intercept = \bar{y} - b_1\bar{x} = \frac{y}{n} - b_1\left(\frac{\sum_{i=1}^{n} x_i}{n}\right) = 23.714 - (5.860)(4.314) = -1.569$$

The regression equation is:

$$\hat{y} = b_0 + b_1 x = -1.569 + 5.860x$$

$$r = \frac{\sum_{i=1}^{n}(x_i - \bar{x})(y_i - \bar{y})}{\left(\left(\sum_{i=1}^{n}(x_i - \bar{x})^2\right)\left(\sum_{i=1}^{n}(y_i - \bar{y})^2\right)\right)^{\frac{1}{2}}} = \frac{SCP_{xy}}{\left((SS_x)(SS_y)\right)^{\frac{1}{2}}} \qquad \text{(Eq. 9.124)}$$

$$r = \frac{236.929}{\left((40.429)(1513.429)\right)^{1/2}} = 0.9578$$

2. From equation of line the slope and intercept are

$$\text{Intercept} = b = -1.569$$

$$\text{Slope} = m = 5.860$$

3. Variance of the error for the model

$$SS_E = Sum\ of\ squares\ of\ error = \sum d^2 = \sum_{i=1}^{n}(y_i - \hat{y}_i)^2 = SS_Y - \frac{(SCP_{XY})^2}{SS_X}$$

$$= 1513.43 - \frac{(236.929)^2}{40.429} = 124.927$$

$$Variance:\ s^2 = Mean\ square = MS = \frac{SS_E}{n-2} = \frac{124.927}{7-2} = 24.985$$

4. Computer design outputs is illustrated in Table 9.56
   ANOVA

	df	SS	MS	F	Significance F
Regression (R)	1	1388.502	1388.502	55.573	0.00069
Residual-(Error)	5	124.927	24.985		
Total	6	1513.429			

Table 9.56. Component FRX analysis computer summary output

Computer summary output	
Regression statistics	
Multiple R	0.957838466
R Square	0.917454526
Adjusted R Square	0.900945431
Standard Error	4.998533354
Observations	7

	Coefficients	Std. error	t Stat	P-value	Lower 95%	Upper 95%
Intercept (b_0)	−1.569	3.8823	−0.404	0.703	−11.549	8.411
X Variable 1(b_1)	5.860	0.786	7.455	0.00069	3.840	7.881

9.9.5 INFERENCES ON THE SLOPE β_1, CONCEPT: t-TEST

In simple linear regression, there is only one estimator variable (x). Inferences in linear regression is normally concentrated on the slope β_1 the coefficient of x, which influences how the response variable y depends on the on the independent variable (or estimator) x. Thus, we have an interest in testing hypothesis around the slope of a regression equation of line. Further we can also test hypothesis about the y-intercept β_0 of a line. However, often this is not an interest. An inference procedure for null and alternative hypothesis about the slope β_1 of the regression line can be on the following steps.

t-test for the population slope β_1: To find if there is a linear relationship between x and y.

1. Null and alternate hypothesis
   H_0: $\beta_1 = 0$ there is no linear relationship
   H_a: $\beta_1 \neq 0$ there is a linear relationship
   The alternate can be one-sided or two-sided. This depends on the application.
2. Test condition:
   For H_0: $\beta_1 = 0$ and H_a: $\beta_1 \neq 0$

   Reject null hypothesis H_0 if $t_{calculated} > t_{\alpha/2, df = n-2}$. Or, $t_{calculated} < -t_{\alpha/2, df = n-2}$.

   If H_0 is true, then t-test statistic supports a t-distribution on $n-2$ degrees of freedom. For H_0: $\beta_1 \leq 0$ and H_a: $\beta_1 > 0$

   Reject null hypothesis H_0 if $t_{calculated} > t_{\alpha, df = n-2}$.

For H_0: $\beta_1 \geq 0$ and H_α: $\beta_1 < 0$

Reject null hypothesis H_0 if $t_{calculated} < -t_{\alpha, df = n-2}$.

3. Test statistics with degree of freedom equal to $df = n - 2$.

$$t = \frac{b_1}{S_{b_1}}$$

where $b1$ = sample regression slope coefficient of x, and S_{b1} is estimator of the standard error of the slope.

Notice that the degrees of is $n - 2$ and not $n - 1$. The loss of two degrees of freedom is due to estimating the intercept and the slope of the line.

4. Make a conclusion statement on the decision of reject null hypothesis of fail to reject null hypothesis test.

As shown in the Table 9.57 the summary of the steps.

Table 9.57. Test of hypothesis on the slope of regression line

Two-tailed test

H_0: $\beta_1 = 0$ No linear relationship

H_α: $\beta_1 \neq 0$ Linear relationship does exist

Test Statistic

$$t = \frac{b_1}{S_{b_1}} = b_1 \left(S_{b_1} \right)^{-1}$$

where, $S_{b_1} = s \left(SS_X \right)^{-1/2}$ and $df = n - 2$.

Test: reject if $|t_{calculated}| > t_{\alpha/2, df = n-2}$

One-tailed Test

H_0: $\beta_1 \leq 0$	H_0: $\beta_1 \geq 0$
H_α: $\beta_1 > 0$	H_α: $\beta_1 < 0$

Test statistic for both one tailed are the same:

$$t = \frac{b_1}{S_{b_1}} = b_1 \left(S_{b_1} \right)^{-1}$$

where $S_{b_1} = s \left(SS_X \right)^{-1/2}$ and $df = n - 2$.

Test: reject H_0 if $t_{calculated} > t_{\alpha, df = n-2}$ Test: reject H_0 if $t_{calculated} < -t_{\alpha, df = n-2}$

Example 9.86

Inferences on the Slope β_1, Concept: A nutritionist was interested in the amount of weight loss caused by a specific diet product called Dr-weight-gone. In a controlled group with 18 volunteers, the amount of weight loss was recorded after eighteen days of daily portion of Dr-weight-gone. Let X is the amount, in grams, of Dr-weight-gone given. Let Y be the weight loss in pounds.

X (daily portion of Dr-weight-gone)	Y (weight loss in pounds)
0.10	0.05
0.10	0.08
0.15	0.11
0.15	0.13
0.20	0.19
0.20	0.21
0.25	0.35
0.25	0.31
0.30	0.41
0.30	0.42
0.35	0.43
0.35	0.42
0.40	0.44
0.40	0.47
0.45	0.51
0.45	0.52
0.50	0.54
0.50	0.53

Is there sufficient evidence to conclude that a significance positive relationship exists between the daily portion of Dr-weight-gone (X) given and weight loss (Y)? Use a significance level of 10% ($\alpha = 0.10$).

Solution

x	y	xy	x^2	y^2	$x - \bar{x}$	$(x - \bar{x})^2$	$y - \bar{y}$	$(y - \bar{y})^2$
0.10	0.05	0.01	0.01	0.0025	−0.20	0.04	−0.29	0.08
0.10	0.08	0.01	0.01	0.0064	−0.20	0.04	−0.26	0.07

(Continued)

(Continued)

x	y	xy	x^2	y^2	$x - \bar{x}$	$(x - \bar{x})^2$	$y - \bar{y}$	$(y - \bar{y})^2$
0.15	0.11	0.02	0.0225	0.0169	−0.15	0.02	−0.23	0.05
0.15	0.13	0.02	0.0225	0.0169	−0.15	0.02	−0.21	0.04
0.20	0.19	0.04	0.04	0.0361	−0.10	0.01	−0.15	0.02
0.20	0.21	0.04	0.04	0.0441	−0.10	0.01	−0.13	0.02
0.25	0.35	0.09	0.0625	0.1225	−0.05	0.00	0.010	0.00
0.25	0.31	0.08	0.0625	0.0961	−0.05	0.00	−0.03	0.00
0.30	0.41	0.12	0.09	0.1681	0.00	0.00	0.07	0.00
0.30	0.42	0.13	0.09	0.1764	0.00	0.00	0.08	0.01
0.35	0.43	0.15	0.1225	0.1849	0.05	0.00	0.09	0.01
0.35	0.42	0.15	0.1225	0.1764	0.05	0.00	0.08	0.01
0.40	0.44	0.18	0.16	0.1936	0.10	0.010	0.10	0.01
0.40	0.47	0.19	0.16	0.2209	0.10	0.010	0.13	0.02
0.45	0.51	0.23	0.2025	0.2601	0.15	0.02	0.17	0.03
0.45	0.52	0.23	0.2025	0.2704	0.15	0.02	0.18	0.03
0.50	0.54	0.27	0.25	0.2916	0.20	0.04	0.20	0.04
0.50	0.53	0.27	0.25	0.2809	0.20	0.04	0.19	0.04
5.4	6.12	2.203	1.92	2.56	0.00	0.30	0.00	0.4792

$n = 18$, $\Sigma x = 5.4$, $\Sigma y = 6.12$, $\Sigma xy = 2.203$, $\Sigma x^2 = 1.92$, $\Sigma y^2 = 2.56$, $\bar{x} = 0.30$, $\bar{y} = 0.34$

$$b_1 = \frac{\text{Sum of cross product for } xy}{\text{Sum of squares of } x} = \frac{SCP_{xy}}{SS_x} = \frac{\sum_{i=1}^{n}(x_i - \bar{x})(y_i - \bar{y})}{\sum_{i=1}^{n}(x_i - \bar{x})^2} = \frac{\sum_{i=1}^{n}x_i y_i - \dfrac{\sum_{i=1}^{n}x_i \sum_{i=1}^{n}y_i}{n}}{\sum_{i=1}^{n}x_i^2 - \dfrac{\left(\sum_{i=1}^{n}x_i\right)^2}{n}}$$

$$SCP_{xy} = \sum_{i=1}^{n}(x_i - \bar{x})(y_i - \bar{y}) = 0.367$$

$$SS_x = \sum_{i=1}^{n}(x_i - \bar{x})^2 = 0.30$$

$$SS_y = \sum_{i=1}^{n}(y_i - \bar{y})^2 = 0.4792$$

$$b_1 = \text{slope} = \frac{SCP_{xy}}{SS_x} = \frac{0.367}{0.30} = 1.2233$$

$$b_0 = \text{intercept} = \bar{y} - b_1\bar{x} = \frac{y}{n} - b_1\left(\frac{\sum_{i=1}^{n}x_i}{n}\right) = 0.34 - (1.2233)(5.4/18) = -0.02699$$

The regression equation is

$$\hat{y} = b_0 + b_1 x = -0.02699 + 1.2233x, \text{ this is equation of line } (y = b + mx)$$

Sum of the squares of error is calculated by

$$SS_E = Sum \ of \ squares \ of \ errors = \sum_{i=1}^{n} d_i^2 = \sum_{i=1}^{n} (y_i - \hat{y}_i)^2$$

$$= SS_Y - \frac{(SCP_{XY})^2}{SS_X} = 0.4792 - \frac{(0.367)^2}{0.30} = 0.030237$$

And the sample variance is the sum of the squares of errors divided by degrees of freedom

$$s^2 = \frac{SS_E}{n-2} = \frac{0.030237}{18-2} = 0.00189$$

The standard deviation is square root of sample variance:

$$s = \left(\frac{SS_E}{n-2} \right)^{-1/2} = \left(\frac{0.030237}{18-2} \right)^{-1/2} = 0.043472$$

Now we can apply the hypothesis testing as stated in the problem statement. This a one-tailed test.

1. The hypothesis testing indicated here for one-tailed test are as follows:
2. $H_0: \beta_1 \leq 0$
   $H_\alpha: \beta_1 > 0$
3. Test statistic is

$$t = \frac{b_1}{S_{b_1}} = b_1 \left(S_{b_1} \right)^{-1}$$

where $S_{b_1} = s \left(SS_X \right)^{-1/2}$ and $df = n - 2$.

$$t = \frac{b_1}{S_{b_1}} = \frac{b_1}{s \left(SS_X \right)^{-1/2}} = \frac{b_1 \left(SS_X \right)^{1/2}}{s} = \frac{(1.2233)(0.30)^{1/2}}{0.043472} = 15.407$$

4. The testing procedure is to

$$\text{Reject } H_0 \text{ if } (t_{\text{calculated}}) > (t_{\alpha = 0.10, \ df = n - 2 = 18 - 2 = 16})$$

$$\text{Reject } H_0 \text{ since } (t_{\text{calculated}} = 15.407) > (t_{\alpha = 0.10, \ df = n - 2 = 18 - 2 = 16} = 1.377 \text{ from table})$$

Figure 9.101 visualizes this concept.

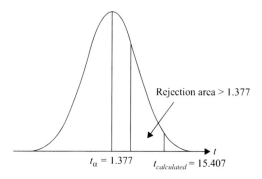

Figure 9.101. t-curve with $df = 16$ rejection area.

Table 9.58. Computer analysis summary output

Regression Statistics	
Multiple R	0.967936868
R Square	0.936901781
Adjusted R Square	0.932958142
Standard Error	0.043471734
Observations	**18**

5. In conclusion positive relationship exists.
6. The computer ANOVA analysis is presented in Table 9.58

ANOVA	df	SS	MS	F	Significance F
Regression	1	0.448963333	0.448963333	237.5729247	5.07898E-11
Residual	16	0.030236667	0.001889792		
Total	17	0.4792			

	Coefficients	Standard error	t Stat	p-value	Lower 95%	Upper 95%
Intercept	−0.027	0.0259	−1.0416	0.3131	−0.0820	0.0279
X Variable 1	1.2233	0.0794	15.4134	5.079E-11	1.0551	1.3916

9.9.6 CONFIDENCE INTERVAL FOR β_1 SLOPE

We now use the t-distribution of the test statistics as previously described to define the confidence interval for population coefficient β_1. The closer this confidence is, the more trust we have in our estimate of population coefficient, and in our equation as a concrete, valid predictor of the dependent variable (y). A $(1 - \alpha) \cdot 100\%$ confidence interval for β_1:

$$b_1 \pm t_{\frac{\alpha}{2}, n-2} S_{b_1} \qquad \text{(Eq. 9.125)}$$

Or

$$b_1 - t_{\frac{\alpha}{2}, n-2} S_{b_1} \text{ to } b_1 + t_{\frac{\alpha}{2}, n-2} S_{b_1}$$

Example 9.87

Determine a 90% confidence interval for the slope of the regression equation used to predict Y (weight loss) in Example 9.86 (weight loss).

$$b_1 \pm t_{\frac{\alpha}{2}, n-2} S_{b_1}$$

$$b_1 - t_{\frac{\alpha}{2}, n-2} S_{b_1} \text{ to } b_1 + t_{\frac{\alpha}{2}, n-2} S_{b_1}$$

$$1.2233 \pm (1.746)(0.0794)$$

$$1.2233 \pm 0.1386$$

$$1.085 \text{ to } 1.362$$

9.9.7 PREDICTION BY REGRESSION ANALYSIS: CONFIDENCE INTERVAL FOR AN INDIVIDUAL Y, GIVEN X

Regression analysis is widely used for *prediction* and *forecasting*. One may need to predict the values of dependent variable (y) based on the values of independent variable (x). Let's consider the following case in Example 9.88.

Example 9.88

The median price of the apartments in Los Angeles increases as the demand goes up. Assuming 11 selected areas of the city prices recorded as shown below.

Areas	Demand rate (X)	Average cost (Y)*10
A1	68.0	98.78
A2	62.80	101.32
A3	68.40	87.16

(Continued)

(Continued)

Areas	Demand rate (X)	Average cost (Y)*10
A4	70.9	109.96
A5	84.20	141.59
A6	75.40	122.94
A7	76.80	173.61
A8	83.60	144.60
A9	62.70	72.80
A10	73.30	125.73
A11	66.80	102.60

1. Find the least square line, identify the values of the intercept, and the slope.
2. Determine the variance of the error for the model.
3. Determine a 95% confidence interval and compare that to prediction *interval* for the cost of an apartment if the demand rate is 72%.

Solution

$n = 11$

1. First we will find the least square line by generating all values in Table 9.59.
   $\bar{x} = 72.0818$, $\bar{y} = 116.4627$, $x_0 = 72\%$
   $n = 11$, $\Sigma x = 792.70$, $\Sigma y = 1{,}281.09$, $\Sigma xy = 94{,}085.46$, $\Sigma x^2 = 54{,}701.63$, $\Sigma y^2 = 157{,}557.50$

Table 9.59. Least square calculation values

(x)	(y)	xy	x^2	y^2	$x-\bar{x}$	$(x-\bar{x})^2$	$y-\bar{y}$	$(y-\bar{y})^2$
68.0	98.78	6,717.04	4,624.49	9,757.49	−4.08	16.66	−17.68	312.68
62.80	101.32	5,961.74	4,678.56	10,265.74	−9.28	86.15	−15.14	229.30
68.40	87.16	5,961.74	4,678.56	7,596.87	−3.68	13.56	−29.30	858.65
70.9	109.96	7,796.16	5,026.81	12,091.20	−1.18	1.40	−6.50	42.29
84.20	141.59	11,921.88	7,089.64	20,047.73	12.12	146.85	25.13	631.38
75.40	122.94	9,269.68	5,685.16	15,114.24	3.32	11.01	6.48	41.96
76.80	173.61	13,333.25	5,898.24	30,140.43	4.72	22.26	57.15	3,265.81
83.60	144.60	12,088.56	6,988.96	20,909.16	11.52	132.67	28.14	791.71
62.70	72.80	4,564.56	3,391.29	5,299.84	−9.38	88.02	−43.66	1,906.43
73.30	125.73	9,216.01	5,372.89	15,808.03	1.22	1.48	9.27	85.88
66.80	102.60	6,853.68	4,462.24	10,526.76	−5.28	27.90	−13.86	192.18
792.7	1,281.09	94,085.46	54,701.63	157,557.50	0.00	547.956	0.00	8,358.26

$$b_1 = \frac{Sum\ of\ cross\ product\ for\ xy}{Sum\ of\ squares\ of\ x} = \frac{SCP_{xy}}{SS_x} = \frac{\sum_{i=1}^{n}(x_i - \bar{x})(y_i - \bar{y})}{\sum_{i=1}^{n}(x_i - \bar{x})^2} = \frac{\sum_{i=1}^{n}x_i y_i - \dfrac{\sum_{i=1}^{n}x_i \sum_{i=1}^{n}y_i}{n}}{\sum_{i=1}^{n}x_i^2 - \dfrac{\left(\sum_{i=1}^{n}x_i\right)^2}{n}}$$

$$SCP_{xy} = \sum_{i=1}^{n}(x_i - \bar{x})(y_i - \bar{y}) = 1{,}742.16$$

$$SS_x = \sum_{i=1}^{n}(x_i - \bar{x})^2 = 547.96$$

$$SS_y = \sum_{i=1}^{n}(y_i - \bar{y})^2 = 8{,}355.26$$

$$b_1 = slope = \frac{SCP_{xy}}{SS_x} = \frac{0.367}{0.30} = 3.18$$

$$b_0 = intercept = \bar{y} - b_1 \bar{x} = \frac{y}{n} - b_1\left(\frac{\sum_{i=1}^{n}x_i}{n}\right) = 116.46 - (3.18)(792.70/11) = -112.71$$

The regression equation is

$$\hat{y} = b_0 + b_1 x = -112.71 + 3.18x,\ this\ is\ equation\ of\ line\ \hat{y} = b + mx = b_0 + b_1 x$$

2. The variance of the error for the model.

$$SS_E = Sum\ of\ squares\ of\ errors = \sum_{i=1}^{n}d_i^2 = \sum_{i=1}^{n}(y_i - \hat{y}_i)^2$$

$$= SS_Y - \frac{(SCP_{XY})^2}{SS_X} = 8{,}358.26 - \frac{(1{,}742.16)^2}{547.96} = 2{,}819.28$$

And the sample variance is the sum of the squares of errors divided by degrees of freedom

$$s^2 = \frac{SS_E}{n-2} = \frac{2{,}819.28}{11-2} = 313.25$$

The standard deviation is square root of sample variance

$$s = \left(\frac{SS_E}{n-2}\right)^{-1/2} = \left(\frac{2{,}819.28}{11-2}\right)^{-1/2} = 17.70$$

3. $\hat{y}$ at $x_0 = 72\%$ and a 95% prediction interval for the cost of an apartment if the demand rate is 72% applied as

$\hat{y} = b_0 + b_1x = -112.71 + 3.18x$, this is equation of line $\hat{y} = b + mx = b_0 + b_1x$

$$\hat{y} = -112.71 + 3.18(72)$$

$$\hat{y} = 11.18$$

$$\hat{y} \pm t_{\frac{\alpha}{2},(df=n-2)} s \left(\frac{1}{n} + \frac{(x_0 + \bar{x})^2}{SS_x} \right)^{1/2} \qquad \text{(Eq. 9.126)}$$

$$116.20 \pm (2.26)(17.73) \left(\frac{1}{11} + \frac{(0.72 + 0.720818)^2}{547.96} \right)^{1/2}$$

$$116.20 \pm (2.26)(17.73)(0.0909 + 0.00)^{1/2}$$

$$116.20 \pm (2.26)(17.73)(0.3015)$$

$$116.20 \pm 12.08$$

$$104.12 \text{ to } 128.28$$

The *prediction interval* for confidence interval estimate of an *individual value* of y given a particular x_0 as shown in Figure 9.95 is

$$\hat{y} \pm t_{\frac{\alpha}{2},(df=n-2)} s \left(1 + \frac{1}{n} + \frac{(x_0 + \bar{x})^2}{SS_x} \right)^{1/2} \qquad \text{(Eq. 9.127)}$$

where
  $\hat{y}$ = Point estimate of the dependent variable (y)
  t = Critical value with degrees of freedom $df = n - 2 = 11 - 2 = 9$
  x_0 = Particular value of independent variable (x) in the sample
  $\bar{x}$ = Average of the independent variable (x) in the sample
  s = Standard error of the estimate

$$116.20 \pm (2.26)(17.73) \left(1 + \frac{1}{11} + \frac{(0.72 + 0.720818)^2}{547.96} \right)^{1/2}$$

$$116.20 \pm (2.26)(17.73)(1 + 0.0909 + 0.00)^{1/2}$$

$$116.20 \pm 41.88$$

$$74.32 \text{ to } 158.08.0$$

Note that by comparing this to the confidence interval of "the mean of y given $x = x_0$" we can see that the individual prediction are significantly less accurate than estimate for the mean cost of an apartment.

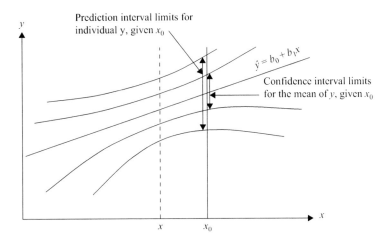

Figure 9.102. Interval estimate for different values of x.

The value of Intercept = −112.7124
The value of Slope = 3.179
Correlation between x and y

$$r = \frac{SCP_{xy}}{\left(\left(SS_x\right)\left(SS_y\right)\right)^{1/2}} = \frac{1,742.16}{\left(\left(548.33\right)\left(8,358.26\right)\right)^{1/2}} = 0.8138$$

Interval estimate for different values of x is featured in Figure 9.102.

9.10   MULTIPLE REGRESSION ANALYSIS

Multiple regression modeling is a powerful methodology in science used for forecasting, predicting, the unknown variables (response y) by applying one or more known independent variables (predictor x).

In a simple linear regression model, a single dependent variable (response y) is function of a single independent variable (predictor x) for each inspection measurement. So the critical estimation of the model is linear function for n observation is

$$y_1 = \beta_0 x_0 + \beta_1 x_1 + e$$

$$y_2 = \beta_0 x_0 + \beta_1 x_1 + e$$

$$\dots \quad \dots \quad \dots \quad \dots$$

$$\dots \quad \dots \quad \dots \quad \dots$$

$$y_n = \beta_0 x_0 + \beta_1 x_n + e$$

However, in most cases, more than one independent variable (predictors) and dependent variable is involved in the problem statement questions and this includes multiple regression analysis for n observations can be written as:

$$y_1 = \beta_0 x_0 + \beta_1 x_{11} + \beta_2 x_{12} + \beta_3 x_{13} + \cdots + \beta_{(m-1)} x_{n(m-1)} + \beta_m x_{nm} + e$$

$$y_2 = \beta_0 x_0 + \beta_1 x_{21} + \beta_2 x_{22} + \beta_3 x_{23} + \cdots + \beta_{(m-1)} x_{n(m-1)} + \beta_m x_{nm} + e$$

$$y_3 = \beta_0 x_0 + \beta_1 x_{31} + \beta_2 x_{32} + \beta_3 x_{33} + \cdots + \beta_{(m-1)} x_{n(m-1)} + \beta_m x_{nm} + e$$

$$\cdots \quad \cdots \quad \cdots \quad \cdots \quad \cdots \quad \cdots \quad \cdots \quad ..$$

$$\cdots \quad \cdots \quad \cdots \quad \cdots \quad \cdots \quad \cdots \quad \cdots \quad ..$$

$$y_n = \beta_0 x_0 + \beta_1 x_{n1} + \beta_2 x_{n2} + \beta_3 x_{n3} + \cdots + \beta_{(m-1)} x_{n(m-1)} + \beta_m x_{nm} + e_m$$

Due to the complexity of multiple regression models, in general the equations are solved by computer software. This would save time and eliminate human calculation errors as well.

9.10.1 MULTIPLE LINEAR REGRESSION MODEL BUILDING

Let's consider n independent variable observations from a random sample of $(x_1, x_2, x_3, \ldots, x_{n-2}, x_{n-1}, x_n)$. So, the relationship between one response y (dependent variable) and predictors two or more x_i (independent variables) for population is

$$y_i = \beta_0 x_0 + \beta_1 x_1 + \beta_2 x_2 + \beta_3 x_3 + \cdots + \beta_{n-2} x_{n-2} + \beta_{n-1} x_{n-1} + \beta_n x_n + e_i \quad \text{(Eq. 9.128)}$$

where β_0 is y-intercept; β_1, β_2, $\ldots$, β_n population slopes, x_0 a dummy variable, is always equal to unity; $x_0 = 1.0$, and "e_i" is random error.

For the estimated multiple regression (Figure 9.104), the above equation becomes

$$\hat{y} = b_0 x_0 + b_1 x_1 + b_2 x_2 + b_3 x_3 + \cdots + b_{n-2} x_{n-2} + b_{n-1} x_{n-1} + b_n x_n \quad \text{(Eq. 9.129)}$$

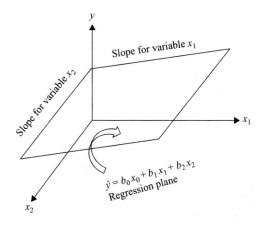

Figure 9.103. Two independent variables hyperplane.

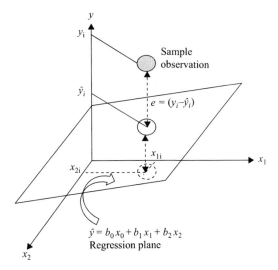

Figure 9.104. Two variable model and hyperplane.

where b_0 is estimated y-intercept; b_1, b_2, ... , b_n estimated slope coefficient or least squares predictors of β_1, β_2, ... , β_n.

The best fit (estimate) equation $\hat{y}$ is achieved by minimizing the sum of the squared errors (Figure 9.104), $\sum e^2$. The multiple linear regression models can be pictured in Figure 9.105.

A residual is defined the difference between the actual response value and the value of its estimate that is $e = (y - \hat{y})$. In other terms, the goal is to find the values of y intercept and slopes that minimize the sum of squares of errors.

$$SS_E = \sum e^2 = \sum_{i=1}^{n}(y_i - \hat{y}_i)^2 \qquad \text{(Eq. 9.130)}$$

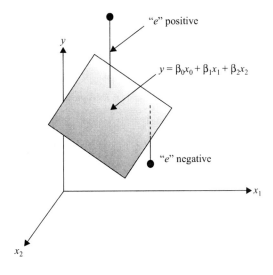

Figure 9.105. Multiple linear regression model.

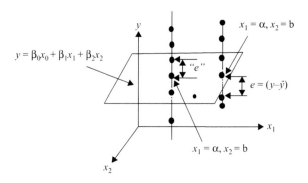

Figure 9.106. Errors in multiple linear regressions with two independent variables.

Note that the errors (or residuals Figure 9.106) from the regression model are normally distributed, centered at zero, with common variance. The mean of the error is zero and the model errors are independent.

Example 9.89

Multiple Linear Regression Forecasting Model: The prediction model $\hat{y} = 3.0 - 2.1x_1 + 7x_2 + 1.4x_3$ was determined from a sample data set with $n = 19$ observations.

Find error degrees of freedom (df). If $x_1 = 2$ value goes up by 2 and $x_2 = 1$ value goes up by 1 with $x_3 = 0$ value constant. Find the net change in prediction model $\hat{y}$.

Solution

The degrees of freedom is

$$df = n - (k+1) = n - k - 1 = 19 - (3 + 1) = 15$$

We substitute the independent variables values in the multiple linear equations given in the problem statement

$$\hat{y} = \beta_0 x_0 + \beta_1 x_1 + \beta_2 x_2 + \cdots + \beta_k x_k; \text{ where } x_0 = 1.$$

$$\hat{y} = 3.0 - 2.1x_1 + 7x_2 + 1.4x_3$$

Substituting the independent variables values to find the net change with the slope we have $\hat{y} = -2.1(2) + 7(1) + 1.4(0)$, net change

$$\hat{y} = 2.8$$

Estimate of σ_e^2 An estimate of σ_e^2 is

$$S_e^2 = \hat{\sigma}_e^2 = \frac{SS_E}{df} = \frac{SS_E}{n - (k+1)} = \frac{SS_E}{n - k - 1} \qquad \text{(Eq. 9.131)}$$

n = Sample size

k = Predictors variables (independent variables)

And the square root of the variance is equal to standard error (residual standard deviation)

$$S_e = \left(\frac{SS_E}{n-k-1} \right)^{1/2} = (MSE)^{1/2}$$

Example 9.90

Suppose for a given problem the sample data is provided as below:

x_1	11	14	17	17	11	12	14	13
x_2	3.1	5.1	5.1	4.4	3.5	3.8	4.9	4.0
y_i	5.4	7.7	9.1	8.8	6.2	7.0	8.4	7.1

Further, the prediction model given is $\hat{y} = 0.214 + 0.341x_1 + 0.608x_2$. Find the estimate of the variance of the error element (S_e^2) of the equation.

Solution

As given in the problem statement the number of predictors are two (k = 2) and number of dependent variables are eight (n = 8) (Table 9.60).

As previously described the general estimate model is

$$\hat{y} = \beta_0 x_0 + \beta_1 x_1 + \beta_2 x_2 + \cdots + \beta_k x_k$$

Table 9.60. Computation of dependent variables

i	y_i	x_1	x_2	$\hat{y} = 0.214 + 0.341x_1 + 0.608x_2$	$e = y_i - \hat{y}$
1	5.4	11	3.1	$\hat{y} = 0.214 + 0.341(11) + 0.608(3.1)$	
2	7.7	14	5.1	$\hat{y} = 0.214 + 0.341(14) + 0.608(5.1)$	
3	9.1	17	5.1	$\hat{y} = 0.214 + 0.341(17) + 0.608(5.1)$	
4	8.8	17	4.4	$\hat{y} = 0.214 + 0.341(17) + 0.608(4.4)$	
5	6.2	11	3.5	$\hat{y} = 0.214 + 0.341(11) + 0.608(3.5)$	
6	7.0	12	3.8	$\hat{y} = 0.214 + 0.341(12) + 0.608(3.8)$	
7	8.4	14	4.9	$\hat{y} = 0.214 + 0.341(14) + 0.608(4.9)$	
8	7.1	13	4.0	$\hat{y} = 0.214 + 0.341(13) + 0.608(4.0)$	

And the estimate model for this problem

$$\hat{y} = 0.214 + 0.341x_1 + 0.608x_2$$

$$Residual = SS_E = \sum e^2 = \sum_{i=1}^{n}(y_i - \hat{y}_i)^2 = \sum(y - 0.214 - 0.341x_1 - 0.608x_2)$$

$$SS_E = \sum_{i=1}^{n}(y_i - \hat{y}_i)^2 = \sum_{i=1}^{n}(y_i - (0.214 + 0.341x_{i1} + 0.608x_{i2}))^2$$

$$= \sum_{i=1}^{8}((y_1 - 0.214 - 0.341x_{11} - 0.608x_{22}) + (y_2 - 0.214 - 0.341x_{21} - 0.608x_{22})$$

$$+ \ldots + (y_8 - 0.214 - 0.341x_{81} - 0.608x_{82}))^2$$

$$SS_E = \sum_{i=1}^{8}((5.4 - 0.214 - 0.341(11) - 0.608(3.1)) + (7.7 - 0.214 - 0.341(14) - 0.608(5.1))$$

$$+ \ldots + (7.1 - 0.214 - 0.341(13) - 0.608(4))^2$$

$$= 0.7505$$

$$df = n - (k + 1) = 8 - (2 + 1) = 5$$

$$S_e^2 = \frac{SS_E}{df} = \frac{SS_E}{n - (k+1)} = \frac{0.7505}{8 - (2+1)} = 0.1501$$

9.10.2 HYPOTHESIS TESTING AND CONFIDENCE INTERVAL

Naturally, one way to evaluate the success of the regression is the nearness of these fitted y values namely $\hat{y}_1, \hat{y}_2, \hat{y}_3, \ldots, \hat{y}_{n-1}, \hat{y}_n$ to the actual observed y values $y_1, y_2, y_3, \ldots, y_{n-1}, y_n$. So, we need to assess whether the overall model is acceptable. Thus, to do that we need to develop regression analysis model using analysis of variances as we previously studied. Let's recall the parameters

$$SS_T = SS_Y = SS_E + SS_R$$

where
SS_T = Total sum of the squares = $\sum_{i=1}^{n}(y_i - \bar{y})^2 = \sum_{i=1}^{n}y_i^2 - \frac{\left(\sum_{i=1}^{n}y_i\right)^2}{n}$ measures the varia-

tion of the y_i values around their mean y.
SS_E = Sum of the squares of errors = $\sum_{i=1}^{n}(y_i - \hat{y}_i)^2$ attributable to elements (factors) other than x and y relationship.

SS_R = Sum of the squares of regression = $\sum_{i=1}^{n}(\hat{y}_i - \bar{y})^2$ attributable to dependent y and independent variables x relationship. Thus,

$$SS_Y = \sum_{i=1}^{n}(y_i - \bar{y})^2 = \sum_{i=1}^{n}(y_i - \hat{y}_i)^2 + \sum_{i=1}^{n}(\hat{y}_i - \bar{y})^2$$

where

y = Actual values of dependent variables

$\bar{y}$ = Mean values of the dependent variable

$\hat{y}$ = Predicted (estimated) values of y for the given x values

And the summary of ANOVA Table

Source	df	SS	MS	F
Regression	k	SS_R	MS_R	$F = MS_R/MS_E$
Residuals	$n - (k+1)$	SS_E	MS_E	
Total	$n - 1$	SS_T		

Test for null hypothesis H_0: all β's = 0

STEP 1

Hypothesis procedure

H_0: $\beta_1 = \beta_2 = \beta_3 = 0$ doesn't contribute to the prediction of y.

H_a: at least one β is not equal to zero. Contribute to the prediction.

STEP 2

Test statistics $F = MS_R / MS_E$ which has F-distribution with n sample size and $k + 1$ number of coefficients and degrees of freedom $df = n - (k + 1)$

STEP 3

Define the rejection area

$$\text{Reject null hypothesis } H_0 \text{ if } F_{\text{Computed}} > F_{\alpha, k, df}$$

where $df = n - k - 1$, k = number of predictors (independent variables), n = sample size. If you *reject* H_0, the outcome is that independent variable doesn't contribute enough to the prediction of y. However, if we *fail to reject* H_0 (Figure 9.107) then we are unable to produce the evidence that predictor (independent variables) making any impact in performance of the dependent variable (prediction y).

STEP 4

Calculate the F value.

STEP 5

Make a statement of conclusion.

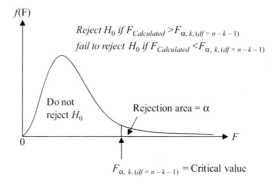

Figure 9.107. *F*-distribution with *k* (*number of predictors*) and $df = n - k - 1$.

Test for null hypothesis H_0: $\beta_i = 0$ versus *Ha*: $\beta_i \neq 0$

When evaluating the individual variables in the multiple regression model, the hypothesis testing involves the following steps.

STEP 1

Define the hypothesis testing procedure

H_0: $\beta_i = 0$

H_a: $\beta_i \neq 0$ at least one β is not equal to zero

STEP 2

Define the test statistic

$$t_{calculated} = \frac{b_i}{S_{b_i}}$$
(Eq. 9.132)

where b_i is the estimates of population coefficient β_i, S_{bi} is the estimated standard error of b_i, and degree of freedom for the test statistic is $df = n - k - 1$.

STEP 3

The *rejection region* (Figure 9.108) for the test of H_0 versus H_a is defined as

Reject null hypothesis H_0 if $| t_{Computed} | > t_{\alpha/2, df = n - k - 1}$.

This term ($t_{\alpha/2, df = n - k - 1}$) is obtained from *t*-table in the appendix.

Reject H_0 if $| t_{Computed} | > t_{\alpha/2, \, df = n - k - 1}$

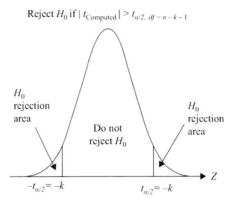

Figure 9.108. t-distribution with and $df = n - k - 1$.

Confidence interval for β_i: One can construct $(1 - \alpha) \cdot 100\%$ confidence interval for β_i based on the t statistic.

$$b_i \pm t_{\frac{\alpha}{2}, \, df = n - (k+1)} S_{b_i}$$

$$b_i - t_{\frac{\alpha}{2}, \, n-k-1} S_{b_i} \text{ to } b_i + t_{\frac{\alpha}{2}, \, n-k-1} S_{b_i} \qquad \text{(Eq. 9.133)}$$

where k is number of independent (predictor) variables used to predict β_i, S_{b_i} is standard error of b_i.

The correlation coefficient and determination The total variation of dependent variable

is driven by SS_T = Total sum of the squares = $\sum_{i=1}^{n} \left(y_i - \overline{y} \right)^2 = \sum_{i=1}^{n} y_i^2 - \dfrac{\left(\sum_{i=1}^{n} y_i \right)^2}{n}$, measures

the variation of the y_i values around their mean y where, n is the sample size. The coefficient of determination (R^2) is required in evaluating the percentages of the above variation which was described by the independent variable in the regression model.

$$R^2 = 1 - (SS_E / SS_T) \qquad \text{(Eq. 9.134)}$$

where, the limit for R^2 is $0 \le R^2 \le 1.0$. In the case of $R^2 = 1$, then 100% of the total variation is concluded, due to the fact that in this case $SS_E = \sum_{i=1}^{n} \left(y_i - \overline{y} \right)^2 = 0$, and so $y_i = \hat{y}_i$ for each inspection in the sample, that is, the equation presents the perfect predictor. However, in reality this may not occur in practice.

Example 9.91

Hypothesis Testing and Confidence Interval for the Parameter β. Bio-Future Company would like to predict the annual income for a chemistry technician based on years of education x_1, quickness (result oriented) x_2, and experience x_3. The following data were collected:

1. Find the least squares prediction model.
2. Determine the residual standard deviation S_e
3. Find out if all the variables help to predict the annual incomes using 10% significance level.
4. Find a 90% confidence interval for β_1
5. Test null hypothesis for $\beta_1 = 0$ at the significance level $\alpha = 0.10$
6. Explain the hypothesis test.
7. Test null hypothesis for $\beta_2 = 0$ at the significance level $\alpha = 0.05$
8. Test null hypothesis for $\beta_3 = 0$ at the significance level $\alpha = 0.05$
9. Repeat part 7 and 8 when $\alpha = 0.10$

Row i	Education x_1	Quickness x_2	Experience x_3	Annual Salary y
1	2	75	7	22,500
2	0	55	11	19,000
3	3	75	6	21,800
4	4	85	1	16,000
5	2.5	60	2	15,800
6	1	45	2	12,500
7	0	65	1	15,000
8	3.5	85	9	26,000
9	0	45	0.50	12,500
10	1	70	3	19,600
11	0	60	0.50	12,400
12	2	65	2	15,120

Solution

The analysis of given data by the computer is shown in the Table 9.61 for regression statistics.

Table 9.61. Summary output of computer analysis

Regression Statistics	
Multiple R	0.942980904
R Square (R^2)	0.889212985
Adjusted R Square	0.847667854
Standard Error (s)	1726.318567
Observations	12

The ANOVA is illustrated in Table 9.62.

Table 9.62. ANOVA table

	df	*SS*	*MS*	*F*	*Significance...F*
Regression (R)	3	191,358,960.3	63,786,320.1	21.403543	0.0003539
Residual (E)	8	23,841,406.35	2,980,175.794		
Total (T)	11	215,200,366.7			

	Coeffi-cients	Standard error	*t*-Stat	*p*-value	Lower 95%	Upper 95%	Lower 90.0%	Upper 90%
Intercept	3,458.36	3,385.23	1.02	0.3369	–4347.99	11264.71	–2836.64	9753.35
X_1 Variable	100.46	556.13	0.181	0.8611	–1181.98	1382.90	–933.698	1134.610
X_2 Variable	164.13	62.31	2.64	0.0299	20.44	307.83	48.257	280.005
X_3 Variable	799.30	151.94	5.26	0.0008	448.91	1149.68	516.747	1081.845

The residual analysis computer output is shown in Table 9.63.

1. The predictor equation is $\hat{y} = 3,458.36 + 100.46x_1 + 164.13x_2 + 799.30x_3$.
2. From the computer output in Table 9.58 the standard deviation $= s_e = 1726.32$.
3. This requires the use of *F*-test so we apply the ANOVA table and the following steps:
   (a) The *hypothesis procedure* is
   $H_0: \beta_1 = \beta_2 = \beta_3 = 0$
   H_a: at least one β is not equal to zero.

Table 9.63. Residuals output

Observation	Predicted Y	Residuals
1	15926.36748	–806.3675
2	12543.29406	–43.2941
3	24954.74071	1045.2593
4	21277.81066	–2277.8107
5	18610.60086	–2610.6009
6	14926.15963	73.8404
7	11243.89416	1256.1059
8	15155.94108	644.0589
9	17445.86189	2154.1381
10	20865.31607	934.6839
11	13705.85727	–1305.8573
12	21564.15611	935.8439

(b) Using information from Table 9.59 the *test statistics* to determine F-ratio is

$$MS_R = SS_R/df = SS_R/k = (191{,}358{,}960.3)/3 = 63{,}786{,}320.1$$

$$MS_E = SS_E/df = SS_E/(n-k-1) = (23{,}841{,}406.35)/(12-3-1) = 2{,}980{,}175.79$$

$$F = MS_R/MS_E$$

$$F_{\text{calculated}} = (63{,}786{,}320.10)/2{,}980{,}175.79 = 21.04$$

(c) The *rejection region* is defined as (Figure 9.109)

Reject null hypothesis H_0 if $F_{\text{Computed}} > F_{\alpha, k, df = n-k-1}$. This term $(F_{\alpha, k, df = n-k-1})$ is obtained from F-Table

$$F_{\text{Computed}} (= 21.04) > F(0.10, 3, 8) = 2.92$$

Thus, in conclusion we *reject* H_0 at least one variable is a significant predictor of annual income for chemistry technician.

4. At 90% **confidence interval** for β_1 can be calculated as

In general we have

$$b_i \pm t_{\frac{\alpha}{2}, df = n - (k+1)} S_{b_i}$$

We now apply the above equation to fond the confidence interval for β_1.

$$b_1 \pm t_{\frac{0.10}{2}, df = 12 - (3+1)} S_{b_1}$$

At $\alpha/2 = 0.05$ and $df = 12 - 3 - 1 = 8$ from t-table we have $t_{0.05} = 1.860$, and from Table 9.61 the standard error is $S_{b1} = 556.13$. Substituting the values in the confidence internal equation we get

$$100.46 \pm (1.860)(556.13)$$

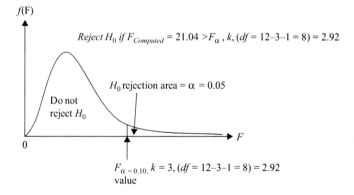

Figure 9.109. *F*-distribution with *k* (*number of predictors*) and $df = n - k - 1$.

$$100.46 - (1.860)(556.13) \; to \; 100.46 + (1.860)(556.13)$$

$$100.46 - 1034.40 \; to \; 100.46 + 1034.40$$

$$-933.70 \; to \; 1134.61$$

5. To test null hypothesis for $\beta_1 = 0$ at the significance level $\alpha = 0.10$ we will also use the ANOVA Table in finding the fact of parameters by using the following steps:
   (a) Let's consider the *hypothesis test* procedure
   $H_0: \boldsymbol{\beta}_1 = 0$
   $H_a: \boldsymbol{\beta}_1 \neq 0$ at least one β is not equal to zero
   (b) The test statistic according to the values obtained from the computer $b_1 = 100.46$, and $S_{b1} = 556.13$.

   $$t_{calculated} = b_1/S_{b1} = 100.46/556.13 = 0.1807$$

   (c) The ***rejection region*** is defined as
   Reject null hypothesis H_0 if $|t_{Computed}| > t_{\alpha/2, \; df \; = \; n \; - \; k \; - \; 1}$. This term $(t_{\alpha/2, \; df \; = \; n \; - \; k \; - \; 1})$ is obtained from t-table
   However the $t_{Calculated} \, (= 0.1807) < t_{(0.05, \; 8)} = 1.860$
   Further, since the value of zero falls in the 90% *confidence interval for* β_1 found in part (4), thus, it will fail to reject null hypothesis test.
6. Not enough sufficient prove to support X_1 significantly contributes for the prediction of annual incomes.
7. Solution is left for students to solve it
8. Solution is left for students to solve it.
9. Solution is left for students to solve it.

9.10.3 POLYNOMIAL AND NONLINEAR REGRESSION MODEL BUILDING

The polynomial regression (or nonlinear) is an alternative way in solving problems when you can't transform model into a linearized form. The multiple regression with powers of independent multiple variables such as one variable quadratic model

$$y = \beta_0 x_0 + \beta_1 x_1 + \beta_2 x^2 + \varepsilon \qquad \text{(Eq. 9.135)}$$

where $x_0 = 0$.
  In general polynomial model looks like

$$y = \beta_0 + \sum_{i=1}^{n} \beta_i x^i + \varepsilon \qquad \text{(Eq. 9.136)}$$

  The ordinary least squares are applied in estimating of linear parameters. So, how the polynomial model is formed?

Let the value of the dependent variable response y changes as a function of independent variables $x_1, x_2, x_3, ..., x_n$ of the n quantitative factors such that

$$y = f(x_1, x_2, x_3, ..., x_n) \tag{Eq. 9.137}$$

where f is called the response function.

The polynomial representation of Equation (9.137) at any point $(x_1, x_2, x_3, ..., x_n)$ in the factor space, or n-dimensional space, can be represented by a regression equation form:

$$Y = \beta_0 + \beta_1 x_1 + \beta_2 x_2 + \cdots + \beta_{11} x_1^2 + \cdots + \beta_{12} x_1 x_2 + \cdots + \beta_{111} x_1^3 + \cdots + \varepsilon$$

And in the summation notation form:

$$Y = \beta_0 x_0 + \sum_{i=1}^{n} \beta_i x_i + \sum_{i=1}^{n} \beta_{ii} x_i^2 + \sum_{i=1}^{n-1}\sum_{j=2}^{n} \beta_{ij} x_i x_j + \sum_{i=1}^{n-2}\sum_{j=2}^{n-1}\sum_{l=3}^{n} \beta_{ijl} x_i x_j x_l + \cdots + \text{etc.} \tag{Eq. 9.138}$$

where x_0, a dummy variable, is equal to unity; $x_1, x_2, x_3, ..., x_n$ are the independent (predictor) variables which effect the y values; β_0, β_i ($i = 1, 2, ..., n$), β_{ij} ($i = 1, 2, ..., n; j = 1, 2, ..., n$), β_{ijl} ($i = 1, 2, ..., n; j = 1, 2, ..., n; l = 1, 2, ..., n$),..., etc. are unknown independent coefficients.

In the case of three factors x_1, x_2, x_3, the corresponding polynomial equation becomes

$$\begin{aligned} y = \beta_0 x_0 + \beta_1 x_1 + \beta_2 x_2 + \beta_3 x_3 & \quad \text{linear terms} \\ + \beta_{11} x_1^2 + \beta_{22} x_2^2 + \beta_{33} x_3^2 & \quad \text{quadratic terms} \\ + \beta_{12} x_1 x_2 + \beta_{13} x_2 x_3 + \beta_{23} x_2 x_3 + e & \quad \text{binary interaction terms} \end{aligned} \tag{Eq. 9.139}$$

The coefficients are

β_0 the level of response
$\beta_1, \beta_2, \beta_3$ measurement of the linear effect
$\beta_{11}, \beta_{22}, \beta_{33}$ measurement of the quadratic effects
$\beta_{12}, \beta_{13}, \beta_{23}$ measurement of the binary interaction effects

This will provide you much more powerful modeling tools than first order equation will. It will be described much more detail in the next chapter (improve) experimental design section.

The summary of hypothesis testing relations is shown in Table 9.64.

9.11 TOLLGATE REVIEW AND DELIVERABLES FOR ANALYSIS PHASE

As expected, before moving to the next phase (improve) you must confirm that you have concluded all the gate meeting conditions for the *Analysis Phase*. You may tailor Figure 9.110 to your product/services processes and come up with solid project analysis techniques. That includes but is not limited to the following list.

Table 9.64. Summary of hypothesis testing relations

H_0: all β_i's $= 0$

H_α: at least one $\beta \neq 0$

$$F = \frac{MS_R}{MS_E} = \frac{R^2 k^{-1}}{\left(1 - R^2\right)\left(n - k - 1\right)^{-1}}$$

$df = n - k - 1$ and $k =$ number of predictors

H_0: $\beta_i = 0$

H_α: $\beta_i \neq 0$ or, H_α: $\beta_i > 0$, or H_α: $\beta_i < 0$

$$t_{calculated} = \frac{b_i}{S_{b_i}}$$

$df = n - k - 1$

Confidence interval for coefficient of population β_i

$$b_i \pm t_{\frac{\alpha}{2}, \, df = n - (k+1)} S_{b_i}$$

Coefficient of determination

$$R^2 = 1 - \left(SS_E\right)\left(SS_T\right)^{-1}$$

where $SS_E =$ Sum of the squares of errors $= \sum_{i=1}^{n}\left(y_i - \hat{y}_i\right)^2$

$$SS_T = \sum_{i=1}^{n}\left(y_i - \bar{y}\right)^2 = \sum_{i=1}^{n}\left(y_i - \hat{y}_i\right)^2 + \sum_{i=1}^{n}\left(\hat{y}_i - \bar{y}\right)^2$$

9.11.1 ANALYSIS PHASE DELIVERABLES AND CHECKLIST

(a) Apply as many as Analysis tools and methodologies such as
    i. Statistical analysis: Hypothesis testing, Confidence Interval, ANOVA (Analysis of Variance), Linear Regression analysis, Multiple regression analysis
    ii. FMEA (Failure mode effect analysis)
(b) Conduct data and process analysis
    i. Conduct value-added and non-value-added steps.
    ii. Develop a model to describe the output variable including the input variables.
(c) Carry out root cause analysis
    i. Develop a Pareto chart (80/20 rule).
    ii. Create a cause and effect (fish bone) diagram.
    iii. Identify the improvement factors and timeline.
    iv. What are the vital few causes in the process and source of variation(s)
(d) Compute the process performance gap and explore the opportunity in closing the gap.

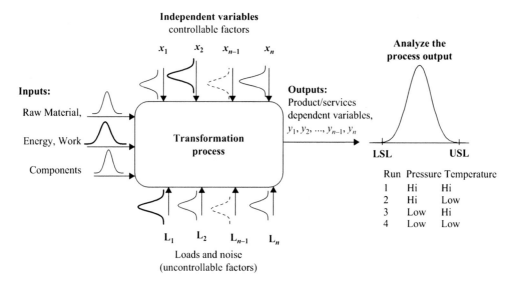

Figure 9.110. Lean Six Sigma transformation process model "Analyze phase."

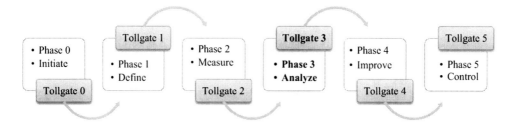

Figure 9.111. Lean Six Sigma Process Improvement projects: Tollgate review—"Analyze phase."

(e) Communicate process outcome using statistical analysis.
(f) Identify the deliverable to be achieved with end-to-end timeline and target:
   i. Has the team analyzed and concluded the potential bottlenecks?
   ii. Can the problem fixed in short term (quick wins) and what are the benefits or long term what are the impacts in process and product availability?
(g) Identify the financial business requirements and resources for completion of this phase.
(h) Create and define RACI-Matrix. Include team members, gate keepers, and anyone who should be part of RACI-Chart.
(i) Concentrate on the Voice-of-Customer (VOC).

Once the checklist is completed, the project must to go through the Tollgate review (Figure 9.111) as explained in Chapter 6 for gate review team approval.

CHAPTER 10

IMPROVE CONCEPTS AND STRATEGIES

EXPERIMENTAL ROBUST ENGINEERING DESIGNS AND ANALYSIS AND FAILURE MODE EFFECT ANALYSIS

The goal of the improve phase is to optimize, design, or redesign the existing process, benchmark the best practices, brainstorm ideas, and develop a breakthrough solution. Normally, carrying out a pilot implementation to improve the new Sigma level is necessary. Some of the tools utilized here are Kaizen events, 5S principles, Lean tools (eliminate non-value-added activities), design of experiment, robust engineering design and analysis, response surface methods, regression, and statistical tools discussed in analysis.

10.1 ADVANCED LEAN SIX SIGMA EXPERIMENTAL DESIGN

Experimental design is the part of statistics that takes place before carrying out an experiment. It is a scientific process that includes designing, implementing, experimenting, analyzing, and translating controlled experiments to understand the behavior of independent factors (inputs) that changes the outcome (dependent factors) of a parameter(s).

If a well-designed experiment is executed, it produces statistically useful results on the response variable based on one or two and even more independent factors. The less the number of the design factors the more expensive and time-consuming the experimental process, however, simpler to model the process. On the other hand, the more independent design factors the more efficient the experimental process but the higher the complexity of the process to model. This may require computer to tackle the computation. Key concepts in designing an experimental design are blocking, randomization, and replication.

A strategically planned experiment should have particular statistical test in mind. The experimental goal is to investigate new phenomena or solve a practical problem (or a process). What we learn from an experiment depends on "where" we look, "how" we look, and the scope of our view. Further, the outcome of an experiment should be reported on the following:

1. What was done (the scope of the investigation)?
2. How was it done (the methodologies applied during the investigation)?
3. What was result (the scientific description of independent and dependent variables relationship)?

10.1.1 EXPERIMENTAL DESIGN TERMINOLOGY

Most frequently used terminologies in design of experiment (DOE) are the following:

Blocking variable—The experimental variables cannot be randomized. Normally, the randomization is carried out during experimental runs for each level of variables.

Center point—Points at the center value of all independent variables.

Coding factor levels—Converting the values of measurement for a factor so that the low value becomes "–1," medium "0," and the high value "+1."

Controllable variable—The quantities that scientists want them to remain constant throughout the experiment.

Data—The sample values obtained from experimental outcome.

Design of Experiments (DOE)—Or experimental design is the plan for collection of sample.

Experiments—Process of collecting sample data which is used to study the effect of parameters as they are set at various levels.

Experimental unit—Object upon which the response (y) is measured.

Error—Unexplained variations in the measured values.

Factors ("x" Manipulated (independent) variables)—Experimental input factors or independent variables ($x_{(s)}$ continuous or discrete) that a researcher manipulates to obtain any changes in the response of the process. In other words, this is the process input that one can adjust to affect the response (e.g., pressure, temperature, speed, rate, viscosity).

Independent variable—The variable that experimenter can adjust to obtain the desired outcome.

Dependent variable—The outcome or response value of the experiment which depends on the independent variable condition.

Interaction—Occurs when the effect of one factor on a response depends on the level of another factor. Further effect of one input (i.e., $x_1, x_2, ..., x_k$) factor depends on another input factor.

Level—The scale of low, medium, and high estimated by a factor in an experiment.

Levels (high/low)—Specific values of factors (inputs) are available when an experiment is carried out at two levels normally denoted by low (–1) or high (+1). Or three levels that are denoted by low (–1), medium (0), and high (+1). This is the value that experimenter will test.

Noise—Uncontrollable factors or experimental error. This can be reduced by assigning treatments to experimental units.

Replication—Completely repeat the experiment with the same conditions and inputs.

Response variable (y)—Dependent variable or output values measured to describe the output response of the process. This is the value that we want to predict or optimize (e.g., total execution time, capacity, down time, quality, part dimension).

Treatment combination (run)—Performing the same experimental combination more than once.

Volume—Size of information in an experiment.

Uncontrollable variable—The experimental noise or error factor.

Experimentation—An experimental design is a method of systematically planning a detailed experiment before carrying out the experiment. It is also a technique in reducing cost of investigation in a timely manner and understanding the process by avoiding trial and error procedures.

Trial and error process—Collecting data through the trial and error is basically build-test-fix procedure. It is impossible to know if the true optimum process is achieved or not. It would be a very inefficient and ineffective attempt to understand and optimize product design and processes, either manufacturing or transactional. It consistently requires optimization over and over, repairs, fire-fighting, and perhaps even luck. There is no science behind it.

Design of Experiment (DOE)—Unlike trial and error, data obtained from design of experiment can be converted into a mathematical model, which could be used for process optimization and simulation using a computer. The theory of DOE normally starts with the idea of process models illustrated in Figure 10.1. However, most empirical models (or equations) of experimental data are either in a linear (first degree polynomial) or quadratic or in a nonlinear (second degree polynomial) form and often it includes third degree (Cubic) as shown in Section 10.6.

There are numerous experimental design methods available. One of the most popular is central composite design (CCD), particularly for second-degree polynomials. The CCD uses full factorial treatment. It is an accurate design with all the possible runs (trials) for specified variables. However, economically, for any treatment higher than three variables, it gets to be

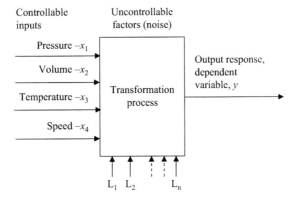

Figure 10.1. General model of transformation process.

a more expensive process because full factorial uses all the levels of each factor. Thus, other methods, such as fractional factorial come into the picture, which use fewer runs than full factorial. The choice or type of the experiment depends on the financial viability for experimenting, as well as time and resources.

The success of DOE depends on the methodology of each design and factor selection. To accelerate the factor selection, one may design a fishbone or cause-and-effect diagram with the team that is working on the project or the process. Or conduct a DOE with multiple factors and determine the most influential factors after the analysis. Then, by using the factors that had the highest impact in the process, design a new experiment to determine the effects of each factor on the output of the process. However, any design should have as many unknown runs as possible. In any experiment, the higher the number of runs, the higher the accuracy of the experiment and of the prediction or response model.

Figure 10.1 shows the general model of transformation process, which includes the mathematical model of a process in which the dependent variable response (y) varies as a function of independent variables $x_1, x_2, x_3, ... x_{n-2}, x_{n-1}, x_n$ of n quantitative variables such that

$$\text{Output (dependent variable)} = f(\text{Input variables})$$

Or,

$$y = f(x) = f(x_1, x_2, x_3, ... x_{n-2}, x_{n-1}, x_n) \qquad \text{(Eq. 10.1)}$$

where f is called response function and y is a dependent variable (in Six Sigma terms strategic objective measurement values or symptom) and x is an independent variable (or cause).

In the case of pressure, temperature, volume, and speed condition, Equation 10.1 becomes

$$y = f(x_1, x_2, x_3, x_4) \qquad \text{(Eq. 10.2)}$$

where y is the response, product performance, or yield and x_1, x_2, x_3, x_4 represent pressure, temperature, volume, and speed condition, respectively.

10.1.2 ELEMENTS OF AN EXPERIMENTAL DESIGN

The essential principles of a statistically experimental design include the following.

1. *Experimental goals and objectives*: Understand the process relationship between process input variables ($x_{(s)}$) and process output variables ($y_{(s)}$). This is an efficient technique of experimentation that identifies key process input variables and their optimum settings that effect process mean and response variations with minimum trials. Additionally the following should be done.

   (a) Create critical and noncritical variables selection list.

   (b) Distinguish dominant process performance improving variables.

   (c) Introduce the identified variables that contribute to the process mean improvement.

   (d) Apply linear and nonlinear multivariable regression modeling and analysis.

   (e) Design an experiment to improve the stated problem and develop a mathematical process model using the data obtained from the experiment. This model will be used in further optimizing and simulating the process.

(f) Omit the process steps that are not essential. Convert the steps to those that will achieve Lean Six Sigma objectives.

(g) Use the statistical tool to optimize your process such as: full factorial design, fractional factorial design, central composite design, response surface analysis, Taguchi, and ANOVA. Use the model obtained from the experimental results to optimize the process through simulation and other techniques.

(h) Implement the optimized process in the production line.

2. *Experimental conditions*: The randomization treatments are more accurate, unbiased, and uniform blocks in an experiment. In other words, if all the experimental trials run in a random order, then statistical analysis of output will be unbiased. Another step is the replication that is the repetition of entire experiment to estimate the experimental error or increase the accuracy.

3. *Specify the methods of analysis*: Statistically one needs to identify the analysis method in achieving the objectives.

10.2   ONE-FACTOR-AT-A-TIME DESIGN (OFATD) $X_1, X_2, ..., X_K$

One factor at a time (OFAT) is a method of experimental design that includes two levels of high and low values and one factor only. It runs all the factors on one condition and holds the other factors constant. Further, just like trial and error it requires more runs, is unable to predict interactions, can miss the optimal points, and is a slow, very expensive, very inefficient, ineffective, and time-consuming process. So, OFAT is appropriate when you are interested in developing a functional relationship between a factor and the response, and you know that interactions between that factor and others are unimportant. This method applies when there is only one factor of interest or importance as shown in Figure 10.2.

On the other hand, other methods in which statistically the experimental design changes multiple factors simultaneously are far more efficient and effective when using two or more factors.

Example 10.1

A manufacturing engineer realized that one of the dimensions in a part is out of the tolerance (actual dimension = 1.677). He asked the process engineer to make the appropriate changes to bring the tolerance within the specification limit.

Solution

The process engineer decided to do one factor at a time design as illustrated in Table 10.1 and 10.2. The experiment is consist of the following: (1) design 2 levels (high and low) temperature, T_1 and T_2, and two level of pressures, P_1 and P_2: testing the part at both T_1 and T_2, holding the pressure constant at P_1.

Figure 10.2 One factorial
$2^{k-1} = 2^1 = 2$ design.

Table 10.1. One factor-at-a-time-design: Round 1

Run	Plastic pressure	Temperature	Response (y)
1	P_1	T_1	1.673
2	P_1	T_2	1.675

Table 10.2. One factor-at-a-time-design: Round 2

Run	Plastic pressure	Plastic temperature	Response (y)
1	P_1	T_2	1.675
2	P_2	T_2	1.676

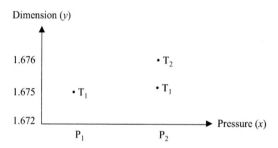

Figure 10.3. Pressure and temperature interaction at "first glance".

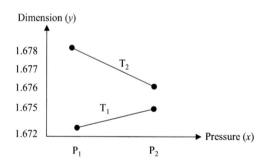

Figure 10.4. Pressure and temperature interaction.

Since the objective is to reduce the variation, from the experiment it seems that T_2 is closer to the desired dimension. So, next we hold T_1 constant and vary the pressure.

At first one may think that pressure and temperature at settings P_2 and T_2 are improving the part dimension (Figure 10.3). Nevertheless, this experimental design is referred as one factor at a time, which do not includes the interactions between the pressure and temperature.

Let's assume that Figure 10.4 represents the untested combination of trial P_2T_1, which would have resulted if it was tested. Figure 10.4 illustrates that interaction relationship between

y and x and has a different slope depending on which temperature, either T_1 or T_2, is considered. Thus, based on Figure 10.4 pressure and temperature interact such that as pressure changes from P_1 to P_2, the variation in part dimension is also changes, depending on the setting values of temperatures T_1 and T_2.

The choice of design of experiment depends on the goal and objectives of the experiment as well as the number of the variables (factors) to be investigated. If you have a few factors to investigate the full factorial design might be the best choice. One of the advantages of the full factorial is that it includes all the attainable combination.

10.3   FULL FACTORIAL DESIGN

A full factorial design is the simplest design to develop, but economically it is not feasible (extremely inefficient) because as the number of the factors increases, it requires more resources. Since each factor is tested at each condition of the factor, only the highest factors that make the biggest impact in the process are included. Moreover, it is efficient and effective if the number of the factors remains within three or four and number of the levels within two, because it covers all the possibilities. This is why the vast majority of the full factorial designs are limited to two factor–two level or three factor–two level. The design becomes more complicated once the number of the level exceeds two and it requires computer software for analysis.

Full factorial design allows the researcher to evaluate the effect of each factor (x) on the output (y) or dependent variable, as well as the effects of interactions (i.e., $x_1 x_2$, $x_1 x_3$, or $x_2 x_3$) of the independent variables on the dependent variable (response y_s) as discussed in Chapter 9. Thus, full factorial design is important for screening to see which effects and interactions are essential for further investigation but it cannot identify nonlinear effects. Moreover, the interactions takes place naturally because of the levels you set individually (i.e., for A and B the interaction AB occurs automatically).

Furthermore, if the number of factors and combinations is too high to be economically feasible, a fractional factorial design (see Section 10.4) may be carried out, in which the number of experimental runs are reduced.

Number of experimental runs: The experimental runs is determined using the equation

$$n = L^k \qquad \text{(Eq. 10.3)}$$

where n is number of test trials
  L = number of design level or process condition (i.e., low, medium, high)
  k = number of design factors

For instance, in a design of two levels ($L = 2$) with three factors ($k = 3$) the number of runs equals to

$$n = L^k = 2^3 = 8$$

A sample of full factorial design with three factors and two level is displayed in Table 10.3.

Table 10.3. DOE: Full factorial design $n = L^k = 2^3 = 8$ layout without response results

	Factors			Factors coded values			Expanded layout interactions coded values				Response value
Trial	x_1	x_2	x_3	x_1	x_2	x_3	$x_1 x_2$	$x_1 x_3$	$x_2 x_3$	$x_1 x_2 x_3$	y
1	Low	Low	Low	−1	−1	−1	+1	+1	+1	−1	
2	Low	Low	High	−1	−1	+1	+1	−1	−1	+1	
3	Low	High	Low	−1	+1	−1	−1	+1	−1	+1	
4	Low	High	High	−1	+1	+1	−1	−1	+1	−1	
5	High	Low	Low	+1	−1	−1	−1	−1	+1	+1	
6	High	Low	High	+1	−1	+1	−1	+1	−1	−1	
7	High	High	Low	+1	+1	−1	+1	−1	−1	−1	
8	High	High	High	+1	+1	+1	+1	+1	+1	+1	

Note that each factor is experimented at each level "high" and "low" four times.

In Table 10.3 "−1" means set factor to low level in this run and "+1" means set factor to high level in this run; run order should be randomized. Failure to randomize is very risky for factor x_1, since it has runs 1 to 4 at low level and 5 to 8 at high level. The researcher always should perform the experiments in "random" order. The randomization will make any variable possibly to contribute to random uncertainty rather than orderly error.

In experimental design, levels do not have to be numeric data. Factors can be either variable data (mathematical numbers) or attribute data (i.e., on/off, go/no go, high/low, red/blue). Once the experiment is completed and the response data collected, then one needs to compute the factor effects.

10.3.1 HOW TO CALCULATE THE EFFECTS

Follow the steps.

1. Calculate the *"effects of response"* of each variable (factor) and the interactions as well.
   (a) Single replication: compute the following:
      i. Sum of all response values when the factor is at the high level (+1)

$$\sum_{i=1}^{n_{(-1)}} y_{i_{(-1)}} = y_{1_{(-1)}} + y_{2_{(-1)}} + y_{3_{(-1)}} + \ldots + y_{n_{(-1)}}$$ (Eq. 10.4)

      ii. Sum of all response values when the factor is at the low level (−1)

$$\sum_{i=1}^{n_{(-1)}} y_{i_{(-1)}} = y_{1_{(-1)}} + y_{2_{(-1)}} + y_{3_{(-1)}} + \ldots + y_{n_{(-1)}}$$ (Eq. 10.5)

      iii. Difference ("Δ" delta) of high and low levels (difference = High total − Low total). Or, you can compute the difference by multiplying each high (+1) or low

(−1) level by the response for its row, and then summing all the values in the column.

$$\Delta_{Effect} = \sum_{i=1}^{n_{(-1)}} y_{i_{(-1)}} - \sum_{i=1}^{n_{(-1)}} y_{i_{(-1)}} \qquad \text{(Eq. 10.6)}$$

iv. Effect is equal to the difference divided by the number of runs at each level or

$$Effect = \frac{High\ level\ total - Low\ level\ total}{number\ of\ runs\ at\ each\ level} = \frac{\Delta_{Effect}}{n_{level_{(high\ or\ low)}}} \qquad \text{(Eq. 10.7)}$$

(b) Multiple replications

  i. Sum of all response values when the factor is at the high level (+1)

$$\overline{\overline{y}}_{(+1)} = \frac{\sum_{i=1}^{n_{(-1)}} \overline{y}_{i_{(-1)}}}{n_{(+1)}} = \frac{(\overline{y}_{1_{(-1)}} + \overline{y}_{2_{(-1)}} + \overline{y}_{3_{(-1)}} + \ldots + \overline{y}_{n_{(-1)}})}{n_{(+1)}} \qquad \text{(Eq. 10.8)}$$

where $\overline{y}_{1_{(-1)}} = \dfrac{\sum_{i=1}^{n} y_{i_{(-1)}}}{n}$, $\overline{y}_{2_{(-1)}} = \dfrac{\sum_{i=1}^{n} y_{i_{(-1)}}}{n}, \ldots, \overline{y}_{n_{(-1)}} = \dfrac{\sum_{i=1}^{n} y_{n_{(-1)}}}{n}$,

n = number of response replicates.

  ii. Sum of all response values when the factor is at the low level (−1)

$$\overline{\overline{y}}_{(-1)} = \frac{\sum_{i=1}^{n_{(-1)}} \overline{y}_{i_{(-1)}}}{n_{(-1)}} = \frac{(\overline{y}_{1_{(-1)}} + \overline{y}_{2_{(-1)}} + \overline{y}_{3_{(-1)}} + \ldots + \overline{y}_{n_{(-1)}})}{n_{(-1)}} \qquad \text{(Eq. 10.9)}$$

where $\overline{y}_{1_{(-1)}} = \dfrac{\sum_{i=1}^{n} y_{i_{(-1)}}}{n}$, $\overline{y}_{2_{(-1)}} = \dfrac{\sum_{i=1}^{n} y_{i_{(-1)}}}{n}, \ldots, \overline{y}_{n_{(-1)}} = \dfrac{\sum_{i=1}^{n} y_{n_{(-1)}}}{n}$,

n = number of response replicates.

  iii. Difference of high and low levels (difference = high total − low total). Or, you can compute the difference by multiplying each high (+1) or low (−1) level by the response for its row, and then summing all the values in the column.

$$\Delta_{Effect} = \overline{\overline{y}}_{(+1)} - \overline{\overline{y}}_{(-1)} \qquad \text{(Eq. 10.10)}$$

2. Calculate the "*effects of standard deviation*" of each variable (factor) and the interactions as well.

   (a) First method of multiple replications:

   i. Sum of all standard deviation values when the factor is at the high level (+1)

$$\sum_{i=1}^{n_{(-1)}} s_{i_{(-1)}} = s_{1_{(-1)}} + s_{2_{(-1)}} + s_{3_{(-1)}} + \ldots + s_{n_{(-1)}} \qquad \text{(Eq. 10.11)}$$

ii. Sum of all standard deviation values when the factor is at the low level (−1)

$$\sum_{i=1}^{n_{(-1)}} s_{i_{(-1)}} = s_{1_{(-1)}} + s_{2_{(-1)}} + s_{3_{(-1)}} + \ldots + s_{n_{(-1)}} \qquad \text{(Eq. 10.12)}$$

iii. Difference of high and low levels (difference = high total − low total). Or, you can compute the difference by multiplying each high (+1) or low (−1) level by the response for its row, and then summing all the values in the column.

$$\Delta_{Effect} = \sum_{i=1}^{n_{(+1)}} s_{i_{(+1)}} - \sum_{i=1}^{n_{(-1)}} s_{i_{(-1)}} \qquad \text{(Eq. 10.13)}$$

iv. Effect is equal to the difference divided by the number of runs at each level or

$$Effect = \frac{High\ level\ total - Low\ level\ total}{number\ of\ runs\ at\ each\ level} = \frac{\Delta_{Effect}}{n_{level_{(high\ or\ low)}}}$$

(b) Second method of multiple replications:

i. Sum of all response values when the factor is at the high level (+1)

$$\bar{s}_{(+1)} = \frac{\sum_{i=1}^{n_{(+1)}} s_{i_{(+1)}}}{n_{(+1)}} = \frac{(s_{1_{(+1)}} + s_{2_{(+1)}} + s_{3_{(+1)}} + \ldots + s_{n_{(+1)}})}{n_{(+1)}} \qquad \text{(Eq. 10.14)}$$

ii. Sum of all response values when the factor is at the low level (−1)

$$\bar{s}_{(-1)} = \frac{\sum_{i=1}^{n_{(-1)}} s_{i_{(-1)}}}{n_{(-1)}} = \frac{(s_{1_{(-1)}} + s_{2_{(-1)}} + s_{3_{(-1)}} + \ldots + s_{n_{(-1)}})}{n_{(-1)}} \qquad \text{(Eq. 10.15)}$$

iii. Difference of high and low levels (difference = high total − low total). Or, you can compute the difference by multiplying each high (+1) or low (−1) level by the standard deviation for its row, and then summing all the values in the column.

$$\Delta_{Effect} = \bar{s}_{(+1)} - \bar{s}_{(-1)} \qquad \text{(Eq. 10.16)}$$

3. State an interpretation of the result as a conclusion, such that by increasing factor value or reducing the factor value it has little or more effect (or is effective) on the yield or response.
4. Then decide which effects are the highest impact in the process.
5. Design another multilevel experiment focusing on the factors that has the highest impact in the process output.

Let's review this concept by an example.

Example 10.2: Full Factorial Design: Two-Level and Three-Factor Design Analysis

An engineer needs to improve the thickness of the camera lens by reducing the variation about the target of 2 mm. The project team concluded that the following factors as shown in Table 10.4 can make the highest impact in the variation of lens.

And the output responses is given in Table 10.5.

Create a full factorial design (2^3) and carry out a complete full analysis to determine the predicted model. Considering that all the other parameters are fixed.

Table 10.4. Independent adjustable variable (factors)

Independent factors	Low level (coded value = −1)	High level (coded value = +1)
Plastic melt temperature	230°C	250°C
Mold temperature	70°C	90°C
Hold time	1.5 sec	3 sec

Table 10.5. Output responses

Run:	1	2	3	4	5	6	7	8
Thickness:	22.60	20.48	22.50	22.70	28.59	26.50	28.70	26.01

Replicated:	9	10	11	12	13	14	15	16
Thickness:	22.71	20.05	22.98	20.99	28.59	26.40	28.78	26.20

Solution

We use two level full factorial design with (Level)^{Factor} = L^k = 2^k = 2^3 = 8 corner points of cube. For replicated design, it will require another eight runs for corner points (four points at low level and another four points at high level) and with no center point.

STEP 1: DESIGN LAYOUT

A full factorial design contains all possible combination treatments of attribute levels. Figure 10.5 presents a cube representing all possible combinations for an experiment with three factors (x_1, x_2, x_3) at two levels each (−1, +1), thus with $2^k = 2^3 = 8$ combination where (±1, 1, 1), (1, ±1, 1), and (1, 1, ±1) are the coded values of the corners points.

Table 10.6 is the detailed format of design illustrated in Figure 10.5 with replications of each run in the design.

Table 10.7 shows the actual values of the design with response (y) value listed.

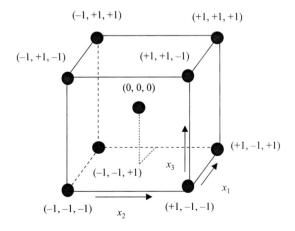

Figure 10.5. Two-level three-factor design coded values.

Table 10.6. Experimental matrix coded values of two level full factorial designs without zero center ($2^k = 2^3 = 8$ plus replicate equal to 16 experiments)

Design type	Experiment run	Plastic melt temperature (x_1)	Mold temperature (x_2)	Hold time (x_3)
Factorial	1 (Corner)	−1	−1	−1
design:	2 (Corner)	+1	−1	−1
	3 (Corner)	−1	+1	−1
	4 (Corner)	+1	+1	−1
	5 (Corner)	−1	−1	+1
	6 (Corner)	+1	−1	+1
	7 (Corner)	−1	+1	+1
	8 (Corner)	+1	+1	+1
Corner	9 (Replicated run 1)	−1	−1	−1
replicated	10 (Replicated run 2)	+1	−1	−1
design:	11 (Replicated run 3)	−1	+1	−1
	12 (Replicated run 4)	+1	+1	−1
	13 (Replicated run 5)	−1	−1	+1
	14 (Replicated run 6)	+1	−1	+1
	15 (Replicated run 7)	−1	+1	+1
	16 (Replicated run 8)	+1	+1	+1

Table 10.7. Experimental matrix and response actual (uncoded) values of two level full factorial designs without zero center ($2^k = 2^3 = 8$ plus replicate equal to 16 experiments (runs))

Design type	Experiment run	Plastic melt temperature (x_1)	Mold temperature (x_2)	Hold time (x_3)	Lens thickness (y)
Factorial	1 (Corner)	230	70	1.50	22.60
design:	2 (Corner)	250	70	1.50	20.48
	3 (Corner)	230	90	1.50	22.50
	4 (Corner)	250	90	1.50	22.70
	5 (Corner)	230	70	3.00	28.59
	6 (Corner)	250	70	3.00	26.50
	7 (Corner)	230	90	3.00	28.70
	8 (Corner)	250	90	3.00	26.01

(Continued)

Table 10.7. (*Continued*)

Design type	Experiment run	Plastic melt temperature (x_1)	Mold temperature (x_2)	Hold time (x_3)	Lens thickness (y)
Corner	9 (Replicated run 1)	230	70	1.50	22.71
replicated	10 (Replicated run 2)	250	70	1.50	20.05
design:	11 (Replicated run 3)	230	90	1.50	22.98
	12 (Replicated run 4)	250	90	1.50	20.99
	13 (Replicated run 5)	230	70	3.00	28.59
	14 (Replicated run 6)	250	70	3.00	26.40
	15 (Replicated run 7)	230	90	3.00	28.78
	16 (Replicated run 8)	250	90	3.00	26.20

The process model from the design chart is

$$y = b_0 + b_1 x_1 + b_2 x_2 + b_3 x_3 + b_{12} x_1 x_2 + b_{13} x_1 x_3 + b_{23} x_2 x_3 + b_{123} x_1 x_2 x_3 \qquad \text{(Eq. 10.17)}$$

STEP 2—MAIN EFFECTS

Figure 10.6 represents the main effects that possibility of plastic melt temperature and hold time had the biggest effect.

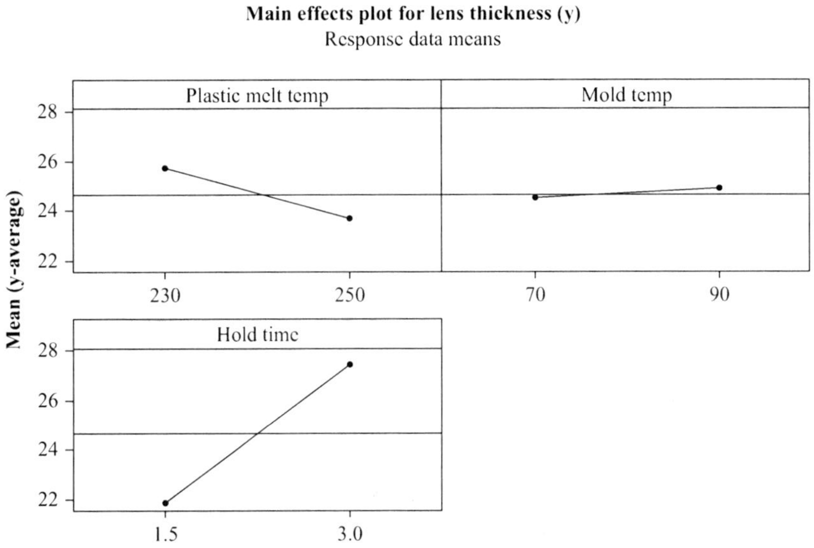

Main effects plot for lens thickness (y)
Response data means

Figure 10.6. Factors (independent variables): Main effects.

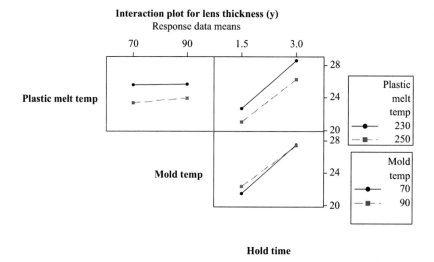

Figure 10.7. Interactions of independent variables.

Interactions: If the plots have steepest lines they represent largest effects. However, here nothing looks significant, but again, these are only graphs (Figure 10.7). Interaction graphs looks similar; moreover the analysis showed that the *p*-values were 0.000. So, one should never make assumptions on graphs alone.

STEP 3—PARETO CHART, RESIDUALS

We now can examine the Pareto chart (Figure 10.8), residuals, and perform the analysis.

Figures 10.6 through Figure 10.8 provide a hint. These three graphs confirm the same observation. Which main effects and interactions are significant? The solid line in the Pareto is your level of significance as selected *a* = 0.05. This means any bar greater than or to the *right of the line will have a p-value lower than* 0.05 ($p < 0.05$ significant) and any bar lower or to the *left of the line will have a p-value greater than* 0.05 ($p > 0.05$ not significant).

The residuals are the difference between the actual *y* value and the *y* value predicted by the regression equation. Residuals should be randomly and normally distributed about a mean of zero, not correlate with the predicted *y*, and not exhibit trends over time (if data is chronological).

STEP 4—NORMAL PROBABILITY OF RESIDUALS (FIGURE 10.9), RESIDUALS AGAINST FITS, AND RESIDUALS AGAINST ORDER

Any trends or patterns in the residual plots indicate inadequacies in the regression model, such as missing *x*(s) or nonlinear relationship. Again, you want them to be normal, with no patterns. Patterns in the residual versus the fitted or predicted value also called fits, which the transfer function or model produce for a given *x* or a set of *x*(s). For instance $y = 5 + 3x$, the fitted value for $x = 5$ is $y = 5 + (3) \cdot (5) = 20$.

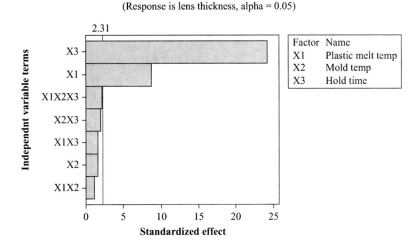

Figure 10.8. Pareto chart of the standardized effects.

Note: One can predict if your model will overpredict or underpredict by looking at the residuals at the target value in Figure 10.9. Since residuals are the values you measured (observed) and fitted, if your target is *Fitted Value* $= y = 24$, and your residuals are near zero at $y = 24$, you can expect your transfer function to confirm; however, if your residuals have a large positive value at $y = 24$, you can expect your transfer function or model to under predict, and the reverse is true too. If your residuals have a large negative value at *Fitted Value* $= 24$, you can expect your model to over predict.

The *normal probability plot of standardized residual* (Figure 10.9) looks good. You're looking for a linear line that basically goes from the lower left to the upper right (including

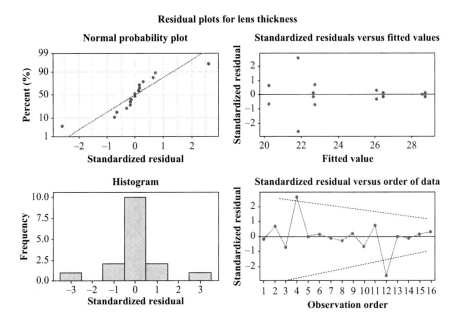

Figure 10.9. Normal probability of standardized residuals.

two outliers at the two ends). The *p-value* should determine if our residuals are normal, but the graphs suggest there should have no problem.

The *residuals vs. fitted values*: This graph suggests there is more variability at the shorter distances than at larger ones. If my target was around 22," I would expect more variability than if my target was around 28"; however, the design only had two repetitions, so let's see what happens when we confirm.

The *Residuals vs. the Order of Data*: It is just a random pattern. This doesn't look too bad. However, if you take your time and look hard enough, you'll find a pattern. You don't want small variability at the beginning with steadily increasing variance toward the end so that the pattern looks like a funnel. This unequal variance is called *heteroscedasticity*; we're looking for homoscedasticity (a sequence or a vector of random variables) or uniform variance throughout the experiment.

On the other hand, if the residuals start in one corner and proceed to the opposite corner, you may have a missing variable (factor x) or your process is changing over time. If the residuals increase or decrease as the inspection increases, you may need to transform the variable (square root or natural log). If you have a nonlinear shape or even curve, you may be missing nonlinear factors.

Step 5—Design Analysis: Start Residuals Normality Test

Apply computer analysis tools of "statistical normality test" to determine if the residuals are normal (see Figure 10.9). You want them to follow a normal distribution. If residuals do not follow a normal probability distribution, then your model may not predict the results very well. Here the *p*-values and normality test showed that the residuals are normal. Anderson-Darling normality test indicates that if the *p*-value is less than 0.05 ($p < 0.05$) then it is not normal. Moreover, if the *p*-value is more than or equal to 0.05 ($p \geq 0.05$) then it is normal. Table 10.8 represents the estimated effects and coefficients for the response (Lens thickness).

Table 10.8. Computer output factorial fit: Lens thickness versus hold time, mold temperature, plastic temperature

Estimated effects and coefficients for lens thickness (coded units)

Term	Effect	Coefficient	Coefficient SE	T	*p*-value
Constant (b_0)		24.674	0.1153	213.97	0.000
Plastic melt temperature (x_1)	-2.015	-1.007	0.1153	-8.74	0.000
Mold temperature (x_2)	0.367	0.184	0.1153	1.59	0.150
Hold time (x_3)	5.595	2.797	0.1153	24.26	0.000
Plastic temperature • Mold temperature ($x_1 • x_2$)	0.250	0.125	0.1153	1.08	0.310
Plastic melt temperature • Hold time ($x_1 • x_3$)	-0.372	-0.186	0.1153	-1.62	0.145
Mold temperature • Hold time ($x_2 • x_3$)	-0.465	-0.232	0.1153	-2.02	0.079
Plastic melt temperature • Mold temperature • Hold time ($x_1 • x_2 • x_3$)	-0.497	-0.249	0.1153	-2.16	0.063

Table 10.9. Analysis of variance for lens thickness (coded units)

S = 0.461248 press = 6.808 R^2=98.84% R^2 (predicted) = 95.35% R^2 (adjusted) = 97.82%

Source	df	Seq SS	Adj SS	Adj MS	F	p-value
Main effects	3	141.997	141.997	47.332	222.48	0.000
Plastic temperature (x_1)	1	16.241	16.241	16.241	76.34	0.000
Mold temperature (x_2)	1	0.540	0.540	0.540	2.54	0.150
Hold time (x_3)	1	125.216	125.216	125.216	588.56	0.000
Two-way interactions	3	1.670	1.670	0.557	2.62	0.123
$x_1 \cdot x_2$	1	0.250	0.250	0.250	1.18	0.310
$x_1 \cdot x_3$	1	0.555	0.555	0.555	2.61	0.145
$x_2 \cdot x_3$	1	0.865	0.865	0.865	4.07	0.079
Three-way interactions	1	0.990	0.990	0.990	4.65	0.063
$x_1 \cdot x_2 \cdot x_3$	1	0.990	0.990	0.990	4.65	0.063
Residual error	8	1.702	1.702	0.213		
Pure error	8	1.702	1.702	0.213		
Total	15	146.359				

Analysis

We want to keep anything with a low *p-value* (Tables 10.8 and 10.9) and in this case $p < 0.05$. The *p*-value (0.000) in Table 10.8 and Figure 10.6 provide sufficient evidence that the plastic melt temperature (x_1) and hold time (x_3) are different and have the most effects.

However, if we kept the three-way interaction then we need to keep all the main effects (including the mold temperature) that are in the three-way interaction regardless whether all the main effects (x_1, x_2, x_3) are significant or not. This is called the hierarchy rule. That is modeling dictates that if a higher order term (quadratic, interaction, etc.) is included in the model, then all linear effects represented in the higher order term will be included in the model regardless of their significance.

The coefficients of actual value of the model are listed in Table 10.10.

Table 10.10. Estimated coefficients for lens thickness using data in uncoded units

Term	Coefficient
Constant	187.281000
Plastic temperature (x_1)	–0.741875
Mold temperature (x_2)	–2.002870
Hold time (x_3)	–51.510000
Plastic temperature • Mold temperature $(x_1 \cdot x_2)$	0.008713
Plastic temperature • Hold time $(x_1 \cdot x_3)$	0.240500
Mold Temperature • Hold time $(x_2 \cdot x_3)$	0.765000
Plastic temperature • Mold temperature • Hold time $(x_1 \cdot x_2 \cdot x_3)$	–0.003317

After Screening Test

STEP 6—COMPUTER DESIGN ANALYSIS: START *"NORMALITY TEST"* AFTER SCREENING

Here we're taking the nonsignificant terms (Tables 10.9 and 10.10) out of our transfer function or model. Remove two- and three-way interactions; you can see the terms and their *p-values*. Remember that you'll need to reanalyze your residuals again with your new model.

After screening we get new values for the process model as shown in Table 10.11.

ANOVA analysis after removing the nonsignificance term is illustrated in Table 10.12.

The coefficients of actual value of the model are given in Table 10.13.

If you add terms that had a high *p-value* back into available terms and remove them from the model, those terms go back into the error term. Notice the degrees of freedom increased from 8 to 13 since we removed one main effect, three two-way interactions, and one three-way interactions. Also notice the residual error is divided into pure error (from the fact we had two repetitions) and lack of fit (LOF) (those one main effect, three two-way interactions, and one three ways interaction we got rid of; notice lack of fit has one degree of freedom).

Table 10.11. Computer output factorial fit: Lens thickness versus plastic melt temperature, hold time

Estimated effects and coefficients for lens thickness (coded units)

Term	Effect	Coefficient	Coefficient SE	T	*p*-value
Constant		24.674	0.1535	160.72	0.000
Plastic melt temperature (x_1)	−2.015	−1.007	0.1535	−6.56	0.000
Hold Time (x_3)	5.595	2.797	0.1535	18.22	0.000

$y = 24.674 - 1.007\, x_1 + 2.797\, x_3$ **Coded-model**

Table 10.12. Analysis of variance for lens thickness (coded units)

S = 0.214618, Press = 1.07196, R^2 = 99.68%, R^2 (predicted) = 99.32%, R^2 (adjusted) = 99.56%

Source	*df*	Seq SS	Adjusted SS	MS	Adjusted F-test	*p*-value
Main effects	2	141.457	141.457	70.728	187.56	0.000
Plastic temperature (x_1)	1	16.241	16.241	16.241	43.07	0.000
Hold time (x_3)	1	125.216	125.216	125.216	332.06	0.000
Residual error	13	4.902	4.902	0.377		
Lack of fit	1	0.555	0.555	0.555	1.53	0.239
Pure error	12	4.347	4.347	0.362		
Total	15	146.359				

Table 10.13. Estimated coefficients for lens thickness using data in uncoded units

Term	Coefficient
Constant b_0	40.4613
Plastic temperature (x_1)	−0.10075
Hold time (x_3)	3.73000
$y = 40.4613 − 0.10075\,x_1 + 3.7300\,x_3$	Uncoded (actual) model

We do not want the lack of fit to be significant. In this case it is not significant (Table 10.12) since the *p-value* = 0.239. However, if you throw too many terms out or you throw just one significant term out, your lack of fit will be significant, which means your model does not adequately fit your data.

Computer gives you two equations, a coded equation where all factors have coded values and an uncoded equation where the student puts in the actual level of x. For instance, if you use the uncoded model and your hold time is 2 seconds, you would multiply the coefficient for hold time by 2. That is (3.73) • (2). This model is intuitive for the students; however, it has limitations. It is not as accurate as the coded model because of round-off errors and the x's are not orthogonal.

The new Pareto chart (Figure 10.10) and residual results (Figure 10.11) show significant difference after elimination of low-impact terms and binary interaction effects.

The "*coded model*" from Table 10.11 is

$$y = 24.674 − 1.007\,x_1 + 2.797\,x_3 \text{ coded model} \quad \text{or,}$$

$$y = 24.674 − 1.007 \text{ (plastic temperature)} + 2.797 \text{ (hold time)}$$

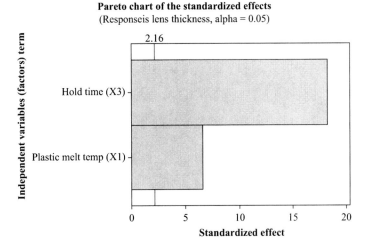

Figure 10.10. Pareto chart of the standardized effects.

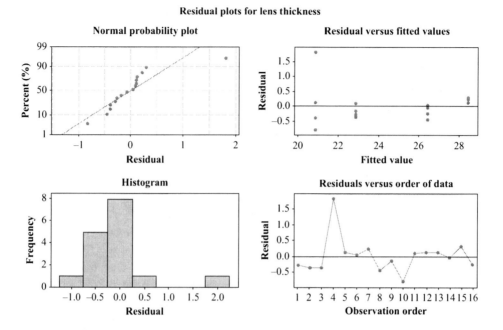

Figure 10.11. Normal probability of residuals.

Let's just arbitrarily set hold time $x_3 = 0$ (coded value) to some value since it is a discrete value and $y = 24$. Now solve for plastic temperature x_1.

$$24 = 24.674 - 1.007\,x_1 + 2.918\,(0)$$
$$24 = 24.674 - 1.007\,x_1 + 0$$
$$24 - 24.674 = -1.007\,x_1$$
$$-0.674 = -1.007\,x_1$$
$$0.669 = x_1 = \text{Plastic temperature coded value when } x_3 = 0.$$

Converting from the coded units:

	Low level		High level
Plastic melt temperature	230°C	?	250°C
	−1	0.669	+1

Now let's apply to "*uncoded model*" from Table 11.13 and that is

$$y = 40.4613 - 0.10075\,x_1 + 3.7300\,x_3 \text{ uncoded model or,}$$

$$y = 40.4613 - 0.10075\,(\text{plastic temperature}) + 3.7300\,(\text{hold time}) \quad \textit{uncoded model}$$

Let's just arbitrarily set x_3 to some value since it is discrete number

- Set hold time to $x_3 = 2$ second and $y = 24$
- Let's solve for plastic melt temperature x_1

Substitute the values in the uncoded model we have

$$24 = 40.4613 - 0.10075\,x_1 + 3.73(2)$$

simplify the equation

$$24 = 40.4613 - 0.10075\,x_1 + 7.46$$

$$24 = 47.9213 - 0.10075\,x_1$$

$$-23.9213 = -0.10075\,x_1$$

$$237.432 = x_1 = \text{Plastic temperature}$$

$$\text{Or, } x_1 = 237.432°C$$

Converting from the coded units:

	Low level		High level
Plastic melt temperature	230°C	237.432°C	250°C
	−1	0.669	+1

10.4   FRACTIONAL (REDUCED) FACTORIAL DESIGN (FFD)

The fractional factorial design is defined as less than full or the reduced experimental runs (trials) to save time, resources, and make quick decisions. Condition combinations are chosen to support the information obtained in concluding the factor effects. Fractional factorial design is more efficient; however, there is a possible risk of missing interactions.

Half-factorial design: A half-factorial design means the researcher will run the design in half a trials (i.e., $2^{3-1} = 2^2 = 4$) rather than full (i.e., $2^3 = 8$) as shown in Figure 10.12 and Table 10.14. This is because the design has $2^{Factor}/2$ runs and it is referred to as $2^{Factor-1}$. Thus, the design requires only half the data that a full factorial design needs. The main effects and all the two interactions are designed (and modeled) using this approach.

Formula for fractional factorial ($n = L^{(k-p)}$): The formula is $n = L^{(k-p)}$ where n = number of experimental runs, L is the design level, k = factor, and p = fractional factor. The letters L, k, and p are positive integer number and k is always greater than p. In $L^{(k-p)}$ the $1/L^{(p)} = L^{(-p)}$ is called the degree of fractionation, since it represents the fraction of observation from full factorial design L^k that are required. For instance in the case of L = 2 we have

- If $p = 0$; $2^{k-p} = 2^{k-0}$ then; $p = 0$ full factorial design, where $2^k \cdot 2^{-p} = 2^{k-p}$
- If $p = 1$; $2^{k-p} = 2^{k-1}$ then; $p = 1$ half fractional factorial design
- If $p = 2$; $2^{k-p} = 2^{k-2}$ then; $p = 2$ quarter fractional factorial design
- Etc.

Furthermore, if there are six factors ($k = 6$), two levels (L = 2), and three fractional factor ($p = 3$), then a full 2^k design would require $2^6 = 64$ observations, while the $2^{(k-p)}$ fractional factorial design would require only $2^{6-3} = 8$ observations.

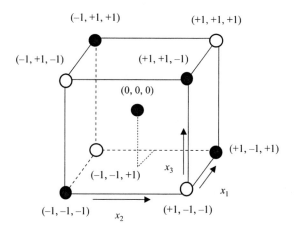

Figure 10.12. Fractional Factorial Design (FFD) or
half-factorial design ($2^{3-1} = 2^2 = 4$).

A sample of fractional factorial design (recall the expanded full factorial design with three factors) with seven factors and two levels is displayed in Table 10.14.

In Table 10.14 "−1" means set factor to low level in this run and "+1" means set factor to high level in this run; run order should be randomized. Failure to randomize is very risky for factor x_1.

Additional examples of half fraction are tabulated in Table 10.15.

To interpret the results follow the effects computation steps as discussed before. The calculation of the effects as well as ANOVA are based on the response values and only not coded values. Even though the coded values are not appeared in the analysis, the values of the levels are very important. Too small difference of high and low means not much change in the response to observe and too much difference may be result in a failed experiment.

Table 10.14. Tabular representation of the design for fractional factorial design ($n = L^{k-p} = 2^{7-4} = 2^3 = 8$) layout without response results

	Factors			Expanded layout fractional factors coded values							
	A	B	C	A	B	C	D = AB	E = AC	F = CD	G = ABC	Response value
Trial	x_1	x_2	x_3	x_1	x_2	x_3	$x_1 x_2$	$x_1 x_3$	$x_2 x_3$	$x_1 x_2 x_3$	y
1	Low	Low	Low	−1	−1	−1	+1	+1	+1	−1	
2	Low	Low	High	−1	−1	+1	+1	−1	−1	+1	
3	Low	High	Low	−1	+1	−1	−1	+1	−1	+1	
4	Low	High	High	−1	+1	+1	−1	−1	+1	−1	
5	High	Low	Low	+1	−1	−1	−1	−1	+1	+1	
6	High	Low	High	+1	−1	+1	−1	+1	−1	−1	
7	High	High	Low	+1	+1	−1	+1	−1	−1	−1	
8	High	High	High	+1	+1	+1	+1	+1	+1	+1	

Note that each factor is experimented at each level "high" and "low" four times.

Table 10.15. Tabular representation of the fractional factorial design for $n = L^{k-1}$

	Sample design 1 $n = 2^{4-1} = 8$				Sample design 2 $n = 2^{4-1} = 8$				Sample design 3 $n = 2^{5-2} = 8$				
	A	B	C	D	A	B	C	D	A	B	C	D	E
Run	x_1	x_2	x_3	x_4	x_1	x_2	x_3	x_4	x_1	x_2	x_3	x_4	x_5
1	−1	−1	−1	−1	+1	−1	−1	−1	−1	−1	−1	−1	+1
2	−1	−1	−1	+1	+1	−1	−1	+1	−1	−1	−1	+1	−1
3	−1	−1	+1	−1	+1	−1	+1	−1	−1	−1	+1	−1	−1
4	−1	−1	+1	+1	+1	−1	+1	+1	−1	−1	+1	+1	+1
5	−1	+1	−1	−1	+1	+1	−1	−1	−1	+1	−1	−1	−1
6	−1	+1	−1	+1	+1	+1	−1	+1	−1	+1	−1	+1	+1
7	−1	+1	+1	−1	+1	+1	+1	−1	−1	+1	+1	−1	+1
8	−1	+1	+1	+1	+1	+1	+1	+1	−1	+1	+1	+1	−1

Note: Use the column product of "A through D coded value" to compute E algebraically.

Example 10.3

Fractional factorial at 2 levels Part shrinkage is one of the main concerns in the injection molding industries. A major injection molding company just built a brand new mold. The company requested the process engineer to sample the mold for qualification and quality assurance engineer to make sure that dimensions are within the customer's guideline (as defined by the 5.580 dimension).

Process engineer would like to find the effect of the following parameters: injection velocity, cooling time, nozzle temperature, mold temperature, pack pressure, and back pressure on the part-XZ shrinkage.

Note: Use only Acetal raw material, no regrind, and run the machine until process becomes stable. At least make 10 runs before collecting parts.

Solution

Fractional factorial process design (Figure 10.13) at two levels.

Independent controllable factors			Low level	High level
A	Pack pressure (hydraulic)	(x_1)	800 psi	1200 psi
B	Injection velocity	(x_2)	1.5	1.75
C	Cooling time	(x_3)	35 sec	40 sec
D	Inject time	(x_4)	1.5 sec	2 sec
E	Mold temperature	(x_5)	80°C	90°C
F	Nozzle temperature	(x_6)	220°C	240°C

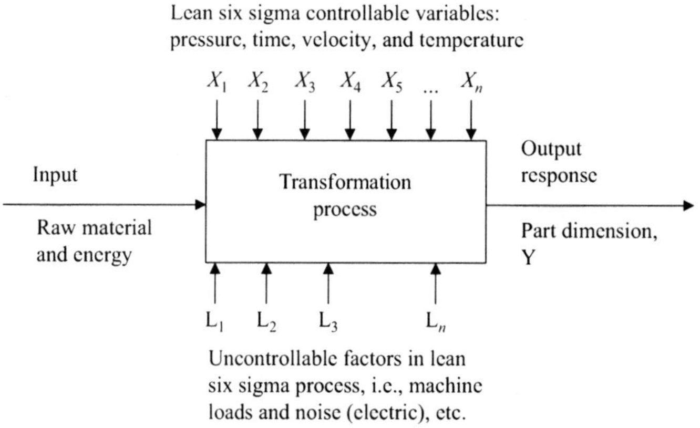

Figure 10.13. Process model.

STEP 1: FRACTIONAL FACTORIAL: $2^{k-p} = 2^{6-3} = 8$

Develop a fractional factorial design to determine the effects of all the independent variables in the process demonstrated in Tables 10.16 as coded variables and Table 10.17 with actual design values.

Let's recall Equations 10.4 through 10.6. Table 10.18 was developed using these three equations.

$$\overline{\overline{y}}_{(+1)} = \frac{\sum_{i=1}^{n} \overline{y}_{(+1)_i}}{n} = \frac{\left(\overline{y}_{(+1)_1} + \overline{y}_{(+1)_2} + \overline{y}_{(+1)_3} + \ldots + \overline{y}_{(+1)_n}\right)}{n}$$

$$\overline{\overline{y}}_{(+1)} = \frac{\sum_{i=1}^{n=4} \overline{y}_{(+1)_i}}{4} = \frac{\left(5.576 + 5.643 + 5.545 + 5.631\right)}{4} = 5.599$$

$$\overline{\overline{y}}_{(-1)} = \frac{\sum_{i=1}^{n} \overline{y}_{(-1)_i}}{n} = \frac{\left(\overline{y}_{(-1)_1} + \overline{y}_{(-1)_2} + \overline{y}_{(-1)_3} + \ldots + \overline{y}_{(-1)_n}\right)}{n}$$

$$\overline{\overline{y}}_{(-1)} = \frac{\sum_{i=1}^{n=4} \overline{y}_{(-1)_i}}{4} = \frac{\left(5.558 + 5.584 + 5.588 + 5.663\right)}{4} = 5.573$$

Table 10.16. Coded values of fractional factorial experimental matrix with random runs

Run	A	B	C	D	E	F	y_1	y_2	y_3	y_4	y_5	y (average)
1	−1	−1	−1	−1	−1	−1	5.575	5.560	5.570	5.585	5.590	**5.576**
2	−1	−1	−1	+1	+1	+1	5.650	5.640	5.640	5.640	5.645	**5.643**
3	−1	+1	+1	−1	−1	+1	5.545	5.545	5.545	5.550	5.540	**5.545**
4	−1	+1	+1	+1	+1	−1	5.630	5.630	5.625	5.635	5.635	**5.631**
5	+1	−1	+1	−1	+1	−1	5.555	5.560	5.560	5.555	5.560	**5.558**
6	+1	−1	+1	+1	−1	+1	5.580	5.580	5.585	5.590	5.585	**5.584**
7	+1	+1	−1	−1	+1	+1	5.600	5.580	5.585	5.590	5.585	**5.588**
8	+1	+1	−1	+1	−1	−1	5.565	5.560	5.565	5.565	5.560	**5.563**
Overall Average												**5.586**

Table 10.17. Actual values of experimental matrix with random runs

Run	A	B	C	D	E	F	y_1	y_2	y_3	y_4	y_5	y_{mean}	Std. dev.
1	800	1.50	35	1.5	80	220	5.575	5.560	5.570	5.585	5.590	5.576	0.0119
2	800	1.50	35	2.0	90	240	5.650	5.560	5.640	5.640	5.645	5.643	0.0045
3	800	1.75	40	1.5	80	240	5.545	5.545	5.545	5.550	5.540	5.545	0.0035
4	800	1.75	40	2.0	90	220	5.630	5.630	5.625	5.635	5.635	5.631	0.0042
5	1200	1.50	40	1.5	90	220	5.555	5.560	5.560	5.555	5.560	5.558	0.0027
6	1200	1.50	40	2.0	80	240	5.580	5.580	5.585	5.590	5.585	5.584	0.0042
7	1200	1.75	35	1.5	90	240	5.600	5.580	5.585	5.590	5.585	5.588	0.0076
8	1200	1.75	35	2.0	80	220	5.565	5.560	5.565	5.565	5.560	5.563	0.0027
Overall average												5.586	0.0052

$$\Delta_{Effect} = \overline{\overline{y}}_{(+1)} - \overline{\overline{y}}_{(-1)}$$

$$\Delta_{Effect} = 5.599 - 5.573 = 0.0255$$

Similarly Table 10.19 developed using standard deviation effect equations.

Table 10.18. Marginal mean analysis of response effects of factors

Cause/Interaction	y− Ave.	y+ Ave.	Δ Effect	Δ |Effect|	Δ½ Effect	Δ½ |Effect|
A = Pack pressure (hydraulic) (x_1)	5.599	5.573	0.0255	0.0255	0.0130	0.0130
B = Injection velocity (x_2)	5.590	5.582	0.0085	**0.0085**	0.0043	0.0043
C = Cooling time (x_3)	5.593	5.580	0.0130	0.0130	−0.0650	0.0650
D = Inject time (x_4)	5.567	5.605	−0.0385	0.0385	−0.0193	0.0193
E = Mold temperature (x_5)	5.567	5.605	0.0380	0.0380	0.0190	0.0190
F = Nozzle temperature (x_7)	5.582	5.590	−0.0080	**0.0080**	0.0040	0.0040

Table 10.19. Response effects of factors (Average of low and high standard deviations)

Cause/Interaction	s− Ave.	s+ Ave.	Δ Effect	Δ |Effect|	Δ½ Effect	Δ½ |Effect|
A = Pack pressure (hydraulic) (x_1)	0.0462	0.0150	−0.0312	0.0312	−0.0156	0.0156
B = Injection velocity (x_2)	0.0368	0.0373	0.0005	**0.0005**	0.00023	0.00023
C = Cooling time (x_3)	0.0352	0.0380	0.0028	0.0028	0.00139	0.00139
D = Inject time (x_4)	0.0190	0.0379	0.0189	0.0189	0.00947	0.00947
E = Mold temperature (x_5)	0.0170	0.0392	0.0222	0.0222	0.01110	0.01110
F = Nozzle temperature (x_7)	0.0335	0.0403	0.0068	**0.0068**	0.00339	0.00339

Interpretation of Results:

A = Pack Pressure (hydraulic) (x_1) High

B = Injection velocity (x_2) High. We will remove this factor due to its minimum effect compare to other terms

C = Cooling time (x_3) High

D = Inject time (x_4) Low

E = Mold temperature (x_5) Low

F = Nozzle temperature (x_6) Low. We will remove this factor due to its less effect compare to other terms

STEP 2: FRACTIONAL FACTORIAL WITH INTERACTION $2^{k-p} = 2^{4-1} = 8$

In an attempt to get the dimension of the part to 5.580, the experimenter would like to look for interaction between the top four main effects on dimension. Create a predictive model for dimension. What are the best settings to achieve 5.580 targets for dimension?

Independent controllable factors			Low level	High level
A	Pack Pressure (hydraulic)	(x_1)	800 psi	1200 psi
B	Cooling time	(x_2)	35 sec	40 sec
C	Inject time	(x_3)	1.5 sec	2 sec
D	Mold temperature	(x_4)	80°C	90°C

Using Tables 10.20 and 10.21 the effect calculations are left for students as an assignment.

Table 10.20. Coded values of full factorial experimental matrix with random runs

Run	A	B	C	D =ABC	y_1	y_2	y_3	y_4	y_5	$y_{(mean)}$
1	−1	−1	−1	−1	5.578	5.560	5.570	5.585	5.590	5.577
2	−1	−1	+1	+1	5.650	5.640	5.612	5.680	5.645	5.645
3	−1	+1	−1	+1	5.545	5.325	5.545	5.550	5.580	5.509
4	−1	+1	+1	−1	5.630	5.534	5.525	5.635	5.535	5.572
5	+1	−1	−1	+1	5.545	5.560	5.620	5.555	5.560	5.568
6	+1	−1	+1	−1	5.580	5.580	5.585	5.480	5.585	5.562
7	+1	+1	−1	−1	5.620	5.545	5.585	5.526	5.585	5.572
8	+1	+1	+1	+1	5.565	5.560	5.565	5.565	5.561	5.563
Overall Average										**5.571**

Table 10.21. Actual values of fractional factorial experimental matrix with random runs

	Actual design values				Response replications						Std.
Run	A	B	C	D = ABC	y_1	y_2	y_3	y_4	y_5	$y_{(mean)}$	dev. (s)
1	800	35	1.50	80	5.578	5.560	5.570	5.585	5.590	5.577	0.0119
2	800	35	2.00	90	5.650	5.640	5.612	5.680	5.645	5.645	0.0243
3	800	40	1.50	90	5.545	5.325	5.545	5.550	5.580	5.509	0.1039
4	800	40	2.00	80	5.630	5.534	5.525	5.635	5.535	5.572	0.0556
5	1200	35	1.50	90	5.545	5.560	5.620	5.555	5.560	5.568	0.0297
6	1200	35	2.00	80	5.580	5.580	5.585	5.480	5.585	5.562	0.0459
7	1200	40	1.50	80	5.620	5.545	5.585	5.526	5.585	5.572	0.0370
8	1200	40	2.00	90	5.565	5.560	5.565	5.565	5.561	5.563	0.0025
Overall Average										5.571	0.0389

10.5   ROBUST ENGINEERING DESIGN AND ANALYSIS

Robust engineering design concept is the process of evaluating competing technologies to develop an optimum new product and processes free of defects for optimum performance. It selects control factors and determines the optimal levels (see Figure 10.14) for each of the factors to response for developing the design specification limits. It focuses on recognizing the ideal function(s) for specific scientific know-how on product/process design. Further, it determines the optimum nominal values of the design parameters that maximize reliability effectiveness.

Dr. Genichi Taguchi defines "robustness" as "the state where the technology, product, or process performance is minimally sensitive to factors causing variability (either in the manufacturing or user's environment) and aging at the lowest unit manufacturing cost."

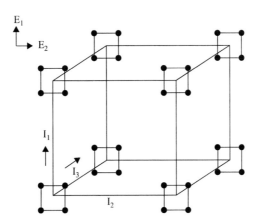

Figure 10.14. Inner $L^k = 2^3$ and outer $L^k = 2^2$ arrays for robust design with "I" the inner array, "E" the outer array.

Taguchi proposed experimental designs that utilize two-, three-, and mixed-level fractional factorial designs particularly large screening design favored by the applicants. Taguchi refers to experimental design as "off-line quality control" (Figure 10.14) since it is a technique of making sure the maximum performance in the early stages of product or processes design.

In Figure 10.14 the "I" is controllable factors with two-level three-factor $L^k = 2^3 = 8$ experimental settings, or runs, and "E" is noise factors (uncontrollable, i.e., ambient temperature, moisture content) with two-level two-factor $L^k = 2^2$. This is also called off-line quality control experimental design, controllable only in the closed system, i.e., research laboratory that requires supplementary resources.

Taguchi approach to experimental design: From every design run, one can obtain an average response ($\bar{y}$) level and a measure of standard deviation (S, $i = 1, 2, ..., n$), which is variation from the response mean, as shown in Figure 10.15 ($\bar{y}_{in}$, s_{in}) in tabular format. These values can be applied in selecting the combination of factor levels that offer the most robust process design.

How to Measure the Robustness (maximum quality, reliable, and enduring summed in efficiency): When and how an engineer knows that a process or product design is robust? Unlike design metrics such as upper/lower specification limits, reliability data, capability performance, scrap rate, rework volume, percent defect rate, and productivity, the problem with depending on these measurement systems to determine the robustness is that they come too late in the product development cycle. However, Taguchi method offers different measurements.

Signal-to-Noise (S/N) ratio calculation: Taguchi states the only measure of the robustness and effectiveness of an existing system is the signal-to-noise (*s/n*) ratio. The following three equations compute the signal-to-noise ratio for maximizing the response, minimizing the response, and meeting the response target desired values.

1. Maximize the response using the signal-to-noise ratio model:

$$\frac{S}{N} = \frac{Signal}{Noise} = -10\log_{10}\frac{1}{n}\sum_{i=1}^{n}\frac{1}{y_i^2}; \text{ for maximizing the response} \qquad \text{(Eq. 10.18)}$$

2. Minimize the response using the following signal-to-noise ratio equation:

$$\frac{S}{N} = \frac{Signal}{Noise} = -10\log_{10}\frac{1}{n}\sum_{i=1}^{n}\left(y_i^2\right); \text{ for minimizing the response} \qquad \text{(Eq. 10.19)}$$

3. For a target response signal-to-noise ratio becomes

$$\frac{S}{N} = \frac{Signal}{Noise} = 10\log_{10}\left(\frac{\bar{y}^2}{S^2}\right); \text{ for a target response} \qquad \text{(Eq. 10.20)}$$

The concept of signal-to-noise ratio is illustrated in Example 10.4. To achieve a robust process, the process of experimenting and analyzing steps are the following.

Initiating steps:

1. Problem identification
2. Brainstorming session: Identify factors, factor settings, possible interactions, and objectives
3. Experimental design: Select orthogonal arrays, and experimental design

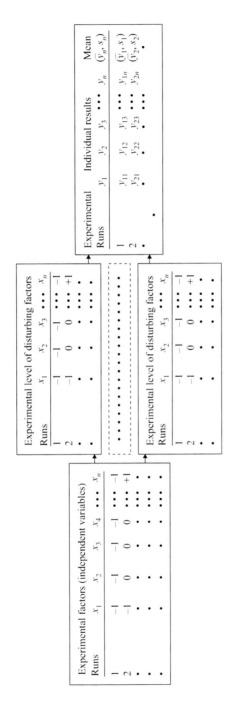

Figure 10.15. Factorial (i.e., 2^k) or fractional factorial design (i.e., 2^{k-p}) with level of noise factor design.

Experimentation steps and analysis:

4. Carry out experiment
5. Conduct analysis of the experimental response
6. Confirm the runs

Example 10.4 Taguchi Application

An injection molding process engineer has been facing issues related with part shrinkage and wrap, which occurs during curing and after part ejected. Normally, plastic parts shrink for 72 hours (slow rate) after being molded. The part shrinkage has contributed to increased part variability and customer dissatisfaction. Mathematically, variation and reproducibility are inversely related to each other (e.g., as variation increases, producibility decreases due to increase of nonconformance probability, which has negative impact in company's profitability).

Thus, the process engineer has been assigned to find the independent variables that contribute to variation in part shrinkage and come up with settings to reduce the shrinkage (difference between cavity and actual part dimension).

As a design of experiment (DOE) expert you are assigned to help process engineer in designing and analyzing the response data. This is visualized in Figure 10.16.

Independent controllable factors		Low level	High level
I_1: Pack pressure (hydraulic)	(x_1)	800 psi	1200 psi
I_2: Injection velocity	(x_2)	1.5	1.75
I_3: Cooling time	(x_3)	35 sec	40 sec
I_4: Inject time	(x_4)	1.5 sec	2 sec
I_5: Mold temperature	(x_5)	80°C	90°C
I_6: Nozzle temperature	(x_6)	220°C	240°C
I_7: Gate size	(x_7)	0.003	0.004

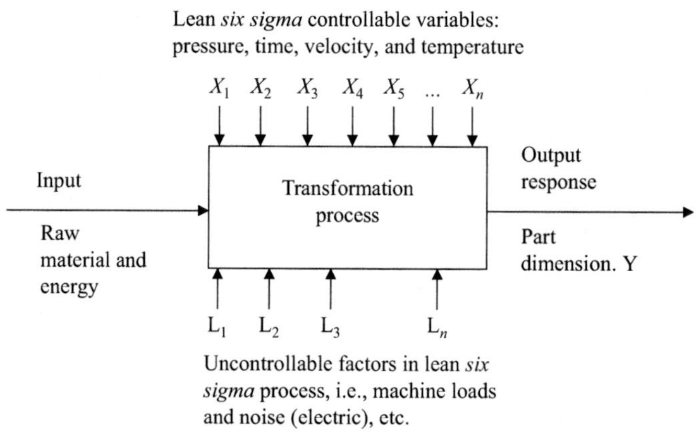

Lean *six sigma* controllable variables:
pressure, time, velocity, and temperature

X_1 X_2 X_3 X_4 X_5 ... X_n

Input

Raw
material and
energy

Transformation
process

Output
response

Part
dimension. Y

L_1 L_2 L_3 L_n

Uncontrollable factors in lean *six
sigma* process, i.e., machine loads
and noise (electric), etc.

Figure 10.16. Process transformation.

Independent uncontrollable factors		Low level	High level
E_1: Percent resin humidity	(x_1)	0.5	1.0
E_2: Room temperature	(x_2)	70°F	75°F
E_3: Percent regrind distribution	(x_3)	5%	10%

Solution

The robust design layout is shown in Figure 10.17.

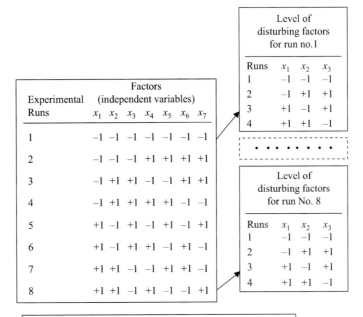

Figure 10.17. Robust design layout $k = 7$.

STEP 1

Apply for inner controllable array (Table 10.22 design-coded values) $2^{k-p} = 2^{7-4} = 2^3 = 8$ or Taguchi method L_8 Fractional Factorial screening matrix and for inner array apply $2^{k-p} = 2^{3-1} = 2^2 = 4$ fractional factorial robust design.

Table 10.22. Inner controllable and outer controllable design—Coded values

Outer matrix 2^{3-1} uncontrollable variables

Run	E_1	E_2	E_3
1	−1	−1	−1
2	−1	+1	+1
3	+1	−1	+1
4	+1	+1	−1

							Outer uncontrollable design 2^{3-1}				
						E_3	−1	+1	+1	−1	
Coded experimental factor values						E_2	−1	+1	−1	+1	Outer
Inner controllable experimental design 2^{7-4}:						E_1	−1	−1	+1	+1	mean
Run	I_1	I_2	I_3	I_4	I_5	I_6	I_7	y_1	y_2	y_3	y_4	$y_{(mean)}$
1	−1	−1	−1	−1	−1	−1	−1	3.5	3.6	3.7	3.7	3.625
2	−1	−1	−1	+1	+1	+1	+1	3.7	3.9	1.7	4.1	3.350
3	−1	+1	+1	−1	−1	+1	+1	4.5	1.9	4.2	1.8	3.100
4	−1	+1	+1	+1	+1	−1	−1	3.4	3.3	3.4	3.2	3.325
5	+1	−1	+1	−1	+1	−1	+1	4.4	4.5	4.4	4.5	4.450
6	+1	−1	+1	+1	−1	+1	−1	3.5	5.6	4.5	2.4	4.000
7	+1	+1	−1	−1	+1	+1	−1	5.4	3.3	3.6	6.0	4.575
8	+1	+1	−1	+1	−1	−1	+1	3.3	3.4	3.2	3.3	3.300
Overall Average												3.716

Note that there are four outputs measured (y_1, y_2, y_3, y_4) on each row. These correspond to the four "outer array" design points at each corner of the "outer array" box. The uncontrollable (difficult to control) design is not required, but as an extreme condition, we can control them during the experiment by additional effort (see Table 10.23 for actual design values). Their use in the design implies more robust results.

Response effects are listed in Table 10.24.

Standard deviation effects are confirmed in Table 10.25.

Table 10.23. Inner controllable and outer controllable design—Actual values

Outer matrix 2^{3-1} uncontrollable variables

Run	E_1	E_2	E_3
1	0.5%	70	5%
2	0.5%	75	10%
3	1%	70	10%
4	1%	75	5%

							Outer uncontrollable design 2^{3-1}:				
						E_3	**5%**	**10%**	**10%**	**5%**	
Actual experimental factor values						E_2	**70**	**75**	**70**	**75**	Standard
Inner controllable experimental design 2^{7-4}:						E_1	**0.5%**	**0.5%**	**1%**	**1%**	deviation

Run	I_1	I_2	I_3	I_4	I_5	I_6	I_7	y_1	y_2	y_3	y_4	(s)
1	800	1.50	35	1.50	80	220	0.003	**3.5**	**3.6**	**3.7**	**3.7**	0.096
2	800	1.50	35	2.00	90	240	0.004	**3.7**	**3.9**	**1.7**	**4.1**	1.112
3	800	1.75	40	1.50	80	240	0.004	**4.5**	**1.9**	**4.2**	**1.8**	1.449
4	800	1.75	40	2.00	90	220	0.003	**3.4**	**3.3**	**3.4**	**3.2**	0.096
5	1200	1.50	40	1.50	90	220	0.004	**4.4**	**4.5**	**4.4**	**4.5**	0.058
6	1200	1.50	40	2.00	80	240	0.003	**3.5**	**5.6**	**4.5**	**2.4**	1.369
7	1200	1.75	35	1.50	90	240	0.003	**5.4**	**3.3**	**3.6**	**6.0**	1.328
8	1200	1.75	35	2.00	80	220	0.004	**3.3**	**3.4**	**3.2**	**3.3**	0.082
Overall Average												0.698

Table 10.24. Marginal mean analysis of response effects of factors

Cause/Interaction	$y-$ Avg	$y+$ Avg	Δ Effect	Δ |Effect|	$\Delta^{1/2}$ Effect	$\Delta^{1/2}$ |Effect|
I_1: Pack pressure (hydraulic) x_1	3.350	4.081	0.7313	**0.7313**	0.3656	0.3656
I_2: Injection velocity (x_2)	3.850	3.575	−0.2813	0.2813	−0.1406	0.1406
I_3: Cooling time (x_3)	3.713	3.719	0.0063	0.0063	0.0031	0.0031
I_4: Inject time (x_4)	3.938	3.494	−0.4438	**0.4438**	−0.2219	0.2219
I_5: Mold temperature (x_5)	3.506	3.925	0.4188	**0.4188**	0.2094	0.2094
I_6: Nozzle temperature (x_6)	3.675	3.756	0.0813	0.0813	0.0406	0.0406
I_7: Gate size (x_7)	3.881	3.550	−0.3313	0.3313	−0.1656	0.1656

Taguchi's three signal-to-noise ratio originates from his loss functions with the following experimental targets.

4. Maximize the response using the following signal-to-noise ratio equation and Table 10.26.

Table 10.25. Response effects of factors

Cause/Interaction	$s-$ Avg	$s+$ Avg	Δ Effect	Δ |Effect|	$\Delta\frac{1}{2}$ Effect	$\Delta\frac{1}{2}$ |Effect|
I_1: Pack pressure (hydraulic) (x_1)	0.6882	0.7089	0.0207	0.0207	0.0104	0.0104
I_2: Injection velocity (x_2)	0.6586	0.7385	0.0800	0.0800	0.0400	0.0400
I_3: Cooling time (x_3)	0.6543	0.07428	0.0886	0.0886	0.0442	0.0442
I_4: Inject time (x_4)	0.7326	0.6645	−0.0680	0.0680	−0.0340	0.0340
I_5: Mold temperature (x_5)	0.7488	0.6483	−0.1005	0.1005	−0.0502	0.0502
I_6: Nozzle temperature (x_6)	0.0827	1.3144	1.2317	1.2317	0.6158	0.6158
I_7: Gate size (x_7)	0.7219	0.6751	−0.0468	0.0468	−0.0234	0.0234

Table 10.26. Maximizing the response signal-to-noise ratio design

Outer matrix 2^{3-1}: Uncontrollable variables

Run	E_1	E_2	E_3
1	−1	−1	−1
2	−1	+1	+1
3	+1	−1	+1
4	+1	+1	−1

							Outer uncontrollable design 2^{3-1}:							Maximize
Coded experimental values of inner controllable experimental/ design 2^{7-4}:							E_3	−1	+1	+1	−1			Response signal/ noise
							E_2	−1	+1	−1	+1	Outer mean	Std. dev.	
							E_1	−1	−1	+1	+1			
Run	I_1	I_2	I_3	I_4	I_5	I_6	I_7	y_1	y_2	y_3	y_4	$y_{(mean)}$	(S)	S/N
1	−1	−1	−1	−1	−1	−1	−1	3.5	3.6	3.7	3.7	3.620	0.096	8.17
2	−1	−1	−1	+1	+1	+1	+1	3.7	3.9	1.7	4.1	3.350	1.112	5.65
3	−1	+1	+1	−1	−1	+1	+1	4.5	1.9	4.2	1.8	3.100	1.449	4.61
4	−1	+1	+1	+1	+1	−1	−1	3.4	3.3	3.4	3.2	3.325	0.096	7.42
5	+1	−1	+1	−1	+1	−1	+1	4.4	4.5	4.4	4.5	4.450	0.058	9.96
6	+1	−1	+1	+1	−1	+1	−1	3.5	5.6	4.5	2.4	4.000	1.369	7.74
7	+1	+1	−1	−1	+1	+1	−1	5.4	3.3	3.6	6.0	4.575	1.328	9.37
8	+1	+1	−1	+1	−1	−1	+1	3.3	3.4	3.2	3.3	3.300	0.082	7.35
Overall Average												3.716	0.698	7.53

$$\frac{S}{N} = \frac{Signal}{Noise} = -10\log_{10}\frac{1}{n}\sum_{i=1}^{n}\frac{1}{y_i^2}; \text{ for maximizing the response}$$

$$= -10\,Log_{10}\frac{1}{4}\sum_{1}^{4}\left(\frac{1}{3.5^2}+\frac{1}{3.6^2}+\frac{1}{3.7^2}+\frac{1}{3.7^2}\right) = -10\,Log_{10}\left(\frac{\frac{1}{3.5^2}+\frac{1}{3.6^2}+\frac{1}{3.7^2}+\frac{1}{3.7^2}}{4}\right)$$

$$= 8.1689 \cong 8.17$$

Signal-to-noise ratios for low and high effects are calculated as follows:

$$\textbf{S/Ns} = \Sigma \textbf{ (S/Ns for I}_1\textbf{)}^- = 8.17 + 5.65 + 4.61 + 7.42 = 6.462$$
$$\textbf{S/Ns} = \Sigma \textbf{ (S/Ns for I}_1\textbf{)}^+ = 9.96 + 7.74 + 9.37 + 7.35 = 8.605$$

Table 10.27 summarizes the S/Ns values for low and high effects.

5. Minimize the response using the following signal noise ratio equation and Table 10.28:

$$\frac{S}{N} = \frac{Signal}{Noise} = -10\log_{10}\frac{1}{n}\sum_{i=1}^{n}\left(y_i^2\right); \text{ for minimizing the response}$$

$$= -10\log_{10}\frac{1}{4}\sum_{1}^{4}\left(3.5^2+3.6^2+3.7^2+3.7^2\right) = -10\log_{10}\left(\frac{3.5^2+3.6^2+3.7^2+3.7^2}{4}\right)$$

$$= -14.1987 \cong -14.20$$

Signal-to-noise ratio for low and high effects are calculated as follows:

$$\textbf{S/Ns} = \Sigma \textbf{ (S/Ns for I}_1\textbf{)}^- = (-14.2) + (-13.9) + (-14.0) + (-13.4) = -13.75$$
$$\textbf{S/Ns} = \Sigma \textbf{ (S/Ns for I}_1\textbf{)}^+ = (-16.0) + (-15.4) + (-16.5) + (-13.4) = -15.32$$

Etc.

Table 10.27. Signal-to-noise ratio for maximizing response

Cause/Interaction	S/Ns– Avg	S/Ns+ Avg	Δ Effect	Δ |Effect|	Δ½ Effect	Δ½ |Effect|
I_1: Pack Pressure (hydraulic) (x_1)	6.462	8.605	2.143	2.143	1.0716	1.0716
I_2: Injection Velocity (x_2)	7.879	7.189	–0.6902	0.6902	–0.3451	0.3451
I_3: Cooling time (x_3)	7.637	7.431	–0.2060	0.2060	–0.1030	0.1030
I_4: Inject time (x_4)	8.027	7.041	–0.9862	0.9862	–0.4931	0.4931
I_5: Mold temperature (x_5)	6.968	8.099	1.1308	**1.1308**	0.5654	0.5654
I_6: Nozzle temperature (x_6)	8.224	6.844	–1.3798	**1.3798**	–0.6899	0.6899
I_7: Gate size (x_7)	8.174	6.893	–1.2819	**1.2819**	–0.6409	0.6409

Table 10.28. Minimizing the response signal-to-noise ratio

Outer matrix 2^{3-1}: Uncontrollable variables

Run	E_1	E_2	E_3
1	−1	−1	−1
2	−1	+1	+1
3	+1	−1	+1
4	+1	+1	−1

								Outer uncontrollable design 2^{3-1}:						**Minimize**
Coded experimental values of							E_3	−1	+1	+1	−1			**Response**
inner controllable experimental/							E_2	−1	+1	−1	+1	**Outer**	**Std.**	**signal/**
design 2^{7-4}:							E_1	−1	−1	+1	+1	**mean**	**dev.**	**noise**
Run	I_1	I_2	I_3	I_4	I_5	I_6	I_7	y_1	y_2	y_3	y_4	$y_{(mean)}$	(S)	S/N
1	−1	−1	−1	−1	−1	−1	−1	3.5	3.6	3.7	3.7	3.620	0.096	−14.2
2	−1	−1	−1	+1	+1	+1	+1	3.7	3.9	1.7	4.1	3.350	1.112	−13.9
3	−1	+1	+1	−1	−1	+1	+1	4.5	1.9	4.2	1.8	3.100	1.449	−14.0
4	−1	+1	+1	+1	+1	−1	−1	3.4	3.3	3.4	3.2	3.325	0.096	−13.4
5	+1	−1	+1	−1	+1	−1	+1	4.4	4.5	4.4	4.5	4.450	0.058	−16.0
6	+1	−1	+1	+1	−1	+1	−1	3.5	5.6	4.5	2.4	4.000	1.369	−15.4
7	+1	+1	−1	−1	+1	+1	−1	5.4	3.3	3.6	6.0	4.575	1.328	−16.5
8	+1	+1	−1	+1	−1	−1	+1	3.3	3.4	3.2	3.3	3.300	0.082	−13.4
Overall Average												**3.716**	**0.698**	**−14.6**

Table 10.29 Summarizes the S/Ns values for low and high effects.

Table 10.29. Signal-to-noise ratio for minimizing response

Cause/Interaction	S/Ns− Avg	S/Ns+ Avg	Δ Effect	Δ |Effect|	Δ½ Effect	Δ½ |Effect|
I_1: Pack Pressure (hydraulic) (x_1)	−13.75	−15.32	−1.4390	**1.4390**	−0.7198	0.7198
I_2: Injection Velocity (x_2)	−14.86	−14.33	0.5336	0.5336	0.2668	0.2668
I_3: Cooling time (x_3)	−14.48	−14.71	−0.2306	0.2306	−0.1153	0.1153
I_4: Inject time (x_4)	−15.17	−14.03	1.1390	**1.1390**	0.5695	0.5695
I_5: Mold temperature (x_5)	−14.25	−14.94	−0.6922	0.6922	−0.3461	0.3461
I_6: Nozzle temperature (x_6)	−14.25	−14.94	−0.6873	0.6873	−0.3436	0.3436
I_7: Gate size (x_7)	−14.89	−14.30	0.5830	0.5830	0.2915	0.2915

6. For a target response using the following signal-to-noise ratio equation and Table 10.30,

$$\frac{S}{N} = \frac{Signal}{Noise} = 10 \log_{10}\left(\frac{\bar{y}^2}{S^2}\right); \text{ for a target response}$$

$$= 10 \log_{10}\left(\frac{3.625^2}{0.096^2}\right) = 4.459 \cong 4.46$$

Signal-to-Noise ratios for low and high effects are calculated as follows:

$$\text{S/Ns} = \Sigma \ (\text{S/Ns for I}_1)^- = 4.46 + 4.52 + 4.39 + 4.46 = 4.458$$
$$\text{S/Ns} = \Sigma \ (\text{S/Ns for I}_1)^+ = 4.46 + 4.46 + 4.45 + 4.46 = 4.455$$

etc.

Table 10.31 summarizes the S/Ns values for low and high effects.

Table 10.30. Signal-to-noise ratio for target response design—coded values

Outer matrix 2^{3-1}: Uncontrollable variables

Run	E_1	E_2	E_3
1	−1	−1	−1
2	−1	+1	+1
3	+1	−1	+1
4	+1	+1	−1

							Outer uncontrollable design 2^{3-1}:						Target
Coded experimental values of inner controllable experimental/ design 2^{7-4}:							E_3	−1	+1	+1	−1			Response signal/ noise
							E_2	−1	+1	−1	+1	Outer mean	Std. dev.	
							E_1	−1	−1	+1	+1			
Run	I_1	I_2	I_3	I_4	I_5	I_6	I_7	y_1	y_2	y_3	y_4	$y_{(mean)}$	(S)	S/N
1	−1	−1	−1	−1	−1	−1	−1	3.5	3.6	3.7	3.7	3.620	0.096	4.46
2	−1	−1	−1	+1	+1	+1	+1	3.7	3.9	1.7	4.1	3.350	1.112	4.52
3	−1	+1	+1	−1	−1	+1	+1	4.5	1.9	4.2	1.8	3.100	1.449	4.39
4	−1	+1	+1	+1	+1	−1	−1	3.4	3.3	3.4	3.2	3.325	0.096	4.46
5	+1	−1	+1	−1	+1	−1	+1	4.4	4.5	4.4	4.5	4.450	0.058	4.46
6	+1	−1	+1	+1	−1	+1	−1	3.5	5.6	4.5	2.4	4.000	1.369	4.45
7	+1	+1	−1	−1	+1	+1	−1	5.4	3.3	3.6	6.0	4.575	1.328	4.46
8	+1	+1	−1	+1	−1	−1	+1	3.3	3.4	3.2	3.3	3.300	0.082	4.45
Overall Average												**3.716**	**0.698**	**4.45**

Table 10.31. Signal-to-noise ratio for target response

Cause/Interaction	S/Ns− Avg	S/Ns+ Avg	Δ Effect	Δ |Effect|	Δ½ Effect	Δ½ |Effect|
I_1: Pack Pressure (hydraulic) (x_1)	4.458	4.455	−0.0042	0.0042	−0.0021	0.0021
I_2: Injection Velocity (x_2)	4.469	4.442	−0.0267	0.0267	−0.0134	0.0134
I_3: Cooling time (x_3)	4.472	4.439	−0.0328	0.0328	−0.0164	0.0164
I_4: Inject time (x_4)	4.442	4.469	0.0269	0.0269	0.0135	0.0135
I_5: Mold temperature (x_5)	4.457	4.454	−0.0035	0.0035	−0.0018	0.0018
I_6: Nozzle temperature (x_6)	4.457	4.454	−0.0035	0.0035	−0.0018	0.0018
I_7: Gate size (x_7)	4.457	4.453	−0.0036	0.0036	−0.0018	0.0018

10.6 RESPONSE SURFACE DESIGNS AND PROCESS/ PRODUCT OPTIMIZATION

The response surface methodology (RSM) and design was originally introduced by Box and Wilson (1951) and later developed by Box and Hunter (1957). It is useful for modeling and analysis of processes in which the response is influenced by numerous factors and the goal is to optimize the response. Thus, the related experiment is planned to allow us to predict the interactions and even quadratic effects, as well as binary effects. As a result, the analysis gives us the response surface structure we are researching for. For this purpose they are called *response surface method (RSM) design*. RSM designs are applied to:

- determine the optimal process settings to obtain the best values of the response;
- develop a robust product or process counter for uncontrollable factors. In other terms, establishing a maximum optimal process that even if the process are relatively insensitive to noncontrollable variables; and
- troubleshoot process issues.

For instance, in order to find the levels of experimental inputs of temperature (x_1) and pressure (x_2) and to maximize their transformation into the response (y) of a process we have

$$y = f(x_1, x_2) + \varepsilon \qquad \text{(Eq. 10.21)}$$

The approximation of the response function $y = f(x_1, x_2) + e$ is referred to *response surface methodology*, where the factors x_1 and x_2 are independent input variables, the response y is the output depending upon the x_1 and x_2 conditions. Variable y is a function of x_1, x_2, and the experimental error factor e. If the response is defined by the first-order design that fits the linear function of independent factors approximation, then first-order model with two variables can be

$$y = \beta_0 + \beta_1 x_1 + \beta_2 x_2 + \epsilon \qquad \text{(Eq. 10.22)}$$

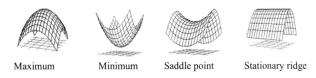

Maximum   Minimum   Saddle point   Stationary ridge

Figure 10.18. Quadratic model forms of response.

and if the response surface has a curving or nonlinear, then a higher degree of polynomial should be applied. In the case of two variables it is called second-order model:

$$y = \beta_0 + \beta_1 x_1 + \beta_2 x_2 + \beta_{11} x_1^2 + \beta_{22} x_2^2 + \beta_{12} x_1 x_2 + \epsilon \qquad \text{(Eq. 10.23)}$$

the polynomial representation of above equation (10.23) at any point $(x_1, x_2, x_3, ..., x_n)$ can be represented by an equation form:

$$y = \beta_0 + \beta_1 x_1 + \beta_2 x_2 + ... + \beta_{11} x_1^2 + \beta_{22} x_2^2 + ... + \beta_{12} x_1 x_2 + ... + \beta_{111} x_1^3 + \beta_{222} x_2^3 + ... \qquad \text{(Eq. 10.24)}$$

The response surface designs purpose is for fitting response surface (Figure 10.18). Thus, the goal is to find the area that optimal response takes place and efficiently in the direction of maximum or minimum response so that the perfect response is achieved.

Application of RSM and central composite design (CCD) will be illustrated in the next section.

10.7   CENTRAL COMPOSITE DESIGN (CCD): OPTIMUM DESIGN

A Box–Wilson composite design usually called central composite design (CCD), which is the most well-known design for estimating the coefficients in the second-order model. It is a very effective and efficient full factorial second order (modeling) design and consists of the following:

1. Factorial 2^k or a suitable fractional factorial design 2^{k-p} (when $k \geq 5$) with center points $(0, 0, ..., 0)$. The values of the coded variables in the factorial design are $(x_1, x_2, x_3, ..., x_n) = (\pm 1, \pm 1, ..., \pm 1)$.
2. The factorial design is improved (reinforced) with a group of "star points" that provide estimation of curvature. If the range from the center of the design space to a factorial point (in the case of square or cube it is called corner points) is ± 1 unit for each factor, the range from the center of the design space to a "start point" is $\pm \alpha$ with $|\alpha| > 1$. The actual value of α is based on the certain conditions of design and the number of the factors included. That is $2k$ vertices $(\pm \alpha, 0, 0, ..., 0)$, $(0, \pm \alpha, 0, ..., 0)$, ..., $(0, 0, ..., 0, \pm \alpha)$.
3. The center point replicate $(x_1, x_2, x_3, ..., x_n) = (0, 0, ..., 0)$, that is $n_0 \geq 1$.

The total number of design points is $n = 2^k + 2k + n_0$ or $n = 2^{k-p} + 2k + n_0$ when a fractional factorial replicate is used. A CCD is designed rotatable for α. The rotatability of α depends upon the number of trials in the factorial segment of CCD. Rotatability has advantages in designing of RSM, which is an optimization technique. The optimum point of any design is unknown in advance of experimental runs. The value of α for CCD with factorial design is

$$\alpha = (L^k)^{1/4} \qquad \text{(Eq. 10.25)}$$

Or for fractional factorial design it converts to

$$\alpha = (L^{k-p})^{1/4} \qquad \text{(Eq. 10.26)}$$

where L^k is the factorial portion design (factorial runs). In the case of two level L = 2, and two factor k = 2 composite design (Figures 10.19 and 10.20) with factorial "α" equal to

$$\alpha = (L^k)^{1/4} = (2^2)^{1/4} = 2^{1/2} = 1.414$$

Similarly the full quadratic model is shown in Figure 10.20.

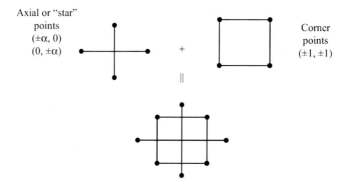

Figure 10.19. CCD for k = 2 variables.

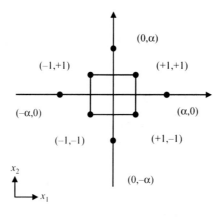

Figure 10.20. Central composite design for k = 2.

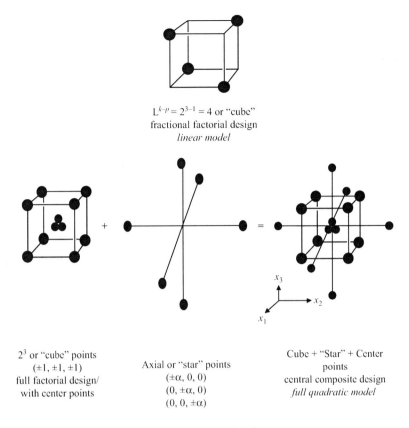

$L^{k-p} = 2^{3-1} = 4$ or "cube"
fractional factorial design
linear model

+

=

2^3 or "cube" points
($\pm 1, \pm 1, \pm 1$)
full factorial design/
with center points

Axial or "star" points
($\pm \alpha, 0, 0$)
($0, \pm \alpha, 0$)
($0, 0, \pm \alpha$)

Cube + "Star" + Center
points
central composite design
full quadratic model

Figure 10.21. Central composite design for $L^k = 2^3$.

Central composite design (Figure 10.21) for two-level (L= 2) and three-factor ($k = 3$) α becomes

$$\alpha = (L^k)^{1/4} = (2^3)^{1/4} = 1.682$$

Center point calculation: The following formula will calculate the approximate center point replications.

$$\text{Center experiment} = 4 \, (\text{Corner experiments} + 1)^{0.50} - 2k \qquad \text{(Eq. 10.27)}$$

For $k = 2$: Center experiment = $4 \, (4 + 1)^{0.50} - 2(2) = 4.94 \cong 5.0$
For $k = 3$: Center experiment = $4 \, (8 + 1)^{0.50} - 2(3) = 6$

Sample CCD design for $k = 2$ and 3 at L = 2 is given in Table 10.32.

Calculating coded and un-coded values: The coded and uncoded values can be calculated using the following equation:

$$\text{Uncoded value (actual value) of } x_i = \left(\frac{\text{Actual high value} + \text{Actual low value}}{2} \right)$$
$$+ \left(\frac{\text{Actual high value} - \text{Actual low value}}{2} \right) \cdot \text{Coded values}$$

$$\text{(Eq. 10.28)}$$

Table 10.32. Composite design settings

Design type	Run	x_1	x_2	Run	x_1	x_2	x_3
		k = 2			**k = 3**		
Factorial	1	−1	−1	1	−1	−1	−1
Design:	2	+1	−1	2	+1	−1	−1
Corner	3	−1	+1	3	−1	+1	−1
Points	4	+1	+1	4	+1	+1	−1
				5	−1	−1	+1
				6	+1	−1	+1
				7	−1	+1	+1
				8	+1	+1	+1
Axial (± α)	5	−1.414	0	9	−1.682	0	0
Design:	6	+1.414	0	10	+1.682	0	0
(Star points)	7	0	−1.414	11	0	−1.682	0
	8	0	+1.414	12	0	+1.682	0
				13	0	0	−1.682
				14	0	0	+1.682
Center	9	0	0	15	0	0	0
Center	10	0	0	16	0	0	0
Design	11	0	0	17	0	0	0
Replicated	12	0	0	18	0	0	0
	13	0	0	19	0	0	0
				20	0	0	0

The following example illustrates this concept.

Example 10.5

Apply two-level composite designs and three-level design analysis.

A design engineer needs to improve the thickness of the camera lens by reducing the variation about the target of 2 mm. The project team concluded that the following factors can make the highest impact in the variation of lens.

Independent factors	Low level	Mid level	High level
Plastic melt Temperature	230°C	240°C	250°C
Mold temperature	70°C	80°C	90°C
Hold time	1.5 sec	2.5 sec	3 sec

1. Carry out two-level full factorial design experiment using high and low levels:
   • Generate the prediction equation from the central composite design.
2. Conduct RSA using central composite design.
3. Carry out three level full factorial design experiments using high, medium, and low levels.
   • Generate the prediction equation from the orthogonally coded design (this section is left for students to be challenged and carry out the full analysis).
4. Compare the predicted models.

Solution

Use two-level full factorial design $(\text{Level})^{\text{Factor}} = 2^k = 2^3 = 8$ corner points of the cube. For composite design, it will require another six runs for star points and plus center point.

STEP 1 DESIGN LAYOUT

The eight coded values of the design (Figure 10.22) for corner points are $(\pm 1, 1, 1)$, $(1, \pm 1, 1)$, and $(1, 1, \pm 1)$. Likewise, the coded values for the star or axial points include $(\pm 1.682, 0, 0)$, $(0, \pm 1.682, 0)$, and $(0, 0, \pm 1.682)$. The process model from the design chart is

$$
\begin{aligned}
y = b_0 + b_1 x_1 + b_2 x_2 + b_3 x_3 \qquad &\text{Linear terms}\\
+ b_{11} x_1^2 + b_{22} x_2^2 + b_{33} x_3^2 \qquad &\text{Quadratic effects}\\
+ b_{12} x_1 x_2 + b_{13} x_1 x_3 + b_{23} x_2 x_3 \qquad &\text{Binary interaction effects}
\end{aligned}
$$

Let's take a look at fractional factorial design presented in Table 10.33 and compare it to full factorial design provided in Table 10.34.

Now we will design a full factorial for the three factors and two levels (Table 10.34) as stated in the problem.

The actual values of design are illustrated in Table 10.35.

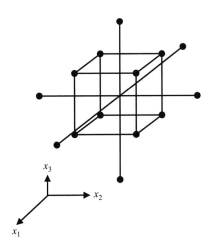

Figure 10.22. Three-factor nonorthogonal composite design.

Table 10.33. Coded values of central composite design: Two levels fractional factorial design $2^{k-1} = 2^{3-1} = 4$

Design type	Plastic melt experiment	Mold temperature	Hold temperature	Time
Fractional	1	−1	−1	−1
Factorial	2	+1	+1	−1
Design	3	−1	−1	+1
(Corner points)	4	+1	+1	+1
Axial	5	−α = −1.682	0	0
Design	6	+α = +1.682	0	0
(Star points)	7	0	−α = −1.682	0
	8	0	+α = +1.682	0
	9	0	0	−α = −1.682
	10	0	0	+α = +1.682
Center	11	0	0	0
Center	12	0	0	0
Design Replicated	13	0	0	0

Table 10.34. Coded values of central composite design: Two level full factorial design $2^{k} = 2^{3} = 8$

Design type	Experiment	Plastic melt temperature	Mold temperature	Hold time
Factorial	1	−1	−1	−1
Design:	2	+1	−1	−1
(Corner	3	−1	+1	−1
Points of	4	+1	+1	−1
Cube)	5	−1	−1	+1
	6	+1	−1	+1
	7	−1	+1	+1
	8	+1	+1	+1
Axial (star)	9	−α = −1.682	0	0
Design	10	+α = +1.682	0	0
(Star points)	11	0	−α = −1.682	0
	12	0	+α = +1.682	0

(Continued)

Table 10.34. (*Continued*)

Design type	Experiment	Plastic melt temperature	Mold temperature	Hold time
	13	0	0	$-\alpha = -1.682$
	14	0	0	$+\alpha = +1.682$
Center	15	0	0	0
Center	16	0	0	0
Design	17	0	0	0
Replicated	18	0	0	0
	19	0	0	0
	20	0	0	0

Table 10.35. Actual values of central composite design

Design type	Experiment	Plastic melt temperature	Mold temperature	Hold time	Lense thickness
Factorial	1	230	70	1.50	11.999
Design	2	250	70	1.50	12.006
(Corner	3	230	90	1.50	12.025
Point of	4	250	90	1.50	12.035
Cube)	5	230	70	3.00	11.955
	6	250	70	3.00	11.983
	7	230	90	3.00	11.974
	8	250	90	3.00	11.995
Axial (star)	9	$-\alpha = 223.18$	80	2.25	11.985
Design	10	$+\alpha = 256.82$	80	2.25	12.056
	11	240	$-\alpha = 63.18$	2.25	11.955
	12	240	$+\alpha = 96.82$	2.25	12.065
	13	240	80	$-\alpha = 0.9886$	11.995
	14	240	80	$+\alpha = 3.5113$	11.975
Center	15	240	80	2.25	11.996
Center	16	240	80	2.25	11.996
Design	17	240	80	2.25	11.995
Replicated	18	240	80	2.25	11.993
	19	240	80	2.25	11.997
	20	240	80	2.25	11.999

The central composite design summaries are

Factors:	3	Replicates:	1
Base runs:	20	Total runs:	20
Base blocks:	1	Total blocks:	1

Two-level factorial: Full factorial

Cube points:	8
Center points in cube:	6
Axial points:	6
Center points in axial:	0
Alpha: $\alpha = 1.68179$	

Figure 10.23 illustrates Pareto chart of variables effects.
Table 10.36 displays estimated effects and coefficients for lens thickness.

Table 10.36. Computer output of factorial fit: Lens thickness versus hold time, mold temperature, plastic temperature

Estimated effects and coefficients for lens thickness (coded units)

Term	Effect	Coefficient
Constant		11.9965
Plastic temperature (x_1)	0.0165	0.0082
Mold temperature (x_2)	0.0215	0.0107
Hold time (x_3)	−0.0395	−0.0198
Plastic temperature • Mold temperature ($x_1 \cdot x_2$)	−0.0010	−0.0005
Plastic temperature • Hold time ($x_1 \cdot x_3$)	0.0080	0.0040
Mold temperature • Hold time ($x_2 \cdot x_3$)	−0.0060	−0.0030
Plastic temperature • Mold temperature • Hold time ($x_1 \cdot x_2 \cdot x_3$)	−0.0025	−0.0013

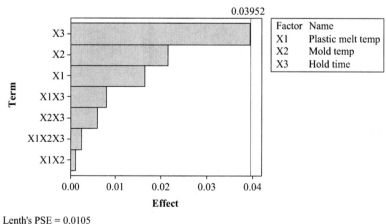

Pareto chart of the effects
(response (y) is lens thickness, alpha = 0.05)

Lenth's PSE = 0.0105

Figure 10.23. Pareto chart of variable the effects.

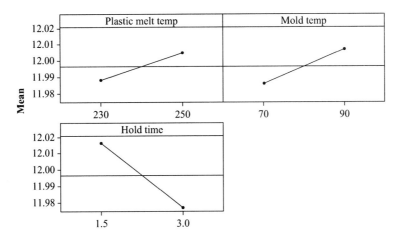

Figure 10. 24. Plot of the response average for each effect.

Figure 10.24 presents main effect plots for Lens thickness.

Table 10.37 provides information of the analysis of variance obtained from computer output on the process response.

Figure 10.25 shows the interaction plot for the lens thickness response average.

STEP 2

Response surface analysis using central composite design is shown in Table 10.38.

Table 10.37. Analysis of variance for lens thickness (coded units)

S = 0.188564, press = 1.1378, R^2 = 99.82% R^2 (pred) = 99.28% R^2 (adj) = 99.66%				
Source	df	Seq SS	Adj SS	Adj MS
Main effects				
Plastic temperature (x_1)	1	0.00458950	0.00458950	0.00152983
Mold temperature (x_2)	1	0.00092450	0.00092450	0.00092450
Hold time (x_3)	1	0.00312050	0.00312050	0.00312050
Two-way interactions	3	0.00020200	0.00020200	0.00006733
$x_1 \bullet x_2$	1	0.00000200	0.00000200	0.00000200
$x_1 \bullet x_3$	1	0.00012800	0.00012800	0.00012800
$x_2 \bullet x_3$	1	0.00007200	0.00007200	0.00007200
Three-way interactions	1	0.00001250	0.00001250	0.00001250
$x_1 \bullet x_2 \bullet x_3$	1	0.00001250	0.00001250	0.00001250
Residual error	0			
Total	7	0.00480400		

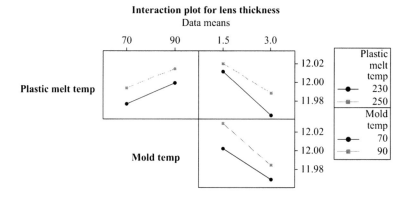

Figure 10.25. Plot of response averages for each interaction.

The model for the process using response surface analysis is

$$y = b_0 + b_1x_1 + b_2x_2 + b_3x_3 \qquad \text{Linear terms}$$
$$+ b_{11}x_1^2 + b_{22}x_2^2 + b_{33}x_3^2 \qquad \text{Quadratic effects}$$
$$+ b_{12}x_1x_2 + b_{13}x_1x_3 + b_{23}x_2x_3 \qquad \text{Binary interaction effects} \qquad \text{(Eq. 10.29)}$$

Table 10.38. Response surface regression: Thickness versus hold time, mold temperature, and plastic melt temperature.

Estimated regression coefficients for thickness using uncoded units				
Term	**Coefficient**	**SE Coef**	**T**	**p-value**
Constant	15.6816	2.99844	5.230	0.000
***Linear terms*:**				
Plastic temperature (x_1)	−0.0317	0.02281	−1.388	0.195
Mold temperature (x_2)	−0.0007	0.01674	−0.043	0.966
Hold time (x_3)	−0.0680	0.21262	−0.320	0.756
***Quadratic effects*:**				
Plastic temperature • Plastic temperature ($x_1 • x_1$)	0.0001	0.00005	1.453	0.177
Mold temperature • Mold temperature ($x_2 • x_2$)	0.0000	0.00005	0.650	0.530
Hold time • Hold time ($x_2 • x_2$)	−0.0104	0.00822	−1.262	0.235
***Binary interaction effects*:**				
Plastic temperature • Mold temperature ($x_1 • x_2$)	−0.0000	0.00006	−0.081	0.937
Plastic temperature • Hold time ($x_1 • x_3$)	0.0005	0.00083	0.645	0.534
Mold temperature • Hold time ($x_2 • x_3$)	−0.0004	0.00083	−0.484	0.639

Substituting the coefficient values for the linear, quadratic, and binary terms from Table 10.17 we have

$$y = 15.6816 - 0.0317x_1 - 0.0007x_2 - 0.0680x_3 \qquad \text{Linear terms}$$
$$+ 0.0001x_1^2 + 0.0000x_2^2 - 0.0104x_3^2 \qquad \text{Quadratic effects}$$
$$- 0.0000\, x_1x_2 + 0.0005x_1x_3 - 0.0004x_2x_3 \qquad \text{Binary interaction effects}$$

The computer output response surface analysis is illustrated in Table 10.39 and Figures 10.26 through Figure 10.30.

STEP 3

A three-level full factorial design (3^3) design schematic presented in Figure 10.31and layout in Table 10.40.

The detailed analysis left for students as an assignment.

Table 10.39. Analysis of variance for lens thickness (coded units)

S = 0.0175492, Press = 0.0232800, R^2 = 79.81%, R^2 (pred.) = 0.00%, R^2 (adj.) = 61.65%						
Source	df	Seq SS	Adj SS	Adj MS	F	p-value
Regression	9	0.012177	0.012177	0.001353	4.39	0.015
Linear:	3	0.010584	0.000632	0.000211	0.68	0.582
Plastic temperature (x_1)	1	0.002517	0.000594	0.000594	1.93	0.195
Mold temperature (x_2)	1	0.005378	0.000001	0.000001	0.00	0.966
Hold time (x_3)	1	0.002689	0.000032	0.000032	0.10	0.756
Square:	3	0.001391	0.001391	0.000464	1.51	0.272
$x_1 \cdot x_1$	1	0.000713	0.000650	0.000650	2.11	0.177
$x_2 \cdot x_2$	1	0.000187	0.000130	0.000130	0.42	0.530
$x_3 \cdot x_3$	1	0.000491	0.000491	0.000491	1.59	0.235
Interactions	3	0.000202	0.000202	0.000067	0.22	0.881
$x_1 \cdot x_2$	1	0.000002	0.000002	0.000002	0.01	0.937
$x_1 \cdot x_3$	1	0.000128	0.000128	0.000128	0.42	0.534
$x_2 \cdot x_3$	1	0.000072	0.000072	0.000072	0.23	0.639
Residual error	10	0.003080	0.003080	0.000308		
Lack of fit	5	0.003060	0.003060	0.000612	152.99	0.000
Pure error	5	0.000020	0.000020	0.000004		
Total	19	0.015257				

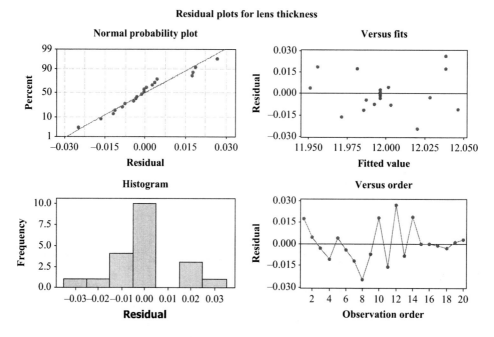

Figure 10.26. Residual plots of response.

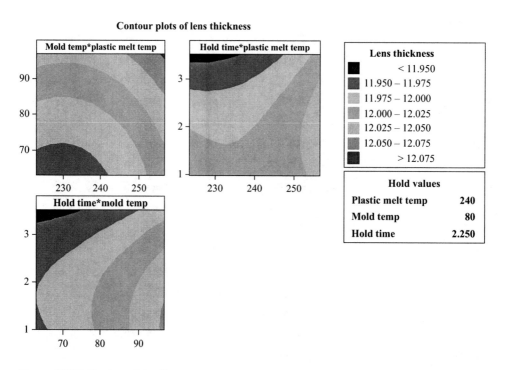

Figure 10.27. Contour plots of response.

Surface plot of lens thickness vs mold temp, plastic melt temp

Hold time = 2.250

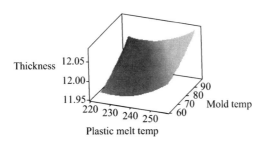

Figure 10.28. Response surface for output (*y*) as a function of independent variables plastic temperature (x_1) and mold temperature (x_2).

Surface plot of lens thickness vs hold time, mold temp

Plastic melt temperature = 240

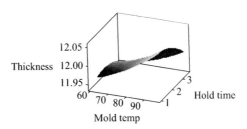

Figure 10.29. Response surface for output (*y*) as a function of independent variables mold temperature (x_2) and hold time (x_3).

Surface plot of lens thickness vs hold time, plastic melt temp

Mold temperature = 80

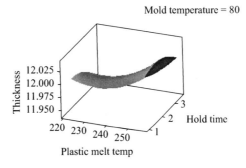

Figure 10.30. Response surface for output (*y*) as a function of independent variables plastic temperature (x_1) and hold time (x_3).

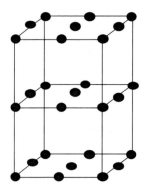

Figure 10.31. A full factorial 3^3 design schematic.

Table 10.40. Three-level full factorial design without composite design actual values of three-level full factorial design $(level)^{factor} = 3^k = 3^3 = 27$

Design type	Experiment	Plastic melt temperature	Mold temperature	Hold time	Lens thickness
Factorial	1	230	70	1.50	11.999
Design:	2	230	70	2.5	11.970
3 Level	3	230	70	3	11.955
	4	230	80	1.50	11.015
	5	230	80	2.50	11.008
	6	230	80	3	11.005
	7	230	90	1.50	12.025
	8	230	90	2.5	12.015
	9	230	90	3	12.010
	10	250	70	1.50	12.006
	11	250	70	2.50	11.995
	12	250	70	3	11.990
	13	250	80	1.50	12.012
	14	250	80	2.50	12.008
	15	250	80	3	11.995
	16	250	90	1.50	12.036
	17	250	90	2.50	12.025
	18	250	90	3	12.020
	19	260	70	1.50	12.015
	20	260	70	2.50	12.017

(Continued)

Table 10.40. (*Continued*)

Design type	Experiment	Plastic melt temperature	Mold temperature	Hold time	Lens thickness
	21	260	70	3	12.013
	22	260	80	1.50	12.022
	23	260	80	2.50	12.014
	24	260	80	3	12.009
	25	260	90	1.50	12.028
	26	260	90	2.50	12.021
	27	260	90	3	12.019

STEP 4

The three-level full factorial design would create the best model, followed by the RSA, and then the two-level full factorial.

10.8   FAILURE MODE EFFECT ANALYSIS (FMEA)

FMEA is an analytical engineering "reliability tools" for preventing defects by prioritizing recognized process or product problems, identifying and eliminating potential failure modes, design, or manufacturing process before customers receive the product. In addition, it identifies actions that could eliminate or reduce the probability or severity of the potential failure and effects. Finally, it is a proactive quality planning and improvement process, associated with risk analysis and the calculation of reducing risks.

FMEA Terminologies

Failure modes: Specific loss of a product functionality that is a concise description of how a particular part, system, of manufacturing product/process may conceivably fail to performed it desired functions.

Failure mode "Effect": It is a consequence (symptom) of a system or part failure.

Severity rating: Severity is a seriousness of the "*effects.*" It is numerical rating of the impact.

Failure mode "Cause": Cause is a description of design or process variables that cause the failure mode. A process or design may have multiple causes. The investigator should focus on causes not the symptoms.

Occurrence rating: It is an approximate number of failure frequencies that will happen for specific cause over the period of life of product.

Failure mode "control": It is a control system that will prevent cause of the failure mode or detect the failure mode from occurring, for instance, mechanism, methodology, procedure, testing, and continuous experimenting.

Detection rating: It is a numerical probability rating that a designed control system will detect specific failure modes before the packaging.

What FMEA can do for you?

1. Identifies the failure modes before they occur
2. Evaluates the effects and severity of failure modes
3. Identifies causes and probability of occurrence of the failure modes
4. Recognizes the controls and their effectiveness
5. Develops action plan to reduce risks
6. Applies the contingency plan of action.

FMEA is continuous in risk analysis and identifies the ways in which a product, process, or service can fail. It estimate the risks associated with specific causes. Then, it takes into consideration the severity of product failure, which may range from hazardous—without warning—to moderate, to none. Eventually, it prioritizes the actions that should be taken to reduce the risks. Where do risks come from? Some of the process risk factors are shown in Figure 10.32.

Managing process and product risks is one of the important aspects of this subject. This involves the probability and reliability, as discussed in previous chapters. In brief, from a management point of view, risk management is equal to identification, quantification, response, and lessons learned during process design. The three steps process includes the FMEA and risk prioritization number (RPN).

1. Identify failure modes and effect (s)

For identifying failure mode effects, you may use a diagram of subsystem, subprocess activities, a tree-diagram to capture linkage and connectivity reason, brainstorming techniques to accumulate areas of potential failure modes, and collecting data to capture areas of actual failure modes.

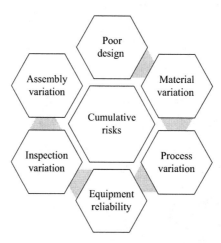

Figure 10. 32. FMEA cumulative risk factor.

2. Decide on the failure modes and effects

Knowing that failures cause defects, one may ask, what are defects? Products and services are defective because they deviate from the desired specification limit from the supplier. They are unsafe due to design defects even though they meet requirements or they are incapable of meeting their designed performance expected. Therefore, they are dangerous because they lack adequate warning and instructions.

Where to find such failure modes? You can find in the area of manufacturing operation for opportunities as product deformed, leaking, misaligned, cracked, broken, eccentric, loose, jams, sticking, imbalance, short circuit, wear, tear, and many more. The typical failure effects may come from process noise, erratic, machine stops, poor appearance, unstable operation, ineffectual equipment, and irregular processes.

A sample of failure mode effects analysis (FMEA) is given in Tables 10.41 and 10.42.

FMEA computes the effects of failures in terms of *risk prioritization or priority number* (RPN), which is also function of risk factors as severity, occurrence, and detection.

$$RPN = f(\text{Severity} \times \text{Occurrence} \times \text{Detection}) \qquad \text{(Eq. 10.30)}$$

Or, it is the product of severity, occurrence, and detection.

$$\text{Risk} = RPN = \text{Severity} \times \text{Occurrence} \times \text{Detection} \qquad \text{(Eq. 10.30)}$$

Table 10.41. Failure mode effect analysis sample table

Processes function	Failure mode	Effects of failure	Severity of failure	Cause of failure	Occurrence (Probability of failure)	Existing control	Detection	Risk prioritization number (RPN)

Table 10.42. Expanded failure mode effect analysis: Sample table

Processes function	Failure mode	Effects of failure	Severity of failure	Cause of failure	Occurrence (Probability of failure)	Existing control	Detection	Risk prioritization number (RPN)	Recommended action plans	Responsibility	Target date	Severity	Occurrence	Detection	Calculate the new RPN

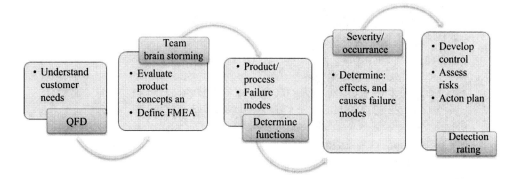

Figure 10.33. FMEA preparation and planning.

what needs to be done before FMEA? The flowchart (Figure 10.33) will guide one to start planning for a FMEA and improvement actions.
3. Application of RPN rating: The RPN ratings are given in Tables 10.43 through Table 10.45.
   Furthermore, RPN rating in terms of process capability is given in Table 10.44.

Improving the RPN

Action plan should be taken very seriously and focus must be on reducing the severity rating, occurrence rating, and the detection rating. To minimize severity, design must be changed. Similarly, to minimize occurrence, design or process change is needed. Concurrently, to minimize detection, one should move to less operator-dependent or verification/inspections.

And an additional analysis regarding inspection rating of Risk Prioritization Number is given in Table 10.45.

Summary of FMEA Steps

1. Identify the product, process, or system components.
2. Brainstorm with process owner potential failure modes.
3. Categorize potential effects of each failure mode.
4. Specify a severity rating for each effect.
5. Specify an occurrence rating for each effect.
6. Specify a detection rating for each failure mode or effect.
7. Perform the RPN calculation before any improvement.
8. Prioritize the failure mode improvement list.
9. Minimize the high-risk failure modes.
10. Compute the resulting risk prioritization number (RPN).

Once the process data has been analyzed, the results will indicate the dominant factors that cause process defects. Then, the process will be simulated or optimized through optimization theory and techniques such as: statistical methods, analysis of variance (ANOVA), design of experiment, and other methods (refer to process improvement methods).

Table 10.43. Rating of severity (risk priority number)

Rating	Description	Definition
10	Dangerously High	Failure may injure the customer or an employee
9	Extremely High	Failure would create noncompliance with federal regulations
8	Very High	Failure renders the unit inoperable or unit for use. Further will create loss of function
7	High	Failure causes a high degree of repair and customer dissatisfaction
6	Moderate	Failure results in disruption of production. Possible scrap. Major loss of Takt-time (cycle time)
5	Low	Failure creates enough of a performance loss to cause the customer to complain. Product must be repaired
4	Very Low	Failure can be overcome with modifications to the customer's process or product, but there is a minor performance loss
3	Minor	Failure would create a minor nuisance to the customer, but the customer can overcome it in the process or product without performance loss
2	Very Minor	Failure may not be readily apparent to the customer, but would have minor effects on the customer's process or product
1	None	Failure will not be observable to the customer and would not affect the customer's process or product. Thus, no effect on production

Table 10.44. Occurrence rating in process capability

Rating	Description	Definition
10	Very high: Failure is almost inevitable	More than one occurrence per day. Or a probability of more than three occurrences in 10 events ($C_{pk} < 0.33$)
9	Process is not in statistical control	One occurrence every 3 to 4 days. Or a probability of occurrence in 10 events. ($C_{pk} < 0.67$).
8	High: Repeated failures	One occurrence per week. Or a probability of five occurrences in 100 events ($C_{pk} < 0.83$)
7	Slightly high to moderate	One occurrence every month. Or one occurrence in 100 events ($C_{pk} < 0.83$)
6	Moderate: Occasional failure	One occurrence every three months. Or three occurrences in 1,000 events ($C_{pk} = 1.00$)
5	Failure or out-of-control condition	One occurrence every six months to one year. Or three occurrences in 10,000 events ($C_{pk} = 1.17$)

(*Continued*)

Table 10.44. (*Continued*)

Rating	Description	Definition
4	Process is in statistical control	One occurrence per year. Or six occurrences in 100,000 events (C_{pk} = 1.33)
3	Low: Relatively few failures	One occurrence every one to three years. Or six occurrences in 10 million events (C_{pk} = 1.67)
2	Only isolated failure	One occurrence every three to five years. Or two occurrences in one billion events. (C_{pk} = 2.00)
1	Remote: Failure is unlikely	One occurrence in greater than five years. Or less than two occurrences in one billion events (C_{pk} > 2.00)

Table 10.45. Detection (Inspection) rating RPN

Rating	Description	Definition
10	Absolute uncertainty	The product is not inspected or the defect caused by failure is not detectable. No known control available.
9	Very remote	Product is sampled, inspected, and released on Acceptable Quality Level (AQL) sampling plan
8	Remote	Product is accepted based on no defectives in a sample
7	Very low	Product is 100% manually inspected in the process
6	Low	Product is 100% manually inspected using go/no-go or mistake-proofing gauges
5	Moderate	Some Statistical Process Control (SPC) is used in the process and the product is final inspected off-line
4	Moderately High	SPC is used and there is immediate reaction to out-of-control conditions
3	High	An effective SPC program is in place with process capabilities (C_{pk}) greater than 1.33
2	Very low	All product is 100% automatically inspected
1	Almost certain	The defect is obvious or there is 100% automatic inspection with regular calibration and preventive maintenance of the inspection equipment.

10.9 POKA-YOKE (JAPANESE TERM FOR MISTAKE PROOFING, PRONOUNCED POH-KAH YOH-KAY).

Poke-yoke is defined as process of analyzing and establishing a system to build quality into an assembly or manufacturing process with simple low-cost devices and methods. The phrase *Poka-Yoke* originates from the Japanese words: *yokeru = to avoid* and *poka = inadvertent* (*careless*) *errors.*

It is a methodology that is used to strive toward zero defects by either preventing or automatically detecting defects. *Poka-Yoke* also called mistake proofing. It is a process design that is robust to mistakes. The mistake proofing steps are identifying problem, prioritizing the problem, finding the root cause, creating solutions, and measuring the results.

10.10   5S KAIZEN PRINCIPLES

Kaizen originates from Japanese terms, kai, which means changer and zen, meaning "for better"; thus Kaizen means continuous improvement. It was coined in Japan by Masaaki Imai. Figure 10.34 illustrates 5S steps that encompass a disciplined technique for establishing a highly clean workplace, safe, orderly, and high performance. It is part of Lean principles that supports waste reduction. So, what does 5S stand for?

10.10.1   SI = SORT (SEIRI)

Sort out what you do not need. Separate items into "necessary" and "unnecessary." Examples include

(a) Sort out filling, documents, electronics, equipment, basically your entire workplace.
(b) Covert all the documents to digital or paperless as much as you can.
(c) Keep what you need and discard the unneeded material.
(d) Eliminate anything that is obsolete.

10.10.2   SII = SET IN ORDER (SEITON)

Organize (what, where, and how). Organize the left over from the sorting process using efficient and effective storage techniques. Just a few examples are the following.

(a) One strategy is painting the floor for process arrangement.
(b) Just like filing cabinets arrange items in alphabetical or by application order. Make it easy to locate and use.
(c) Items needed most or frequently used must be reachable.

10.10.3   SIII = SHINE (SEISO)

Once sorting (S_I) and organizing (S_{II}) have been completed, the next step is cleaning the area; for example, bookshelves, equipment, storage area, conference room, offices rooms, under table, etc.

Figure 10.34. 5S Principles.

(a) What to clean?
Make a checklist of what to clean.
(b) How to clean?
Cleaning procedure also needs to be included.
(c) Who will clean?
The functional people and their responsibilities must be clear and determined. Regularly, cleaning areas needs to be inspected.
(d) How much cleaning is necessary?
Make cleaning a daily habit.

10.10.4 S_{IV} = STANDARDIZE (SEIKETSU)

It has the improved and perhaps the optimum process along with its procedures that have been standardized (or documented to be standardized). Now, S_I through S_{III} have been completed; to keep it clean we need to standardize by making it part of a daily routine job function.

10.10.5 S_V = SUSTAIN (SHITSUKE)

Control and sustain the standardized 3S (S_{IV}) process and implement it with corporate culture.

10.11 TOLLGATE REVIEW AND DELIVERABLES FOR IMPROVE PHASE

Once again before moving to the next phase, control project team must assure that all the deliverables for the *Improve Phase* have been achieved. This is necessary for the gate meeting review. Figure 10.35 can be adjusted to any project that includes product or process improvement.

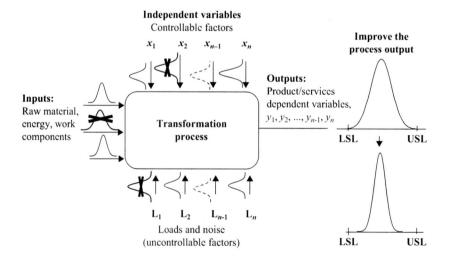

Figure 10.35. Lean Six Sigma transformation process model "Improve phase."

Project team must come up with a firm Six Sigma methodology for improving the projects. That includes but is not limited to the following checklist.

10.11.1 IMPROVE PHASE DELIVERABLES AND CHECKLIST

(a) Apply as many as Improve tools and methodologies such as
    i. Design of Experiment, Robust Engineering, Failure Mode Effect Analysis, Risk Prioritization number
    ii. 5S Kaizen principles, Poka-Yoke
    iii. Complete the Six Sigma Scorecard.
(b) Create the best solution
    i. What techniques are used to develop potential solution?
    ii. Was any screening method utilized in the process of developing solution?
    iii. Is the solution short-term or long-term and what are the financial benefits
(c) Are the changes standardized?
(d) Compute the new Sigma level and compare to process baseline before improvement.
(e) Communicate process results and changes with improvement techniques to people who are involved with the project.
(f) Identify the deliverable to be sustain the improvement.
(g) Create and define RACI-Matrix. Include team members, gate keepers, and anyone who should be part of RACI-Chart.
(h) Concentrate on the Voice-of-Customer (VOC).
(i) Will the improvement meet the organization's financial target?

Once the checklist has been completed, the project must to go through the Tollgate review (Figure 10.36) as explained in Chapter 6 for gate review team's approval.

Figure 10.37 visualizes the process distribution curve before and after the improvement phase. That is before improvement the process variation ranges are wide and not consistent. While after applying the Six Sigma tools the process variation ranges have been minimized and consistency has been established. For example, if we go back to Figure 8.1 cause and effect for pizza delivery, the goal is to deliver the pizza to the customer in a timely fashion. Variability is caused by the particular specialized offered, day of the week, time of a day, traffic outlook, staff experience, and perhaps many other factors. We can visualize the inconsistency in the pizza delivery with statistical distribution curve before the improvement. In the distribution curve the horizontal axis symbolizes the time customers waited to receive their

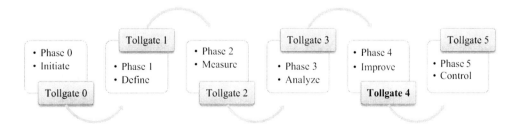

Figure 10.36. Lean Six Sigma process improvement projects: Tollgate review—"Improve phase."

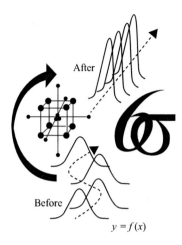

Figure 10.37. Lean Six Sigma
process vision: Before and after
process improvement model

pizza. The vertical axis represents how many times pizza were delivered. The graph before the improvement visualizes that the time to deliver the pizza ranges are wide. On the other hand, the graphs after the improvement reflects that pizzas are almost always were consistent in delivery service.

CHAPTER 11

CONTROL CONCEPTS AND STRATEGIES

The control phase focuses on maintaining and ensuring the gains achieved in the improve phase. This applies for the life of the product. Thus, it is essential to standardize, document the standardized operating procedure (SOP), train the new process owner, and design a process monitoring system that would not deviate from its target. The details of 10 essential process and quality control tools are discussed in this chapter to meet process monitoring requirements.

11.1   PROCESS CONTROL STRATEGY

Process control is defined to ensure that the process will continue to produce the same products all the time, i.e., the articles having the same physical properties and functionality to meet the customer's specification. It requires a process and control plan design, team development, and advanced process control system. In addition, the following important factors must be taken seriously:

- Customer *needs* and *wants* should be the top priority.
- Primary focus must be on the process improvement to support customer satisfaction.
- Watch for system failure that causes most variations (shift changes) in the process. Note that as the variation increases, producability decreases due to increase of nonconformance probability.
- Teamwork is integral and extremely important to quality management.
- Organizational changes and transformation are necessary.
- Lean Six Sigma concepts produce higher quality products at the lowest manufacturing cost.

11.2   PROCESS CONTROL OBJECTIVES

Now that you have completed the improvement phase, the control plan needs to be implemented to keep the sustainability of the process. Some of the details of the key elements given below were described in the previous chapters as well.

1. *Develop a process control plan.* This ensures that customer "needs" and "wants" are met. In this step you establish a process flow, document process measurement and team responsibilities, and determine action plan in the case of unexpected events.
2. Apply Lean tools and Lean enterprise.
3. Execute control charts for essential factors.
4. Utilize mistake proofing techniques. This is an effective tool that improves mistakes detected in the process improvement phase or in the design phase. It is a simple cost-effective procedure that minimizes defect rate and improves quality.
5. Develop risk prioritization number before and after the improvement.
6. Describe the control charts outcome.
7. Set up a documentation plan. All the activity and results must be documented for future analysis purposes.
8. Create performance monitoring metrics.
    i. Develop input/output process measurement system including control charts.
    ii. Start training, if any, required for control charts interpretation.
    iii. Record the most recent process output.
    iv. Does the process performance meet the customer's requirement?

11.3   SUSTAINING THE IMPROVED PROCESS

As described in the previous chapters, process output is a function of process input (Figure 11.1) or independent variable, i.e., Output $= f$ (input) or $y = f(x)$ where y is called dependent, output, effect, symptom, or monitor and x is called independent, input, cause, problem, or control.

Some of the key day-to-day monitoring responsibilities are as follows:

Where will the process occur and location of the workstation?
How the measuring process will take place?
What specific factors of the process need to be controlled?
Who is responsible for maintaining the inspection, measurement, checking, replacing, documentation, and applying mistake proof concepts?
What techniques, gauges, procedures, and devices are used in the process?
When are the time intervals, what is the number of pieces, and what is the frequency of measurement?

Throughout the course of process monitoring and sustaining the improved process, the input (cause) and output (symptoms) must be reviewed continuously. Questions such as the following should be answered:

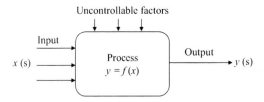

Figure 11.1. Process model.

What are the most effective causes (*xs*) in the process?
What is the best method of measuring the *xs*?
What are the optimum values of *xs*?
What steps should be taken if the *xs* shifts from their nominal?
How often the team or process owner monitor the output values (*ys*)?
What steps should be taken if the process deviates from the desired values?
What process design should be implemented if the uncontrollable factors required to be measured?

11.4   TEN ESSENTIAL PROCESS/QUALITY CONTROL TOOLS

Seven basic tools need to be implemented throughout the process monitoring and should be documented as well.

1. *Process Flowcharts (or Process Mapping)*
Map out the process (Figure 11.2) to better visualize and understand the opportunities for improvement at each step. The benefits of this exercise are as follows:

(a) Provides a clear vision of the process
(b) Distinguishes the value-added and non-value-added steps
(c) Assists in the progress of team and communication

2. *Brainstorming of Ideas*
Brainstorming is a process of collecting ideas for developing a creative optimum solution to problems. Brainstorming is effective and works by concentrating and engaging the team on the problem. The new ideas emerging out of the brainstorming is refined and categorized as an affinity diagram (see Chapter 7).

Brainstorming steps:

1. Prepare a summary description of the problem.
2. Form a team from wide variety range of the disciplines.
3. Ensure that everyone is clear with the problem statement.
4. Write down all the ideas and solutions that come to mind, and there should no criticism on solutions. Welcome as many ideas as as possible.
5. Keep the team focused on the subject matter. Do not deviate from the problem.

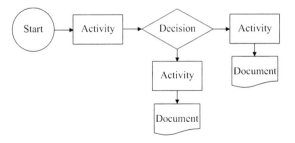

Figure 11.2. Process flow chart.

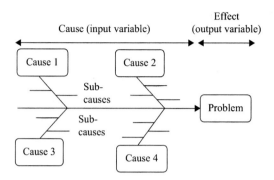

Figure 11.3. Cause and effect.

6. Encourage the team to come up with practical solutions.
7. Once all the ideas collected, evaluate the list using an affinity diagram and actions in solving the problem.

3. *Fishbone Diagram (cause and effect, Figure 11.3)*

The cause-and-effect diagram identifies all the possible causes for an effect or a problem. It can also be used in a brainstorming session to structure a fishbone diagram. Just like brainstorming idea starts with a problem statement, make sure that team members have understood the effect. Then write all the categories of causes as main branches from the main arrow. Begin brainstorming session with the team members and write down all the possible causes on the appropriate branch for each category. Continue asking questions, i.e., ask five whys (see Chapter 8 for details) and post the answer as subcauses.

4. *Histogram*

A histogram is a graphical representation of data using bars of different heights. Basically the purpose is to graphically summarize the distribution of univariate (*function*) data set. The histogram graphically illustrates the following:

1. Center, i.e., the location of the data
2. Spread (variation from data-the scale) of the data
3. Skewness of the inspection data
4. Proof of multiple modes if any
5. Proof of outliers if any

The above features show that a strong evidence exists for proper distributional model for the observation. The most used histogram is obtained by dividing the range of the observations into equal classes, i.e., frequency (vertical axis) and response variable (horizontal axis) as shown in Figure 11.4.

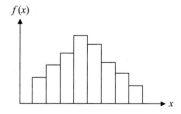

Figure 11.4. Histogram sample plot.

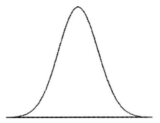

Figure 11.5. Distribution curve.

5. *Distribution Curve*

A distribution curve (Figure 11.5) is basically a symmetrical "bell-shaped" curve that demonstrates the occurrence between two variables. The most common type of distribution is the continuous probability distributions described by the normal equation. The distribution curve suggests whether the data set distribution is skewed to the left or right, or, is it symmetrically oriented. Furthermore, the normality tests, i.e., Anderson-Darling normality test are used to evaluate if the data set is well modeled by distribution curve.

6. *Pareto Chart*

Pareto chart as shown in Figure 11.6 is a bar chart that represents the frequency and designed with the highest bars on the left and the shortest to the right. The chart visually describes which direction is most substantial and tangible; for example 80% of the component scraps in a process are caused by 20% of the factors. This also called 80-20 rule.

Application of Pareto chart: Pareto chart is used in analyzing the data such as the following:

1. When you are analyzing a data and looking for a specific cause that makes the most impact in the process outcome?
2. When you have too many causes and trying to look at their pattern of effect?
3. When you want to report out the responses visually?

7. *Trend Charts: Identify and Interpret the Trend*

By definition, it is a graphical illustration of the data in a sequence order over time. The curve in Figure 11.7 represents a change in the general pattern of the graph.

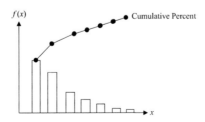

Figure 11.6. Pareto chart.

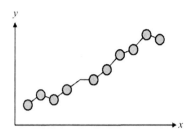

Figure 11. 7. Trent chart sample plot.

8. Scatter (x–y) Plots—Examine the Relationships

The scatter plot affirms relationships between two dependent and independent variables. Different common types of plots are shown in Figure 11.8. The plots in Figure 11.8 represent linear relationship between the two variables.

A scatter plot is a plot of y response variable (vertical axis) versus the corresponding value of x manipulated variable (horizontal axis). Scatter plots can respond to questions like

1. Is x variable related to y variable?
2. Is the relationship linear or not linear?
3. Do the symptoms in the y response changes depend on the x?
4. Is the pattern shows any outliers?

9. Check Sheet

Check sheet (Table 11.1) is the form that is used to collect data in real time at the location where the process is carried out. The collected data can be either qualitative or qualitative depending on the type of the process.

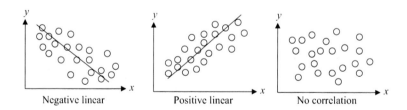

Figure 11.8. Scatter diagram.

Table 11.1. Check sheet/data collection

Cause	Week 1	Week 2	Total number of occurrences
Factor 1	||||| ||||| ||	||||| ||	19
Factor 2	||	|||	5
Factor 3	||||| |	|||	9

Figure 11.9. Statistical process control.
(UCL = Upper control limit, CL = Process average or center line,
LCL = Lower control limit).

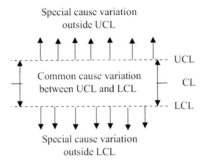

Figure 11.10. Statistical process control.

10. *Statistical Process Control Charts—Examine the Performance of a Process Over the Time.*

Statistical process control (SPC) charts are used in the system to monitor process behavior or product properties. A SPC chart as shown in Figure 11.9 records the data history and illustrates the process variations compared to an optimum process performance.

Total process variation: Variation is natural, inherent in the world. The inherent variation can be reduced but not eliminated. These variations can arise from manpower, machine malfunction, material lot variation, different methodologies, gage R&R measurement, and of course, due to uncontrollable factors in our environment.

Common cause variation: It is a natural variation with the process that takes place in the visual observation or when the parts are being measured. These types of variations are normally expected.

Special cause variation: In general, it is recognized by data points falling outside of upper control limit and lower control limit as shown in Figure 11.10. This is called abnormal or unexpected variations. On the other hand, there are cases that data points falls within the control limit and still considered special cases such as shift in the chart pattern (trends) which process influenced by some special cause variation.

In both cases, it may not affect the quality of parts and perhaps it is due to variations such as in machine, operator, or others.

11.5   CONTROL CHART TYPES

Categorically there are two types of control charts: the ones that exhibit attributes data and variable data. The attribute data presents the information raised from number of occurrences in a single class, i.e., go/no-go, pass/fail, and blue/green. On the other hand, the variable data represents the response values measured from a continuous process. While the above categories cover various types of control charts, there are only three types that are mostly utilized by data analysis.

11.5.1   X-BAR ($\bar{x}$) AND R-CHART

In a process capability study, either manufacturing or transactional, the most essential factor is "time." The x-bar ($\bar{x}$) chart presents when the process is centered; if the center fluctuates, the $\bar{x}$ order (pattern) also fluctuates. If the distribution center moves up and down, the $\bar{x}$ order also follows the same course. Thus, the $\bar{x}$ chart changes assist us in evaluating and taking action for keeping the process in control.

The purpose of control chart (also known as "Shewhart" chart) is to enable the inspectors to identify between *random variation* and *variations due to transactional causes* and control the process efficiency over time for constancy. It also supports in identifying the opportunities for further improvement.

The common types of control charts depend on the circumstance and type of data available to determine and construct the following charts.

The model for $\bar{x}$ center line (CL) and control chart limits using R are defined as

$$UCL = \bar{\bar{x}} + A_2 \bar{R} \qquad \text{(Eq. 11.1)}$$

$$Center \; line\left(CL\right) = \bar{\bar{x}} \qquad \text{(Eq. 11.2)}$$

$$LCL = \bar{\bar{x}} - A_2 \bar{R} \qquad \text{(Eq. 11.3)}$$

where upper control limit (UCL),lower control limit (LCL), and A_2 value as given in Table 11.3, and x-double bar ($\bar{\bar{x}}$ the mean of all the means) is determined by

$$\bar{\bar{x}} = \frac{\sum_{i=1}^{n} \bar{x}_i}{n}, \; i = 1, 2, 3, \ldots, \; n \; and \; n = sample \; number \qquad \text{(Eq. 11.4)}$$

where the process mean ($\bar{x}$) is an estimator of the population mean (μ), and R-bar ($\bar{R}$) the mean of range in each sample group. The range of the sample is simply the difference between the largest value and the smallest observation. Example 11.1 clarifies this concept.

Example 11.1

A quality control engineer was assigned to evaluate the product dimension of an experiment with 15 runs and each run with 5 samples. The sample measurements are given in Table 11.2. Determine if the process is in control by calculating appropriate upper and lower control limits of $\bar{x}$ and R-charts.

Table 11.2. Sample measurements for $\bar{X}$ and R calculations

Sample number	Individual measurement				
	Trial 1	Trial 2	Trial 3	Trial 4	Trial 5
1	28.842	29.125	27.746	27.956	29.254
2	25.892	28.674	29.562	30.253	31.625
3	28.050	26.658	28.768	29.684	30.422
4	30.565	31.897	27.644	29.212	26.337
5	26.323	28.475	29.125	30.598	28.254
6	31.846	28.784	30.786	31.597	27.488
7	29.548	27.369	30.478	29.985	29.647
8	28.254	26.985	30.284	30.756	30.284
9	27.654	30.147	31.896	30.547	27.854
10	30.274	26.852	27.654	28.456	30.951
11	27.854	27.968	28.397	29.654	29.684
12	26.753	27.854	30.862	27.984	28.674
13	29.852	28.654	28.759	30.758	29.684
14	30.456	30.756	28.778	30.984	30.874
15	27.964	29.989	29.897	27.884	28.675

Solution

STEP 1

Calculate the sample mean ($\bar{x}$) of subgroups (m), sample range (R) of subgroup, mean of means ($\bar{\bar{x}}$), and mean of ranges ($\bar{R}$). Let's recall Equation 11.4 and expand it:

$$\bar{x}_i = \frac{\sum_{i=1}^{i=m} x_i}{m} = \frac{x_1 + x_2 + \cdots + x_m}{m} \qquad \text{(Eq. 11.5)}$$

$$\bar{R} = \frac{\sum_{i=1}^{i=n} R_i}{n} = \frac{R_1 + R_2 + \cdots + R_n}{n} \qquad \text{(Eq. 11.6)}$$

using Equation 11.5 for $\bar{x}_1$ we have

$$\bar{x}_1 = \frac{\sum_{i=1}^{i=5} x_i}{m} = \frac{x_1 + x_2 + \cdots + x_5}{m} = \frac{28.842 + 29.125 + \cdots + 29.254}{5} = 28.585$$

The calculations for all the other means are listed in Table 11.3.
Now, the x-double bar ($\bar{\bar{x}}$) or mean of means is calculated by

$$\bar{\bar{x}} = \frac{\sum_{i=1}^{i=n} \bar{x}_i}{n} = \frac{\bar{x}_1 + \bar{x}_2 + \cdots + \bar{x}_n}{n} = \frac{28.585 + 29.201 + \cdots + 28.882}{15} = 29.160$$

Table 11.3. $\overline{X}$ and R chart calculation

Sample number	Individual measurement					Subgroup measures	
	Trial 1	Trial 2	Trial 3	Trial 4	Trial 5	Mean ($\overline{x}$)	Range (R)
1	28.842	29.125	27.746	27.956	29.254	28.585	1.508
2	25.892	28.674	29.562	30.253	31.625	29.201	5.733
3	28.050	26.658	28.768	29.684	30.422	28.716	3.764
4	30.565	31.897	27.644	29.212	26.337	29.131	5.560
5	26.323	28.475	29.125	30.598	28.254	28.555	4.275
6	31.846	28.784	30.786	31.597	27.488	30.100	4.358
7	29.548	27.369	30.478	29.985	29.647	29.405	3.109
8	28.254	26.985	30.284	30.756	30.284	29.313	3.771
9	27.654	30.147	31.896	30.547	27.854	29.620	4.242
10	30.274	26.852	27.654	28.456	30.951	28.837	4.099
11	27.854	27.968	28.397	29.654	29.684	28.711	1.830
12	26.753	27.854	30.862	27.984	28.674	28.425	4.109
13	29.852	28.654	28.759	30.758	29.684	29.541	1.198
14	30.456	30.756	28.778	30.984	30.874	30.370	2.096
15	27.964	29.989	29.897	27.884	28.675	28.882	2.105
					Average $\overline{\overline{x}}$ = 29.160		$\overline{R}$ = 3.450

STEP 2

Using Table 11.4 at sample number $n = 5$, the control limit factor is $A_2 = 0.577$, and from Table 11.3 the $\overline{\overline{x}} = 29.160$, $\overline{R} = 3.450$. Now by substituting the Equation 11.1 we get

$$UCL = \overline{\overline{x}} + A_2\overline{R} = 29.160 + (0.577)(3.450) = 31.151$$

where
$$A_2 = \frac{3}{d_2\sqrt{n}} \qquad \text{(Eq. 11.7)}$$

$$Center\,line\,(CL) = \overline{\overline{x}} = 29.160$$

Likewise we apply the same values in Equation 11.3.

$$LCL = \overline{\overline{x}} - A_2\overline{R} = 29.160 - (0.577)(3.450) = 27.169$$

11.5.2 R-CHART LIMITS MODELS

The model for R control chart limits are defined as in Equations 11.8 through 11.10:

$$UCL = D_4\overline{R} \qquad \text{(Eq. 11.8)}$$

Table 11.4. Control factors for constructing limits $\bar{X}$ and R charts

n	A_2	D_3	D_4	d_2	d_3
2	1.880	0.000	3.267	1.128	0.853
3	1.023	0.000	2.574	1.693	0.888
4	0.729	0.000	2.282	2.059	0.880
5	0.577	0.000	2.115	2.326	0.864
6	0.483	0.000	2.004	2.534	0.868
7	0.419	0.076	1.924	2.704	0.833
8	0.373	0.136	1.864	2.847	0.820
9	0.337	0.184	1.816	2.970	0.808
10	0.308	0.223	1.777	3.078	0.797
11	0.285	0.256	1.744	3.173	0.787
12	0.266	0.284	1.716	3.258	0.777
13	0.249	0.308	1.692	3.336	0.769
14	0.235	0.329	1.671	3.407	0.762
15	0.223	0.348	1.652	3.472	0.754
16	0.212	0.364	1.636	3.532	0.749
17	0.203	0.379	1.621	3.588	0.743
18	0.194	0.392	1.608	3.640	0.738
19	0.187	0.404	1.596	3.689	0.733
20	0.180	0.414	1.586	3.735	0.729

$$Center\ line\left(CL\right)=\bar{R} \qquad \text{(Eq. 11.9)}$$

$$LCL = D_3\bar{R} \qquad \text{(Eq. 11.10)}$$

Values of D_3 and D_4 are defined as

$$D_3 = 1-3\frac{d_3}{d_2} \text{ and } D_4 = 1+3\frac{d_3}{d_2} \qquad \text{(Eq. 11.11)}$$

where d_2 is the factor for estimating sample standard deviation.

$$s = \frac{\bar{R}}{d_2} \qquad \text{(Eq. 11.12)}$$

Values of d_3 and d_2 are provided in Table 11.3. Since sample range values are not negative, thus D_3 values are zero whenever Equation 11.11 is negative, that is, when $n = 2, 3, 4, 5,$ and 6. In addition, values of D_3 and D_4 also are given in Table 11.3.

Example 11.2

Referring to Example 11.1, average range is $\bar{R} = 3.450$ and from Table 11.3 at $n = 5$ the values for $D_3 = 0.00$, and $D_4 = 2.115$, so the resultant control limits will be

$$UCL = D_4\bar{R} = (2.115)(3.450) = 7.297$$

$$Center\ line(CL) = \bar{R} = 3.450$$

$$LCL = D_3\bar{R} = (0.00)(3.450) = 0.00$$

Likewise, control models for the $\bar{x}$ center line and control chart limits using S are defined as

$$UCL = \bar{\bar{x}} + A_3\bar{S} \qquad \text{(Eq. 11.13)}$$

$$Center\ line(CL) = \bar{\bar{x}}$$

$$LCL = \bar{\bar{x}} - A_3\bar{S} \qquad \text{(Eq. 11.14)}$$

11.5.3 STEPS FOR DEVELOPING $\bar{X}$ AND R CHARTS

1. Collect n samples of data n sets, each size of m, i.e., $n = 15$ data sets, each set equal to sample size of $m = 5$, like in Example 11.1 (Table 11.1).
2. Compute the mean of each subgroup i.e., $\bar{x}_1, \bar{x}_2, \ldots, \bar{x}_n$.
3. Compute the range of each subgroup i.e., $R_1, R_2, R_3, .., R_n$.
4. Find the overall mean for step 2, $\bar{\bar{x}}$, where $\bar{\bar{x}}$ is the mean of n sets of $\bar{x}$.
5. Find the overall mean for step 3, $\bar{R}$, where $\bar{R}$ is the mean of n sets of R.
6. Compute sample standard deviation $s = \dfrac{\bar{R}}{d_2}$, to estimate s, where values of d_2 are listed in Table 11.3.
7. Compute the three Sigma control limits for $\bar{x}$ control chart:

$$UCL = \bar{\bar{x}} + 3\frac{s}{\sqrt{n}} = \bar{\bar{x}} + 3\frac{\left(\dfrac{\bar{R}}{d_2}\right)}{\sqrt{n}} = \bar{\bar{x}} + \frac{3}{d_2\sqrt{n}}\bar{R} = \bar{\bar{x}} + A_2\bar{R}$$

$$Center\ line\ (CL) = \bar{\bar{x}}$$

$$LCL = \bar{\bar{x}} - 3\frac{s}{\sqrt{n}} = \bar{\bar{x}} - 3\frac{\left(\dfrac{\bar{R}}{d_2}\right)}{\sqrt{n}} = \bar{\bar{x}} - \frac{3}{d_2\sqrt{n}}\bar{R} = \bar{\bar{x}} - A_2\bar{R}$$

8. Calculate the three Sigma control limits for $\bar{x}$ control chart:

$$UCL = D_4\bar{R} \quad \text{(Refer to Table 11.3)}$$

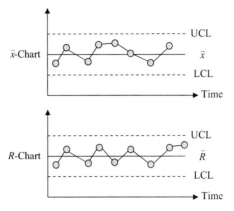

Figure 11.11. $\bar{x}$-chart and R-chart.

$$Center\ line\ (CL) = \bar{R}$$

$$LCL = D_3\bar{R} \quad \text{(Refer to Table 11.3)}$$

9. Plot control chart using $\bar{x}$ and R points for each subgroup on the same vertical line for sample numbers are given in Figure 11.11.

11.6 *P*-CHART: ATTRIBUTE CONTROL CHART

P-chart is control chart for percentage or proportions and it is an attribute chart. Its objective is to monitor the pattern of the process over time and take the necessary steps to resolve the process issues if it needed. p is the symbol for proportion and simply for percent defective components. Note that p-chart is less effective than $\bar{x} - R$ chart since it follows only one variable. The p-chart depends more on practical expertise than $\bar{x} - R$ chart. Thus, it is more applied in the areas where the workforce or process owners are familiar with the highest impact variables.

The p-chart illustrates the proportion of nonconforming items, not the proportion of defects. For instance, count the number of the cell phones and divide by total cell phones inspected. Cell phone is either defective or not defective. So, finding a defective phone can be classified as a "success." *P-chart* is applied for equal or unequal sample sizes and it should have $np > 5$ and $n(1-p) > 5$ as discussed in Chapter 9.

Steps for creating *p*-chart: For constant sample size the steps are as follows:

1. Accumulate m samples of data, usually each sample size of $n = 20$ to 25.
2. Compute the proportion nonconforming items (p_i) for each sample size (m) using Equation 11.15.

$$p_i = \frac{W_i}{n} \qquad \text{(Eq. 11.15)}$$

where W_i is the number of defective items in sample i and n is the sample size.

3. Determine the overall proportion of nonconforming items.

(a) If equal sample size

$$\bar{p} = \frac{\sum_i^m w_i}{nm} = \frac{\sum p_i}{m} = \frac{\textit{total number of nonconforming items}}{\textit{Total number of sample size}} \quad \text{(Eq. 11.16a)}$$

where $\bar{p}$ is the mean of subgroup proportion (m), p_i = sample proportion for subgroup i, m = number of subgroups of size n.

(b) If unequal sample size

$$\bar{p} = \frac{\sum n_i p_i}{\sum n_i} \quad \text{(Eq. 11.16b)}$$

where
n_i = number of items in sample i, $\sum n_i$ = total number of items sampled in m samples.

4. Calculate the upper and lower control limits for 3-sigma using Equations 11.17 through 11.19:

$$UCL = \bar{p} + 3\left(\frac{\bar{p}(1-\bar{p})}{n}\right)^{1/2} = \bar{p} + 3\sigma \quad \text{(Eq. 11.17)}$$

$$CL = \bar{p} \quad \text{(Eq. 11.18)}$$

$$LCL = \bar{p} - 3\left(\frac{\bar{p}(1-\bar{p})}{n}\right)^{\frac{1}{2}} = \bar{p} - 3\sigma \quad \text{(Eq. 11.19)}$$

5. Using the lines to illustrate the control limits on the p-charts.

A numerical example will now be given to illustrate aforementioned principles.

Example 11.3

A mechanical engineer was asked to examine the petrochemical valve whether it contains leak due to improper design of cape and valve body. Twenty runs and on each set of runs 150 sample parts were measured and nonconforming item (number of success) is defined in case of misalignment of the cape and valve body. The data for this experiment is recorded in Table 11.5.

Solution

We use Equation 11.16 to calculate the proportion mean.

$$\bar{p} = \frac{\sum_i^m w_i}{nm} = \frac{\sum p_i}{m} = \frac{\textit{total number of nonconforming items}}{\textit{Total number of sample size}}$$

$$= \frac{p_1 + p_2 + \cdots + p_{20}}{m} = \frac{1.0}{20} = 0.05$$

Table 11.5. Experimental results for petrochemical valve

Subgroup number (m)	Sample size (n)	Number of nonconforming items (successes) w_i	Proportions (p_i)
1	100	3	0.03
2	100	11	0.11
3	100	4	0.04
4	100	9	0.09
5	100	8	0.08
6	100	6	0.06
7	100	4	0.04
8	100	12	0.12
9	100	6	0.06
10	100	5	0.05
11	100	2	0.02
12	100	4	0.04
13	100	9	0.09
14	100	1	0.01
15	100	3	0.03
16	100	4	0.04
17	100	2	0.02
18	100	1	0.01
19	100	7	0.07
20	100	2	0.02
	Total = 2000	$\sum W_i = 103$	$\sum p_i = 1.00$
		Average subgroup proportion $(\bar{p})$ =	0.052

or, we got the overall nonconformity equal to 103 and 20 runs with 100 samples equal to $(20) \cdot (100) = 2000$ items. Thus

$$\bar{p} = \frac{w_1 + w_2 + \cdots + w_{20}}{(20)(100)} = \frac{103}{2000} = 0.052$$

The control limits are then derived using Equations 11.7 through 11.19:

$$UCL = \bar{p} + 3\left(\frac{\bar{p}(1-\bar{p})}{n}\right)^{1/2} = 0.052 + \left(\frac{0.052(1-0.52)}{100}\right)^{1/2} = 0.074$$

$$CL = 0.052$$

$$LCL = \bar{p} - 3\left(\frac{\bar{p}(1-\bar{p})}{n}\right)^{\frac{1}{2}} = 0.052 - \left(\frac{0.052(1-0.52)}{100}\right)^{\frac{1}{2}} = 0.029$$

11.7   c–CHART

The p-chart focused on the proportion of nonconforming items in a sample of n units. However, it is possible for a unit to have more than one nonconforming item. If this is the case then we should also take into account the number of nonconformities (defects) per unit. Therefore, the c-chart comes into effect to be used for controlling a single type of defects or controlling all types of defects without classifying between the types. For instance, imagine a plastic part product with contamination and sink mark; or a camera lens with air bubble and under sized dimension and so on.

The assumption is that the number of defects occurring in a unit holds normal approximation to the Poisson process distribution. Thus, symmetry of the control limits. Further, both the mean and the variance of the Poisson random variable are the same ($\sigma^2 = \mu$) and standard deviation is equal to the square root of the mean ($\sigma = \sqrt{\mu}$).

Steps for Constructing A c-Chart

1. Accumulate m samples of data, usually each sample size of $n = 20$ to 25 units
2. Find the number of nonconforming (c_i) for the ith unit.
3. Compute the average number of defects per unit ($\bar{c}$).

$$\bar{c} = \frac{\sum_{i}^{m} c_i}{m} = \frac{c_1 + c_2 + \cdots + c_m}{m} \qquad \text{(Eq. 11.20)}$$

4. Calculate the upper and lower control limits for 3-sigma using the Equations 11.21 through 11.23:

$$UCL = \bar{c} + 3(\bar{c})^{1/2} = \bar{c} + 3s \qquad \text{(Eq. 11.21)}$$

$$Center\ Line\ (CL) = \bar{c} \qquad \text{(Eq. 11.22)}$$

$$LCL = \bar{c} - 3(\bar{c})^{\frac{1}{2}} = \bar{c} - 3s \qquad \text{(Eq. 11.23)}$$

where $s = \sqrt{\bar{c}}$ = standard deviation for c-chart.
5. Using the lines to illustrate the control limits on the c-charts and by plotting the values of c_i.

A numerical example will be given to illustrate aforementioned principles.

Example 11.4

A tire manufacturing company interested in inspecting the tire surface for minor air bubbles. The following data were obtained using a 25 tires for mid-sized automobile:

Tire (1 through 15)	1	2	3	4	5	6	7	8	9	10	11	12	13	14	15
Air Bubbles	3	1	0	2	1	3	2	0	5	0	2	1	2	0	8

Tire (16 through 25)	16	17	18	19	20	21	22	23	24	25
Air Bubbles	4	0	1	4	2	1	0	1	3	2

Construct control chart and determine if all the data points are within the control charts

Solution

The average number of defects (air bubbles) for the sample of 25 tires is $\bar{c}$, where

$$\bar{c} = \frac{3+1+0+2+\cdots+3+2}{25} = \frac{48}{25} = 1.92$$

The limits (UCL and LCL) and center line (CL) for the resulting c-chart are

$$UCL = \bar{c} + 3(\bar{c})^{1/2} = 1.92 + 3(1.92)^{1/2} = 6.08$$

$$Center\ Line\ (CL) = 1.92$$

$$LCL = \bar{c} - 3(\bar{c})^{\frac{1}{2}} = 1.92 - 3(1.92)^{\frac{1}{2}} = -2.24,\ (set\ LCL = 0)$$

Since LCL < 0 and the Poisson variable is never negative, so the LCL is set equal to zero and is inactive.

11.8   CONTROL LIMITS VERSUS SPECIFICATION LIMITS

Frequently there is a mix-up or misunderstanding between the control limit and specification limit. Actually, not only are they not related, they are also different as well. Particularly the control limits are dependent on the experimental data response; however, the specification limits are assigned by the customer "needs" and "wants" or product functionality limits. For any product to have lower cost and higher return on investment, specification limits need to be outside the control limits. The comparison of control limits versus specification limits is given in Table 11.6.

Table 11.6. Control limits and customer specification limits
(UCL, CL ($\bar{\bar{x}}$), LCL versus USL, mean, LSL)

Control Limits (CL)	Specification Limits (SL)
Process data (Voice of process)	Customer specification (Voice of customer)
UCL, CL (average), LCL values calculated using observed data from the process	USL, (mean), LSL are defined based on customer's requirements.
Control chart	Generated from histogram
Data is utilized to see if the process statistically in control	Data is utilized to see if the process is outside the specification limit
If the data points are outside the graph then process is out of control	If the data points are outside the specification limit (i.e., not capable), however within the control limits, then statistically process is in control, but it may not meet the customer's "needs" and "wants"

11.9   PROCESS CAPABILITY RATIO, C_P AND C_{PK}

Capability ratio is the repeatability of a manufacturing or transactional process based on the customer lower specification limit (LSL) and upper specification limit (USL). The capability ratio C_p does not take into account how the data is centered. On the other hand, C_{pk} is applied to define the actual tolerance with specified limits. If the process is in statistical process control as discussed earlier and the process data is concentrated on the target so that the bell-shaped curve is symmetrical, then the capability index C_p can be determined using Equation 11.24.

$$C_p = \frac{USL - LSL}{6\sigma} = \frac{Specification\ width}{Process\ width} = \frac{``allowable''\ process\ variation}{``actual''\ process\ variation} \qquad \text{(Eq. 11.24)}$$

In general C_p has the following conditions:

$C_p < 1.0$	Poor process; the process has more variations and defects.
$C_p = 1.0$	Adequate (Ok); the process is acceptable just meeting the required specifications.
$1.33 \le C_p \le 1.50$	Good; the process variation is less than specified, however, possible defects and the process is required to be centered or to be sustained in control limits.
$C_p = 2.0$	Excellent, this is Six Sigma process

Figures 11.12 through 11.14 demonstrate the preceding concepts.

In this case, C_p does not include the distribution mean in the process review. Therefore, it does not support how well the process mean is centered to the target value. On the other hand, C_{pk} does include the distribution mean into account. So, C_{pk} is the more realistic capability ratio.

When determining the initial observation (C_p), it is assumed that

1. the process response is at the target and within customer specification limits;
2. the process is normally distributed and symmetric; and
3. the process is in control as discussed before.

The C_p measurement is basically an observation of process capability with the specification limits and it is true if the above statement of assumptions holds. To get an idea of the value of

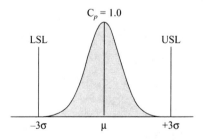

Figure 11.12. Standard normal distribution centered at the target.

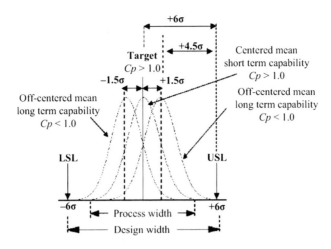

Figure 11.13. Standard normal distribution curve shifted ±1.5 sigma to the right or to the left of target.

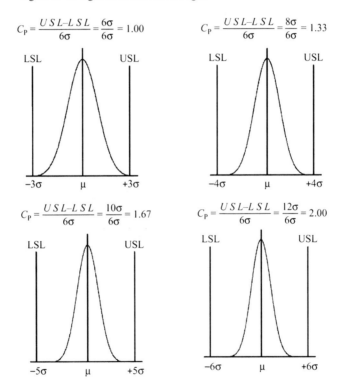

Figure 11.14. Standard normal distribution plots showing C_p for varying process widths.

the capability index C_p statistic for changing process specification widths, consider the following distribution plots.

Figure 11.14 can be stated numerically by Table 11.7.

Rejects plots are based on the assumptions that normal distribution is centered at mean (μ). Example 11.5 reviews this concept.

Table 11.7. Interpreting capability into defects

USL-LSL (width)	6	8	10	12
C_p	1.00	1.33	1.67	2.00
Defects	0.27%	64 ppm*	0.60 ppm*	2 ppb*
% of specification used	100	75	60	50

*ppm = parts per million, ppb = parts per billion.

Example 11.5

A manufacturing company produces a product that has a part with a mean diameter of 3 inches ($\mu = 3$) and a standard deviation (variation from the mean) of 0.035 inches. The lower and upper design specification limits, LSL = 3.05 and USL = 2.85 inches:

$$C_p = \frac{USL - LSL}{6\sigma} = \frac{3.05 - 2.85}{6(0.035)} = \frac{0.20}{0.21} = 0.95$$

Since $C_p < 1.0$, process is not capable.

Process Capability and C_{pk} Location: When computing the C_p index, we compared the width of normal distribution curve (the process spread) to the width of the lower and upper specification limit (the difference between the LSL and USL), with the assumption that the distribution curve shifting to the left or right of the center. The C_p index is a measure of potential capability.

Unlike C_p, the process capability index C_{pk} not only measures the width of the normal distribution curve compared to the width of the specification limit; in addition it goes further to specify whether the normal distribution curve is "on" or "off" the center line. The ratio determines the variation (distance) from the process center to the closest specification limit (either LSL or USL), as shown in Figure 11.14.

Just like C_p when using C_{pk} the following assumptions are made:

1. The process spread is normally distributed and symmetric.
2. The process may or may not be centered between lower and upper specification limit.
3. The process is in control, and control charts will be applied to observe the process progress over time.

Steps to compute the C_{pk}: C_{pk} is the smallest value of Equation 11.25 or 11.26. That is

$$C_{pk} = \min\left(C_{pk}(LSL), C_{PK}(USL)\right)$$

$$C_{pk}(LSL) = \frac{\mu - LSL}{3\sigma} \qquad \text{(Eq. 11.25)}$$

$$C_{pk}(USL) = \frac{USL - \mu}{3\sigma} \qquad \text{(Eq. 11.26)}$$

where

$$\sigma = \frac{USL - u}{z} \qquad \text{(Eq. 11.27)}$$

and the value of C_{pk} will always be less than or equal to C_p, that is $C_{pk} \leq C_p$. Normally the acceptance value of $C_{pk} = 1.0$. Furthermore, if the $C_{pk} > 1.0$ means process is capable then

1. monitor and sustain the process;
2. keep on continuous process improvement;

If the process is not capable which means $C_{pk} \leq 1.0$, then do the following:

1. Monitor and sustain the process
2. Keep on continuous process improvement
3. Apply Lean Six Sigma tools (define, measure, analyze, improve, and control) to reduce process variation.

As C_{pk} increases, the Sigma (σ) level also increases (see Tables 11.7 and 11.8), i.e., the variation from the mean reduces, product quality improves, and market share increases. This means cycle time (Takt time) is reduced, quality inspector check is minimized, operating cost goes down, customer satisfaction is achieved and their "need and wants" are met, and 80% of lead time is reduced, which is caused by 20% workstation time traps.

Example 11.6

Let's recall Example 11.5, in which a manufacturing company produces a product that has a part with a mean diameter of 3 inches ($\mu = 3$) and a standard deviation (variation from the mean) of 0.035 inches. The lower and upper design specification limits, LSL = 2.85 and USL = 3.05 inches.

$$C_{pk} = \min\left(\frac{\mu - LSL}{3\sigma}, \frac{USL - \mu}{3\sigma}\right) = \min\left(\frac{3.0 - 2.85}{3(0.035)}, \frac{3.05 - 3.0}{3(0.035)}\right) = \min(1.43, 0.48) = 0.48$$

Table 11.8. Short term process capability (C_{pk}) without shift changes when the process is centered on the target

C_{pk}	Nonconformance [parts per million (ppm)]	Sigma (σ) level
0.33	317,310	1.00
0.50	133,614	1.50
0.67	45,500	2.00
0.75	24,448	2.25
1.000	2,700	3.00
1.167	465	3.50
1.333	63	4.00
1.500	6.8	4.50
1.667	0.6	5.00
2.000	0.002	6.00 (World class manufacturing)

Table 11.9. Long-term process capability (C_{pk}) when the process is shifted by 1.5-sigma from the target

C_{pk}	Nonconformance [parts per million (ppm)]	Sigma (σ) Level
0.00	500,000	1.5
0.17	308,300	2.0
0.33	158,650	2.5
0.50	66,807	3.0
0.67	22,700	3.5
0.83	6,220	4.0
1.00	1,350	4.5
1.17	233	5.0
1.33	32	5.5
1.50	3.4	6.0 (World class manufacturing)

Since $C_{pk} < 1.0$, the process is not capable. The variation is too big and the process mean is not on the target.

Tables 11.8 and 11.9 were generated by applying Equations 11.24 and 11.27. These tables illustrate the short-term and long-term process capabilities.

11.10 TOLLGATE REVIEW AND DELIVERABLES FOR CONTROL PHASE

Control and sustain is the last phase of the Lean Six Sigma methodology. By now, you have completed the launch objectives and trying to organize a process management system that will maintain the fulfillments of deployment efficiently and effectively.

The team must ensure that all the deliverables for the *Control phase* have been standardized and is in process before presenting the results in the gate meeting. Figure 11.15 can be adjusted to any project that includes product or processes improvement control system. Project team must come up with Lean Six Sigma methodology to control and maintain the process improvement. That includes but is not limited to the checklist discussed below.

11.10.1 CONTROL PHASE DELIVERABLES AND CHECKLIST

(a) Apply as many as Control tools and methodologies such as
   i. X-bar ($\bar{x}$) and R-Chart
   ii. R-Chart Limits Models
   iii. *P*-Chart Attribute Control Chart, *P*-Chart
   iv. Continuously monitor and analyze the process capability ratios C_p and C_{pk}
   v. Complete the Six Sigma Scorecard.
   vi. Establish SOP (Standard Operating Procedures)

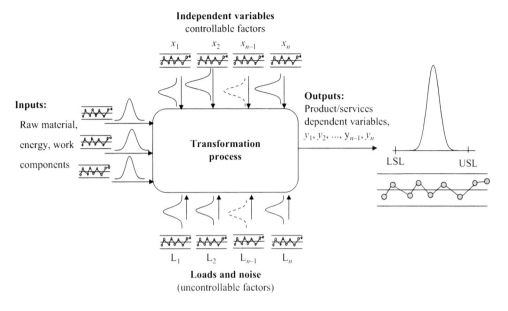

Figure 11.15. Lean Six Sigma transformation process model "Control phase."

(b) Identify the deliverable to be sustain the improvement.

(c) Have the financial benefits gained based on the define phase project charter.

(d) Have the goals and objectives are achieved according to project charter.

(e) Are the changes standardized and controlled?

(f) Document the lessons learned from the process.

(g) Communicate the lessons learned from the process to team members.

(h) Train and educate the new ownership of the process on the following:

    i. Input and output variables must be monitored and analyzed continuously

    ii. Control charts must be implemented effectively

    iii. Monitor the process Sigma level

(i) Create and define RACI-Matrix. Include team members, gate keepers, and anyone who should be part of RACI-Chart.

(j) Does the Voice-of-Customer (VOC) meets the process performance.

(k) Will the improvement meet the organization's financial target or are further improvements required?

    Once the checklist has been completed, the project must to go through the Tollgate review (Figure 11.16) as explained in Chapter 6 for gate review team approval.

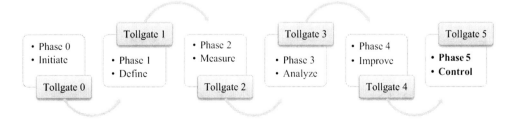

Figure 11.16. Lean Six Sigma process improvement projects: Tollgate review—"Control phase."

CHAPTER 12

CASE STUDIES: LEAN SIX SIGMA APPLICATIONS

12.1 DEFECT REDUCTION IN INJECTION MOLDING PRODUCTION COMPONENTS

Scope: The objective of Lean Six Sigma Black Belt project, component production optimization (CPO), is to establish and maintain optimized process settings for molding that will reduce the amount of component scrap and number of dimensional inspections. The scope of the project included creating a model process to be used for any molded parts. The data was reviewed on the most problematic molds that were measured using optical gaging products (OGP) and narrowed the focus to the XF product line for molds to target. The goals of the project are the following:

1. Reduce dimensional inspection requirements by 25%.
2. Reduce cycle time (Takt time) by 10%.
3. Reduce scrap rate by 25%, which will ultimately reduce Non-conforming Material Report (NCMR).
4. Reduce machine downtime by 10%.
5. Ensure specification requirements are met with critical-to-quality (CTQ) capability of $C_{pk} \geq 1.33$.
6. Prove out the optimized process on 5 parts using Lean Six Sigma methodologies.

12.1.1 DEFINE PHASE

The define phase utilized the following Lean Six Sigma tools:

1. Project charter
2. Affinity diagrams
3. VOC ("Voice of the Customer") analysis
4. Quality Function Deployment (QFD) analysis

5. SIPOC (Supplier, Input, Process, Output, and Customer) Chart
6. SWOT (Strength, Weaknesses, Opportunities, and Threats) analysis.

The first step was to complete the project charter. The elements included in the project charter were as follows:

1. Project need—why is the project important
2. Project scope—what's included and not included
3. Assumptions, constraints, and risks associated with the project
4. Start and finish date (end-to-end boundaries)
5. Roles and responsibilities
6. Resource availabilities
7. Goals and objectives
8. Milestones

Once the project charter was completed by the team, the next step was to create a high-level process map, or SIPOC. The results of this process are shown in Figure 12.1.

Once the SIPOC was established, we surveyed the customers to find out what the key drivers of customer satisfaction were. The key customers to this process were identified as Manufacturing Engineers and Molding Process Specialists. The customers were surveyed to determine what the key drivers of this process would be and how well the current system met these drivers. The results of this survey are illustrated in Table 12.1 "Voice of Customer" (VOC) analysis.

Soon after the VOC was analyzed, team needed to get a better understanding of the problem in order to create breakthrough solutions. For this part of the project, team utilized an affinity diagram. This allowed the team to brainstorm and group the issues and ideas that

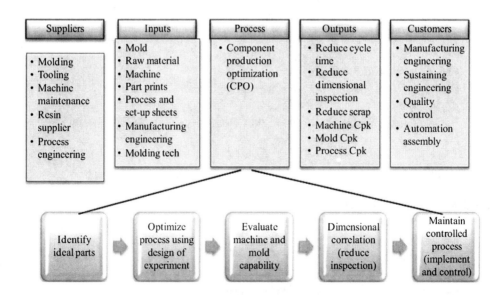

Figure 12.1. SIPOC Chart (Supplier, Input, Process, Output, and Customer).

Table 12.1. VOC (voice of customer) analysis

Customer	Primary needs	Secondary needs	Importance rating (1–5)	Self-rating rating (1–5)
Automation	Part quality	Correct part	5	2.5
		Correct material	5	4
		Meets dimensional tolerance	5	3
		Meets visual requirements	5	3.5
		Meet destructive testing requirements	5	3.5
	Part availability	Enough parts	5	3.5

Customer	Primary needs	Secondary needs	Rating (1–5)	Rating (1-5)
Molding	Material requirements	Correct material	5	5
		Correct colorant	4.4	4.5
		Correct mix	4.2	4.5
		Dried material	3.2	4.5
		No contamination	4.4	4.5
	Optimized process	Largest molding window	4.2	3.5
		Established for machine type	4.6	3.5
		Shortest cycle time	3.6	4.5
	Machine selection	Correct barrel size	4.2	4.5
		Correct screw type	4	4.5
		Correct nozzle tip	4.2	3.5
		Correct machine type	4.4	4
	Mold setup	Short setup time	4	4.5
		Correct water connections	4.6	5

impacted meeting the customer's "needs" and "wants." The results of this session are listed in Figure 12.2, and were grouped into four main subsystems.

To get a better understanding of why the customers rated the current system as they did (see Table 12.1) a SWOT analysis was performed. With this tool, the team explored the Strengths, Weaknesses, Opportunities, and Threats of the existing system. A breakdown of each of these areas is described in Figure 12.3.

Now that the strengths, weaknesses, opportunities, and threats of the current system have been identified; the customer's needs, self-rating, and issues can be assessed in the "house of quality" or quality function deployment (QFD) in Figure 12.4.

Mold design	Machine capability	Material	Processing
• Material shrinkage • Cooling system design • Steel type • Surface finish • Gate size & location • Venting • Runner design (cold vs. hot) • Mold maintenance • Ejection	• Machine type • Calibration (mech. & elec.) • Barrel size • Screw type • Nozzle size • Machine size • Maintenance • Controller • Die protection • Tie bar wear • Clamping • Platen deflection • Thermolator • Sensors	• Material properties • Material additives • Material variation • Lot to lot consistency • Material drying • Material handling • Material contamination • Blending consistency • Colorant ratio	• Scientific molding • DOE • Fill time • Cooling time • Hold time • Cycle time • Shot size • Mold setup • Water flow rate • Clamp force • Screw speed • Injection pressure • Pack pressure • Hold pressure • Back pressure • Mold temperature • Melt temperature • Process control

Figure 12.2. Affinity diagram.

Strengths
1. Using Lean six sigma structure
2. Machines are maintained by qualified engineers
3. On-site mold designers and builders
4. Full control over material handling
5. Highly motivated teamgy
6. Easy access to data

Weaknesses
1. Molding machines have varying capabilities
2. Linking scrap due to material causes
3. Having the proper equipment to measure machine and mold capability
4. Many variables to explore
5. Each shop operates differently

Opportunities
1. Robust processing techniques are widely known
2. Incorporating material, machine, mold, and process to drive part quality
3. New NCMR process provides more visible scrap
4. Target XF product line (highest volume of scrap parts)

Threats
1. Willingness to adopt change
2. Molding does not have documented or standardized procedures to address part production issues
3. Capital spending will be tight
4. Accountability for maintaining project gains
5. Peak production season limits available time on machines

Figure 12.3. SWOT (Strength, Weaknesses, Opportunities, and Threads) analyses.

Ultimate goal: Customer attributes, needs, requirements, or demanded quality — Primary goal	Secondary goal	Relative importance or weight	Part quality: Ppk>1.0	Part quality: Visual inspection	Part quality: Functional testing	Material requirements: Moisture content	Material requirements: Material certification	Material requirements: Material ratio	Material requirements: Material containment	Material requirements: Material purge	Material requirements: Hopper clean	Optimized process: Process settings	Optimized process: Stabilization time	Optimized process: Process audit time	Machine selection: Machine type	Machine selection: Nozzle size	Machine selection: Machine status	Mold setup: Water flow	Mold setup: Setup time	Mold setup: Qualification time	Mold setup: Mold configuration	Mold setup: Mold status	Mold setup: Mold location
Part quality	Correct part	5		3																	9		
	Correct material	5	3	3	3	9	9	9	9	3	3												
	Meets dimensional tolerances	5	9	1	1	1	3	3	1			9	3		3	3		9			3		
	Meets visual requirements	5		9		3		3	3	1	1	9	3					3			3		
	Meets destructive testing requirements	5	9	1	9	3	9	3	3	1	1	9	1		1	1		3					
Part availability	Enough parts	5															9					9	
Material requirements	Correct material	5					9		3														
	Correct colorant	4.7	1	9	1			9	3	1	1												
	Correct mix	4.5	1	3	3			9	3	1	1												
	Dried	4.1	1	3	1	9			9		3												
	No contamination	4.7		9	3					9	9												
Optimized process	Largest molding window	4.5	3	3	3		1	1	1			9			9	3		9			3		
	Established for machine type	4.7	1	1	1							9			9	1		9			1		
	Shortest cycle time	4.2	1	1	3							9			9	1		9			1		
Machine selection	Correct barrel size	4.5	3	3	1																9		
	Correct screw type	4.4										3			9								
	Correct nozzle	4.5	3	3	3											9					9		
	Correct machine type	4.6	3	3	3							3			9						9		
Mold setup	Short set up time	4.4											1						9		9		3
	Correct water connections	4.8	9	9	9													9	3		9		
	Direction of improvement																						
	Measurement units		Ppk>1.0	go/no go	go/no go	go/no go	percent	percent	correct/not	prged/not	clean/not	go/no go	minutes	10 minutes	go/no go	inches	Available/ Not Available	gpm	under an hour	go/no go	Available/ Not Available	there/not there	0
	Our value		2	1	2	3	2	3	2	3	2	1	3	3	3	4	4	4	4	3	4	4	4
	Competitor X		0	0	0	0	0	0	0	0	0	0	0	0	0	0	0	0	0	0	0	0	
	Competitor Y		0	0	0	0	0	0	0	0	0	0	0	0	0	0	0	0	0	0	0	0	
	Delighted or ideal target specification values																						
	Disgusted or marginal target specification values																						
	Technical importance absolute		225	302	207	117	155	177	164	77	89	283	39	0	222	83	45	239	54	0	303	45	13
	Technical importance relative		**8%**	**11%**	**7%**	4%	5%	**6%**	6%	3%	3%	**10%**	1%	0%	**8%**	3%	2%	**8%**	2%	0%	**11%**	2%	0%

Figure 12.4. House of quality (or quality function deployment).

The QFD would allow the team to gain more insight on the relationship between the customer's needs and the engineering requirements. The results of this exercise identified the following key metrics in order of technical importance:

1. Visual inspection—11%
2. Mold configuration—11%
3. Process condition settings—10%
4. Water flow cooling system temperature—8%
5. Dimensional capability requirements to be $C_{pk} > 1.00$ with 90% confidence interval (CI)
6. Machine type and its control system—8%
7. Molded parts functionality testing—7%
8. Plastics material mixture ratio—6%

If the company addresses the key metrics with the most technical importance, the customer will have greater satisfaction. In summary, these areas include the molding process condition, mold design, machine type, and material properties. Actually the process performance is function of all these conditions, such that *Process performance = f (mold design, material properties, machine, process condition).*

The last step of the "define phase" was to complete a detailed process flow chart that would be used for value stream map analysis. This diagram can be found in Figure 12.5.

12.1.2 MEASURE PHASE

During the "Measure phase," the goal was to collect baseline data on the defects and calculate the process sigma for the target parts. The first step was to identify the ideal parts for the project. One year's worth of scrap data was analyzed from computer database. The team brainstormed to select the defects that would be impacted by the molding process. This was done to filter out what molding-related scrap team wanted from the report. Next, the data was filtered according to the scrap being produced in the factory. Finally, once the data was compiled, it was then analyzed using a Pareto chart (the 80:20 rule) shown in Figure 12.6.

From the top scrapped part number's (PNs), five were selected with a preference of being able to use an automated vision system (OGP) to perform the dimensional inspection. The result of this analysis created the following list of ideal PNs:

1. 30001—XF1
2. 30002—XF2
3. 30003—XF3
4. 30004—XF4
5. 30005—XF5

Once the PNs were identified along with the amount of scrap produced, the next step was to understand the impact of the molding process on mold repair and machine downtime. The mold repair data came from computer data base and the machine downtime was compiled from manufacturing monitoring system, as shown in Figures 12.7 and 12.8.

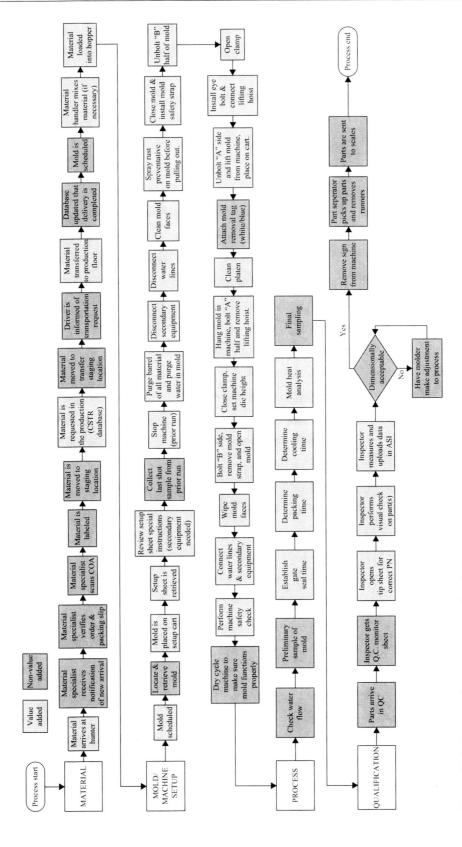

Figure 12.5. Process value stream map and analysis.

Pareto chart of dimensional molds

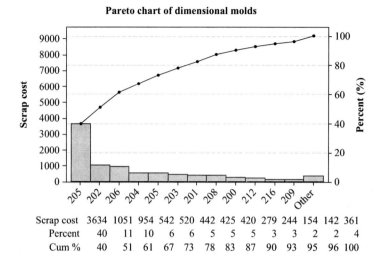

Scrap cost	3634	1051	954	542	520	442	425	420	279	244	154	142	361
Percent	40	11	10	6	6	5	5	5	3	3	2	2	4
Cum %	40	51	61	67	73	78	83	87	90	93	95	96	100

Figure 12.6. Pareto chart of top scrapped part number's based on molding production nonconformance's for one year.

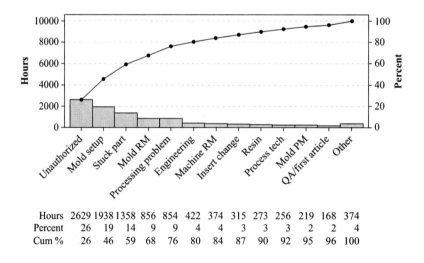

Hours	2629	1938	1358	856	854	422	374	315	273	256	219	168	374
Percent	26	19	14	9	9	4	4	3	3	3	2	2	4
Cum %	26	46	59	68	76	80	84	87	90	92	95	96	100

Figure 12.7. Pareto chart of mold repair causes for the period of year.

The results of the Pareto analysis for mold repair and machine downtime as impacted by the molding process are shown in Table 12.2.

The Pareto analysis has shown that there are more than enough impacts from the top causes of both mold repair and machine downtime to achieve the project goals. The analysis of the scrap reporting from computer shows 70% of all scraps is related to visual defects. Cycle time reduction will be proved as the molding process optimization takes place. A summary of goals and plans to address these objectives are listed in Table 12.3.

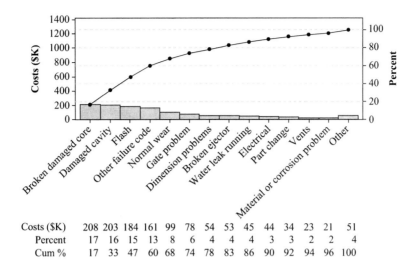

Costs ($K)	208	203	184	161	99	78	54	53	45	44	34	23	21	51
Percent	17	16	15	13	8	6	4	4	4	3	3	2	2	4
Cum %	17	33	47	60	68	74	78	83	86	90	92	94	96	100

Figure 12.8. Pareto chart of machine downtime causes for the period of 9 months of machine monitoring system data.

Table 12.2. Mold repair and machine downtime causes related to molding process

Mold repair	Machine downtime
1. Flash: 15%	1. Mold setup: 19%
2. Gate problem: 6%	2. Stuck part: 14%
3. Dimensional problem: 4%	3. Processing problem: 9%
Total impact = 25%	Total impact = 42%

Table 12.3. Project goal and plan of action

Project goals	Project plan
Reduce scrap and inspection by 25%	More emphasis on visual inspection instead of dimension
Reduce mold repair and machine downtime by 10%	Control the molding process to eliminate shorting and flashing
Reduce molding process cycle time by 10%	Optimize process condition using DOE

The next step of the "Measure phase" was to create a baseline for the current process. Each of the five PNs was evaluated based on the existing capability in the existing system. The main response to gauge the system's performance was based on the dimensional capability of the parts. An example of the baseline study for Mold-205 can be found in Table 12.4.

Before evaluating the current process, the team first qualified the equipment using three basic tests. These tests measured the speed linearity, shot repeatability, and load sensitivity of the machine used to mold the parts. The speed linearity assessment tests the speed control over

Table 12.4. Example of Dimensional Capability Baseline

PN	Mold No.	Dimen- sion	Equip- ment	LSL	Target	USL	Mean	St. dev.	Cpk
30001	205	A	OGP	0.1352	0.1370	0.1390	0.1350	0.00007	0.09
		B	OGP	0.1620	0.1635	0.1650	0.1623	0.00007	0.16

the full injection speed setting in the press. The shot repeatability test is a comparison of shots against each other at different levels to ensure that the variation from shot to shot is at a minimum. Lastly, the load sensitivity inspection determines how the injection flow rate is altered by changes in molding conditions, such as material viscosity. The team employs these tests to confirm that the machine is capable and can control the process. Without capable equipment, it is difficult to have control of the process.

In addition to the equipment qualification, team utilized the Six-Step Study to determine the most optimized process settings. The Six-Step Study takes into account the following factors of injection molding:

1. *Viscosity curve*: Ensures filling stage of process will stay consistent. Shot-to-shot variations should be reduced in order to achieve repeatable quality of parts.
2. *Cavity balance*: Checks the percent variation between maximum and minimum fill cavities. The percent variation should not be greater than 5% or even 3% for tight tolerance parts.
3. *Pressure drop*: Ensure the tool is not pressure-limiting the machine's capability.
4. *Process window*: Indicates how much you can vary the process and still make an acceptable part. An ideal situation is to have a wide process window.
5. *Gate seal*: Determines the holding time for the part, such that the holding time is just beyond the time where the part weight is constant.
6. *Cooling*: Cycle time is the most important factor that makes the bottom line profit. If the process is capable at lower cooling times, we can achieve the same dimensions at lower cycle times

12.1.3 ANALYZE PHASE

During the "Analyze phase," the goal is to develop a focused problem statement, explore potential causes, collect data, quantify causes, and effect relationships. In order to develop a focused problem statement and explore potential causes, a fishbone (cause and effect) diagram was utilized. The team brainstormed all of the possible causes that would create an out of control molding process. The team then decided to put the primary focus on causes that were a result of plastic processing conditions. The results of the fishbone diagram are listed in Figure 12.9.

Once the primary focus was established, the team needed to explore the plastic processing causes in detail. In order to effectively accomplish this, each mold was to be evaluated using design of experiments (DOEs). The goal of the experiment was to establish an optimized molding process that would allow for the study of dimensional correlation while also reducing the

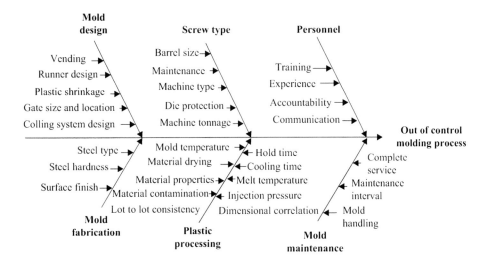

Figure 12.9. Cause and effect diagram (or also called fishbone).

cycle time. During the experiment, a direct relationship between the molding parameters and the part characteristics would be established.

A full factorial central composite design (CCD) with three factors and two levels would be utilized. This design would also include one center point, which would closely represent the setting used to produce production parts in the current system, for a total of 15 runs. The design was set up to allow for the quickest transitions between runs by minimizing the necessary time to stabilize each run, which was dictated by melt temperature. Once each run was stabilized, five shots would be collected for dimensional and visual inspection and one shot was collected for backup. The three factors chosen were holding pressure for dimensional stability, along with mold and melt temperature for visual appearance as well as dimensional stability. The ranges for each of these factors can be found in Table 12.5.

Once the DOE was completed and parts were measured, visually inspected, and functionally tested if needed, the next step was to evaluate the correlation of the responses to one another in order to determine which response was the key indicator of the parts quality. After the key indicator, or predictor dimension, had been identified then a regression analysis was performed for every dimension. The regression analysis included 90% confidence intervals for both the upper and lower bounds of the best fit of the data. Each response was evaluated against the predictor dimension. The goal of this exercise was to determine the operating range for the predictor dimension that would ensure all other responses met specifications. An example of this analysis can be seen in Figure 12.10.

Table 12.5. Factors for Design of Experiment (DOE)

Factor	Levels (Low–High)	Impact
Hold pressure	± 20–30% of nominal	Dimensional stability
Mold temperature	± 10–20% of nominal	Dimensional stability, Visual
Melt temperature	± 10–20% of nominal	Dimensional stability, Visual

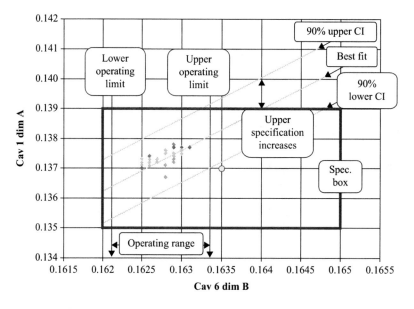

Figure 12.10 Regression analysis of Cavity 6, Dimension B versus Cavity 1, Dimension A for PN30001-XF1.

As shown in Figure 12.10, Cavity number 6, dimension B was the predictor dimension and Cavity 1, dimension A was the predicted dimension. There was a strong positive correlation between these two dimensions, so as dimension B increased, dimension A would increase as well. The area bounded by the upper and lower confidence intervals represented what the process was able to do to these dimensions. The intersection of the upper confidence interval and the top of the specification box represented the upper operating limit and the lower operating limit was determined by the intersection between the lower confidence interval and the bottom of the specification box. The distance between the lower operating limit and upper operating limit represented the range in which the predictor dimension must remain to keep the predicted dimension in specification.

The next piece of information represented in this plot was the necessary changes to the upper specification limit needed to run at nominal. This was found by determining the offset from the best fit line and the center of the specification box in the *y*-axis, as shown in Figure 12.11, Cavity 1 dimension A would need to be adjusted by 0.001.

As each dimension was evaluated, the team needed to determine the necessary steps to increase the operating window large enough to be able to maintain a process capability of 1.33 (P_{pk}>1.33). This will be accomplished by modifying one or a combination of the following areas:

1. Control process in the ideal spot
2. Adjust tolerances (modify upper specification limit)
3. Eliminate dimensions from correlation

All of these solutions would be simulated through the regression analysis prior to implementation.

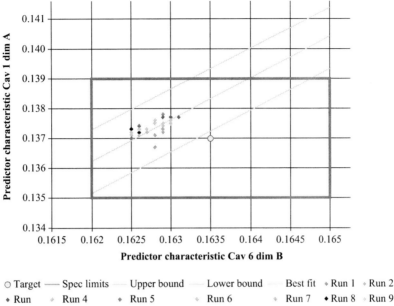

Figure 12.11. Regression analysis with centered process.

12.1.4 IMPROVE PHASE

During the Improve phase, the possible solutions needed to be identified, selected, implemented, and verified. In order to identify the solutions, a regression analysis (Figure 12.11) was analyzed based on maximizing the operating range of the predictor dimension/response. The areas considered are listed above and examples are shown below:

1. *Control process in the ideal spot*:

    In order to control this dimension near nominal, the process would need to be similar to the settings used for run 2, 3, or 4 in the design of experiment trials.

2. *Adjust tolerances*:

    In order to remove the constraints of this dimension the tolerances could be relaxed. In this example, as shown previously in Figure 12.10, the upper specification limit was increased by +0.001. Now the intersection of the upper confidence intervals and the specification box have widened the operating range.

3. *Eliminate dimensions from correlation*:

    If a predicted dimension correlates strongly with the determined predictor dimension, then there is a possibility for removing that dimension's inspection requirements. Monitoring the predictor dimension can be shown to sufficiently determine whether or not changes are occurring in other cavities or dimensions.

    In all cases, the process should be analyzed to improve cycle time without negatively impacting part quality, and if needed, reestablished to improve the dimensional capability. To determine these settings, the DOE needs to be analyzed. The first step is to analyze the effects of each of the process parameters to the dimensional capability of

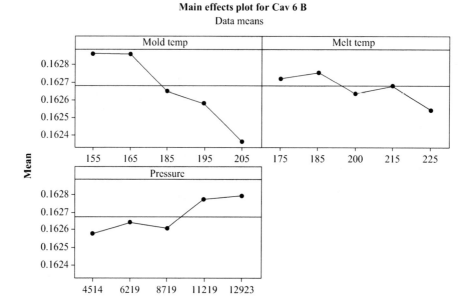

Figure 12.12. Main effects for PN30001, cavity 6, dimension B.

the part. Once this is determined, the optimum parameters for the most efficient process producing a quality part may be established.

The results of the analysis, for PN30001—XF1, are shown starting with Figure 12.12.

As can be seen in Figure 12.12, mold temperature is the main parameter that drives dimension B. During the regression analysis of Cavity 1, dimension A (see Figure 12.13), it was also noted that large variation existed with respect to mold temperature. To ensure more control over each dimension, mold temperature must be stabilized using a water temperature controller called Thermolator.

To determine the acceptable ranges of each of these parameters, the predictive equations from the DOE analysis can be used. The predictive equations for cavity 6, dimension B and cavity 1, dimension A are given in Equations 12.1 and 12.2.

$$\text{Cav 6 Dim B} = 0.1648 - 9.607\text{E}{-}06\bullet(\text{Mold Temp})$$
$$- 3.049\text{E}{-}06\bullet(\text{Melt Temp}) + 2.607\text{E}{-}08\bullet(\text{Hold pressure}) \qquad (12.1)$$

$$\text{Cav 1 Dim A} = 0.1425 - 2.033\text{E}{-}05\bullet(\text{Mold Temp})$$
$$- 7.443\text{E}{-}06\bullet(\text{Melt Temp}) - 1.648\text{E}{-}08\bullet(\text{Hold pressure}) \qquad (12.2)$$

Any combination of the options listed above would represent a possible solution. The team must consider the best solution based on quality, cost, and time. The next step of the process was to evaluate the risk associated with each subprocess. The areas evaluated included

1. Mold design
2. Machine capability
3. Material handling
4. Molding process conditions

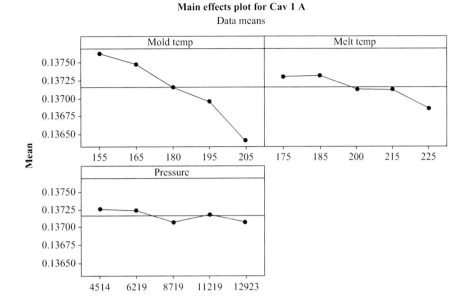

Figure 12.13. Main effects for PN30001, cavity 1, dimension A.

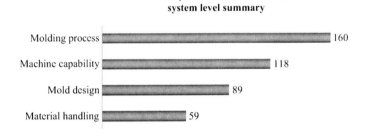

Figure 12.14. FMEA subsystem summary.

Each of these subsystems was evaluated to determine what were the possible risks, their severity, occurrence, and detection. This was best captured with the use of the failure mode effect analysis (FMEA). Each one of the potential risks in each subsystem would be evaluated and assigned a risk prioritization number (RPN). The RPNs would be ranked and the goal is to go after the top RPNs and create action items to lower the RPN to an acceptable level. The results of this exercise are summarized in Figure 12.14.

As determined through the course of the entire project, the emphasis will be placed on the molding process. As one controls the molding process, this will lead to improvements in the areas of machine capability, mold design, and material handling.

12.1.5 CONTROL PHASE

The control phase will require changes in "Tipsheets" or summary of engineering design print, molding setup sheets, and procedures for M-205, PN-30001 (XF1). It will also include monitored performance for each of the project goals in order to create the before and after results. The savings related to mold repair, scrap, and machine downtime will take roughly one year to evaluate. With the removal of production monitoring system, it may be difficult to calculate the machine downtime with high confidence.

To control the molding process for M-205, it is critical to keep an eye on the injection/hold pressure for the critical machine types. The visual inspection on the parts is extremely important and the main role in the control system. This increase in frequency will help reduce the amount of visual defects found in inventory and/or automation assembly. The increased frequency of inspection will also allow for more feedback on the control of the molding process.

Next Steps

1. Implement process for mold qualification.
2. Complete out the optimization on the remaining PNs.
3. Continue to monitor performance of optimized PNs.
4. Continue efforts with molding production optimization and control (MPOC).

12.2 OVERALL EQUIPMENT EFFECTIVENESS: A PROCESS ANALYSIS

Purpose: Equipment effectiveness is extremely important to XM1 Corporation. The majority of manufacturing assembly processes contains at least partial machine automation, and many processes are completely automated. Manufacturing automation allows for greater output of finished goods with minimal associated labor cost. Automation machines, however, are complex and must be constantly maintained to ensure the efficiency levels promised through process. Performance indicators are identified and constantly measured to diagnose problems and classify defective procedures. The existing analytics tools and data collection abilities allow seeing a machine's overall effectiveness in 8-hour blocks of time, determined by shift. However, these automated counters are manually downloaded, placed in manually created spreadsheets, manually manipulated for purpose, and manually distributed to the various end users of information. By the time this process is complete, the data received is up to 7 days old, one-dimensional and ambiguous. A new way must be found to collect the data and analyze it in real time. Thus, the team set out to analyze the existing data collection capabilities, assess their effectiveness, and considered areas for improvement. In completing this, the project team find that implementation of new software built upon a more robust data collection tool can potentially save $80K in *non-value-added* work and $250K in *direct labor cost* over a one-year period. Development and implementation of a new process would also add greater visibility to inefficient practices and allow the company to direct improvement costs with greater focus.

12.2.1 DEFINE PHASE

In Defining this project, team will identify business financial drivers, determine critical to quality (CTQ) processes, define specification limits, and identify goals and values. Once identified, the team can develop a path toward completion of the project. In order to identify drivers for machines efficiency team must define OEE as it pertains to XM1-Corporation. This is shown in Equations 12.3 through 12.6.

$$Availability \ \% = \frac{Run \ time}{Run \ time + down \ time} \qquad (12.3)$$

$$Quality \ \% = \frac{Good \ parts}{Good \ parts + Bad \ parts} \qquad (12.4)$$

$$Performance \ \% = \left(\frac{\frac{(Design \ cycles \ per \ second) \ Actual \ cycles}{Run \ time \ in \ seconds}}{(Good \ parts + Bad \ parts)} \right) \qquad (12.5)$$

$$Overall \ Equipment \ Effectiveness = Availability + Quality + Performance \qquad (12.6)$$

Overall equipment effectiveness (OEE) represents a hierarchy of metrics that evaluates and indicates how effectively a manufacturing operation is utilized. Company uses data collection software to capture faults, waiting, pauses, and good and bad assemblies on machines used to assemble various products. The counts associated with these occurrences represent key performance indicators (KPI) used to determine the effectiveness of automation processes. The above mathematical formulas (Equations 12.3–12.6) show how the key performance indicators of run time, down time, good parts, bad parts, cycles, and design cycle makeup availability, performance, and quality. These indicators make up company's automation effectiveness criteria.

Voice of Customer (VOC)

The ability to obtain accurate and timely efficiency analysis in manufacturing is of great importance to many of the company's business units. Engineering teams of various responsibilities can use this information to improve processes, lower cost, enhance the quality of goods, and troubleshoot nonconformance in production. A critical to satisfaction (CTS) analysis shows that end users of automation analysis data define satisfaction as accurate, current, easily digestible, and comprehensive. In completing OEE project, the team will address and improve the current process and increase the satisfaction of end users.

Process

The company currently employs its machine data interface (MDI) software to collect data from numerous programmable logic controllers (PLC). These PLCs are installed on the machines to control the machine's actions as well as capture time (in 0.01 seconds) according to the machine's index/dwell settings, as well as counts of occurrence of stoppages or faults. This information is compiled over the duration of the shift (usually 8 hours). At the completion of the shift, the data from the PLCs are captured in an xx.csv file type and placed in the appropriate machine folder labeled in a date/shift format.

Shift data is continually placed in the respective machine's MDI folder at the completion of each shift. For instance, on Monday of the following work week, a data entry clerk downloads the numerous files for each machine into an Excel spreadsheet for analysis. The following list depicts the steps taken by the clerk in the creation of analyzable weekly data:

1. Open MDI Calc. desktop application
2. Open OEE report from "Share-point" library
3. Select machine type, week beginning date, and week ending date in MDI Calc.
4. Run MDI Calc. application to create data table
5. Export MDI Calc. data table to Excel
6. Cut and paste exported data to existing OEE report
7. Report 3–6 for all other machines
8. Hide rows in OEE report show existing month plus two months
9. Email report to recipients
10. Download OEE values into automation scorecard
11. Email link to automation scorecard to recipients

The process of gathering weekly machine efficiency data and putting it into an analyzable format currently takes one person 20 minutes to accomplish per machine. Data is usually compiled on the Monday following the previous week. Consolidation of the previous week's data is completed by the mid-Tuesday.

Figure 12.15 represents a Value Stream Map (VSM) of the process. It shows a 20-minute task for each of the automation machines. The inefficiency of this process and the marginal benefits derived from it led the team to consider the contributing factors to efficiency in relation to the data collecting process. Team developed a SIPOC chart to identify and categorize the process suppliers, inputs, outputs, and customers in a chronological order from left to right. The completion of this task defines the question at the root of the group's project: How can the team enhance the data collection and analysis capabilities through improvement of data collection inputs and transformation of outputs? The goal is that a change to this process will result in a significant advancement in automation effectiveness visibility and understanding in the company.

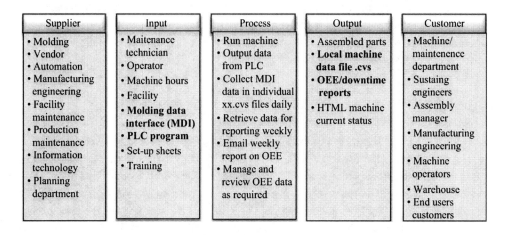

Figure 12.15. Value Stream Map (VSM) of the process.

12.2.2 MEASURE PHASE

In attempting to measure the effectiveness of the data collection process, the team decided to focus on a specific automation machine. XF-parts subassembly automation was chosen due to its level of integration with the company's MDI system as well as its historically complete data set. The goal is to measure indicators of efficiency using current methods in order to show gaps in data collection and analysis. The assumption is that the characteristics identified in the measurements can be attributed to all automation machines, and that shortfalls in collection and analysis are universally applicable to various work centers.

The process capability chart (Figure 12.16) is a statistical tool that provides an assessment of the ability to control the process in relation to a standard. This chart was performed on OEE of XF machines, confirming that the process is out of control (indicated by a C_{pk} value of 0.60). What team typically expects is a C_{pk} value of 1.33, which is equivalent to a sigma level of approximately 4 without shift changes and with shift changes C_{pk} value of 1.33 will translate into 5.5 sigma level.

This test was conducted to highlight the lack of visibility in automation processes. Unfortunately the process capability assessment identifies the need for improvement but is not able to be connected to any driver by current analytics. Thus, team ability to see areas for improvement are useless when combined with team inability to identify cause(s).

The disparity between reality and expectation is also evident in XF-automation's distance between actual OEE and effectiveness goals. A histogram of OEE data (Figures 12.17–12.19) for the past three years shows that the average good parts per hour (GPPH) is consistently well below the goal of 725. In relation to data showing only for the past one year GPPH data, one can see that XF machines remain well below the efficiency goals historically as well as in the current year. On average GPPH falls approximately 100 parts below the goal of 725 (differences in machine vary).

The Pareto chart is used to display frequency of occurrences of an item. This particular Pareto showed the fault conditions resulting in machine stoppages from an XF machine for the past two years. When team focused on the top five conditions, they discovered that placement faults and feeder faults were commonly high-occurring failures present in 80% of the machines.

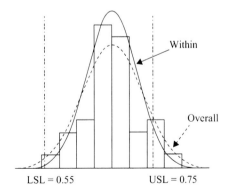

Figure 12.16. Process capability of overall equipment efficiency (OEE), $n = 43$; potential capability: $C_p = 0.69$, $C_{pl} = 0.78$, $C_{pu} = 0.60$, $C_{pk} = 0.60$
$P_p = 0.50$, $P_{pl} = 0.56$, $P_{pu} = 0.43$, $P_{pk} = 0.43$
St. dev. (within) = 0.048, St. dev. (Overall) = 0.048.

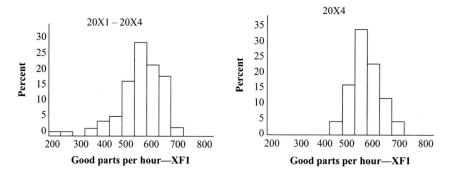

Figure 12.17. Histogram of OEE data—XF1 parts.

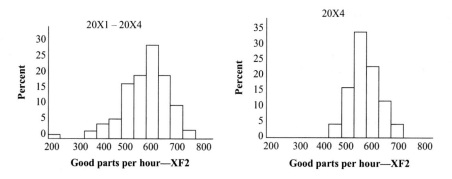

Figure 12.18. Histogram of OEE data—XF2 parts.

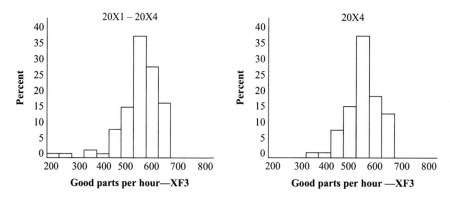

Figure 12.19. Histogram of OEE data—XF3 parts.

It is important to point out that many of the methods used to analyze data on the XF machines were manually created, and are manually sustained. The various histogram, Pareto, and process charts have taken much longer to create than the resulting value. From the existing measurements, the team has only to conclude that certain faults are typical and abundant, and that it falls significantly short of teams expected efficiency goals on XF automation. This measurement, however, does not identify "Why?" In the analysis of efficiency process, the project will identify answers to this question.

12.2.3 ANALYZE PHASE

In the analysis of data collection process, the existing practices compared to a new data collection system that is in the process of testing. This new Machine Data Interface (MDI) software outputs data with greater granularity, and exports into a more user-friendly Sequel Database (SQL). It is proposed that this SQL database can be used in conjunction with existing and future analytics software to produce near real-time analysis of automation KPI combined with greater trend analysis. In comparing the two systems, a month of data was gathered (say October) from the new MDI collection system. For data analysis the team chose the XF1 machine.

This machine was chosen as a test platform for the new MDI software due to its simplicity of use and availability of redundant machines. Thus, the XF1 machine will be used in the analysis as well.

Hypothesis Testing: The purpose of statistical inference is to draw conclusions about a population on the basis of data obtained from a sample of that population. Hypothesis testing is the process used to evaluate the strength of evidence from the sample and provides a framework for making determinations related to the population, i.e., it provides a method for understanding how reliably one can extrapolate observed findings in a sample under study to the larger population from which the sample was drawn. The team used a two-year sample of XF1-OEE to test the hypothesis (Figure 12.20) that the current practices fall short of the machine's goal of 75% OEE. Then the mean OEE was tested against the target of 75% and determined the null hypothesis is rejected. This was concluded through calculations of Z-scores and the calculated Z scores that show that the null hypothesis is not equal to the project objective's target and assists the team in supporting the project objective. The test mathematically shows that the sample data falls short of the OEE goals, and that the same can be said about the entire population.

Boxplot Analysis: This plot (Figure 12.21) shows the same 3-day period of data analyzed using differing levels of granularity. It is proposed that the new MDI system will allow for increased visibility of KPI through greater granularity, and this system has been compared to existing granularity in order to assess the importance of such a capability. This data is representative of an entire day's worth of data for three consecutive days (October 1 to 3). The plot on the left shows the key indicators of availability, quality, and performance as it relates to OEE.

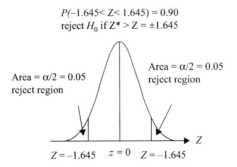

Figure 12.20. Hypothesis test (H_0 = mean OEE = 75%) (H_a = mean of OEE $\neq$ 75%) Normal mean = 0.6872, Standard deviation = 0.1029.

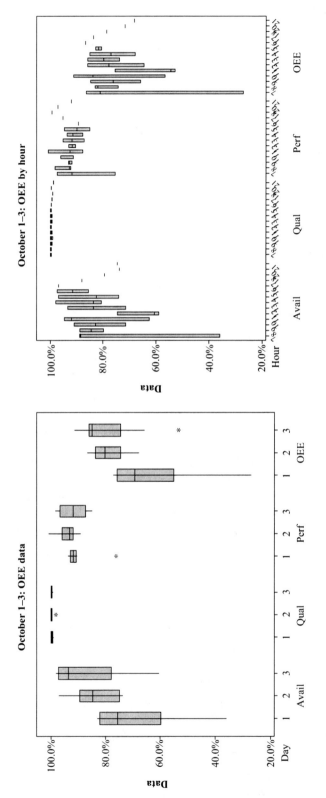

Figure 12.21. Box plots of OEE data.

The left plot shows data received over the course of an 8-hour shift, encompassing two shifts per day. The information derived from this plot shows a substantial decrease in availability and OEE on October 1 (Monday). The plot on the right shows the same KPI over the same 3 days, but is compiled by hour as opposed to shift. When looking at the plot on the right, one can see that the most significant decreases in KPI occur in the 7 am hour and 11 am hour. The left plot allows us to focus efforts on a day of the week, while the right plot allows us to see KPI affects hour by hour.

Design of Experiment

A short-run Design of Experiment (DOE) (Figure 12.22) was created to show the benefits of more granular analysis. A test was created to assess the effectiveness of the machine 250 vibratory bowl settings in relation to part feeding. The test called for one machine operator to adjust the vibratory settings and bowl fill amounts every 10 minutes over the course of a shift. For the purpose of testing, it was assumed that all molded parts were within design specification limits and that all electronic and mechanical components are working as designed. The following results were presented (Figure 12.22).

The DOE results indicate that part feeding mechanisms on the 250 machine are most efficient when vibratory settings are higher and bowl fill amounts are lower. The significance of this test is minor, but illustrates an important point: the current data granularity cannot show this appropriately. This experiment highlights the correlation between feeder settings and machine stoppages, and does so over a 90-minute period. The results of this experiment will lead the team to pursue a different approach to feeder fault reduction. Current methods of collection, grouped by 8-hour shift, would only show that a certain amount of faults occurred at that station for the entire period. A cause and effect analysis for this is also shown in Figure 12.23.

A comparison of current and future methods for data collection describes that the current data collection capabilities are lacking and that existing analytics are ineffective in addressing areas for improvement. Moving forward, the hope is to identify ways to improve collection methods on automation equipment. The new analytical methods also will be given focus and combined with greater collection capabilities. The goal is that the combination of these improvements will bring greater visibility of process variance and lead to more effective solutions and efficiency gains.

12.2.4 IMPROVE PHASE

Software: In using the results to improve the current collection and analysis process, team began to look at commercially made software solutions that would use SQL data to analyze machine efficiency in near-real-time. Then, the team compared possible solutions to the idea of making similar products in-house using existing software, such as SAP Business Objects or Performance Point Dashboard Designer (PPDD). Both have advantages and disadvantages. Commercial software would add initial expenses in the purchase, installation, and integration of the solution. There are also potential costs associated with annual licensing with some options. A solution created in-house would add labor hours to existing employees and requires new technology to be learned prior to development. An addition to departmental duties without addition

	C1	C2	C3	C4	C5	C6	C7	C8	C9	C10	C11	C12	C13	C14	C15
↓	Std order	Run order	Center pt	Blocks	Bowl	Inline	Parts	Good parts	Bad parts	Run time	Downtime	OEE	Avail	Perf	Qual
1	1	1	1	1	60	40	500	46	2	6	4	0.15333	0.6000	0.26667	0.9583
2	2	2	1	1	90	40	500	274	6	10	0	0.91333	1.0000	0.93333	0.9786
3	3	3	1	1	60	80	500	60	6	7	3	0.20000	0.7000	0.31429	0.9091
4	4	4	1	1	90	80	500	282	4	10	0	0.94000	1.0000	0.95333	0.9860
5	5	5	1	1	60	40	1000	27	9	7	3	0.09000	0.7000	0.17143	0.7500
6	6	6	1	1	90	40	1000	214	4	9	1	0.71333	0.9000	0.80741	0.9817
7	7	7	1	1	60	80	1000	35	13	6	4	0.11667	0.6000	0.26667	0.7292
8	8	8	1	1	90	80	1000	304	1	10	0	1.01333	1.0000	1.01667	0.9967

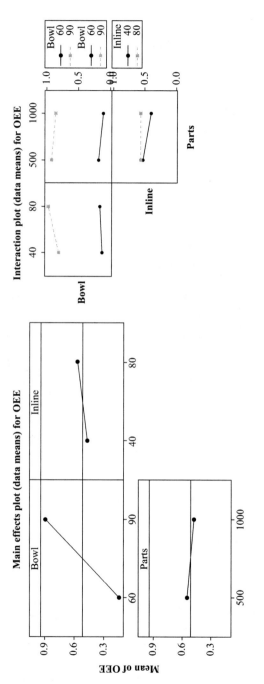

Figure 12.22. Design of Experiment: main effects.

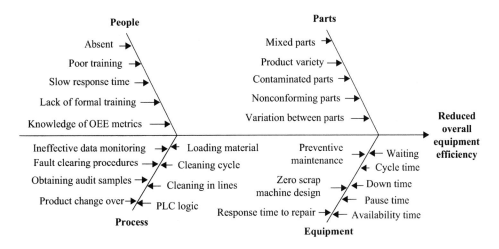

Figure 12.23. Cause and effect diagram.

of labor would keep costs low but result in lost productivity elsewhere. Further examination of possible software solutions will be needed before the true costs and benefits of both options can be assessed.

Process Improvement

In order to obtain the level of granularity needed in future machine analysis, there must be improvements made to the collection and distribution of data from machine to software. The new MDI component, currently in beta phase, will pass information from the station PLC to a Sequel (SQL) database where it can be used by future analytics software. Two of the machines have been converted to SQL database for testing.

Once improvements (Figure 12.24) have been made to software and systems, visibility should be increased with respect to real-time analysis of KPIs (Figure 12.25). Media monitors in each work center will show current analytics in a dashboard format. This will allow all operators, mechanics, engineers, and supervisors to see the same real-time data at the appropriate roll-up level of reporting. These end users will also be able to access these machine analytics from their local work station computer and even tailor their view to show information most pertinent to the user. This means that future information will pass from the machine PLC to a SQL database, then to a program that analyzes defined KPIs in near-real-time, pushed to

Figure 12.24. Process improvement.

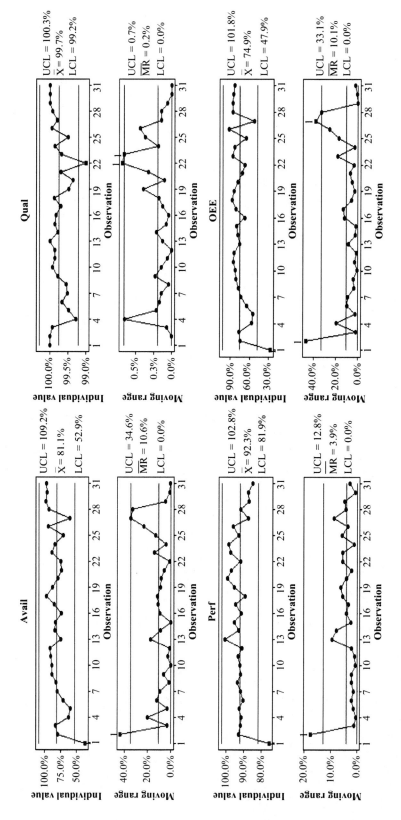

Figure 12.25. Run charts of Key Performance Indicators for XF1 subassembly automation machine.

computerized dashboards that allow for consistent monitoring by operators of machine performance metrics. This tool will be accessible by all end users, and will create intelligence-based numeric and visual reporting as needed by the user.

PDCA (Plan, Do, Check, Act) Cycle

Upon implementation of the new process, a Plan, Do, Check, Act (PDCA) cycle process will be implemented to ensure proper and timely action based on new data analytics. This cycle will be executed in response to action taken in efficiency improvement and will be implemented universally among assembly work centers. This cycle will ensure periodic and systematic review of automation efficiency.

12.2.5   CONTROL PHASE

The primary objective of this phase was to ensure that the gains obtained during "Improve" are maintained long after the project has ended. To that end, it is necessary to standardize and document procedures, make sure all employees are trained, and communicate the project's results. In addition, the project team needs to create a plan for ongoing monitoring of the process and for reacting to any problems that arise.

Standardizing and Documenting

The process control plan describes the level of monitoring and the level of metric reviewing of OEE from the shop floor level to the manager Level. It details the granular review by the shop floor operator hourly with a dashboard review and reporting every two hours. The plan details the steps the operator should take in the event of decreased OEE indicators. This plan also describes the reporting mechanism through Corporate Portal (XM1-CP) and its review cycle. The control plan describes two support groups: Manufacturing Engineering and Assembly Maintenance, for Production. While the process control plan (Figure 12.26) is a key component of the documentation, it also serves as the new escalation process for manufacturing organization.

12.3   POWDER COAT IMPROVEMENT

Executive Summary: A Six Sigma Black Belt team selected for a project at LF-II to reduce WIP (work in progress) scrap in the powder coat department. Additional defect codes were added to the documentation process to clarify the disposition procedure. The powder coat bubble (BB) totaled to the largest defect code in dollars, so the team focused resources on reducing that specific code.

After a screening test and full factorial DOE, the team concluded that two influences had a significant effect on the size of the powder coat bubbles: *curing time* and *curing temperature*. When both time and temperature are reduced, the team saw a 43% reduction in bubbles. If this data were generalized across all similar components, the effect on WIP scrap savings would result in an annual savings of up to 8% of 20X2 WIP scrap.

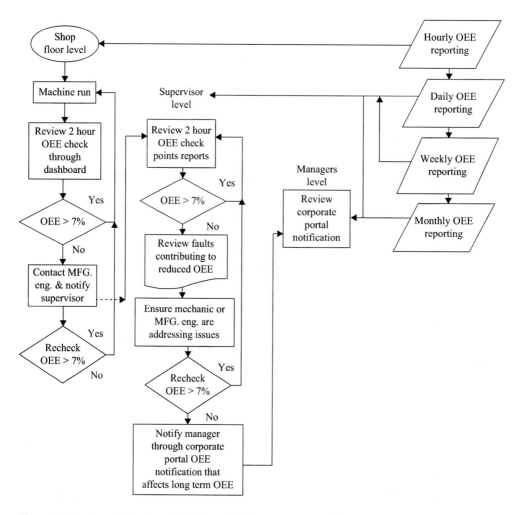

Figure 12.26. Overall Equipment Efficiency (OEE) process control plan.

Purpose: The team selected to perform a Six Sigma project on the powder coat department at LF-II. As the bottleneck of manufacturing facility, this process is considered the heart of production and often is the determining factor on product longevity in the field. Key areas for improvement include: reduce WIP scrap, increase efficiency, and lower overtime and temporary employee labor expense.

12.3.1 DEFINE PHASE

Problem Statement: The volume of powder coat WIP scrap and the velocity of several parts through the process lead to added costs in scrap, overtime, temporary labor needs, and overall strain on the system. Improvement in any one of these areas will result in labor savings, scrap savings, and increased customer satisfaction. The Six Sigma team chose this project with the goal of improving documentation, reducing WIP scrap, and increasing velocity.

Scope: This project will focus on reducing powder-coat-related WIP scrap and increasing the velocity of parts through the powder coat process. Areas of improvement will remain solely within the LF-II powder coat department, and not spread into die-casting, CNC machining, assembly, or any other department.

Voice of Customer (VOC): The glue station and assembly departments sit as the subsequent processes, waiting for available parts prior to beginning. Therefore, those two units are considered the customer of powder coating. Additional lead time due to scrap or slower velocity will cause the assembly lines to fall behind schedule, potentially delaying shipment of the finished product to end users.

SWOT Analysis: A strengths, weaknesses, opportunities, and threats (SWOT) analysis structured a summary of the powder coating department (Figure 12.27). From the analysis, we determined the infrastructure is one of the principle strengths of the process. A continuously moving conveyor transports each part through the entire system reducing a lot of excess waste. The personnel are already trained and can turn around customer orders relatively quickly due to holding on hand popular components that have already been coated. The ability to powder coat each color on site requires less inventory to carry since the facility can adapt raw components to the customer's color of choice.

The process falls short in areas such as consistency, changeover times, velocity, and part/color combinations. The powder coat is manually applied introducing the operator as a variable. Often an operator will apply too little or too much powder resulting in consistency issues and WIP scrap. Also, color changeovers add time, not from changing colors in the spray guns, but from time/speed in the curing oven. The number of parts that can be coated at one time is a limiting factor, leading to increased cycle times. Lastly, there are often several mating parts that require the same powder coat color (e.g., shroud, body, base, etc.). A couple of colors (Desert Granite and Sedona Brown) vary in brightness depending on the thickness applied, leading to mismatch issues.

Opportunities for the process include reduction in WIP scrap as well as a gain in efficiencies. The threats associated with powder coating in any industry include corrosion, adhesion, and defects.

Figure 12.27. SWOT analysis on powder coating.

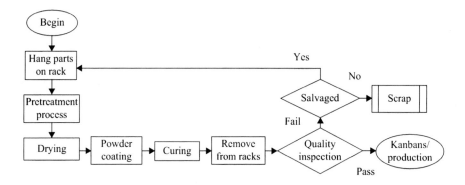

Figure 12.28. Powder coat process map.

Process: Powder coating is a type of coating that is applied as a free-flowing, dry powder. The main difference between a conventional liquid paint and a powder coating is that the powder coating does not require a solvent to keep the binder and filler parts in a liquid suspension form. The coating is typically applied electrostatically and is then cured under heat to allow it to flow and form a "skin." The powder may be a thermoplastic or a thermoset polymer. It is usually used to create a hard finish that is tougher than conventional paint.

Process map: All parts are attached to a moving conveyor that transports the parts throughout the entire process. The process map as shown in Figure 12.28 depicts the components of the powder coating process. Various hooks, racks, and fixtures are designed to connect the parts to the conveyor, carrying each part from beginning to the end of the system. Prior to hanging, some parts require an initial preparation stage. Often various features are covered using masking or plugs to prevent the powder coat from covering threads, holes, or surfaces. Parts are then hung on their respective racks and sent through the system. Initially, the parts enter a three-phase phosphate bath (washing machine), which chemically etches the face of the parts to clean and decontaminate the surface. A clean surface is very important to the powder coat adhesion properties. Next, the parts enter the oven for drying. With a dry surface, the parts are ready for powder coating. An operator manually sprays the powder coat onto each component. A completely coated part reenters the oven for curing. The parts remain in the oven between 10 and 30 minutes depending on the part design, finish, and personnel availability. Lastly, all parts are removed from the racks and prepared for assembly.

SIPOC: The team created a SIPOC diagram (Figure 12.29) to summarize the inputs and outputs of the process. Suppliers range from internal to external, and inputs vacillate between manufacturing supplies to actual components. The process was discussed in previous section, with outputs including WIP, scrap, unused powder, and various metrics. Lastly, the chief customers are personnel at subassembly, final assembly, and metal salvage.

Affinity diagram: Through the affinity diagram presented in Figure 12.30, the team concluded that the problem variables that affect the powder coat department's performance consist of five (5) categories. The part design, the racks, the cure settings, equipment, and personnel. The goal variables are to improve the manufacturing efficiency by reducing the amount of WIP scrap and by improving the productivity of the line.

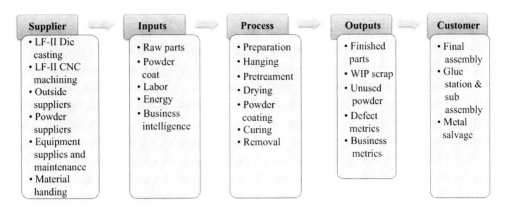

Figure 12.29. Powder coat SIPOC diagram.

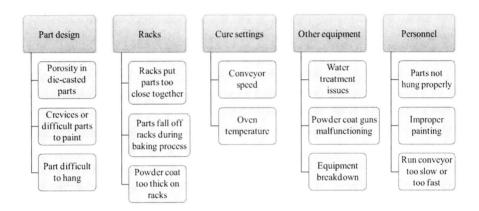

Figure 12.30. Affinity diagram.

12.3.2 MEASURE PHASE

Data collection: An existing process is designed to report the weekly WIP scrap and record that information in computer database. Figure 12.31 displays the amount of WIP reported in 20X2 up to the end of July. The "powder coat" (PC) and "cosmetic defect" (AD) categories total to almost half of the semiannual WIP scrap. Each category has a vague definition and the same defect will often be included in either category depending on the lead reporting the information. Also, AD is seen as an "other" or "miscellaneous" category in the powder coat department, comprising of defects such as bubbles, pin holes, burrs, "fish eyes," divots. The categories require adjustment to accurately classify WIP scrap.

Three additional defect codes were uploaded into database and onto the WIP scrap reporting form: Powder Coat Bubble (BB), Burr (BR), and Divot (DV). Each of these three codes often makes up the majority of the "Other Cosmetic Defect" (AD) code. Three months of data after the addition of the new codes show bubbles (BB) to be the highest scrap category in terms of dollars (Figure 12.32). Categories in the bottom 20% were removed from the pie chart in Figure 12.33 to reduce clutter. The team concluded from this information that bubbles are a key category on which to focus process improvements.

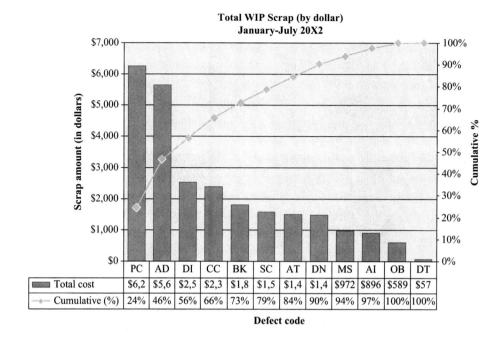

Figure 12.31. January–July dollars scrap by defect code (July–20X2).

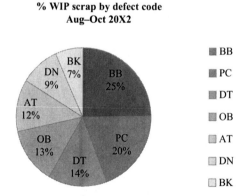

Figure 12.32. WIP scrap category percentages after code adjustments (only top 80% included).

Process capability: The LX faceplate constantly had bubble issues. Many engineering techniques were used in the casting operation to reduce those issues, but to no avail. Eventually the casting of this part was outsourced. The factory still had hundreds of these original faceplates lying around, so it was selected as the ideal part on which to experiment.

Production powder coated an initial run of 31 parts at the existing powder coat settings (20 minutes at 380 degrees). The histogram in Figure 12.33 categories in bins of four (4) the 24 bubble failures. The failure rate is calculated to be 77%.

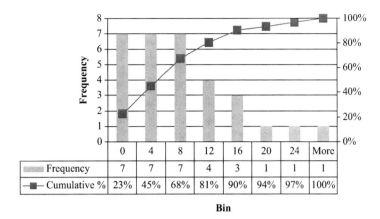

	0	4	8	12	16	20	24	More
Frequency	7	7	7	4	3	1	1	1
Cumulative %	23%	45%	68%	81%	90%	94%	97%	100%

Bin

Figure 12.33. Initial bubble frequency histogram cart on LX-faceplates.

Defect rate = 24/31

Yield = 22.58%

DPMO = 774,194

Process sigma = 0.75

Figure 12.34. Sigma level.

Sigma Level: With data available, a quick check of the Sigma Level was performed, and a result of 0.75 was determined (Figure 12.34). However, this is only a quick reference check solely based off the number of "opportunities" and the number of "defects."

12.3.3 ANALYZE PHASE

Each phase in the DMAIC (Define-Measure-Analyze-Improve-Control) process follows a strict set of procedures. The "analyze" phase focuses on using data and tools to understand the cause-and-effect relationship in the process or system.

Cause and Effect: The Ishikawa (fishbone) diagram is a brainstorming tool used to explore and display sources of variation or influence of a process. In the case of powder coat bubble defects, Figure 12.35 reports three categories of influence (training, materials, and operation), with corresponding causes.

Variation: The bubble type defects are thought to start from void spaces in the metal part. When the part is heated during the curing process, there is potential for the void space to expand and create a bubble or blister on the surface. It was theorized that hot cure temperature and long durations in the oven could result in more bubble defects from high pressure gas trying to escape the void.

Hypothesis Test: To test this hypothesis, two runs with 32 LX parts each were run through the powder coat process. The first run was at standard 380°F for 20 minutes. The second run was for 400°F for 20 minutes. After the run, the team calculated both the number of defective parts as well as the surface area of parts covered by bubble defects.

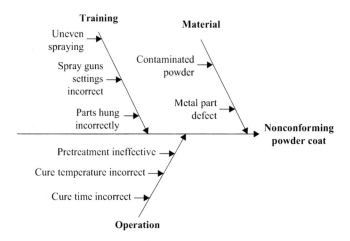

Figure 12.35. Cause-and-effect diagram for powder coat bubble defects.

An ANOVA test ($\alpha = 0.05$) showed there was a significant difference in defective surface area between the two runs. However, there was not a significant difference between the run in the number of defective samples. The team decided to run a Design of Experiments to look further into the effect of temperature and cure time on bubble defects.

12.3.4 IMPROVE PHASE

The team sought to improve the bubble defects through performing a structured experiment changing the curing time and temperature settings.

Design of Experiment (DOE)

A face-centered central composite design (CCD) was selected, creating a 3^2 factorial design. In this design the star points are at the center of each face of the factorial space, so $\alpha = \pm 1$. A model of the experimental design is located in Figure 12.36.

The oven is always set at 380° and never changes for the various parts and colors. The amount of time each part/color cures varies depending on the part, the operator's skill, the number of operators available that day, etc. Material specification sheets were requested from the powder coat vendors. Using all available information, the team developed Table 12.6 for each level.

Table 12.4 was combined to create the coded design found in Table 12.7. This is the format the team followed to conduct the experiment. The design required a block over the temperature variable. The oven requires a significant amount of time to adjust the temperature. Therefore, the run order follows closely to the temperature and then alternates over the time.

Experiment Response Variables

The amount of "bubble" type defect was the primary response variable. The area covered on each powder coated part was measured. Two response variables were obtained: the number of

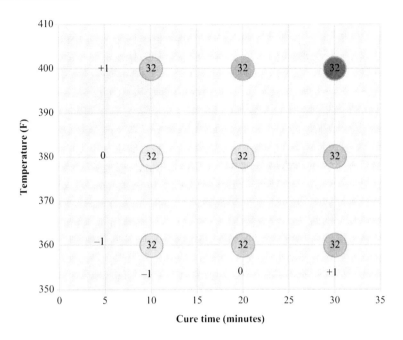

Figure 12.36. Face-centered central composite design points.

Table 12.6. Design level high, medium, and low values for cure time and temperature

Level (code)	Cure time (minutes)	Temperature (degrees) F
−1	10	360
0	20	380
+1	30	400

Table 12.7. Two-Factor Composite Design (Coded Variables) of Cure Time and Temperature

Run	Parts/Test	Parts	Label	Cure Time	Temperature
A	10	32	− −	10	360
B	10	30	− O	10	380
C	10	30	− +	10	400
D	10	30	O −	20	360
E	50	150	OO	20	380
F	10	30	O +	20	400
G	10	30	+ −	30	360
H	10	30	+O	30	380
I	10	30	++	30	400

defective parts and the surface area of bubble defects. Figure 12.37 shows the histogram of the defect areas for each of the nine runs of the DOE. Not surprisingly, the data is not normal and skewed to the right because the defect area cannot be less than zero.

Defect Rate Results

The mean of the area of bubble defects are illustrated in Figure 12.38. To detect if there was a significant difference between the factor levels of temperature and time, an ANOVA was run on the data. A strong effect of both temperature ($p < 0.00001$) and cure time ($p = 0.0062$) on defect area was shown. There was no significant interaction of temperature and cure time ($p = 0.2662$).

The response variable can also be considered to be the binomial measure of whether a bubble defect existed, regardless of size. The number of defective parts in each run was divided by the total number of parts in each run. This gave an estimated defect rate as illustrated in Figure 12.39.

Analysis of DOE: The DOE and ANOVA results indicate that increasing either cure time or cure temperature can lead to larger mean areas of defects, however, if only the presence or absence of defects are considered; only increased cure times appear to affect the defect rate. Separate analysis of log transformed data showed similar effects from cure time and temperature.

12.3.5 CONTROL PHASE

The two most important issues to remember when permanently adjusting temperature and time settings: (1) coating adhesion and (2) part application. The DOE findings show a reduction in defects on the LF faceplate, but that does not necessarily mean other parts will have the same response.

Quality Control and Reliability

In addition to measuring bubble defects, three ASTM tests were used verify if other quality measures of the power coat process were maintained. These accelerated exposure tests are meant to assess the performance of the powder coat over its service life. Three samples from each run were tested.

Salt Spray Corrosion Test (ASTM B117)

A line was scribed to the bare metal on three power-coated parts. The parts were placed in a salt fog chamber. The chamber used an atomizer to create a salt fog from a 5% salt solution. After one week, no noticeable corrosion was noticed along the line. No difference between parts from different runs was observed.

Impact Test (ASTM D2794)

Power coat parts were impacted with a weight from a height of three inches. Tape was applied to the impact site. There was some cracking of the brittle zinc part but no loss of powder coat adhesion was found when the tape was removed.

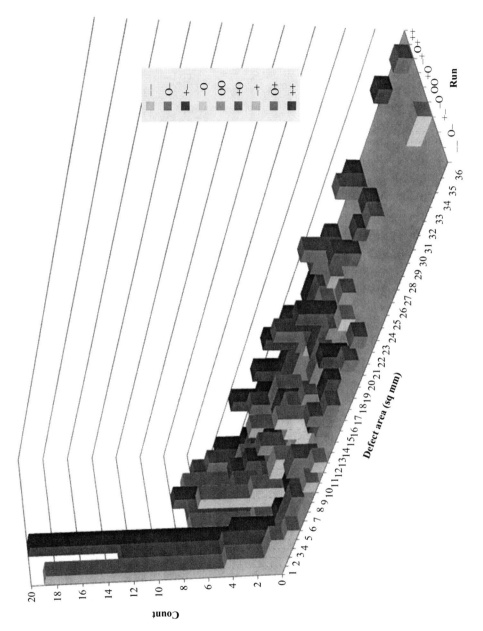

Figure 12.37. Histogram of bubble defect area for DOE.

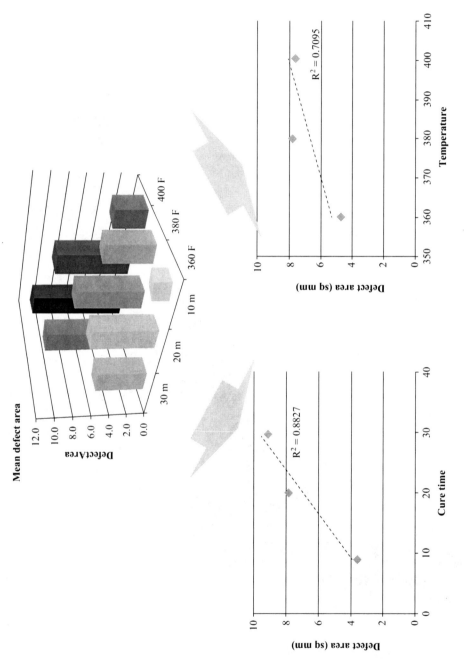

Figure 12.38. DOE Surface area of defect response to temperature and cure time factors.

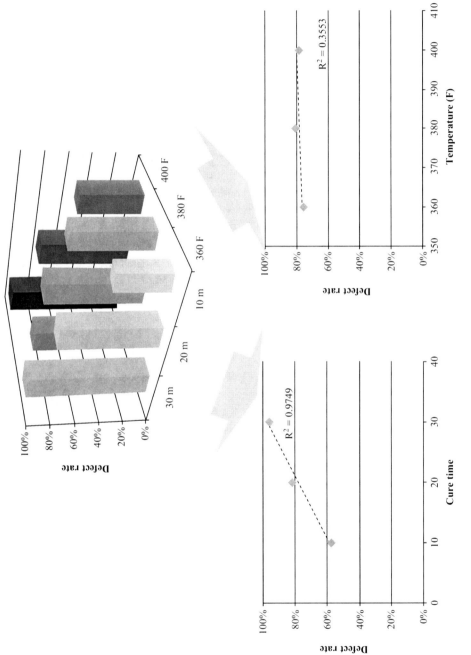

Figure 12.39. DOE defect rate in response to cure time and temperature.

Cross Hatch (ASTM D3359)

Crossing parallel lines were scribed though the coating to the substrate. Sticky tape was applied and removed from the cross hatch area. The amount of powder coat removed between the cross hatched lines is then recorded. No loss of adhesion was observed for the parts tested.

From these ASTM results, it can be expected that powder coat quality and reliability will be maintained with any of the cure time and temperature levels that were tested.

Conclusion

This chapter presented the Define, Measure, Analyze, Improve, and Control phases of a Six Sigma Black Belt project on the LF-II powder coating department. The team established that both curing time and temperature have a significant effect on powder coat bubbles in the MM faceplates. Adjusting the two settings can result in a 43% reduction in powder coat bubbles. If this data were generalized across all zinc components, the effect on WIP scrap savings would result in an annual savings of up to 8% of 20X2 WIP scrap.

Next steps on this project require ensuring the same effects can be applied to all components and all powder coat colors. When the new components pass a visual inspection, the three tests performed in the control phase will be repeated to guarantee the changes in settings do not negatively affect the adhesion or other properties of the powder coat.

APPENDICES

Statistical Tables used for Mastering Lean Six Sigma:

APPENDIX I

HIGHLIGHTS OF SYMBOLS AND ABBREVIATIONS

ANOVA	analysis of variance
b_0	y-intercept in the prediction equation
b_1	the coefficient of linear effect of variable in a regression model
b_{11}	the coefficient of quadratic term for the first variable in a regression model
C_p	process capability (assumes process centered on the target)
C_{pk}	process capability index (does not assume process is centered on the target)
DOE	design of experiment
df	degrees of freedom
f	function (used to describe a function)
$f(x)$	function of x
P_p	process performance
P_{pk}	process performance index
s	sample standard deviation
USL	upper specification limit
LSL	lower specification limit
z	a standardized value
μ	mean
σ	standard deviation of population
$\int$	integral of
$\int_a^b$	integral between the limits a and b
Σx	the sum of all of the values in a set of values
$\sum_j^n x_j$	the sum of values from $j = 1$ to n, which is equivalent to Σx only when n equals total number of values
$\lvert x \rvert$	absolute value of x, if x is negative number the sign is ignored and the outcome will be positive value.
log	common logarithm
ln	natural logarithm
E	exponential notation. Example $1.58E + 2$ is the same as 1.58×10^2

STATISTICAL TABLES AND FORMULAS

Table A. Estimating one population parameter

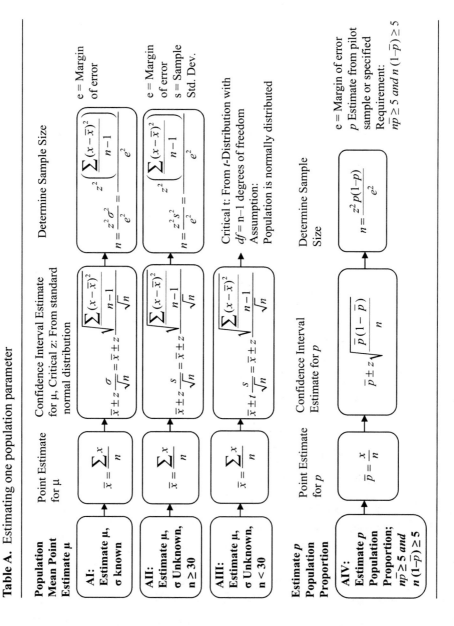

Table B. Estimating two population values

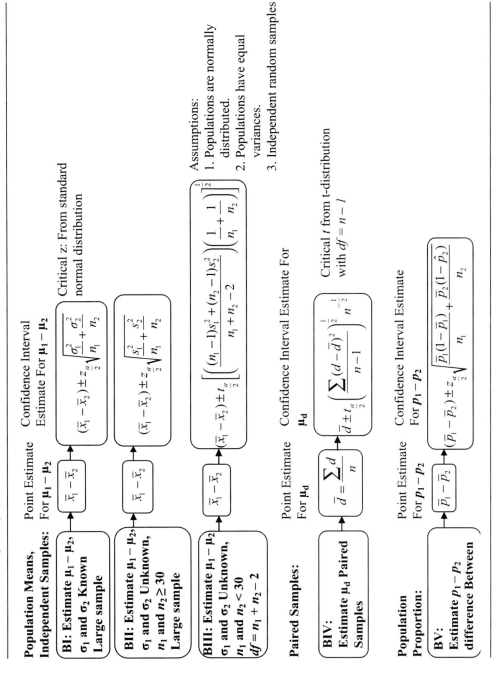

Population Means, Independent Samples:

Point Estimate For $\mu_1 - \mu_2$	Confidence Interval Estimate For $\mu_1 - \mu_2$

BI: Estimate $\mu_1 - \mu_2$, σ_1 and σ_2 Known Large sample

$$\bar{x}_1 - \bar{x}_2$$

$$(\bar{x}_1 - \bar{x}_2) \pm z_{\frac{a}{2}} \sqrt{\frac{\sigma_1^2}{n_1} + \frac{\sigma_2^2}{n_2}}$$

Critical z: From standard normal distribution

BII: Estimate $\mu_1 - \mu_2$, σ_1 and σ_2 Unknown, n_1 and $n_2 \geq 30$ Large sample

$$\bar{x}_1 - \bar{x}_2$$

$$(\bar{x}_1 - \bar{x}_2) \pm z_{\frac{a}{2}} \sqrt{\frac{s_1^2}{n_1} + \frac{s_2^2}{n_2}}$$

BIII: Estimate $\mu_1 - \mu_2$ σ_1 and σ_2 Unknown, n_1 and $n_2 < 30$ $df = n_1 + n_2 - 2$

$$\bar{x}_1 - \bar{x}_2$$

$$(\bar{x}_1 - \bar{x}_2) \pm t_{\frac{a}{2}} \left[\left(\frac{(n_1-1)s_1^2 + (n_2-1)s_2^2}{n_1 + n_2 - 2} \right) \left(\frac{1}{n_1} + \frac{1}{n_2} \right) \right]^{\frac{1}{2}}$$

Critical t from t-distribution with $df = n - 1$

Assumptions:
1. Populations are normally distributed.
2. Populations have equal variances.
3. Independent random samples

Paired Samples:

Point Estimate For μ_d	Confidence Interval Estimate For μ_d

BIV: Estimate μ_d Paired Samples

$$\bar{d} = \frac{\sum d}{n}$$

$$\bar{d} \pm t_{\frac{a}{2}} \left(\frac{\sum (d - \bar{d})^2}{n-1} \right)^{\frac{1}{2}} n^{-\frac{1}{2}}$$

Population Proportion:

Point Estimate For $p_1 - p_2$	Confidence Interval Estimate For $p_1 - p_2$

BV: Estimate $p_1 - p_2$ difference Between

$$\bar{p}_1 - \bar{p}_2$$

$$(\bar{p}_1 - \bar{p}_2) \pm z_{\frac{a}{2}} \sqrt{\frac{\bar{p}_1(1 - \bar{p}_1)}{n_1} + \frac{\bar{p}_2(1 - \hat{p}_2)}{n_2}}$$

Table C. Hypothesis tests for one population value

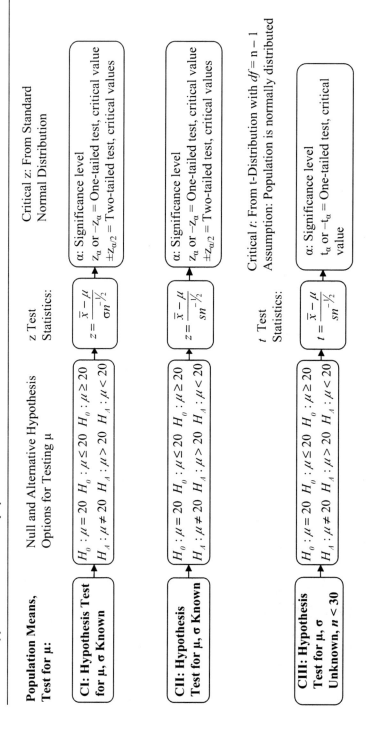

Population Means, Test for μ:

Null and Alternative Hypothesis Options for Testing μ

z Test Statistics:

Critical z: From Standard Normal Distribution

CI: Hypothesis Test for μ, σ Known

$H_0: \mu = 20$ $H_0: \mu \leq 20$ $H_0: \mu \geq 20$
$H_A: \mu \neq 20$ $H_A: \mu > 20$ $H_A: \mu < 20$

$z = \dfrac{\bar{x} - \mu}{\sigma n^{-\frac{1}{2}}}$

α: Significance level
z_α or $-z_\alpha$ = One-tailed test, critical value
$\pm z_{\alpha/2}$ = Two-tailed test, critical values

CII: Hypothesis Test for μ, σ Known

$H_0: \mu = 20$ $H_0: \mu \leq 20$ $H_0: \mu \geq 20$
$H_A: \mu \neq 20$ $H_A: \mu > 20$ $H_A: \mu < 20$

$z = \dfrac{\bar{x} - \mu}{s n^{-\frac{1}{2}}}$

α: Significance level
z_α or $-z_\alpha$ = One-tailed test, critical value
$\pm z_{\alpha/2}$ = Two-tailed test, critical values

t Test Statistics:

Critical t: From t-Distribution with $df = n - 1$
Assumption: Population is normally distributed

CIII: Hypothesis Test for μ, σ Unknown, n < 30

$H_0: \mu = 20$ $H_0: \mu \leq 20$ $H_0: \mu \geq 20$
$H_A: \mu \neq 20$ $H_A: \mu > 20$ $H_A: \mu < 20$

$t = \dfrac{\bar{x} - \mu}{s n^{-\frac{1}{2}}}$

α: Significance level
t_α or $-t_\alpha$ = One-tailed test, critical value

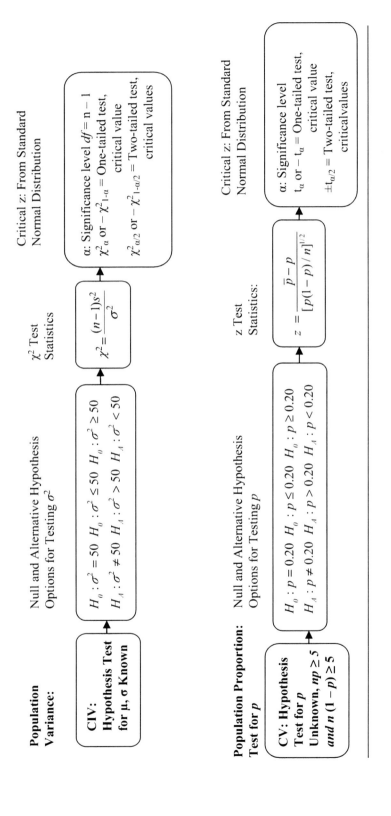

Population Variance:

Critical z: From Standard Normal Distribution

χ^2 Test Statistics

Null and Alternative Hypothesis Options for Testing σ^2

CIV: Hypothesis Test for μ, σ Known

$H_0 : \sigma^2 = 50$ $H_0 : \sigma^2 \leq 50$ $H_0 : \sigma^2 \geq 50$
$H_A : \sigma^2 \neq 50$ $H_A : \sigma^2 > 50$ $H_A : \sigma^2 < 50$

$$\chi^2 = \frac{(n-1)s^2}{\sigma^2}$$

α: Significance level $df = n - 1$
χ^2_α or $-\chi^2_{1-\alpha}$ = One-tailed test, critical value
$\chi^2_{\alpha/2}$ or $-\chi^2_{1-\alpha/2}$ = Two-tailed test, critical values

Population Proportion: Test for p

Critical z: From Standard Normal Distribution

z Test Statistics:

Null and Alternative Hypothesis Options for Testing p

CV: Hypothesis Test for p Unknown, $np \geq 5$ and $n (1 - p) \geq 5$

$H_0 : p = 0.20$ $H_0 : p \leq 0.20$ $H_0 : p \geq 0.20$
$H_A : p \neq 0.20$ $H_A : p > 0.20$ $H_A : p < 0.20$

$$z = \frac{\bar{p} - p}{[p(1-p)/n]^{1/2}}$$

α: Significance level
t_α or $-t_\alpha$ = One-tailed test, critical value
$\pm t_{\alpha/2}$ = Two-tailed test, criticalvalues

Table D. Hypothesis tests for two population values

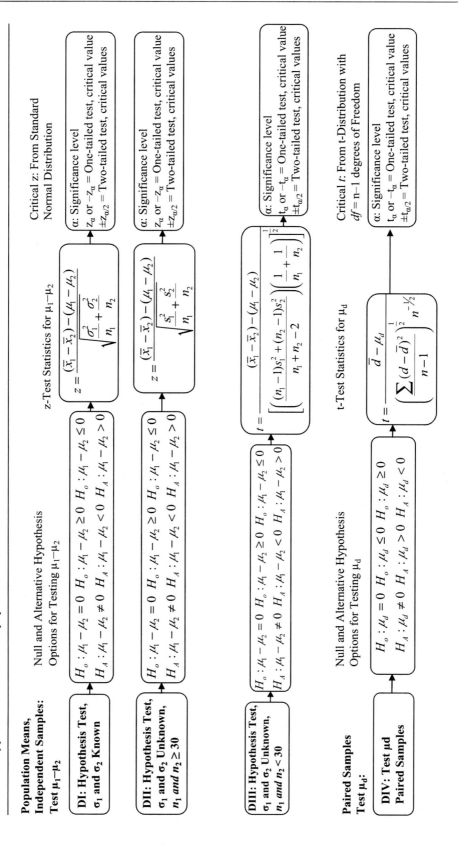

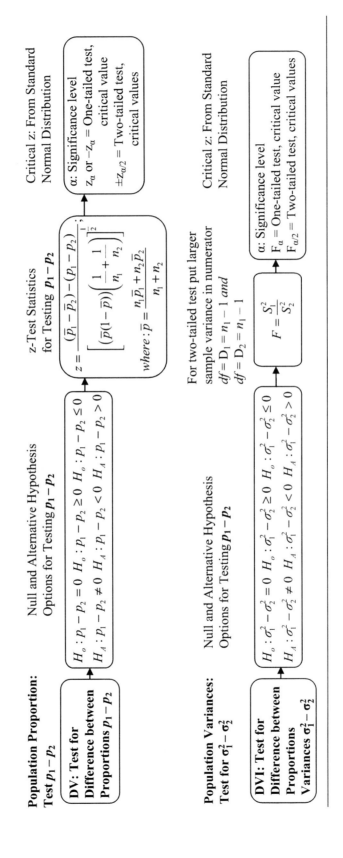

Population Proportion: Test $p_1 - p_2$

DV: Test for Difference between Proportions $p_1 - p_2$

Null and Alternative Hypothesis Options for Testing $p_1 - p_2$

$H_o: p_1 - p_2 = 0$ $H_o: p_1 - p_2 \geq 0$ $H_o: p_1 - p_2 \leq 0$
$H_A: p_1 - p_2 \neq 0$ $H_A: p_1 - p_2 < 0$ $H_A: p_1 - p_2 > 0$

z-Test Statistics for Testing $p_1 - p_2$

$$z = \frac{(\bar{p}_1 - \bar{p}_2) - (p_1 - p_2)}{\left[(\bar{p}(1-\bar{p}))\left(\dfrac{1}{n_1} + \dfrac{1}{n_2}\right) \right]^{\frac{1}{2}}};$$

$where: \bar{p} = \dfrac{n_1\bar{p}_1 + n_2\bar{p}_2}{n_1 + n_2}$

Critical z: From Standard Normal Distribution

α: Significance level
z_α or $-z_\alpha$ = One-tailed test, critical value
$\pm z_{\alpha/2}$ = Two-tailed test, critical values

Population Variances: Test for $\sigma_1^2 - \sigma_2^2$

DVI: Test for Difference between Proportions Variances $\sigma_1^2 - \sigma_2^2$

Null and Alternative Hypothesis Options for Testing $p_1 - p_2$

$H_o: \sigma_1^2 - \sigma_2^2 = 0$ $H_o: \sigma_1^2 - \sigma_2^2 \geq 0$ $H_o: \sigma_1^2 - \sigma_2^2 \leq 0$
$H_A: \sigma_1^2 - \sigma_2^2 \neq 0$ $H_A: \sigma_1^2 - \sigma_2^2 < 0$ $H_A: \sigma_1^2 - \sigma_2^2 > 0$

For two-tailed test put larger sample variance in numerator
$df = D_1 = n_1 - 1$ and
$df = D_2 = n_1 - 1$

$$F = \frac{S_1^2}{S_2^2}$$

Critical z: From Standard Normal Distribution

α: Significance level
F_α = One-tailed test, critical value
$F_{\alpha/2}$ = Two-tailed test, critical values

Table E. Hypothesis tests for 3 or more population means

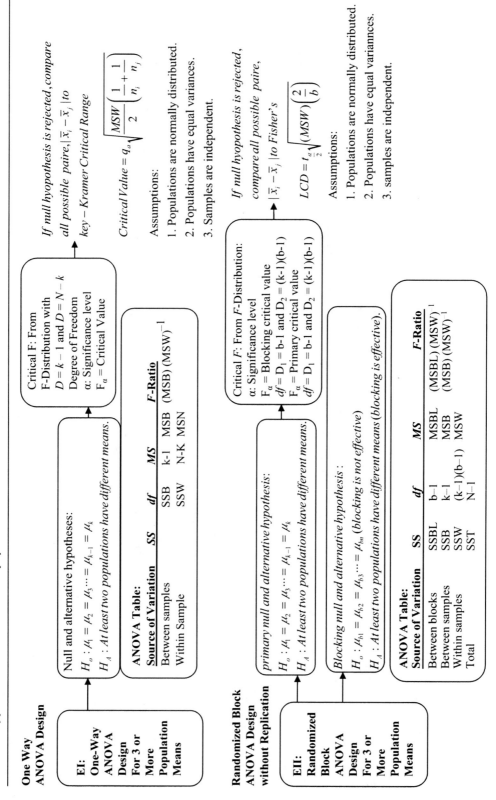

One Way ANOVA Design

EI: One-Way ANOVA Design For 3 or More Population Means

Null and alternative hypotheses:

$H_o : \mu_1 = \mu_2 = \mu_3 \cdots = \mu_{k-1} = \mu_k$

$H_A :$ At least two populations have different means.

ANOVA Table:

Source of Variation	SS	df	MS	F-Ratio
Between samples	SSB	k-1	MSB	$(MSB)(MSW)^{-1}$
Within Sample	SSW	N-K	MSN	

Critical F: From F-Distribution with $D = k - 1$ and $D = N - k$ Degree of Freedom

α: Significance level

F_α = Critical Value

If null hyopothesis is rejected, compare all possible paire, $|\overline{x}_i - \overline{x}_j|$ *to key – Kramer Critical Range*

$Critical\,Value = q_a \sqrt{\dfrac{MSW}{2}\left(\dfrac{1}{n_i}+\dfrac{1}{n_j}\right)}$

Assumptions:

1. Populations are normally distributed.

2. Populations have equal variances.

3. Samples are independent.

Randomized Block ANOVA Design without Replication

EII: Randomized Block ANOVA Design For 3 or More Population Means

primary null and alternative hypothesis:

$H_o : \mu_1 = \mu_2 = \mu_3 \cdots = \mu_{k-1} = \mu_k$

$H_A :$ At least two populations have different means.

Blocking null and alternative hypothesis :

$H_o : \mu_{b1} = \mu_{b2} = \mu_{b33} \cdots = \mu_{bn}$ (blocking is not effective)

$H_A :$ At least two populations have different means (blocking is effective).

ANOVA Table:

Source of Variation	SS	df	MS	F-Ratio
Between blocks	SSBL	b-1	MSBL	$(MSBL)(MSW)^{1}$
Between samples	SSB	k-1	MSB	$(MSB)(MSW)^{1}$
Within samples	SSW	(k-1)(b-1)	MSW	
Total	SST	N-1		

Critical F: From F-Distribution:

α: Significance level

F_α = Blocking critical value

$df = D_1 = $ b-1 and $D_2 = $ (k-1)(b-1)

F_α = Primary critical value

$df = D_1 = $ b-1 and $D_2 = $ (k-1)(b-1)

If null hyopothesis is rejected, compare all possible paire, $|\overline{x}_i - \overline{x}_j|$ *to Fisher's*

$LCD = t_{\frac{a}{2}} \sqrt{(MSW)\left(\dfrac{2}{b}\right)}$

Assumptions:

1. Populations are normally distributed.

2. Populations have equal variannces.

3. samples are independent.

Two-Way ANOVA Design with Replication

EIII: Two-Factor ANOVA Design With Replications

Factor A null and alternative hypothesis:

$H_o: \mu_{A1} = \mu_{A2} = \mu_{A3}\cdots = \mu_{Ak}$

H_A : *No all means are equal.*

Factor B null and alternative hypothesis:

$H_o: \mu_{A1} = \mu_{A2} = \mu_{A3}\cdots = \mu_{Ak}$

H_A : *No all means are equal.*

Null and alternate hypothesis for testing whether the two factors interact.

H_o : *Factor A and B do not interact to affect the mean response.*

H_A : *Factor A and B do not interac.*

α: Significance level

F_α = Factor A critical value; $df = D_1 = a-1$ and $D_2 = N - ab$

F_α = Factor B critical value; $df = D_1 = b-1$ and $D_2 = N - ab$

F_α = Interaction critical value; $df = D_1 = (a-1)(b-1)$ and $D_2 = N - ab$

Assumptions:

1. The Population values for each combination of pair wise factor levels are normally distributed

2. The variances for each population are equal.

3. The observation samples are independent.

ANOVA Table:

Source of Variation	SS	df	MS	F-Ratio
Factor A	SS_A	$a-1$	MS_A	$(MS_A)(MSE)^{-1}$
Factor B	SS_B	$b-1$	MS_B	$(MS_B)(MSE)^{-1}$
AB Interaction	SS_{AB}	$(a-1)(b-1)$	MS_{AB}	$(MS_{AB})(MSE)^{-1}$
Error	SSE	$N-ab$	MSE	
Total	**SST**	**N–1**		

APPENDIX III

VALUES OF Y = EXP($-\eta$)

η	$e^{-\eta}$	η	$e^{-\eta}$	η	$e^{-\eta}$	η	$e^{-\eta}$	η	$e^{-\eta}$	η	$e^{-\eta}$
0.0	1.0000										
0.01	0.9900	0.41	0.6637	0.81	0.4449	1.21	0.2982	1.61	0.1999	2.01	0.1340
0.02	0.9802	0.42	0.6570	0.82	0.4404	1.22	0.2952	1.62	0.1979	2.02	0.1327
0.03	0.9704	0.43	0.6505	0.83	0.4306	1.23	0.2923	1.63	0.1959	2.03	0.1313
0.04	0.9608	0.44	0.6440	0.84	0.4317	1.24	0.2894	1.64	0.1940	2.04	0.1300
0.05	0.9512	0.45	0.6376	0.85	0.4274	1.25	0.2865	1.65	0.1920	2.05	0.1287
0.06	0.9418	0.46	0.6313	0.86	0.4232	1.26	0.2837	1.66	0.1901	2.06	0.1275
0.07	0.9324	0.47	0.6250	0.87	0.4190	1.27	0.2808	1.67	0.1882	2.07	0.1262
0.08	0.9231	0.48	0.6188	0.88	0.4148	1.28	0.2780	1.68	0.1864	2.08	0.1249
0.09	0.9139	0.49	0.6162	0.89	0.4107	1.29	0.2753	1.69	0.1845	2.09	0.1237
0.10	0.9048	0.50	0.6065	0.90	0.4066	1.30	0.2725	1.70	0.1827	2.10	0.1225
0.11	0.8958	0.51	0.6005	0.91	0.4025	1.31	0.2698	1.71	0.1809	2.11	0.1212
0.12	0.8869	0.52	0.5945	0.92	0.3985	1.32	0.2671	1.72	0.1791	2.12	0.1200
0.13	0.8781	0.53	0.5886	0.93	0.3946	1.33	0.2645	1.73	0.1773	2.13	0.1188
0.14	0.8694	0.54	0.5827	0.94	0.3906	1.34	0.2618	1.74	0.1775	2.14	0.1177
0.15	0.8607	0.55	0.5769	0.95	0.3867	1.35	0.2592	1.75	0.1738	2.15	0.1165
0.16	0.8521	0.56	0.5712	0.96	0.3829	1.36	0.2567	1.76	0.1720	2.16	0.1153
0.17	0.8437	0.57	0.5655	0.97	0.3791	1.37	0.2541	1.77	0.1703	2.17	0.1142
0.18	0.8353	0.58	0.5599	0.98	0.3753	1.38	0.2516	1.78	0.1686	2.18	0.1130
0.19	0.8270	0.59	0.5543	0.99	0.3716	1.39	0.2491	1.79	0.1670	2.19	0.1119
0.20	0.8187	0.60	0.5488	1.00	0.3679	1.40	0.2466	1.80	0.1653	2.20	0.1108
0.21	0.8106	0.61	0.5434	1.01	0.3642	1.41	0.2441	1.81	0.1637	2.21	0.1097
0.22	0.8025	0.62	0.5379	1.02	0.3606	1.42	0.2417	1.82	0.1620	2.22	0.1086
0.23	0.7945	0.63	0.5326	1.03	0.3570	1.43	0.2393	1.83	0.1604	2.23	0.1075
0.24	0.7866	0.64	0.5273	1.04	0.3535	1.44	0.2369	1.84	0.1588	2.24	0.1065
0.25	0.7788	0.65	0.5220	1.05	0.3499	1.45	0.2346	1.85	0.1572	2.25	0.1054
0.26	0.7711	0.66	0.5169	1.06	0.3465	1.46	0.2322	1.86	0.1557	2.26	0.1044
0.27	0.7634	0.67	0.5117	1.07	0.3430	1.47	0.2299	1.87	0.1541	2.27	0.1033
0.28	0.7558	0.68	0.5066	1.08	0.3396	1.48	0.2276	1.88	0.1526	2.28	0.1023
0.29	0.7483	0.69	0.5016	1.09	0.3362	1.49	0.2259	1.89	0.1511	2.29	0.1013
0.30	0.7408	0.70	0.4966	1.10	0.3329	1.50	0.2231	1.90	0.1496	2.30	0.1003
0.31	0.7334	0.71	0.4916	1.11	0.3296	1.51	0.2209	1.91	0.1481	2.31	0.0993
0.32	0.7261	0.72	0.4868	1.12	0.3263	1.52	0.2187	1.92	0.1466	2.32	0.0983
0.33	0.7189	0.73	0.4819	1.13	0.3230	1.53	0.2165	1.93	0.1451	2.33	0.0973
0.34	0.7118	0.74	0.4771	1.14	0.3198	1.54	0.2144	1.94	0.1437	2.34	0.0963
0.35	0.7047	0.75	0.4724	1.15	0.3166	1.55	0.2122	1.95	0.1423	2.35	0.0954
0.36	0.6977	0.76	0.4677	1.16	0.3135	1.56	0.2101	1.96	0.1409	2.36	0.0944
0.37	0.6907	0.77	0.4630	1.17	0.3104	1.57	0.2080	1.97	0.1395	2.37	0.0935
0.38	0.6907	0.78	0.4584	1.18	0.3073	1.58	0.2060	1.98	0.1381	2.38	0.0926
0.39	0.6839	0.79	0.4538	1.19	0.3042	1.59	0.2039	1.99	0.1367	2.39	0.0916
0.40	0.6771	0.80	0.4493	1.20	0.3012	1.60	0.2019	2.00	0.1353	2.40	0.0907

DPMO TO *SIGMA* TO YIELD% CONVERSION TABLE

DPMO	*Sigma* Level	1.5 *Sigma*-Shift Yield %	DPMO	*Sigma* Level	1.5 *Sigma*-Shift Yield %
3.40	6.008	99.99966	108.00	5.200	99.98920
5.00	5.914	99.99950	150.00	5.116	99.98500
5.42	5.896	99.99946	159.00	5.101	99.98410
8.00	5.821	99.99920	200.00	5.040	99.98000
8.55	5.802	99.99915	232.00	5.001	99.97680
10.0	5.765	99.99900	233.00	4.999	99.97670
13.35	5.700	99.99867	300.00	4.932	99.97000
15.0	5.672	99.99850	340.00	4.898	99.96600
20.00	5.607	99.99800	400.00	4.853	99.96000
30.00	5.514	99.99700	485.00	4.799	99.95150
31.70	5.500	99.99683	500.00	4.790	99.95000
40.00	5.444	99.99600	600.00	4.739	99.94000
48.10	5.400	99.99519	685.00	4.701	99.93150
50.00	5.391	99.99500	700.00	4.695	99.93000
60.00	5.346	99.99400	800.00	4.656	99.92000
70.00	5.309	99.99300	900.00	4.621	99.91000
72.40	5.300	99.99276	965.00	4.601	99.90350
80.00	5.274	99.99200	1,000.00	4.590	99.90000
90.00	5.246	99.99100	1,350.00	4.500	99.86500
100.00	5.219	99.99000	1,500.00	4.468	99.85000

(Continued)

(*Continued*)

DPMO	*Sigma* Level	1.5 *Sigma*-Shift Yield %	DPMO	*Sigma* Level	1.5 *Sigma*-Shift Yield %
1,870.00	4.399	99.81300	75,000.00	2.940	92.50000
2,000.00	4.378	99.98000	80,800.00	2.900	91.92000
2,550.00	4.301	99.74500	96,800.00	2.800	90.32000
3,000.00	4.248	99.70000	100,000.00	2.782	90.00000
3,470.00	4.200	99.65300	115,100.00	2.700	88.49000
4,000.00	4.152	99.96000	135,700.00	2.600	86.43000
4,660.00	4.100	99.53400	158,700.00	2.500	84.13000
5,000.00	4.076	99.50000	184,100.00	2.400	81.59000
6,210.00	4.000	99.37900	211,900.00	2.300	78.81000
8,200.00	3.900	99.18000	242,100.00	2.200	75.79000
10,000.00	3.826	99.00000	250,000.00	2.174	75.00000
10,700.00	3.801	98.93000	274,400.00	2.100	72.56000
13,900.00	3.700	98.61000	308,700.00	2.000	69.13000
17,900.00	3.599	98.21000	344,700.00	1.900	65.53000
20,000.00	3.554	98.00000	382,000.00	1.800	61.80000
22,750.00	3.500	97.72500	420,900.00	1.700	57.91000
25,000.00	3.460	97.50000	460,300.00	1.600	53.97000
28,700.00	3.400	97.13000	500,000.00	1.500	50.00000
35,900.00	3.300	96.41000	540,000.00	1.400	46.00000
44,600.00	3.200	95.54000	579,200.00	1.300	42.08000
50,000.00	3.145	95.00000	618,000.00	1.200	38.20000
54,800.00	3.100	94.52000	655,600.00	1.100	34.44000
66,800.00	3.000	93.32000	691,500.00	1.000	30.85000

APPENDIX V

STANDARD NORMAL DISTRIBUTION

The Z-table contains the area under the standard normal curve from 0 to z. This can be used to compute the cumulative distribution values for the standard normal distribution.

The entries in Z-table are the probabilities that a standard normal random variable is between 0 and z (the indicated area).

Basically the area given in the z-table is $P(0 \leq Z \leq |a|)$ where a is a constant desired z value (this is shown in the graph below for $a = 1.0$).

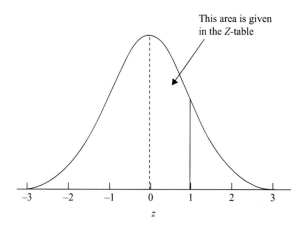

z-**Table.** Critical values for the normal distribution (area under the normal curve from 0 to X)

z	0.00000	0.01000	0.02000	0.03000	0.04000	0.05000	0.06000	0.07000	0.08000	0.09000
0.0	0.00000	0.00399	0.00798	0.01197	0.01595	0.01994	0.02392	0.02790	0.03188	0.03586
0.1	0.03983	0.04380	0.04776	0.05172	0.05567	0.05962	0.06356	0.06749	0.07142	0.07535
0.2	0.07926	0.08317	0.08706	0.09095	0.09483	0.09871	0.10257	0.10642	0.11026	0.11409
0.3	0.11791	0.12172	0.12552	0.12930	0.13307	0.13683	0.14058	0.14431	0.14803	0.15173
0.4	0.15542	0.15910	0.16276	0.16640	0.17003	0.17364	0.17724	0.18082	0.18439	0.18793
0.5	0.19146	0.19497	0.19847	0.20194	0.20540	0.20884	0.21226	0.21566	0.21904	0.22240

(Continued)

542

z-Table. (*Continued*)

z	0.00000	0.01000	0.02000	0.03000	0.04000	0.05000	0.06000	0.07000	0.08000	0.09000
0.6	0.22575	0.22907	0.23237	0.23565	0.23891	0.24215	0.24537	0.24857	0.25175	0.25490
0.7	0.25804	0.26115	0.26424	0.26730	0.27035	0.27337	0.27637	0.27935	0.28230	0.28524
0.8	0.28814	0.29103	0.29389	0.29673	0.29955	0.30234	0.30511	0.30785	0.31057	0.31327
0.9	0.31594	0.31859	0.32121	0.32381	0.32639	0.32894	0.33147	0.33398	0.33646	0.33891
1.0	0.34134	0.34375	0.34614	0.34849	0.35083	0.35314	0.35543	0.35769	0.35993	0.36214
1.1	0.36433	0.36650	0.36864	0.37076	0.37286	0.37493	0.37698	0.37900	0.38100	0.38298
1.2	0.38493	0.38686	0.38877	0.39065	0.39251	0.39435	0.39617	0.39796	0.39973	0.40147
1.3	0.40320	0.40490	0.40658	0.40824	0.40988	0.41149	0.41308	0.41466	0.41621	0.41774
1.4	0.41924	0.42073	0.42220	0.42364	0.42507	0.42647	0.42785	0.42922	0.43056	0.43189
1.5	0.43319	0.43448	0.43574	0.43699	0.43822	0.43943	0.44062	0.44179	0.44295	0.44408
1.6	0.44520	0.44630	0.44738	0.44845	0.44950	0.45053	0.45154	0.45254	0.45352	0.45449
1.7	0.45543	0.45637	0.45728	0.45818	0.45907	0.45994	0.46080	0.46164	0.46246	0.46327
1.8	0.46407	0.46485	0.46562	0.46638	0.46712	0.46784	0.46856	0.46926	0.46995	0.47062
1.9	0.47128	0.47193	0.47257	0.47320	0.47381	0.47441	0.47500	0.47558	0.47615	0.47670
2.0	0.47725	0.47778	0.47831	0.47882	0.47932	0.47982	0.48030	0.48077	0.48124	0.48169
2.1	0.48214	0.48257	0.48300	0.48341	0.48382	0.48422	0.48461	0.48500	0.48537	0.48574
2.2	0.48610	0.48645	0.48679	0.48713	0.48745	0.48778	0.48809	0.48840	0.48870	0.48899
2.3	0.48928	0.48956	0.48983	0.49010	0.49036	0.49061	0.49086	0.49111	0.49134	0.49158
2.4	0.49180	0.49202	0.49224	0.49245	0.49266	0.49286	0.49305	0.49324	0.49343	0.49361
2.5	0.49379	0.49396	0.49413	0.49430	0.49446	0.49461	0.49477	0.49492	0.49506	0.49520
2.6	0.49534	0.49547	0.49560	0.49573	0.49585	0.49598	0.49609	0.49621	0.49632	0.49643
2.7	0.49653	0.49664	0.49674	0.49683	0.49693	0.49702	0.49711	0.49720	0.49728	0.49736
2.8	0.49744	0.49752	0.49760	0.49767	0.49774	0.49781	0.49788	0.49795	0.49801	0.49807
2.9	0.49813	0.49819	0.49825	0.49831	0.49836	0.49841	0.49846	0.49851	0.49856	0.49861
3.0	0.49865	0.49869	0.49874	0.49878	0.49882	0.49886	0.49889	0.49893	0.49896	0.49900
3.1	0.49903	0.49906	0.49910	0.49913	0.49916	0.49918	0.49921	0.49924	0.49926	0.49929
3.2	0.49931	0.49934	0.49936	0.49938	0.49940	0.49942	0.49944	0.49946	0.49948	0.49950
3.3	0.49952	0.49953	0.49955	0.49957	0.49958	0.49960	0.49961	0.49962	0.49964	0.49965
3.4	0.49966	0.49968	0.49969	0.49970	0.49971	0.49972	0.49973	0.49974	0.49975	0.49976
3.5	0.49977	0.49978	0.49978	0.49979	0.49980	0.49981	0.49981	0.49982	0.49983	0.49983
3.6	0.49984	0.49985	0.49985	0.49986	0.49986	0.49987	0.49987	0.49988	0.49988	0.49989
3.7	0.49989	0.49990	0.49990	0.49990	0.49991	0.49991	0.49992	0.49992	0.49992	0.49992
3.8	0.49993	0.49993	0.49993	0.49994	0.49994	0.49994	0.49994	0.49995	0.49995	0.49995
3.9	0.49995	0.49995	0.49996	0.49996	0.49996	0.49996	0.49996	0.49996	0.49997	0.49997
4.0	0.49997	0.49997	0.49997	0.49997	0.49997	0.49997	0.49998	0.49998	0.49998	0.49998

Adapted from the National Institute of Standards and Technology Engineering Statistics Handbook.

CRITICAL VALUES OF T (T-DISTRIBUTION)

This table contains the upper critical values of the student's t-distribution. The upper critical values are computed using the percent point function. Due to the symmetry of the t-distribution, this table can be used for both one-sided (lower and upper) and two-sided tests using the appropriate value of α.

The significance level α is demonstrated with the graph below which plots a t-distribution with 10 degrees of freedom. The most commonly used significance level is $\alpha = 0.05$. For a two-sided test, we compute the percent point function at $\alpha/2$ (0.025). If the absolute value of the test statistic is greater than the upper critical value (0.025), then we reject the null hypothesis. Due to the symmetry of the t-distribution, we only tabulate the upper critical values in the table Appendix VI.

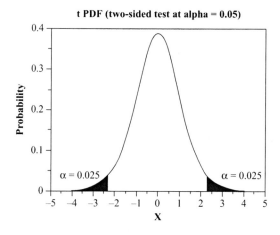

Given a specified value a,

1. For a two-sided test, find the column corresponding to $a/2$ and reject the null hypothesis if the absolute value of the test statistic is greater than the value of $t_{a/2}$ in the table below.
2. For an upper one-sided test, find the column corresponding to α and reject the null hypothesis if the test statistic is greater than the tabled value.
3. For a lower one-sided test, find the column corresponding to α and reject the null hypothesis if the test statistic is less than the negative of tabled value.

Upper critical values of student's *t*-distribution with *df* degrees of freedom probability of exceeding the critical value

df	$t_{0.10}$	$t_{0.05}$	$t_{0.025}$	$t_{0.01}$	$t_{0.005}$	$t_{0.001}$
1.	3.078	6.314	12.706	31.821	63.657	318.313
2.	1.886	2.920	4.303	6.965	9.925	22.327
3.	1.638	2.353	3.182	4.541	5.841	10.215
4.	1.533	2.132	2.776	3.747	4.604	7.173
5.	1.476	2.015	2.571	3.365	4.032	5.893
6.	1.440	1.943	2.447	3.143	3.707	5.208
7.	1.415	1.895	2.365	2.998	3.499	4.782
8.	1.397	1.860	2.306	2.896	3.355	4.499
9.	1.383	1.833	2.262	2.821	3.250	4.296
10.	1.372	1.812	2.228	2.764	3.169	4.143
11.	1.363	1.796	2.201	2.718	3.106	4.024
12.	1.356	1.782	2.179	2.681	3.055	3.929
13.	1.350	1.771	2.160	2.650	3.012	3.852
14.	1.345	1.761	2.145	2.624	2.977	3.787
15.	1.341	1.753	2.131	2.602	2.947	3.733
16.	1.337	1.746	2.120	2.583	2.921	3.686
17.	1.333	1.740	2.110	2.567	2.898	3.646
18.	1.330	1.734	2.101	2.552	2.878	3.610
19.	1.328	1.729	2.093	2.539	2.861	3.579
20.	1.325	1.725	2.086	2.528	2.845	3.552
21.	1.323	1.721	2.080	2.518	2.831	3.527
22.	1.321	1.717	2.074	2.508	2.819	3.505
23.	1.319	1.714	2.069	2.500	2.807	3.485
24.	1.318	1.711	2.064	2.492	2.797	3.467
25.	1.316	1.708	2.060	2.485	2.787	3.450
26.	1.315	1.706	2.056	2.479	2.779	3.435
27.	1.314	1.703	2.052	2.473	2.771	3.421
28.	1.313	1.701	2.048	2.467	2.763	3.408
29.	1.311	1.699	2.045	2.462	2.756	3.396
30.	1.310	1.697	2.042	2.457	2.750	3.385
31.	1.309	1.696	2.040	2.453	2.744	3.375
32.	1.309	1.694	2.037	2.449	2.738	3.365
33.	1.308	1.692	2.035	2.445	2.733	3.356

(Continued)

(*Continued*)

df	$t_{0.10}$	$t_{0.05}$	$t_{0.025}$	$t_{0.01}$	$t_{0.005}$	$t_{0.001}$
34.	1.307	1.691	2.032	2.441	2.728	3.348
35.	1.306	1.690	2.030	2.438	2.724	3.340
36.	1.306	1.688	2.028	2.434	2.719	3.333
37.	1.305	1.687	2.026	2.431	2.715	3.326
38.	1.304	1.686	2.024	2.429	2.712	3.319
39.	1.304	1.685	2.023	2.426	2.708	3.313
40.	1.303	1.684	2.021	2.423	2.704	3.307
41.	1.303	1.683	2.020	2.421	2.701	3.301
42.	1.302	1.682	2.018	2.418	2.698	3.296
43.	1.302	1.681	2.017	2.416	2.695	3.291
44.	1.301	1.680	2.015	2.414	2.692	3.286
45.	1.301	1.679	2.014	2.412	2.690	3.281
46.	1.300	1.679	2.013	2.410	2.687	3.277
47.	1.300	1.678	2.012	2.408	2.685	3.273
48.	1.299	1.677	2.011	2.407	2.682	3.269
49.	1.299	1.677	2.010	2.405	2.680	3.265
50.	1.299	1.676	2.009	2.403	2.678	3.261
51.	1.298	1.675	2.008	2.402	2.676	3.258
52.	1.298	1.675	2.007	2.400	2.674	3.255
53.	1.298	1.674	2.006	2.399	2.672	3.251
54.	1.297	1.674	2.005	2.397	2.670	3.248
55.	1.297	1.673	2.004	2.396	2.668	3.245
56.	1.297	1.673	2.003	2.395	2.667	3.242
57.	1.297	1.672	2.002	2.394	2.665	3.239
58.	1.296	1.672	2.002	2.392	2.663	3.237
59.	1.296	1.671	2.001	2.391	2.662	3.234
60.	1.296	1.671	2.000	2.390	2.660	3.232
61.	1.296	1.670	2.000	2.389	2.659	3.229
62.	1.295	1.670	1.999	2.388	2.657	3.227
63.	1.295	1.669	1.998	2.387	2.656	3.225
64.	1.295	1.669	1.998	2.386	2.655	3.223
65.	1.295	1.669	1.997	2.385	2.654	3.220
66.	1.295	1.668	1.997	2.384	2.652	3.218
67.	1.294	1.668	1.996	2.383	2.651	3.216
68.	1.294	1.668	1.995	2.382	2.650	3.214

(*Continued*)

(*Continued*)

df	$t_{0.10}$	$t_{0.05}$	$t_{0.025}$	$t_{0.01}$	$t_{0.005}$	$t_{0.001}$
69.	1.294	1.667	1.995	2.382	2.649	3.213
70.	1.294	1.667	1.994	2.381	2.648	3.211
71.	1.294	1.667	1.994	2.380	2.647	3.209
72.	1.293	1.666	1.993	2.379	2.646	3.207
73.	1.293	1.666	1.993	2.379	2.645	3.206
74.	1.293	1.666	1.993	2.378	2.644	3.204
75.	1.293	1.665	1.992	2.377	2.643	3.202
76.	1.293	1.665	1.992	2.376	2.642	3.201
77.	1.293	1.665	1.991	2.376	2.641	3.199
78.	1.292	1.665	1.991	2.375	2.640	3.198
79.	1.292	1.664	1.990	2.374	2.640	3.197
80.	1.292	1.664	1.990	2.374	2.639	3.195
81.	1.292	1.664	1.990	2.373	2.638	3.194
82.	1.292	1.664	1.989	2.373	2.637	3.193
83.	1.292	1.663	1.989	2.372	2.636	3.191
84.	1.292	1.663	1.989	2.372	2.636	3.190
85.	1.292	1.663	1.988	2.371	2.635	3.189
86.	1.291	1.663	1.988	2.370	2.634	3.188
87.	1.291	1.663	1.988	2.370	2.634	3.187
88.	1.291	1.662	1.987	2.369	2.633	3.185
89.	1.291	1.662	1.987	2.369	2.632	3.184
90.	1.291	1.662	1.987	2.368	2.632	3.183
91.	1.291	1.662	1.986	2.368	2.631	3.182
92.	1.291	1.662	1.986	2.368	2.630	3.181
93.	1.291	1.661	1.986	2.367	2.630	3.180
94.	1.291	1.661	1.986	2.367	2.629	3.179
95.	1.291	1.661	1.985	2.366	2.629	3.178
96.	1.290	1.661	1.985	2.366	2.628	3.177
97.	1.290	1.661	1.985	2.365	2.627	3.176
98.	1.290	1.661	1.984	2.365	2.627	3.175
99.	1.290	1.660	1.984	2.365	2.626	3.175
100.	1.290	1.660	1.984	2.364	2.626	3.174
∞	1.282	1.645	1.960	2.326	2.576	3.090

Adapted from the *National Institute of Standards and Technology Engineering Statistics Handbook.*

CRITICAL VALUES OF CHI-SQUARE DISTRIBUTION WITH DEGREES OF FREEDOM

This table contains the critical values of the chi-square distribution. Because of the lack of symmetry of the chi-square distribution, separate tables are provided for the upper and lower tails of the distribution.

A test statistic with degrees of freedom is computed from the data. For upper one-sided tests, the test statistic is compared with a value from the table of upper critical values. For two-sided tests, the test statistic is compared with values from both the table for the upper critical value and the table for the lower critical value.

The significance level, α, is demonstrated with the graph below, which shows a chi-square distribution with 3 degrees of freedom for a two-sided test at significance level $\alpha = 0.05$. If the test statistic is greater than the upper critical value or less than the lower critical value, we reject the null hypothesis. Specific instructions are given below.

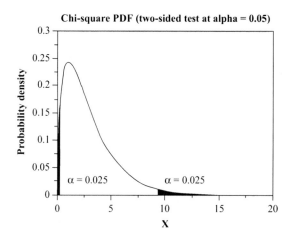

Given a specified value for α:

1. For a two-sided test, find the column corresponding to $\alpha/2$ in the table for upper critical values and reject the null hypothesis if the test statistic is greater than the tabled value. Similarly, find the column corresponding to $1 - \alpha/2$ in the table for lower critical values and reject the null hypothesis if the test statistic is less than the tabled value.
2. For an upper one-sided test, find the column corresponding to α in the upper critical values table and reject the null hypothesis if the test statistic is greater than the tabled value.
3. For a lower one-sided test, find the column corresponding to $1 - \alpha$ in the lower critical values table and reject the null hypothesis if the computed test statistic is less than the tabled value.

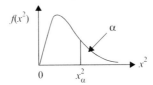

Upper critical values of chi-square distribution with *df* degrees of freedom chi–square $\chi^2_{\alpha,df}$

df \ α	0.9950	0.9900	0.9750	0.9500	0.9000	0.5000	0.1000	0.0500	0.0250	0.0100	0.0050
1	0.0000	0.0001	0.0009	0.0039	0.0157	0.4549	2.7055	3.8414	5.0238	6.6349	7.8794
2	0.0100	0.0201	0.0506	0.1025	0.2107	1.3862	4.6051	5.9914	7.3777	9.2103	10.596
3	0.0717	0.1148	0.2157	0.3518	0.5843	2.3659	6.2513	7.8147	9.3484	11.344	12.838
4	0.2069	0.2971	0.4844	0.7107	1.0636	3.3566	7.7794	9.4877	11.143	13.276	14.860
5	0.4117	0.5542	0.8312	1.1454	1.6103	4.3514	9.2363	11.070	12.832	15.086	16.749
6	0.6757	0.8720	1.2373	1.6353	2.2041	5.3481	10.646	12.591	14.449	16.811	18.547
7	0.9892	1.2390	1.6898	2.1673	2.8331	6.3458	12.017	14.067	16.012	18.475	20.277
8	1.3444	1.6464	2.1797	2.7326	3.4895	7.3441	13.361	15.507	17.534	20.090	21.955
9	1.7349	2.0879	2.7003	3.3251	4.1681	8.3428	14.683	16.919	19.022	21.666	23.589
10	2.1558	2.5582	3.2469	3.9403	4.8651	9.3418	15.987	18.307	20.483	23.209	25.188
11	2.6032	3.0534	3.8157	4.5748	5.5777	10.341	17.275	19.675	21.920	24.725	26.756
12	3.0738	3.5705	4.4037	5.2260	6.3038	11.340	18.549	21.026	23.336	26.217	28.299
13	3.5650	4.1069	5.0087	5.8918	7.0415	12.340	19.811	22.362	24.735	27.688	29.819
14	4.0746	4.6604	5.6287	6.5706	7.7895	13.339	21.064	23.684	26.119	29.141	31.319
15	4.6009	5.2293	6.2621	7.2609	8.5467	14.338	22.307	24.995	27.488	30.577	32.801
16	5.1422	5.8122	6.9076	7.9616	9.3122	15.338	23.541	26.296	28.845	31.999	34.267
17	5.6972	6.4077	7.5641	8.6717	10.085	16.338	24.769	27.587	30.191	33.408	35.718
18	6.2648	7.0149	8.2307	9.3904	10.864	17.338	25.989	28.869	31.526	34.805	37.156

(*Continued*)

(*Continued*)

df \ α	0.9950	0.9900	0.9750	0.9500	0.9000	0.5000	0.1000	0.0500	0.0250	0.0100	0.0050
19	6.8439	7.6327	8.9065	10.117	11.650	18.337	27.203	30.143	32.852	36.190	38.582
20	7.4338	8.2604	9.5908	10.850	12.442	19.337	28.412	31.410	34.169	37.556	39.996
21	8.0336	8.8972	10.282	11.591	13.239	20.337	29.615	32.670	35.478	38.932	41.401
22	8.6427	9.5424	10.982	12.338	14.041	21.337	30.813	33.924	36.780	40.289	42.795
23	9.2604	10.195	11.688	13.090	14.847	22.337	32.006	35.172	38.075	41.638	44.181
24	9.8862	10.856	12.401	13.848	15.653	23.337	33.196	36.415	39.364	42.979	45.558
25	10.519	11.524	13.119	0.9500	16.473	24.336	34.381	37.625	40.646	44.314	46.927
26	11.160	12.198	13.843	15.379	17.291	25.336	35.563	38.885	41.923	45.641	48.289
27	11.807	12.878	14.573	16.151	18.113	26.336	36.741	40.113	43.194	46.963	49.644
28	12.461	13.564	15.307	16.927	18.939	27.336	37.915	41.337	44.460	48.278	50.993
29	13.121	14.256	16.047	17.708	19.767	28.336	39.087	42.556	45.722	49.587	52.335
30	13.786	14.953	16.790	18.492	20.599	29.336	40.256	43.772	46.979	50.892	53.672
31	14.458	15.655	17.539	19.281	21.434	30.336	41.422	44.985	48.232	52.191	55.003
32	15.134	16.362	18.291	20.072	22.271	31.336	42.585	46.194	49.480	53.486	56.328
33	15.815	17.074	19.047	20.867	23.110	32.336	43.745	47.400	50.725	54.776	57.648
34	16.501	17.789	19.806	21.664	23.952	33.335	44.903	48.602	51.966	56.061	58.964
35	17.192	18.509	20.569	22.465	24.797	34.335	46.059	49.802	53.203	57.342	60.275
36	17.887	19.336	23.269	23.269	25.643	35.335	47.212	50.998	54.437	58.619	61.581
37	18.586	19.960	24.075	24.075	26.492	36.335	48.363	52.192	55.668	59.893	62.883
38	19.289	20.691	24.884	24.884	27.343	37.335	49.513	53.384	56.896	61.162	64.181
39	19.996	21.426	25.695	25.695	28.196	38.335	50.660	58.120	58.120	62.428	65.475
40	20.706	22.164	24.433	26.509	29.050	39.335	51.805	55.758	59.341	63.691	66.766
41	21.421	22.906	25.215	27.326	29.907	40.335	52.949	56.942	60.561	64.950	68.053
42	22.138	23.650	25.999	28.144	30.765	41.335	54.090	58.124	61.777	66.206	69.336
43	22.859	24.398	26.785	28.965	31.625	42.335	55.230	59.304	62.990	67.459	70.616
44	23.587	25.148	27.575	29.787	32.487	43.335	56.369	60.481	64.201	68.710	71.892
45	24.311	25.901	28.366	30.612	33.350	44.335	57.505	61.656	65.410	69.957	73.166
46	25.041	26.657	29.160	31.439	34.215	45.335	58.641	62.830	66.617	71.201	74.436
47	25.775	27.416	29.956	32.268	35.081	46.335	59.774	64.001	67.821	72.443	75.704
48	26.510	28.177	30.755	33.098	35.949	47.335	60.907	65.171	69.023	73.683	76.969
49	27.249	28.914	31.555	33.930	36.818	48.334	62.038	66.339	70.222	74.919	78.231
50	27.991	2.5582	32.357	34.764	37.689	49.334	63.167	67.505	71.420	76.154	79.490
51	28.735	30.475	33.162	35.600	38.560	50.335	64.295	68.669	72.616	77.386	80.747
52	29.481	31.246	33.968	36.437	39.433	51.335	65.422	69.832	73.810	78.616	82.001
53	30.230	32.018	34.776	37.276	40.308	52.335	66.548	70.993	75.002	79.843	83.252

(*Continued*)

(*Continued*)

df \ α	0.9950	0.9900	0.9750	0.9500	0.9000	0.5000	0.1000	0.0500	0.0250	0.0100	0.0050
54	30.981	32.793	35.586	38.116	41.183	53.335	67.673	72.153	76.192	81.069	84.502
55	31.735	33.570	36.398	38.958	42.060	54.335	68.796	73.311	77.380	82.292	85.749
56	32.490	34.350	37.212	39.801	42.937	55.335	69.919	74.468	78.567	83.513	86.994
57	33.248	35.027	38.027	40.646	43.816	56.335	71.040	75.624	79.756	84.733	88.236
58	34.008	38.844	38.844	41.492	44.696	57.335	72.160	76.778	80.936	85.950	89.477
59	34.770	36.698	39.662	42.339	45.577	58.335	73.279	77.931	82.117	87.166	90.715
60	35.534	37.485	40.482	43.188	46.459	59.335	74.397	79.082	83.298	88.379	91.952
61	36.300	38.273	41.303	44.038	47.342	60.335	75.514	80.232	84.476	89.591	93.186
62	37.068	39.063	42.126	44.889	48.226	61.335	76.630	81.381	85.654	90.802	94.419
63	37.838	39.855	42.950	45.741	49.111	62.335	77.745	82.529	86.830	92.010	95.649
64	38.610	40.649	43.776	46.595	49.996	63.334	78.860	83.675	88.004	93.217	96.878
65	39.383	41.444	44.603	47.450	50.883	64.334	79.973	84.821	89.177	94.422	98.105
66	40.158	42.240	45.431	48.305	51.770	65.334	81.085	85.965	90.349	95.626	99.330
67	40.935	43.038	46.261	49.162	52.659	66.334	82.197	87.108	91.519	96.828	100.55
68	41.713	43.838	47.092	50.020	53.548	67.334	83.308	88.250	92.689	98.028	101.78
69	42.493	44.639	47.924	50.879	54.438	68.334	84.418	89.391	93.856	99.228	102.99
70	43.275	45.442	48.758	51.739	55.329	69.334	85.527	90.531	95.023	100.42	104.21
71	44.058	46.246	49.592	52.600	56.221	70.334	86.635	91.670	96.189	101.62	105.43
72	44.843	47.051	50.428	53.462	57.113	71.334	87.743	92.808	97.353	102.81	106.65
73	45.629	47.858	51.265	54.325	58.006	72.334	88.850	93.945	98.516	104.01	107.86
74	46.417	48.666	52.103	55.189	58.900	73.334	89.956	95.081	99.678	105.20	109.07
75	47.206	49.475	52.942	56.054	59.795	74.334	91.061	96.217	100.84	106.39	110.28
76	47.996	50.286	53.782	56.920	60.690	75.334	92.166	97.351	101.99	107.58	111.49
77	48.788	51.097	54.623	57.786	61.586	76.334	93.270	98.484	103.16	108.77	112.70
78	49.581	51.910	55.466	58.654	62.483	77.334	94.374	99.617	104.31	109.96	113.91
79	50.376	52.725	56.309	59.522	63.380	78.334	95.476	100.75	105.47	111.14	115.12
80	51.172	53.540	57.153	60.391	64.278	79.334	96.578	101.88	106.63	112.33	116.32
81	51.969	54.357	57.998	61.261	65.176	80.334	97.680	103.01	107.78	113.51	117.52
82	52.767	55.174	58.845	62.132	66.076	81.334	98.780	104.14	108.94	114.69	118.73
83	53.567	55.993	59.692	63.004	66.976	82.334	99.880	105.27	110.09	115.87	119.93
84	54.367	56.813	60.540	63.876	67.876	83.334	100.98	106.40	111.24	117.06	121.13
85	55.170	57.634	61.389	64.749	68.777	84.334	102.08	107.52	112.39	118.23	122.32
86	55.973	58.456	62.239	65.623	69.679	85.334	103.18	108.65	113.54	119.41	123.52
87	56.777	59.279	63.089	66.498	70.581	86.334	104.28	109.77	114.69	120.59	124.72
88	57.582	60.103	63.941	67.373	71.484	87.334	105.37	110.90	115.84	121.76	125.91

(*Continued*)

(*Continued*)

df \ α	0.9950	0.9900	0.9750	0.9500	0.9000	0.5000	0.1000	0.0500	0.0250	0.0100	0.0050
89	58.389	60.928	64.793	68.249	72.387	88.334	106.47	112.02	116.99	122.94	127.11
90	59.196	61.754	65.647	69.126	73.291	89.334	107.57	113.14	118.14	124.11	128.30
91	60.005	62.581	66.501	70.003	74.196	90.334	108.66	114.27	119.28	125.29	129.49
92	60.814	63.409	67.356	70.882	75.100	91.334	109.76	115.39	120.43	126.46	130.68
93	61.625	64.238	68.211	71.760	76.006	92.334	110.85	116.51	121.57	127.63	131.87
94	62.437	65.068	69.068	72.640	76.912	93.334	111.94	117.63	122.71	128.80	133.06
95	63.250	65.898	69.925	73.520	77.818	94.334	113.04	118.75	123.86	129.97	134.25
96	64.063	66.730	70.783	74.401	78.725	95.334	114.13	119.87	125.00	131.14	135.43
97	64.878	67.562	71.642	75.282	79.633	96.334	115.22	120.99	126.14	132.31	136.62
98	65.693	68.396	72.501	76.164	80.541	97.334	116.31	122.11	127.28	133.48	137.80
99	66.510	69.230	73.361	77.046	81.449	98.334	117.41	123.22	128.42	134.64	138.98
100	67.327	70.065	74.222	77.929	82.358	99.334	118.50	124.34	129.56	135.81	140.17

UPPER CRITICAL VALUES OF THE F-DISTRIBUTION

Upper Critical Values of the F-Distribution for df_1 Numerator Degrees of Freedom and df_2 Denominator Degrees of Freedom.

HOW TO USE F-TABLE

The F-Table contains the upper critical values of the F-distribution. This table is used for one-sided F tests at α = 0.05, 0.10, and 0.01 levels.

More specifically, a test statistic is computed with df_1 (df error) and df_2 (df factor) degrees of freedom, and the result is compared to Table Appendix VII. For a one-sided test, the null hypothesis is rejected when the test statistic is greater than the tabled value. This is demonstrated with the graph of an F distribution with df_1 = 10 and df_2 = 10. The shaded area of the graph indicates the rejection region at the "α" significance level. Since this is a one-sided test, we have α probability in the upper tail of exceeding the critical value and zero in the lower tail. Because the F-distribution is asymmetric, a two-sided test requires a set of tables (not included here) that contain the rejection regions for both the lower and upper tails.

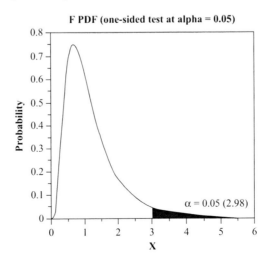

553

Upper critical values of the F distribution for df_1 numerator degrees of freedom and df_2 denominator degrees of freedom 10% significance level $F_{0.10} (df_1, df_2)$

df_2 \ df_1	1	2	3	4	5	6	7	8	9	10
1	39.863	49.500	53.593	55.833	57.240	58.204	58.906	59.439	59.858	60.195
2	8.526	9.000	9.162	9.243	9.293	9.326	9.349	9.367	9.381	9.392
3	5.538	5.462	5.391	5.343	5.309	5.285	5.266	5.252	5.240	5.230
4	4.545	4.325	4.191	4.107	4.051	4.010	3.979	3.955	3.936	3.920
5	4.060	3.780	3.619	3.520	3.453	3.405	3.368	3.339	3.316	3.297
6	3.776	3.463	3.289	3.181	3.108	3.055	3.014	2.983	2.958	2.937
7	3.589	3.257	3.074	2.961	2.883	2.827	2.785	2.752	2.725	2.703
8	3.458	3.113	2.924	2.806	2.726	2.668	2.624	2.589	2.561	2.538
9	3.360	3.006	2.813	2.693	2.611	2.551	2.505	2.469	2.440	2.416
10	3.285	2.924	2.728	2.605	2.522	2.461	2.414	2.377	2.347	2.323
11	3.225	2.860	2.660	2.536	2.451	2.389	2.342	2.304	2.274	2.248
12	3.177	2.807	2.606	2.480	2.394	2.331	2.283	2.245	2.214	2.188
13	3.136	2.763	2.560	2.434	2.347	2.283	2.234	2.195	2.164	2.138
14	3.102	2.726	2.522	2.395	2.307	2.243	2.193	2.154	2.122	2.095
15	3.073	2.695	2.490	2.361	2.273	2.208	2.158	2.119	2.086	2.059
16	3.048	2.668	2.462	2.333	2.244	2.178	2.128	2.088	2.055	2.028
17	3.026	2.645	2.437	2.308	2.218	2.152	2.102	2.061	2.028	2.001
18	3.007	2.624	2.416	2.286	2.196	2.130	2.079	2.038	2.005	1.977
19	2.990	2.606	2.397	2.266	2.176	2.109	2.058	2.017	1.984	1.956
20	2.975	2.589	2.380	2.249	2.158	2.091	2.040	1.999	1.965	1.937
21	2.961	2.575	2.365	2.233	2.142	2.075	2.023	1.982	1.948	1.920
22	2.949	2.561	2.351	2.219	2.128	2.060	2.008	1.967	1.933	1.904
23	2.937	2.549	2.339	2.207	2.115	2.047	1.995	1.953	1.919	1.890
24	2.927	2.538	2.327	2.195	2.103	2.035	1.983	1.941	1.906	1.877
25	2.918	2.528	2.317	2.184	2.092	2.024	1.971	1.929	1.895	1.866
26	2.909	2.519	2.307	2.174	2.082	2.014	1.961	1.919	1.884	1.855
27	2.901	2.511	2.299	2.165	2.073	2.005	1.952	1.909	1.874	1.845
28	2.894	2.503	2.291	2.157	2.064	1.996	1.943	1.900	1.865	1.836
29	2.887	2.495	2.283	2.149	2.057	1.988	1.935	1.892	1.857	1.827
30	2.881	2.489	2.276	2.142	2.049	1.980	1.927	1.884	1.849	1.819
31	2.875	2.482	2.270	2.136	2.042	1.973	1.920	1.877	1.842	1.812
32	2.869	2.477	2.263	2.129	2.036	1.967	1.913	1.870	1.835	1.805
33	2.864	2.471	2.258	2.123	2.030	1.961	1.907	1.864	1.828	1.799

11	12	13	14	15	16	17	18	19	20
60.473	60.705	60.903	61.073	61.220	61.350	61.464	61.566	61.658	61.740
9.401	9.408	9.415	9.420	9.425	9.429	9.433	9.436	9.439	9.441
5.222	5.216	5.210	5.205	5.200	5.196	5.193	5.190	5.187	5.184
3.907	3.896	3.886	3.878	3.870	3.864	3.858	3.853	3.849	3.844
3.282	3.268	3.257	3.247	3.238	3.230	3.223	3.217	3.212	3.207
2.920	2.905	2.892	2.881	2.871	2.863	2.855	2.848	2.842	2.836
2.684	2.668	2.654	2.643	2.632	2.623	2.615	2.607	2.601	2.595
2.519	2.502	2.488	2.475	2.464	2.455	2.446	2.438	2.431	2.425
2.396	2.379	2.364	2.351	2.340	2.329	2.320	2.312	2.305	2.298
2.302	2.284	2.269	2.255	2.244	2.233	2.224	2.215	2.208	2.201
2.227	2.209	2.193	2.179	2.167	2.156	2.147	2.138	2.130	2.123
2.166	2.147	2.131	2.117	2.105	2.094	2.084	2.075	2.067	2.060
2.116	2.097	2.080	2.066	2.053	2.042	2.032	2.023	2.014	2.007
2.073	2.054	2.037	2.022	2.010	1.998	1.988	1.978	1.970	1.962
2.037	2.017	2.000	1.985	1.972	1.961	1.950	1.941	1.932	1.924
2.005	1.985	1.968	1.953	1.940	1.928	1.917	1.908	1.899	1.891
1.978	1.958	1.940	1.925	1.912	1.900	1.889	1.879	1.870	1.862
1.954	1.933	1.916	1.900	1.887	1.875	1.864	1.854	1.845	1.837
1.932	1.912	1.894	1.878	1.865	1.852	1.841	1.831	1.822	1.814
1.913	1.892	1.875	1.859	1.845	1.833	1.821	1.811	1.802	1.794
1.896	1.875	1.857	1.841	1.827	1.815	1.803	1.793	1.784	1.776
1.880	1.859	1.841	1.825	1.811	1.798	1.787	1.777	1.768	1.759
1.866	1.845	1.827	1.811	1.796	1.784	1.772	1.762	1.753	1.744
1.853	1.832	1.814	1.797	1.783	1.770	1.759	1.748	1.739	1.730
1.841	1.820	1.802	1.785	1.771	1.758	1.746	1.736	1.726	1.718
1.830	1.809	1.790	1.774	1.760	1.747	1.735	1.724	1.715	1.706
1.820	1.799	1.780	1.764	1.749	1.736	1.724	1.714	1.704	1.695
1.811	1.790	1.771	1.754	1.740	1.726	1.715	1.704	1.694	1.685
1.802	1.781	1.762	1.745	1.731	1.717	1.705	1.695	1.685	1.676
1.794	1.773	1.754	1.737	1.722	1.709	1.697	1.686	1.676	1.667
1.787	1.765	1.746	1.729	1.714	1.701	1.689	1.678	1.668	1.659
1.780	1.758	1.739	1.722	1.707	1.694	1.682	1.671	1.661	1.652
1.773	1.751	1.732	1.715	1.700	1.687	1.675	1.664	1.654	1.645

(Continued)

(*Continued*)

df_2 df_1	1	2	3	4	5	6	7	8	9	10
34	2.859	2.466	2.252	2.118	2.024	1.955	1.901	1.858	1.822	1.793
35	2.855	2.461	2.247	2.113	2.019	1.950	1.896	1.852	1.817	1.787
36	2.850	2.456	2.243	2.108	2.014	1.945	1.891	1.847	1.811	1.781
37	2.846	2.452	2.238	2.103	2.009	1.940	1.886	1.842	1.806	1.776
38	2.842	2.448	2.234	2.099	2.005	1.935	1.881	1.838	1.802	1.772
39	2.839	2.444	2.230	2.095	2.001	1.931	1.877	1.833	1.797	1.767
40	2.835	2.440	2.226	2.091	1.997	1.927	1.873	1.829	1.793	1.763
41	2.832	2.437	2.222	2.087	1.993	1.923	1.869	1.825	1.789	1.759
42	2.829	2.434	2.219	2.084	1.989	1.919	1.865	1.821	1.785	1.755
43	2.826	2.430	2.216	2.080	1.986	1.916	1.861	1.817	1.781	1.751
44	2.823	2.427	2.213	2.077	1.983	1.913	1.858	1.814	1.778	1.747
45	2.820	2.425	2.210	2.074	1.980	1.909	1.855	1.811	1.774	1.744
46	2.818	2.422	2.207	2.071	1.977	1.906	1.852	1.808	1.771	1.741
47	2.815	2.419	2.204	2.068	1.974	1.903	1.849	1.805	1.768	1.738
48	2.813	2.417	2.202	2.066	1.971	1.901	1.846	1.802	1.765	1.735
49	2.811	2.414	2.199	2.063	1.968	1.898	1.843	1.799	1.763	1.732
50	2.809	2.412	2.197	2.061	1.966	1.895	1.840	1.796	1.760	1.729
60	2.791	2.393	2.177	2.041	1.946	1.875	1.819	1.775	1.738	1.707
70	2.779	2.380	2.164	2.027	1.931	1.860	1.804	1.760	1.723	1.691
80	2.769	2.370	2.154	2.016	1.921	1.849	1.793	1.748	1.711	1.680
90	2.762	2.363	2.146	2.008	1.912	1.841	1.785	1.739	1.702	1.670
100	2.756	2.356	2.139	2.002	1.906	1.834	1.778	1.732	1.695	1.663

11	12	13	14	15	16	17	18	19	20
1.767	1.745	1.726	1.709	1.694	1.680	1.668	1.657	1.647	1.638
1.761	1.739	1.720	1.703	1.688	1.674	1.662	1.651	1.641	1.632
1.756	1.734	1.715	1.697	1.682	1.669	1.656	1.645	1.635	1.626
1.751	1.729	1.709	1.692	1.677	1.663	1.651	1.640	1.630	1.620
1.746	1.724	1.704	1.687	1.672	1.658	1.646	1.635	1.624	1.615
1.741	1.719	1.700	1.682	1.667	1.653	1.641	1.630	1.619	1.610
1.737	1.715	1.695	1.678	1.662	1.649	1.636	1.625	1.615	1.605
1.733	1.710	1.691	1.673	1.658	1.644	1.632	1.620	1.610	1.601
1.729	1.706	1.687	1.669	1.654	1.640	1.628	1.616	1.606	1.596
1.725	1.703	1.683	1.665	1.650	1.636	1.624	1.612	1.602	1.592
1.721	1.699	1.679	1.662	1.646	1.632	1.620	1.608	1.598	1.588
1.718	1.695	1.676	1.658	1.643	1.629	1.616	1.605	1.594	1.585
1.715	1.692	1.672	1.655	1.639	1.625	1.613	1.601	1.591	1.581
1.712	1.689	1.669	1.652	1.636	1.622	1.609	1.598	1.587	1.578
1.709	1.686	1.666	1.648	1.633	1.619	1.606	1.594	1.584	1.574
1.706	1.683	1.663	1.645	1.630	1.616	1.603	1.591	1.581	1.571
1.703	1.680	1.660	1.643	1.627	1.613	1.600	1.588	1.578	1.568
1.680	1.657	1.637	1.619	1.603	1.589	1.576	1.564	1.553	1.543
1.665	1.641	1.621	1.603	1.587	1.572	1.559	1.547	1.536	1.526
1.653	1.629	1.609	1.590	1.574	1.559	1.546	1.534	1.523	1.513
1.643	1.620	1.599	1.581	1.564	1.550	1.536	1.524	1.513	1.503
1.636	1.612	1.592	1.573	1.557	1.542	1.528	1.516	1.505	1.494

Upper critical values of the F distribution for df_1 numerator degrees of freedom and df_2 denominator degrees of freedom 5% significance level $F_{0.05}$ (df_1, df_2)

df_2 \ df_1	1	2	3	4	5	6	7	8	9	10
1	161.448	199.500	215.707	224.583	230.162	233.986	236.768	238.882	240.543	241.882
2	18.513	19.000	19.164	19.247	19.296	19.330	19.353	19.371	19.385	19.396
3	10.128	9.552	9.277	9.117	9.013	8.941	8.887	8.845	8.812	8.786
4	7.709	6.944	6.591	6.388	6.256	6.163	6.094	6.041	5.999	5.964
5	6.608	5.786	5.409	5.192	5.050	4.950	4.876	4.818	4.772	4.735
6	5.987	5.143	4.757	4.534	4.387	4.284	4.207	4.147	4.099	4.060
7	5.591	4.737	4.347	4.120	3.972	3.866	3.787	3.726	3.677	3.637
8	5.318	4.459	4.066	3.838	3.687	3.581	3.500	3.438	3.388	3.347
9	5.117	4.256	3.863	3.633	3.482	3.374	3.293	3.230	3.179	3.137
10	4.965	4.103	3.708	3.478	3.326	3.217	3.135	3.072	3.020	2.978
11	4.844	3.982	3.587	3.357	3.204	3.095	3.012	2.948	2.896	2.854
12	4.747	3.885	3.490	3.259	3.106	2.996	2.913	2.849	2.796	2.753
13	4.667	3.806	3.411	3.179	3.025	2.915	2.832	2.767	2.714	2.671
14	4.600	3.739	3.344	3.112	2.958	2.848	2.764	2.699	2.646	2.602
15	4.543	3.682	3.287	3.056	2.901	2.790	2.707	2.641	2.588	2.544
16	4.494	3.634	3.239	3.007	2.852	2.741	2.657	2.591	2.538	2.494
17	4.451	3.592	3.197	2.965	2.810	2.699	2.614	2.548	2.494	2.450
18	4.414	3.555	3.160	2.928	2.773	2.661	2.577	2.510	2.456	2.412
19	4.381	3.522	3.127	2.895	2.740	2.628	2.544	2.477	2.423	2.378
20	4.351	3.493	3.098	2.866	2.711	2.599	2.514	2.447	2.393	2.348
21	4.325	3.467	3.072	2.840	2.685	2.573	2.488	2.420	2.366	2.321
22	4.301	3.443	3.049	2.817	2.661	2.549	2.464	2.397	2.342	2.297
23	4.279	3.422	3.028	2.796	2.640	2.528	2.442	2.375	2.320	2.275
24	4.260	3.403	3.009	2.776	2.621	2.508	2.423	2.355	2.300	2.255
25	4.242	3.385	2.991	2.759	2.603	2.490	2.405	2.337	2.282	2.236
26	4.225	3.369	2.975	2.743	2.587	2.474	2.388	2.321	2.265	2.220
27	4.210	3.354	2.960	2.728	2.572	2.459	2.373	2.305	2.250	2.204
28	4.196	3.340	2.947	2.714	2.558	2.445	2.359	2.291	2.236	2.190
29	4.183	3.328	2.934	2.701	2.545	2.432	2.346	2.278	2.223	2.177
30	4.171	3.316	2.922	2.690	2.534	2.421	2.334	2.266	2.211	2.165
31	4.160	3.305	2.911	2.679	2.523	2.409	2.323	2.255	2.199	2.153
32	4.149	3.295	2.901	2.668	2.512	2.399	2.313	2.244	2.189	2.142
33	4.139	3.285	2.892	2.659	2.503	2.389	2.303	2.235	2.179	2.133
34	4.130	3.276	2.883	2.650	2.494	2.380	2.294	2.225	2.170	2.123

11	12	13	14	15	16	17	18	19	20
242.983	243.906	244.690	245.364	245.950	246.464	246.918	247.323	247.686	248.013
19.405	19.413	19.419	19.424	19.429	19.433	19.437	19.440	19.443	19.446
8.763	8.745	8.729	8.715	8.703	8.692	8.683	8.675	8.667	8.660
5.936	5.912	5.891	5.873	5.858	5.844	5.832	5.821	5.811	5.803
4.704	4.678	4.655	4.636	4.619	4.604	4.590	4.579	4.568	4.558
4.027	4.000	3.976	3.956	3.938	3.922	3.908	3.896	3.884	3.874
3.603	3.575	3.550	3.529	3.511	3.494	3.480	3.467	3.455	3.445
3.313	3.284	3.259	3.237	3.218	3.202	3.187	3.173	3.161	3.150
3.102	3.073	3.048	3.025	3.006	2.989	2.974	2.960	2.948	2.936
2.943	2.913	2.887	2.865	2.845	2.828	2.812	2.798	2.785	2.774
2.818	2.788	2.761	2.739	2.719	2.701	2.685	2.671	2.658	2.646
2.717	2.687	2.660	2.637	2.617	2.599	2.583	2.568	2.555	2.544
2.635	2.604	2.577	2.554	2.533	2.515	2.499	2.484	2.471	2.459
2.565	2.534	2.507	2.484	2.463	2.445	2.428	2.413	2.400	2.388
2.507	2.475	2.448	2.424	2.403	2.385	2.368	2.353	2.340	2.328
2.456	2.425	2.397	2.373	2.352	2.333	2.317	2.302	2.288	2.276
2.413	2.381	2.353	2.329	2.308	2.289	2.272	2.257	2.243	2.230
2.374	2.342	2.314	2.290	2.269	2.250	2.233	2.217	2.203	2.191
2.340	2.308	2.280	2.256	2.234	2.215	2.198	2.182	2.168	2.155
2.310	2.278	2.250	2.225	2.203	2.184	2.167	2.151	2.137	2.124
2.283	2.250	2.222	2.197	2.176	2.156	2.139	2.123	2.109	2.096
2.259	2.226	2.198	2.173	2.151	2.131	2.114	2.098	2.084	2.071
2.236	2.204	2.175	2.150	2.128	2.109	2.091	2.075	2.061	2.048
2.216	2.183	2.155	2.130	2.108	2.088	2.070	2.054	2.040	2.027
2.198	2.165	2.136	2.111	2.089	2.069	2.051	2.035	2.021	2.007
2.181	2.148	2.119	2.094	2.072	2.052	2.034	2.018	2.003	1.990
2.166	2.132	2.103	2.078	2.056	2.036	2.018	2.002	1.987	1.974
2.151	2.118	2.089	2.064	2.041	2.021	2.003	1.987	1.972	1.959
2.138	2.104	2.075	2.050	2.027	2.007	1.989	1.973	1.958	1.945
2.126	2.092	2.063	2.037	2.015	1.995	1.976	1.960	1.945	1.932
2.114	2.080	2.051	2.026	2.003	1.983	1.965	1.948	1.933	1.920
2.103	2.070	2.040	2.015	1.992	1.972	1.953	1.937	1.922	1.908
2.093	2.060	2.030	2.004	1.982	1.961	1.943	1.926	1.911	1.898
2.084	2.050	2.021	1.995	1.972	1.952	1.933	1.917	1.902	1.888

(*Continued*)

(Continued)

df_2 \ df_1	1	2	3	4	5	6	7	8	9	10
35	4.121	3.267	2.874	2.641	2.485	2.372	2.285	2.217	2.161	2.114
36	4.113	3.259	2.866	2.634	2.477	2.364	2.277	2.209	2.153	2.106
37	4.105	3.252	2.859	2.626	2.470	2.356	2.270	2.201	2.145	2.098
38	4.098	3.245	2.852	2.619	2.463	2.349	2.262	2.194	2.138	2.091
39	4.091	3.238	2.845	2.612	2.456	2.342	2.255	2.187	2.131	2.084
40	4.085	3.232	2.839	2.606	2.449	2.336	2.249	2.180	2.124	2.077
41	4.079	3.226	2.833	2.600	2.443	2.330	2.243	2.174	2.118	2.071
42	4.073	3.220	2.827	2.594	2.438	2.324	2.237	2.168	2.112	2.065
43	4.067	3.214	2.822	2.589	2.432	2.318	2.232	2.163	2.106	2.059
44	4.062	3.209	2.816	2.584	2.427	2.313	2.226	2.157	2.101	2.054
45	4.057	3.204	2.812	2.579	2.422	2.308	2.221	2.152	2.096	2.049
46	4.052	3.200	2.807	2.574	2.417	2.304	2.216	2.147	2.091	2.044
47	4.047	3.195	2.802	2.570	2.413	2.299	2.212	2.143	2.086	2.039
48	4.043	3.191	2.798	2.565	2.409	2.295	2.207	2.138	2.082	2.035
49	4.038	3.187	2.794	2.561	2.404	2.290	2.203	2.134	2.077	2.030
50	4.034	3.183	2.790	2.557	2.400	2.286	2.199	2.130	2.073	2.026
60	4.001	3.150	2.758	2.525	2.368	2.254	2.167	2.097	2.040	1.993
70	3.978	3.128	2.736	2.503	2.346	2.231	2.143	2.074	2.017	1.969
80	3.960	3.111	2.719	2.486	2.329	2.214	2.126	2.056	1.999	1.951
90	3.947	3.098	2.706	2.473	2.316	2.201	2.113	2.043	1.986	1.938
100	3.936	3.087	2.696	2.463	2.305	2.191	2.103	2.032	1.975	1.927

11	12	13	14	15	16	17	18	19	20
2.075	2.041	2.012	1.986	1.963	1.942	1.924	1.907	1.892	1.878
2.067	2.033	2.003	1.977	1.954	1.934	1.915	1.899	1.883	1.870
2.059	2.025	1.995	1.969	1.946	1.926	1.907	1.890	1.875	1.861
2.051	2.017	1.988	1.962	1.939	1.918	1.899	1.883	1.867	1.853
2.044	2.010	1.981	1.954	1.931	1.911	1.892	1.875	1.860	1.846
2.038	2.003	1.974	1.948	1.924	1.904	1.885	1.868	1.853	1.839
2.031	1.997	1.967	1.941	1.918	1.897	1.879	1.862	1.846	1.832
2.025	1.991	1.961	1.935	1.912	1.891	1.872	1.855	1.840	1.826
2.020	1.985	1.955	1.929	1.906	1.885	1.866	1.849	1.834	1.820
2.014	1.980	1.950	1.924	1.900	1.879	1.861	1.844	1.828	1.814
2.009	1.974	1.945	1.918	1.895	1.874	1.855	1.838	1.823	1.808
2.004	1.969	1.940	1.913	1.890	1.869	1.850	1.833	1.817	1.803
1.999	1.965	1.935	1.908	1.885	1.864	1.845	1.828	1.812	1.798
1.995	1.960	1.930	1.904	1.880	1.859	1.840	1.823	1.807	1.793
1.990	1.956	1.926	1.899	1.876	1.855	1.836	1.819	1.803	1.789
1.986	1.952	1.921	1.895	1.871	1.850	1.831	1.814	1.798	1.784
1.952	1.917	1.887	1.860	1.836	1.815	1.796	1.778	1.763	1.748
1.928	1.893	1.863	1.836	1.812	1.790	1.771	1.753	1.737	1.722
1.910	1.875	1.845	1.817	1.793	1.772	1.752	1.734	1.718	1.703
1.897	1.861	1.830	1.803	1.779	1.757	1.737	1.720	1.703	1.688
1.886	1.850	1.819	1.792	1.768	1.746	1.726	1.708	1.691	1.676

Upper critical values of the F distribution for df_1 numerator degrees of freedom and df_2 denominator degrees of freedom 1% significance level $F_{0.01}$ (df_1, df_2)

df_2 \ df_1	1	2	3	4	5	6	7	8	9	10
1	4052.19	4999.52	5403.34	5624.62	5763.65	5858.97	5928.33	5981.10	6022.50	6055.85
2	98.502	99.000	99.166	99.249	99.300	99.333	99.356	99.374	99.388	99.399
3	34.116	30.816	29.457	28.710	28.237	27.911	27.672	27.489	27.345	27.229
4	21.198	18.000	16.694	15.977	15.522	15.207	14.976	14.799	14.659	14.546
5	16.258	13.274	12.060	11.392	10.967	10.672	10.456	10.289	10.158	10.051
6	13.745	10.925	9.780	9.148	8.746	8.466	8.260	8.102	7.976	7.874
7	12.246	9.547	8.451	7.847	7.460	7.191	6.993	6.840	6.719	6.620
8	11.259	8.649	7.591	7.006	6.632	6.371	6.178	6.029	5.911	5.814
9	10.561	8.022	6.992	6.422	6.057	5.802	5.613	5.467	5.351	5.257
10	10.044	7.559	6.552	5.994	5.636	5.386	5.200	5.057	4.942	4.849
11	9.646	7.206	6.217	5.668	5.316	5.069	4.886	4.744	4.632	4.539
12	9.330	6.927	5.953	5.412	5.064	4.821	4.640	4.499	4.388	4.296
13	9.074	6.701	5.739	5.205	4.862	4.620	4.441	4.302	4.191	4.100
14	8.862	6.515	5.564	5.035	4.695	4.456	4.278	4.140	4.030	3.939
15	8.683	6.359	5.417	4.893	4.556	4.318	4.142	4.004	3.895	3.805
16	8.531	6.226	5.292	4.773	4.437	4.202	4.026	3.890	3.780	3.691
17	8.400	6.112	5.185	4.669	4.336	4.102	3.927	3.791	3.682	3.593
18	8.285	6.013	5.092	4.579	4.248	4.015	3.841	3.705	3.597	3.508
19	8.185	5.926	5.010	4.500	4.171	3.939	3.765	3.631	3.523	3.434
20	8.096	5.849	4.938	4.431	4.103	3.871	3.699	3.564	3.457	3.368
21	8.017	5.780	4.874	4.369	4.042	3.812	3.640	3.506	3.398	3.310
22	7.945	5.719	4.817	4.313	3.988	3.758	3.587	3.453	3.346	3.258
23	7.881	5.664	4.765	4.264	3.939	3.710	3.539	3.406	3.299	3.211
24	7.823	5.614	4.718	4.218	3.895	3.667	3.496	3.363	3.256	3.168
25	7.770	5.568	4.675	4.177	3.855	3.627	3.457	3.324	3.217	3.129
26	7.721	5.526	4.637	4.140	3.818	3.591	3.421	3.288	3.182	3.094
27	7.677	5.488	4.601	4.106	3.785	3.558	3.388	3.256	3.149	3.062
28	7.636	5.453	4.568	4.074	3.754	3.528	3.358	3.226	3.120	3.032
29	7.598	5.420	4.538	4.045	3.725	3.499	3.330	3.198	3.092	3.005
30	7.562	5.390	4.510	4.018	3.699	3.473	3.305	3.173	3.067	2.979
31	7.530	5.362	4.484	3.993	3.675	3.449	3.281	3.149	3.043	2.955
32	7.499	5.336	4.459	3.969	3.652	3.427	3.258	3.127	3.021	2.934
33	7.471	5.312	4.437	3.948	3.630	3.406	3.238	3.106	3.000	2.913
34	7.444	5.289	4.416	3.927	3.611	3.386	3.218	3.087	2.981	2.894

11	12	13	14	15	16	17	18	19	20
6083.35	6106.35	6125.86	6142.70	6157.28	6170.12	6181.42	6191.52	6200.58	6208.74
99.408	99.416	99.422	99.428	99.432	99.437	99.440	99.444	99.447	99.449
27.133	27.052	26.983	26.924	26.872	26.827	26.787	26.751	26.719	26.690
14.452	14.374	14.307	14.249	14.198	14.154	14.115	14.080	14.048	14.020
9.963	9.888	9.825	9.770	9.722	9.680	9.643	9.610	9.580	9.553
7.790	7.718	7.657	7.605	7.559	7.519	7.483	7.451	7.422	7.396
6.538	6.469	6.410	6.359	6.314	6.275	6.240	6.209	6.181	6.155
5.734	5.667	5.609	5.559	5.515	5.477	5.442	5.412	5.384	5.359
5.178	5.111	5.055	5.005	4.962	4.924	4.890	4.860	4.833	4.808
4.772	4.706	4.650	4.601	4.558	4.520	4.487	4.457	4.430	4.405
4.462	4.397	4.342	4.293	4.251	4.213	4.180	4.150	4.123	4.099
4.220	4.155	4.100	4.052	4.010	3.972	3.939	3.909	3.883	3.858
4.025	3.960	3.905	3.857	3.815	3.778	3.745	3.716	3.689	3.665
3.864	3.800	3.745	3.698	3.656	3.619	3.586	3.556	3.529	3.505
3.730	3.666	3.612	3.564	3.522	3.485	3.452	3.423	3.396	3.372
3.616	3.553	3.498	3.451	3.409	3.372	3.339	3.310	3.283	3.259
3.519	3.455	3.401	3.353	3.312	3.275	3.242	3.212	3.186	3.162
3.434	3.371	3.316	3.269	3.227	3.190	3.158	3.128	3.101	3.077
3.360	3.297	3.242	3.195	3.153	3.116	3.084	3.054	3.027	3.003
3.294	3.231	3.177	3.130	3.088	3.051	3.018	2.989	2.962	2.938
3.236	3.173	3.119	3.072	3.030	2.993	2.960	2.931	2.904	2.880
3.184	3.121	3.067	3.019	2.978	2.941	2.908	2.879	2.852	2.827
3.137	3.074	3.020	2.973	2.931	2.894	2.861	2.832	2.805	2.781
3.094	3.032	2.977	2.930	2.889	2.852	2.819	2.789	2.762	2.738
3.056	2.993	2.939	2.892	2.850	2.813	2.780	2.751	2.724	2.699
3.021	2.958	2.904	2.857	2.815	2.778	2.745	2.715	2.688	2.664
2.988	2.926	2.871	2.824	2.783	2.746	2.713	2.683	2.656	2.632
2.959	2.896	2.842	2.795	2.753	2.716	2.683	2.653	2.626	2.602
2.931	2.868	2.814	2.767	2.726	2.689	2.656	2.626	2.599	2.574
2.906	2.843	2.789	2.742	2.700	2.663	2.630	2.600	2.573	2.549
2.882	2.820	2.765	2.718	2.677	2.640	2.606	2.577	2.550	2.525
2.860	2.798	2.744	2.696	2.655	2.618	2.584	2.555	2.527	2.503
2.840	2.777	2.723	2.676	2.634	2.597	2.564	2.534	2.507	2.482
2.821	2.758	2.704	2.657	2.615	2.578	2.545	2.515	2.488	2.463

(Continued)

(Continued)

df_2 \ df_1	1	2	3	4	5	6	7	8	9	10
35	7.419	5.268	4.396	3.908	3.592	3.368	3.200	3.069	2.963	2.876
36	7.396	5.248	4.377	3.890	3.574	3.351	3.183	3.052	2.946	2.859
37	7.373	5.229	4.360	3.873	3.558	3.334	3.167	3.036	2.930	2.843
38	7.353	5.211	4.343	3.858	3.542	3.319	3.152	3.021	2.915	2.828
39	7.333	5.194	4.327	3.843	3.528	3.305	3.137	3.006	2.901	2.814
40	7.314	5.179	4.313	3.828	3.514	3.291	3.124	2.993	2.888	2.801
41	7.296	5.163	4.299	3.815	3.501	3.278	3.111	2.980	2.875	2.788
42	7.280	5.149	4.285	3.802	3.488	3.266	3.099	2.968	2.863	2.776
43	7.264	5.136	4.273	3.790	3.476	3.254	3.087	2.957	2.851	2.764
44	7.248	5.123	4.261	3.778	3.465	3.243	3.076	2.946	2.840	2.754
45	7.234	5.110	4.249	3.767	3.454	3.232	3.066	2.935	2.830	2.743
46	7.220	5.099	4.238	3.757	3.444	3.222	3.056	2.925	2.820	2.733
47	7.207	5.087	4.228	3.747	3.434	3.213	3.046	2.916	2.811	2.724
48	7.194	5.077	4.218	3.737	3.425	3.204	3.037	2.907	2.802	2.715
49	7.182	5.066	4.208	3.728	3.416	3.195	3.028	2.898	2.793	2.706
50	7.171	5.057	4.199	3.720	3.408	3.186	3.020	2.890	2.785	2.698
60	7.077	4.977	4.126	3.649	3.339	3.119	2.953	2.823	2.718	2.632
70	7.011	4.922	4.074	3.600	3.291	3.071	2.906	2.777	2.672	2.585
80	6.963	4.881	4.036	3.563	3.255	3.036	2.871	2.742	2.637	2.551
90	6.925	4.849	4.007	3.535	3.228	3.009	2.845	2.715	2.611	2.524
100	6.895	4.824	3.984	3.513	3.206	2.988	2.823	2.694	2.590	2.503

Adapted from the *National Institute of Standards and Technology Engineering Statistics Handbook.*

11	12	13	14	15	16	17	18	19	20
2.803	2.740	2.686	2.639	2.597	2.560	2.527	2.497	2.470	2.445
2.786	2.723	2.669	2.622	2.580	2.543	2.510	2.480	2.453	2.428
2.770	2.707	2.653	2.606	2.564	2.527	2.494	2.464	2.437	2.412
2.755	2.692	2.638	2.591	2.549	2.512	2.479	2.449	2.421	2.397
2.741	2.678	2.624	2.577	2.535	2.498	2.465	2.434	2.407	2.382
2.727	2.665	2.611	2.563	2.522	2.484	2.451	2.421	2.394	2.369
2.715	2.652	2.598	2.551	2.509	2.472	2.438	2.408	2.381	2.356
2.703	2.640	2.586	2.539	2.497	2.460	2.426	2.396	2.369	2.344
2.691	2.629	2.575	2.527	2.485	2.448	2.415	2.385	2.357	2.332
2.680	2.618	2.564	2.516	2.475	2.437	2.404	2.374	2.346	2.321
2.670	2.608	2.553	2.506	2.464	2.427	2.393	2.363	2.336	2.311
2.660	2.598	2.544	2.496	2.454	2.417	2.384	2.353	2.326	2.301
2.651	2.588	2.534	2.487	2.445	2.408	2.374	2.344	2.316	2.291
2.642	2.579	2.525	2.478	2.436	2.399	2.365	2.335	2.307	2.282
2.633	2.571	2.517	2.469	2.427	2.390	2.356	2.326	2.299	2.274
2.625	2.562	2.508	2.461	2.419	2.382	2.348	2.318	2.290	2.265
2.559	2.496	2.442	2.394	2.352	2.315	2.281	2.251	2.223	2.198
2.512	2.450	2.395	2.348	2.306	2.268	2.234	2.204	2.176	2.150
2.478	2.415	2.361	2.313	2.271	2.233	2.199	2.169	2.141	2.115
2.451	2.389	2.334	2.286	2.244	2.206	2.172	2.142	2.114	2.088
2.430	2.368	2.313	2.265	2.223	2.185	2.151	2.120	2.092	2.067

APPENDIX IX

CUMULATIVE POISSON PROBABILITY DISTRIBUTION TABLE

						μ				
x	0.01	0.01	0.02	0.03	0.04	0.05	0.06	0.07	0.08	0.09
0	0.9950	0.9900	0.9802	0.9704	0.9608	0.9512	0.9418	0.9324	0.9231	0.9139
1	1.0000	1.0000	0.9998	0.9996	0.9992	0.9988	0.9983	0.9977	0.9970	0.9962
2	1.0000	1.0000	1.0000	1.0000	1.0000	1.0000	1.0000	0.9999	0.9999	0.9999
3	1.0000	1.0000	1.0000	1.0000	1.0000	1.0000	1.0000	1.0000	1.0000	1.0000

						μ				
x	0.10	0.20	0.30	0.40	0.50	0.60	0.70	0.80	0.90	1.00
0	0.9048	0.8187	0.7408	0.6703	0.6065	0.5488	0.4966	0.4493	0.4066	0.3679
1	0.9953	0.9825	0.9631	0.9384	0.9098	0.8781	0.8442	0.8088	0.7725	0.7358
2	0.9998	0.9989	0.9964	0.9921	0.9856	0.9769	0.9659	0.9526	0.9371	0.9197
3	1.0000	0.9999	0.9997	0.9992	0.9982	0.9966	0.9942	0.9909	0.9865	0.9810
4	1.0000	1.0000	1.0000	0.9999	0.9998	0.9996	0.9992	0.9986	0.9977	0.9963
5	1.0000	1.0000	1.0000	1.0000	1.0000	1.0000	0.9999	0.9998	0.9997	0.9994
6	1.0000	1.0000	1.0000	1.0000	1.0000	1.0000	1.0000	1.0000	1.0000	0.9999
7	1.0000	1.0000	1.0000	1.0000	1.0000	1.0000	1.0000	1.0000	1.0000	1.0000

						μ				
x	1.10	1.20	1.30	1.40	1.50	1.60	1.70	1.80	1.90	2.00
0	0.3329	0.3012	0.2725	0.2466	0.2231	0.2019	0.1827	0.1653	0.1496	0.1353
1	0.6990	0.6626	0.6268	0.5918	0.5578	0.5249	0.4932	0.4628	0.4337	0.4060
2	0.9004	0.8795	0.8571	0.8335	0.8088	0.7834	0.7572	0.7306	0.7037	0.6767

(Continued)

(*Continued*)

					μ					
x	**1.10**	**1.20**	**1.30**	**1.40**	**1.50**	**1.60**	**1.70**	**1.80**	**1.90**	**2.00**
3	0.9743	0.9662	0.9569	0.9463	0.9344	0.9212	0.9068	0.8913	0.8747	0.8571
4	0.9946	0.9923	0.9893	0.9857	0.9814	0.9763	0.9704	0.9636	0.9559	0.9473
5	0.9990	0.9985	0.9978	0.9968	0.9955	0.9940	0.9920	0.9896	0.9868	0.9834
6	0.9999	0.9997	0.9996	0.9994	0.9991	0.9987	0.9981	0.9974	0.9966	0.9955
7	1.0000	1.0000	0.9999	0.9999	0.9998	0.9997	0.9996	0.9994	0.9992	0.9989
8	1.0000	1.0000	1.0000	1.0000	1.0000	1.0000	0.9999	0.9999	0.9998	0.9998
9	1.0000	1.0000	1.0000	1.0000	1.0000	1.0000	1.0000	1.0000	1.0000	1.0000

					μ					
x	**2.10**	**2.20**	**2.30**	**2.40**	**2.50**	**2.60**	**2.70**	**2.80**	**2.90**	**3.00**
0	0.1225	0.1108	0.1003	0.0907	0.0821	0.0743	0.0672	0.0608	0.0550	0.0498
1	0.3796	0.3546	0.3309	0.3084	0.2873	0.2674	0.2487	0.2311	0.2146	0.1991
2	0.6496	0.6227	0.5960	0.5697	0.5438	0.5184	0.4936	0.4695	0.4460	0.4232
3	0.8386	0.8194	0.7993	0.7787	0.7576	0.7360	0.7141	0.6919	0.6696	0.6472
4	0.9379	0.9275	0.9162	0.9041	0.8912	0.8774	0.8629	0.8477	0.8318	0.8153
5	0.9796	0.9751	0.9700	0.9643	0.9580	0.9510	0.9433	0.9349	0.9258	0.9161
6	0.9941	0.9925	0.9906	0.9884	0.9858	0.9828	0.9794	0.9756	0.9713	0.9665
7	0.9985	0.9980	0.9974	0.9967	0.9958	0.9947	0.9934	0.9919	0.9901	0.9881
8	0.9997	0.9995	0.9994	0.9991	0.9989	0.9985	0.9981	0.9976	0.9969	0.9962
9	0.9999	0.9999	0.9999	0.9998	0.9997	0.9996	0.9995	0.9993	0.9991	0.9989
10	1.0000	1.0000	1.0000	1.0000	0.9999	0.9999	0.9999	0.9998	0.9998	0.9997
11	1.0000	1.0000	1.0000	1.0000	1.0000	1.0000	1.0000	1.0000	0.9999	0.9999
12	1.0000	1.0000	1.0000	1.0000	1.0000	1.0000	1.0000	1.0000	1.0000	1.0000

					μ					
x	**3.10**	**3.20**	**3.30**	**3.40**	**3.50**	**3.60**	**3.70**	**3.80**	**3.90**	**4.00**
0	0.0450	0.0408	0.0369	0.0334	0.0302	0.0273	0.0247	0.0224	0.0202	0.0183
1	0.1847	0.1712	0.1586	0.1468	0.1359	0.1257	0.1162	0.1074	0.0992	0.0916
2	0.4012	0.3799	0.3594	0.3397	0.3208	0.3027	0.2854	0.2689	0.2531	0.2381
3	0.6248	0.6025	0.5803	0.5584	0.5366	0.5152	0.4942	0.4735	0.4532	0.4335
4	0.7982	0.7806	0.7626	0.7442	0.7254	0.7064	0.6872	0.6678	0.6484	0.6288
5	0.9057	0.8946	0.8829	0.8705	0.8576	0.8441	0.8301	0.8156	0.8006	0.7851
6	0.9612	0.9554	0.9490	0.9421	0.9347	0.9267	0.9182	0.9091	0.8995	0.8893
7	0.9858	0.9832	0.9802	0.9769	0.9733	0.9692	0.9648	0.9599	0.9546	0.9489
8	0.9953	0.9943	0.9931	0.9917	0.9901	0.9883	0.9863	0.9840	0.9815	0.9786

(*Continued*)

(*Continued*)

					μ					
x	3.10	3.20	3.30	3.40	3.50	3.60	3.70	3.80	3.90	4.00
9	0.9986	0.9982	0.9978	0.9973	0.9967	0.9960	0.9952	0.9942	0.9931	0.9919
10	0.9996	0.9995	0.9994	0.9992	0.9990	0.9987	0.9984	0.9981	0.9977	0.9972
11	0.9999	0.9999	0.9998	0.9998	0.9997	0.9996	0.9995	0.9994	0.9993	0.9991
12	1.0000	1.0000	1.0000	0.9999	0.9999	0.9999	0.9999	0.9998	0.9998	0.9997
13	1.0000	1.0000	1.0000	1.0000	1.0000	1.0000	1.0000	1.0000	0.9999	0.9999
14	1.0000	1.0000	1.0000	1.0000	1.0000	1.0000	1.0000	1.0000	1.0000	1.0000

					μ					
x	4.10	4.20	4.30	4.40	4.50	4.60	4.70	4.80	4.90	5.00
0	0.0166	0.0150	0.0136	0.0123	0.0111	0.0101	0.0091	0.0082	0.0074	0.0067
1	0.0845	0.0780	0.0719	0.0663	0.0611	0.0563	0.0518	0.0477	0.0439	0.0404
2	0.2238	0.2102	0.1974	0.1851	0.1736	0.1626	0.1523	0.1425	0.1333	0.1247
3	0.4142	0.3954	0.3772	0.3594	0.3423	0.3257	0.3097	0.2942	0.2793	0.2650
4	0.6093	0.5898	0.5704	0.5512	0.5321	0.5132	0.4946	0.4763	0.4582	0.4405
5	0.7693	0.7531	0.7367	0.7199	0.7029	0.6858	0.6684	0.6510	0.6335	0.6160
6	0.8786	0.8675	0.8558	0.8436	0.8311	0.8180	0.8046	0.7908	0.7767	0.7622
7	0.9427	0.9361	0.9290	0.9214	0.9134	0.9049	0.8960	0.8867	0.8769	0.8666
8	0.9755	0.9721	0.9683	0.9642	0.9597	0.9549	0.9497	0.9442	0.9382	0.9319
9	0.9905	0.9889	0.9871	0.9851	0.9829	0.9805	0.9778	0.9749	0.9717	0.9682
10	0.9966	0.9959	0.9952	0.9943	0.9933	0.9922	0.9910	0.9896	0.9880	0.9863
11	0.9989	0.9986	0.9983	0.9980	0.9976	0.9971	0.9966	0.9960	0.9953	0.9945
12	0.9997	0.9996	0.9995	0.9993	0.9992	0.9990	0.9988	0.9986	0.9983	0.9980
13	0.9999	0.9999	0.9998	0.9998	0.9997	0.9997	0.9996	0.9995	0.9994	0.9993
14	1.0000	1.0000	1.0000	0.9999	0.9999	0.9999	0.9999	0.9999	0.9998	0.9998
15	1.0000	1.0000	1.0000	1.0000	1.0000	1.0000	1.0000	1.0000	0.9999	0.9999
16	1.0000	1.0000	1.0000	1.0000	1.0000	1.0000	1.0000	1.0000	1.0000	1.0000

					μ					
x	5.10	5.20	5.30	5.40	5.50	5.60	5.70	5.80	5.90	6.00
0	0.0061	0.0055	0.0050	0.0045	0.0041	0.0037	0.0033	0.0030	0.0027	0.0025
1	0.0372	0.0342	0.0314	0.0289	0.0266	0.0244	0.0224	0.0206	0.0189	0.0174
2	0.1165	0.1088	0.1016	0.0948	0.0884	0.0824	0.0768	0.0715	0.0666	0.0620
3	0.2513	0.2381	0.2254	0.2133	0.2017	0.1906	0.1800	0.1700	0.1604	0.1512
4	0.4231	0.4061	0.3895	0.3733	0.3575	0.3422	0.3272	0.3127	0.2987	0.2851
5	0.5984	0.5809	0.5635	0.5461	0.5289	0.5119	0.4950	0.4783	0.4619	0.4457

(*Continued*)

(*Continued*)

	μ									
x	5.10	5.20	5.30	5.40	5.50	5.60	5.70	5.80	5.90	6.00
6	0.7474	0.7324	0.7171	0.7017	0.6860	0.6703	0.6544	0.6384	0.6224	0.6063
7	0.8560	0.8449	0.8335	0.8217	0.8095	0.7970	0.7841	0.7710	0.7576	0.7440
8	0.9252	0.9181	0.9106	0.9027	0.8944	0.8857	0.8766	0.8672	0.8574	0.8472
9	0.9644	0.9603	0.9559	0.9512	0.9462	0.9409	0.9352	0.9292	0.9228	0.9161
10	0.9844	0.9823	0.9800	0.9775	0.9747	0.9718	0.9686	0.9651	0.9614	0.9574
11	0.9937	0.9927	0.9916	0.9904	0.9890	0.9875	0.9859	0.9841	0.9821	0.9799
12	0.9976	0.9972	0.9967	0.9962	0.9955	0.9949	0.9941	0.9932	0.9922	0.9912
13	0.9992	0.9990	0.9988	0.9986	0.9983	0.9980	0.9977	0.9973	0.9969	0.9964
14	0.9997	0.9997	0.9996	0.9995	0.9994	0.9993	0.9991	0.9990	0.9988	0.9986
15	0.9999	0.9999	0.9999	0.9998	0.9998	0.9998	0.9997	0.9996	0.9996	0.9995
16	1.0000	1.0000	1.0000	0.9999	0.9999	0.9999	0.9999	0.9999	0.9999	0.9998
17	1.0000	1.0000	1.0000	1.0000	1.0000	1.0000	1.0000	1.0000	1.0000	0.9999
18	1.0000	1.0000	1.0000	1.0000	1.0000	1.0000	1.0000	1.0000	1.0000	1.0000

	μ									
x	6.10	6.20	6.30	6.40	6.50	6.60	6.70	6.80	6.90	7.00
0	0.0022	0.0020	0.0018	0.0017	0.0015	0.0014	0.0012	0.0011	0.0010	0.0009
1	0.0159	0.0146	0.0134	0.0123	0.0113	0.0103	0.0095	0.0087	0.0080	0.0073
2	0.0577	0.0536	0.0498	0.0463	0.0430	0.0400	0.0371	0.0344	0.0320	0.0296
3	0.1425	0.1342	0.1264	0.1189	0.1118	0.1052	0.0988	0.0928	0.0871	0.0818
4	0.2719	0.2592	0.2469	0.2351	0.2237	0.2127	0.2022	0.1920	0.1823	0.1730
5	0.4298	0.4141	0.3988	0.3837	0.3690	0.3547	0.3406	0.3270	0.3137	0.3007
6	0.5902	0.5742	0.5582	0.5423	0.5265	0.5108	0.4953	0.4799	0.4647	0.4497
7	0.7301	0.7160	0.7017	0.6873	0.6728	0.6581	0.6433	0.6285	0.6136	0.5987
8	0.8367	0.8259	0.8148	0.8033	0.7916	0.7796	0.7673	0.7548	0.7420	0.7291
9	0.9090	0.9016	0.8939	0.8858	0.8774	0.8686	0.8596	0.8502	0.8405	0.8305
10	0.9531	0.9486	0.9437	0.9386	0.9332	0.9274	0.9214	0.9151	0.9084	0.9015
11	0.9776	0.9750	0.9723	0.9693	0.9661	0.9627	0.9591	0.9552	0.9510	0.9467
12	0.9900	0.9887	0.9873	0.9857	0.9840	0.9821	0.9801	0.9779	0.9755	0.9730
13	0.9958	0.9952	0.9945	0.9937	0.9929	0.9920	0.9909	0.9898	0.9885	0.9872
14	0.9984	0.9981	0.9978	0.9974	0.9970	0.9966	0.9961	0.9956	0.9950	0.9943
15	0.9994	0.9993	0.9992	0.9990	0.9988	0.9986	0.9984	0.9982	0.9979	0.9976
16	0.9998	0.9997	0.9997	0.9996	0.9996	0.9995	0.9994	0.9993	0.9992	0.9990
17	0.9999	0.9999	0.9999	0.9999	0.9998	0.9998	0.9998	0.9997	0.9997	0.9996

(*Continued*)

(*Continued*)

x	6.10	6.20	6.30	6.40	6.50	6.60	6.70	6.80	6.90	7.00
					μ					
18	1.0000	1.0000	1.0000	1.0000	0.9999	0.9999	0.9999	0.9999	0.9999	0.9999
19	1.0000	1.0000	1.0000	1.0000	1.0000	1.0000	1.0000	1.0000	1.0000	1.0000
20	1.0000	1.0000	1.0000	1.0000	1.0000	1.0000	1.0000	1.0000	1.0000	1.0000

x	7.10	7.20	7.30	7.40	7.50	7.60	7.70	7.80	7.90	8.00
					μ					
0	0.0008	0.0007	0.0007	0.0006	0.0006	0.0005	0.0005	0.0004	0.0004	0.0003
1	0.0067	0.0061	0.0056	0.0051	0.0047	0.0043	0.0039	0.0036	0.0033	0.0030
2	0.0275	0.0255	0.0236	0.0219	0.0203	0.0188	0.0174	0.0161	0.0149	0.0138
3	0.0767	0.0719	0.0674	0.0632	0.0591	0.0554	0.0518	0.0485	0.0453	0.0424
4	0.1641	0.1555	0.1473	0.1395	0.1321	0.1249	0.1181	0.1117	0.1055	0.0996
5	0.2881	0.2759	0.2640	0.2526	0.2414	0.2307	0.2203	0.2103	0.2006	0.1912
6	0.4349	0.4204	0.4060	0.3920	0.3782	0.3646	0.3514	0.3384	0.3257	0.3134
7	0.5838	0.5689	0.5541	0.5393	0.5246	0.5100	0.4956	0.4812	0.4670	0.4530
8	0.7160	0.7027	0.6892	0.6757	0.6620	0.6482	0.6343	0.6204	0.6065	0.5925
9	0.8202	0.8096	0.7988	0.7877	0.7764	0.7649	0.7531	0.7411	0.7290	0.7166
10	0.8942	0.8867	0.8788	0.8707	0.8622	0.8535	0.8445	0.8352	0.8257	0.8159
11	0.9420	0.9371	0.9319	0.9265	0.9208	0.9148	0.9085	0.9020	0.8952	0.8881
12	0.9703	0.9673	0.9642	0.9609	0.9573	0.9536	0.9496	0.9454	0.9409	0.9362
13	0.9857	0.9841	0.9824	0.9805	0.9784	0.9762	0.9739	0.9714	0.9687	0.9658
14	0.9935	0.9927	0.9918	0.9908	0.9897	0.9886	0.9873	0.9859	0.9844	0.9827
15	0.9972	0.9969	0.9964	0.9959	0.9954	0.9948	0.9941	0.9934	0.9926	0.9918
16	0.9989	0.9987	0.9985	0.9983	0.9980	0.9978	0.9974	0.9971	0.9967	0.9963
17	0.9996	0.9995	0.9994	0.9993	0.9992	0.9991	0.9989	0.9988	0.9986	0.9984
18	0.9998	0.9998	0.9998	0.9997	0.9997	0.9996	0.9996	0.9995	0.9994	0.9993
19	0.9999	0.9999	0.9999	0.9999	0.9999	0.9999	0.9998	0.9998	0.9998	0.9997
20	1.0000	1.0000	1.0000	1.0000	1.0000	1.0000	0.9999	0.9999	0.9999	0.9999
21	1.0000	1.0000	1.0000	1.0000	1.0000	1.0000	1.0000	1.0000	1.0000	1.0000

x	8.10	8.20	8.30	8.40	8.50	8.60	8.70	8.80	8.90	9.00
					μ					
0	0.0003	0.0003	0.0002	0.0002	0.0002	0.0002	0.0002	0.0002	0.0001	0.0001
1	0.0028	0.0025	0.0023	0.0021	0.0019	0.0018	0.0016	0.0015	0.0014	0.0012
2	0.0127	0.0118	0.0109	0.0100	0.0093	0.0086	0.0079	0.0073	0.0068	0.0062
3	0.0396	0.0370	0.0346	0.0323	0.0301	0.0281	0.0262	0.0244	0.0228	0.0212

(*Continued*)

(*Continued*)

					μ					
x	**8.10**	**8.20**	**8.30**	**8.40**	**8.50**	**8.60**	**8.70**	**8.80**	**8.90**	**9.00**
4	0.0940	0.0887	0.0837	0.0789	0.0744	0.0701	0.0660	0.0621	0.0584	0.0550
5	0.1822	0.1736	0.1653	0.1573	0.1496	0.1422	0.1352	0.1284	0.1219	0.1157
6	0.3013	0.2896	0.2781	0.2670	0.2562	0.2457	0.2355	0.2256	0.2160	0.2068
7	0.4391	0.4254	0.4119	0.3987	0.3856	0.3728	0.3602	0.3478	0.3357	0.3239
8	0.5786	0.5647	0.5507	0.5369	0.5231	0.5094	0.4958	0.4823	0.4689	0.4557
9	0.7041	0.6915	0.6788	0.6659	0.6530	0.6400	0.6269	0.6137	0.6006	0.5874
10	0.8058	0.7955	0.7850	0.7743	0.7634	0.7522	0.7409	0.7294	0.7178	0.7060
11	0.8807	0.8731	0.8652	0.8571	0.8487	0.8400	0.8311	0.8220	0.8126	0.8030
12	0.9313	0.9261	0.9207	0.9150	0.9091	0.9029	0.8965	0.8898	0.8829	0.8758
13	0.9628	0.9595	0.9561	0.9524	0.9486	0.9445	0.9403	0.9358	0.9311	0.9261
14	0.9810	0.9791	0.9771	0.9749	0.9726	0.9701	0.9675	0.9647	0.9617	0.9585
15	0.9908	0.9898	0.9887	0.9875	0.9862	0.9848	0.9832	0.9816	0.9798	0.9780
16	0.9958	0.9953	0.9947	0.9941	0.9934	0.9926	0.9918	0.9909	0.9899	0.9889
17	0.9982	0.9979	0.9977	0.9973	0.9970	0.9966	0.9962	0.9957	0.9952	0.9947
18	0.9992	0.9991	0.9990	0.9989	0.9987	0.9985	0.9983	0.9981	0.9978	0.9976
19	0.9997	0.9997	0.9996	0.9995	0.9995	0.9994	0.9993	0.9992	0.9991	0.9989
20	0.9999	0.9999	0.9998	0.9998	0.9998	0.9998	0.9997	0.9997	0.9996	0.9996
21	1.0000	1.0000	0.9999	0.9999	0.9999	0.9999	0.9999	0.9999	0.9998	0.9998
22	1.0000	1.0000	1.0000	1.0000	1.0000	1.0000	1.0000	1.0000	0.9999	0.9999
23	1.0000	1.0000	1.0000	1.0000	1.0000	1.0000	1.0000	1.0000	1.0000	1.0000

					μ					
x	**9.10**	**9.20**	**9.30**	**9.40**	**9.50**	**9.60**	**9.70**	**9.80**	**9.90**	**10.00**
0	0.0001	0.0001	0.0001	0.0001	0.0001	0.0001	0.0001	0.0001	0.0001	0.0000
1	0.0011	0.0010	0.0009	0.0009	0.0008	0.0007	0.0007	0.0006	0.0005	0.0005
2	0.0058	0.0053	0.0049	0.0045	0.0042	0.0038	0.0035	0.0033	0.0030	0.0028
3	0.0198	0.0184	0.0172	0.0160	0.0149	0.0138	0.0129	0.0120	0.0111	0.0103
4	0.0517	0.0486	0.0456	0.0429	0.0403	0.0378	0.0355	0.0333	0.0312	0.0293
5	0.1098	0.1041	0.0986	0.0935	0.0885	0.0838	0.0793	0.0750	0.0710	0.0671
6	0.1978	0.1892	0.1808	0.1727	0.1649	0.1574	0.1502	0.1433	0.1366	0.1301
7	0.3123	0.3010	0.2900	0.2792	0.2687	0.2584	0.2485	0.2388	0.2294	0.2202
8	0.4426	0.4296	0.4168	0.4042	0.3918	0.3796	0.3676	0.3558	0.3442	0.3328
9	0.5742	0.5611	0.5479	0.5349	0.5218	0.5089	0.4960	0.4832	0.4705	0.4579
10	0.6941	0.6820	0.6699	0.6576	0.6453	0.6329	0.6205	0.6080	0.5955	0.5830

(*Continued*)

(*Continued*)

						μ				
x	9.10	9.20	9.30	9.40	9.50	9.60	9.70	9.80	9.90	10.00
11	0.7932	0.7832	0.7730	0.7626	0.7520	0.7412	0.7303	0.7193	0.7081	0.6968
12	0.8684	0.8607	0.8529	0.8448	0.8364	0.8279	0.8191	0.8101	0.8009	0.7916
13	0.9210	0.9156	0.9100	0.9042	0.8981	0.8919	0.8853	0.8786	0.8716	0.8645
14	0.9552	0.9517	0.9480	0.9441	0.9400	0.9357	0.9312	0.9265	0.9216	0.9165
15	0.9760	0.9738	0.9715	0.9691	0.9665	0.9638	0.9609	0.9579	0.9546	0.9513
16	0.9878	0.9865	0.9852	0.9838	0.9823	0.9806	0.9789	0.9770	0.9751	0.9730
17	0.9941	0.9934	0.9927	0.9919	0.9911	0.9902	0.9892	0.9881	0.9870	0.9857
18	0.9973	0.9969	0.9966	0.9962	0.9957	0.9952	0.9947	0.9941	0.9935	0.9928
19	0.9988	0.9986	0.9985	0.9983	0.9980	0.9978	0.9975	0.9972	0.9969	0.9965
20	0.9995	0.9994	0.9993	0.9992	0.9991	0.9990	0.9989	0.9987	0.9986	0.9984
21	0.9998	0.9998	0.9997	0.9997	0.9996	0.9996	0.9995	0.9995	0.9994	0.9993
22	0.9999	0.9999	0.9999	0.9999	0.9999	0.9998	0.9998	0.9998	0.9997	0.9997
23	1.0000	1.0000	1.0000	1.0000	0.9999	0.9999	0.9999	0.9999	0.9999	0.9999
24	1.0000	1.0000	1.0000	1.0000	1.0000	1.0000	1.0000	1.0000	1.0000	1.0000

						μ				
x	11.00	12.00	13.00	14.00	15.00	16.00	17.00	18.00	19.00	20.00
0	0.0000	0.0000	0.0000	0.0000	0.0000	0.0000	0.0000	0.0000	0.0000	0.0000
1	0.0002	0.0001	0.0000	0.0000	0.0000	0.0000	0.0000	0.0000	0.0000	0.0000
2	0.0012	0.0005	0.0002	0.0001	0.0000	0.0000	0.0000	0.0000	0.0000	0.0000
3	0.0049	0.0023	0.0011	0.0005	0.0002	0.0001	0.0000	0.0000	0.0000	0.0000
4	0.0151	0.0076	0.0037	0.0018	0.0009	0.0004	0.0002	0.0001	0.0000	0.0000
5	0.0375	0.0203	0.0107	0.0055	0.0028	0.0014	0.0007	0.0003	0.0002	0.0001
6	0.0786	0.0458	0.0259	0.0142	0.0076	0.0040	0.0021	0.0010	0.0005	0.0003
7	0.1432	0.0895	0.0540	0.0316	0.0180	0.0100	0.0054	0.0029	0.0015	0.0008
8	0.2320	0.1550	0.0998	0.0621	0.0374	0.0220	0.0126	0.0071	0.0039	0.0021
9	0.3405	0.2424	0.1658	0.1094	0.0699	0.0433	0.0261	0.0154	0.0089	0.0050
10	0.4599	0.3472	0.2517	0.1757	0.1185	0.0774	0.0491	0.0304	0.0183	0.0108
11	0.5793	0.4616	0.3532	0.2600	0.1848	0.1270	0.0847	0.0549	0.0347	0.0214
12	0.6887	0.5760	0.4631	0.3585	0.2676	0.1931	0.1350	0.0917	0.0606	0.0390
13	0.7813	0.6815	0.5730	0.4644	0.3632	0.2745	0.2009	0.1426	0.0984	0.0661
14	0.8540	0.7720	0.6751	0.5704	0.4657	0.3675	0.2808	0.2081	0.1497	0.1049
15	0.9074	0.8444	0.7636	0.6694	0.5681	0.4667	0.3715	0.2867	0.2148	0.1565
16	0.9441	0.8987	0.8355	0.7559	0.6641	0.5660	0.4677	0.3751	0.2920	0.2211

(*Continued*)

(*Continued*)

x	11.00	12.00	13.00	14.00	15.00	16.00	17.00	18.00	19.00	20.00
					μ					
17	0.9678	0.9370	0.8905	0.8272	0.7489	0.6593	0.5640	0.4686	0.3784	0.2970
18	0.9823	0.9626	0.9302	0.8826	0.8195	0.7423	0.6550	0.5622	0.4695	0.3814
19	0.9907	0.9787	0.9573	0.9235	0.8752	0.8122	0.7363	0.6509	0.5606	0.4703
20	0.9953	0.9884	0.9750	0.9521	0.9170	0.8682	0.8055	0.7307	0.6472	0.5591
21	0.9977	0.9939	0.9859	0.9712	0.9469	0.9108	0.8615	0.7991	0.7255	0.6437
22	0.9990	0.9970	0.9924	0.9833	0.9673	0.9418	0.9047	0.8551	0.7931	0.7206
23	0.9995	0.9985	0.9960	0.9907	0.9805	0.9633	0.9367	0.8989	0.8490	0.7875
24	0.9998	0.9993	0.9980	0.9950	0.9888	0.9777	0.9594	0.9317	0.8933	0.8432
25	0.9999	0.9997	0.9990	0.9974	0.9938	0.9869	0.9748	0.9554	0.9269	0.8878
26	1.0000	0.9999	0.9995	0.9987	0.9967	0.9925	0.9848	0.9718	0.9514	0.9221
27	1.0000	0.9999	0.9998	0.9994	0.9983	0.9959	0.9912	0.9827	0.9687	0.9475
28	1.0000	1.0000	0.9999	0.9997	0.9991	0.9978	0.9950	0.9897	0.9805	0.9657
29	1.0000	1.0000	1.0000	0.9999	0.9996	0.9989	0.9973	0.9941	0.9882	0.9782
30	1.0000	1.0000	1.0000	0.9999	0.9998	0.9994	0.9986	0.9967	0.9930	0.9865
31	1.0000	1.0000	1.0000	1.0000	0.9999	0.9997	0.9993	0.9982	0.9960	0.9919
32	1.0000	1.0000	1.0000	1.0000	1.0000	0.9999	0.9996	0.9990	0.9978	0.9953
33	1.0000	1.0000	1.0000	1.0000	1.0000	0.9999	0.9998	0.9995	0.9988	0.9973
34	1.0000	1.0000	1.0000	1.0000	1.0000	1.0000	0.9999	0.9998	0.9994	0.9985
35	1.0000	1.0000	1.0000	1.0000	1.0000	1.0000	1.0000	0.9999	0.9997	0.9992
36	1.0000	1.0000	1.0000	1.0000	1.0000	1.0000	1.0000	0.9999	0.9998	0.9996
37	1.0000	1.0000	1.0000	1.0000	1.0000	1.0000	1.0000	1.0000	0.9999	0.9998
38	1.0000	1.0000	1.0000	1.0000	1.0000	1.0000	1.0000	1.0000	1.0000	0.9999
39	1.0000	1.0000	1.0000	1.0000	1.0000	1.0000	1.0000	1.0000	1.0000	0.9999
40	1.0000	1.0000	1.0000	1.0000	1.0000	1.0000	1.0000	1.0000	1.0000	1.0000

CUMULATIVE BINOMIAL PROBABILITY DISTRIBUTION

This table was developed using the following equations and Excel:

$$P(x \le x_1) = \sum_{k}^{x_1} \binom{n}{k} p^k (1-p)^{n-k}$$

$$P(x \le x_1) = \sum_{k}^{x_1} \left(\frac{n!}{k!(n-k!)} \right) p^k (1-p)^{n-k}$$

where probability of x_1 or fewer successes in n trials, or of up to and including x_1 successes in n trials, when each trial has a probability of succeeding of p.

n	x	p 0.01	0.02	0.03	0.04	0.05	0.06	0.07	0.08	0.09	0.10	0.15	0.20	0.25	0.30	0.35	0.40	0.45	0.50
1	0	0.9900	0.9800	0.9700	0.9600	0.9500	0.9400	0.9300	0.9200	0.9100	0.9000	0.8500	0.8000	0.7500	0.7000	0.6500	0.6000	0.5500	0.5000
1	1	1.0000	1.0000	1.0000	1.0000	1.0000	1.0000	1.0000	1.0000	1.0000	1.0000	1.0000	1.0000	1.0000	1.0000	1.0000	1.0000	1.0000	1.0000

n	x	p = 0.55	0.60	0.65	0.70	0.75	0.80	0.85	0.90	0.91	0.92	0.93	0.94	0.95	0.96	0.97	0.98	0.99	1.00
1	0	0.4500	0.4000	0.3500	0.3000	0.2500	0.2000	0.1500	0.1000	0.0900	0.0800	0.0700	0.0600	0.0500	0.0400	0.0300	0.0200	0.0100	0.0000
1	1	1.0000	1.0000	1.0000	1.0000	1.0000	1.0000	1.0000	1.0000	1.0000	1.0000	1.0000	1.0000	1.0000	1.0000	1.0000	1.0000	1.0000	1.0000

n	x	p = 0.01	0.02	0.03	0.04	0.05	0.06	0.07	0.08	0.09	0.10	0.15	0.20	0.25	0.30	0.35	0.40	0.45	0.50
2	0	0.9801	0.9604	0.9409	0.9216	0.9025	0.8836	0.8649	0.8464	0.8281	0.8100	0.7225	0.6400	0.5625	0.4900	0.4225	0.3600	0.3025	0.2500
2	1	0.9999	0.9996	0.9991	0.9984	0.9975	0.9964	0.9951	0.9936	0.9919	0.9900	0.9775	0.9600	0.9375	0.9100	0.8775	0.8400	0.7975	0.7500
2	2	1.0000	1.0000	1.0000	1.0000	1.0000	1.0000	1.0000	1.0000	1.0000	1.0000	1.0000	1.0000	1.0000	1.0000	1.0000	1.0000	1.0000	1.0000

n	x	p = 0.55	0.60	0.65	0.70	0.75	0.80	0.85	0.90	0.91	0.92	0.93	0.94	0.95	0.96	0.97	0.98	0.99	1.00
2	0	0.2025	0.1600	0.1225	0.0900	0.0625	0.0400	0.0225	0.0100	0.0081	0.0064	0.0049	0.0036	0.0025	0.0016	0.0009	0.0004	0.0001	0.0000
2	1	0.6975	0.6400	0.5775	0.5100	0.4375	0.3600	0.2775	0.1900	0.1719	0.1536	0.1351	0.1164	0.0975	0.0784	0.0591	0.0396	0.0199	0.0000
2	2	1.0000	1.0000	1.0000	1.0000	1.0000	1.0000	1.0000	1.0000	1.0000	1.0000	1.0000	1.0000	1.0000	1.0000	1.0000	1.0000	1.0000	1.0000

(Continued)

(Continued)

n	x	p = 0.01	0.02	0.03	0.04	0.05	0.06	0.07	0.08	0.09	0.10	0.15	0.20	0.25	0.30	0.35	0.40	0.45	0.50
3	0	0.9703	0.9412	0.9127	0.8847	0.8574	0.8306	0.8044	0.7787	0.7536	0.7290	0.6141	0.5120	0.4219	0.3430	0.2746	0.2160	0.1664	0.1250
3	1	0.9997	0.9988	0.9974	0.9953	0.9928	0.9896	0.9860	0.9818	0.9772	0.9720	0.9393	0.8960	0.8438	0.7840	0.7183	0.6480	0.5748	0.5000
3	2	1.0000	1.0000	1.0000	0.9999	0.9999	0.9998	0.9997	0.9995	0.9993	0.9990	0.9966	0.9920	0.9844	0.9730	0.9571	0.9360	0.9089	0.8750
3	3	1.0000	1.0000	1.0000	1.0000	1.0000	1.0000	1.0000	1.0000	1.0000	1.0000	1.0000	1.0000	1.0000	1.0000	1.0000	1.0000	1.0000	1.0000

n	x	p = 0.55	0.60	0.65	0.70	0.75	0.80	0.85	0.90	0.91	0.92	0.93	0.94	0.95	0.96	0.97	0.98	0.99	1.00
3	0	0.0911	0.0640	0.0429	0.0270	0.0156	0.0080	0.0034	0.0010	0.0007	0.0005	0.0003	0.0002	0.0001	0.0001	0.0000	0.0000	0.0000	0.0000
3	1	0.4253	0.3520	0.2818	0.2160	0.1563	0.1040	0.0608	0.0280	0.0228	0.0182	0.0140	0.0104	0.0073	0.0047	0.0026	0.0012	0.0003	0.0000
3	2	0.8336	0.7840	0.7254	0.6570	0.5781	0.4880	0.3859	0.2710	0.2464	0.2213	0.1956	0.1694	0.1426	0.1153	0.0873	0.0588	0.0297	0.0000
3	3	1.0000	1.0000	1.0000	1.0000	1.0000	1.0000	1.0000	1.0000	1.0000	1.0000	1.0000	1.0000	1.0000	1.0000	1.0000	1.0000	1.0000	1.0000

n	x	p = 0.01	0.02	0.03	0.04	0.05	0.06	0.07	0.08	0.09	0.10	0.15	0.20	0.25	0.30	0.35	0.40	0.45	0.50
4	0	0.9606	0.9224	0.8853	0.8493	0.8145	0.7807	0.7481	0.7164	0.6857	0.6561	0.5220	0.4096	0.3164	0.2401	0.1785	0.1296	0.0915	0.0625
4	1	0.9994	0.9977	0.9948	0.9909	0.9860	0.9801	0.9733	0.9656	0.9570	0.9477	0.8905	0.8192	0.7383	0.6517	0.5630	0.4752	0.3910	0.3125
4	2	1.0000	1.0000	0.9999	0.9998	0.9995	0.9992	0.9987	0.9981	0.9973	0.9963	0.9880	0.9728	0.9492	0.9163	0.8735	0.8208	0.7585	0.6875
4	3	1.0000	1.0000	1.0000	1.0000	1.0000	1.0000	1.0000	1.0000	0.9999	0.9999	0.9995	0.9984	0.9961	0.9919	0.9850	0.9744	0.9590	0.9375
4	4	1.0000	1.0000	1.0000	1.0000	1.0000	1.0000	1.0000	1.0000	1.0000	1.0000	1.0000	1.0000	1.0000	1.0000	1.0000	1.0000	1.0000	1.0000

n	x	p = 0.55	0.60	0.65	0.70	0.75	0.80	0.85	0.90	0.91	0.92	0.93	0.94	0.95	0.96	0.97	0.98	0.99	1.00
4	0	0.0410	0.0256	0.0150	0.0081	0.0039	0.0016	0.0005	0.0001	0.0001	0.0000	0.0000	0.0000	0.0000	0.0000	0.0000	0.0000	0.0000	0.0000
4	1	0.2415	0.1792	0.1265	0.0837	0.0508	0.0272	0.0120	0.0037	0.0027	0.0019	0.0013	0.0008	0.0005	0.0002	0.0001	0.0000	0.0000	0.0000
4	2	0.6090	0.5248	0.4370	0.3483	0.2617	0.1808	0.1095	0.0523	0.0430	0.0344	0.0267	0.0199	0.0140	0.0091	0.0052	0.0023	0.0006	0.0000
4	3	0.9085	0.8704	0.8215	0.7599	0.6836	0.5904	0.4780	0.3439	0.3143	0.2836	0.2519	0.2193	0.1855	0.1507	0.1147	0.0776	0.0394	0.0000
4	4	1.0000	1.0000	1.0000	1.0000	1.0000	1.0000	1.0000	1.0000	1.0000	1.0000	1.0000	1.0000	1.0000	1.0000	1.0000	1.0000	1.0000	1.0000

n	x	p = 0.01	0.02	0.03	0.04	0.05	0.06	0.07	0.08	0.09	0.10	0.15	0.20	0.25	0.30	0.35	0.40	0.45	0.50
5	0	0.9510	0.9039	0.8587	0.8154	0.7738	0.7339	0.6957	0.6591	0.6240	0.5905	0.4437	0.3277	0.2373	0.1681	0.1160	0.0778	0.0503	0.0313
5	1	0.9990	0.9962	0.9915	0.9852	0.9774	0.9681	0.9575	0.9456	0.9326	0.9185	0.8352	0.7373	0.6328	0.5282	0.4284	0.3370	0.2562	0.1875
5	2	1.0000	0.9999	0.9997	0.9994	0.9988	0.9980	0.9969	0.9955	0.9937	0.9914	0.9734	0.9421	0.8965	0.8369	0.7648	0.6826	0.5931	0.5000
5	3	1.0000	1.0000	1.0000	1.0000	1.0000	0.9999	0.9999	0.9998	0.9997	0.9995	0.9978	0.9933	0.9844	0.9692	0.9460	0.9130	0.8688	0.8125
5	4	1.0000	1.0000	1.0000	1.0000	1.0000	1.0000	1.0000	1.0000	1.0000	1.0000	0.9999	0.9997	0.9990	0.9976	0.9947	0.9898	0.9815	0.9688
5	5	1.0000	1.0000	1.0000	1.0000	1.0000	1.0000	1.0000	1.0000	1.0000	1.0000	1.0000	1.0000	1.0000	1.0000	1.0000	1.0000	1.0000	1.0000

n	x	p = 0.55	0.60	0.65	0.70	0.75	0.80	0.85	0.90	0.91	0.92	0.93	0.94	0.95	0.96	0.97	0.98	0.99	1.00
5	0	0.0185	0.0102	0.0053	0.0024	0.0010	0.0003	0.0001	0.0000	0.0000	0.0000	0.0000	0.0000	0.0000	0.0000	0.0000	0.0000	0.0000	0.0000
5	1	0.1312	0.0870	0.0540	0.0308	0.0156	0.0067	0.0022	0.0005	0.0003	0.0002	0.0001	0.0001	0.0000	0.0000	0.0000	0.0000	0.0000	0.0000
5	2	0.4069	0.3174	0.2352	0.1631	0.1035	0.0579	0.0266	0.0086	0.0063	0.0045	0.0031	0.0020	0.0012	0.0006	0.0003	0.0001	0.0000	0.0000
5	3	0.7438	0.6630	0.5716	0.4718	0.3672	0.2627	0.1648	0.0815	0.0674	0.0544	0.0425	0.0319	0.0226	0.0148	0.0085	0.0038	0.0010	0.0000
5	4	0.9497	0.9222	0.8840	0.8319	0.7627	0.6723	0.5563	0.4095	0.3760	0.3409	0.3043	0.2661	0.2262	0.1846	0.1413	0.0961	0.0490	0.0000
5	5	1.0000	1.0000	1.0000	1.0000	1.0000	1.0000	1.0000	1.0000	1.0000	1.0000	1.0000	1.0000	1.0000	1.0000	1.0000	1.0000	1.0000	1.0000

(*Continued*)

(Continued)

n	x	p = 0.01	0.02	0.03	0.04	0.05	0.06	0.07	0.08	0.09	0.10	0.15	0.20	0.25	0.30	0.35	0.40	0.45	0.50
6	0	0.9415	0.8858	0.8330	0.7828	0.7351	0.6899	0.6470	0.6064	0.5679	0.5314	0.3771	0.2621	0.1780	0.1176	0.0754	0.0467	0.0277	0.0156
6	1	0.9985	0.9943	0.9875	0.9784	0.9672	0.9541	0.9392	0.9227	0.9048	0.8857	0.7765	0.6554	0.5339	0.4202	0.3191	0.2333	0.1636	0.1094
6	2	1.0000	0.9998	0.9995	0.9988	0.9978	0.9962	0.9942	0.9915	0.9882	0.9842	0.9527	0.9011	0.8306	0.7443	0.6471	0.5443	0.4415	0.3438
6	3	1.0000	1.0000	1.0000	1.0000	0.9999	0.9998	0.9997	0.9995	0.9992	0.9987	0.9941	0.9830	0.9624	0.9295	0.8826	0.8208	0.7447	0.6563
6	4	1.0000	1.0000	1.0000	1.0000	1.0000	1.0000	1.0000	1.0000	1.0000	0.9999	0.9996	0.9984	0.9954	0.9891	0.9777	0.9590	0.9308	0.8906
6	5	1.0000	1.0000	1.0000	1.0000	1.0000	1.0000	1.0000	1.0000	1.0000	1.0000	1.0000	0.9999	0.9998	0.9993	0.9982	0.9959	0.9917	0.9844
6	6	1.0000	1.0000	1.0000	1.0000	1.0000	1.0000	1.0000	1.0000	1.0000	1.0000	1.0000	1.0000	1.0000	1.0000	1.0000	1.0000	1.0000	1.0000

n	x	p = 0.55	0.60	0.65	0.70	0.75	0.80	0.85	0.90	0.91	0.92	0.93	0.94	0.95	0.96	0.97	0.98	0.99	1.00
6	0	0.0083	0.0041	0.0018	0.0007	0.0002	0.0001	0.0000	0.0000	0.0000	0.0000	0.0000	0.0000	0.0000	0.0000	0.0000	0.0000	0.0000	0.0000
6	1	0.0692	0.0410	0.0223	0.0109	0.0046	0.0016	0.0004	0.0001	0.0000	0.0000	0.0000	0.0000	0.0000	0.0000	0.0000	0.0000	0.0000	0.0000
6	2	0.2553	0.1792	0.1174	0.0705	0.0376	0.0170	0.0059	0.0013	0.0008	0.0005	0.0003	0.0002	0.0001	0.0000	0.0000	0.0000	0.0000	0.0000
6	3	0.5585	0.4557	0.3529	0.2557	0.1694	0.0989	0.0473	0.0159	0.0118	0.0085	0.0058	0.0038	0.0022	0.0012	0.0005	0.0002	0.0000	0.0000
6	4	0.8364	0.7667	0.6809	0.5798	0.4661	0.3446	0.2235	0.1143	0.0952	0.0773	0.0608	0.0459	0.0328	0.0216	0.0125	0.0057	0.0015	0.0000
6	5	0.9723	0.9533	0.9246	0.8824	0.8220	0.7379	0.6229	0.4686	0.4321	0.3936	0.3530	0.3101	0.2649	0.2172	0.1670	0.1142	0.0585	0.0000
6	6	1.0000	1.0000	1.0000	1.0000	1.0000	1.0000	1.0000	1.0000	1.0000	1.0000	1.0000	1.0000	1.0000	1.0000	1.0000	1.0000	1.0000	1.0000

n	x	p = 0.01	0.02	0.03	0.04	0.05	0.06	0.07	0.08	0.09	0.10	0.15	0.20	0.25	0.30	0.35	0.40	0.45	0.50
7	0	0.9321	0.8681	0.8080	0.7514	0.6983	0.6485	0.6017	0.5578	0.5168	0.4783	0.3206	0.2097	0.1335	0.0824	0.0490	0.0280	0.0152	0.0078
7	1	0.9980	0.9921	0.9829	0.9706	0.9556	0.9382	0.9187	0.8974	0.8745	0.8503	0.7166	0.5767	0.4449	0.3294	0.2338	0.1586	0.1024	0.0625
7	2	1.0000	0.9997	0.9991	0.9980	0.9962	0.9937	0.9903	0.9860	0.9807	0.9743	0.9262	0.8520	0.7564	0.6471	0.5323	0.4199	0.3164	0.2266
7	3	1.0000	1.0000	1.0000	0.9999	0.9998	0.9996	0.9993	0.9988	0.9982	0.9973	0.9879	0.9667	0.9294	0.8740	0.8002	0.7102	0.6083	0.5000
7	4	1.0000	1.0000	1.0000	1.0000	1.0000	1.0000	1.0000	0.9999	0.9999	0.9998	0.9988	0.9953	0.9871	0.9712	0.9444	0.9037	0.8471	0.7734
7	5	1.0000	1.0000	1.0000	1.0000	1.0000	1.0000	1.0000	1.0000	1.0000	1.0000	0.9999	0.9996	0.9987	0.9962	0.9910	0.9812	0.9643	0.9375
7	6	1.0000	1.0000	1.0000	1.0000	1.0000	1.0000	1.0000	1.0000	1.0000	1.0000	1.0000	1.0000	0.9999	0.9998	0.9994	0.9984	0.9963	0.9922
7	7	1.0000	1.0000	1.0000	1.0000	1.0000	1.0000	1.0000	1.0000	1.0000	1.0000	1.0000	1.0000	1.0000	1.0000	1.0000	1.0000	1.0000	1.0000

n	x	p = 0.55	0.60	0.65	0.70	0.75	0.80	0.85	0.90	0.91	0.92	0.93	0.94	0.95	0.96	0.97	0.98	0.99	1.00
7	0	0.0037	0.0016	0.0006	0.0002	0.0001	0.0000	0.0000	0.0000	0.0000	0.0000	0.0000	0.0000	0.0000	0.0000	0.0000	0.0000	0.0000	0.0000
7	1	0.0357	0.0188	0.0090	0.0038	0.0013	0.0004	0.0001	0.0000	0.0000	0.0000	0.0000	0.0000	0.0000	0.0000	0.0000	0.0000	0.0000	0.0000
7	2	0.1529	0.0963	0.0556	0.0288	0.0129	0.0047	0.0012	0.0002	0.0001	0.0001	0.0000	0.0000	0.0000	0.0000	0.0000	0.0000	0.0000	0.0000
7	3	0.3917	0.2898	0.1998	0.1260	0.0706	0.0333	0.0121	0.0027	0.0018	0.0012	0.0007	0.0004	0.0002	0.0001	0.0000	0.0000	0.0000	0.0000
7	4	0.6836	0.5801	0.4677	0.3529	0.2436	0.1480	0.0738	0.0257	0.0193	0.0140	0.0097	0.0063	0.0038	0.0020	0.0009	0.0003	0.0000	0.0000
7	5	0.8976	0.8414	0.7662	0.6706	0.5551	0.4233	0.2834	0.1497	0.1255	0.1026	0.0813	0.0618	0.0444	0.0294	0.0171	0.0079	0.0020	0.0000
7	6	0.9848	0.9720	0.9510	0.9176	0.8665	0.7903	0.6794	0.5217	0.4832	0.4422	0.3983	0.3515	0.3017	0.2486	0.1920	0.1319	0.0679	0.0000
7	7	1.0000	1.0000	1.0000	1.0000	1.0000	1.0000	1.0000	1.0000	1.0000	1.0000	1.0000	1.0000	1.0000	1.0000	1.0000	1.0000	1.0000	1.0000

(Continued)

(Continued)

n	x	p = 0.01	0.02	0.03	0.04	0.05	0.06	0.07	0.08	0.09	0.10	0.15	0.20	0.25	0.30	0.35	0.40	0.45	0.50
8	0	0.9227	0.8508	0.7837	0.7214	0.6634	0.6096	0.5596	0.5132	0.4703	0.4305	0.2725	0.1678	0.1001	0.0576	0.0319	0.0168	0.0084	0.0039
8	1	0.9973	0.9897	0.9777	0.9619	0.9428	0.9208	0.8965	0.8702	0.8423	0.8131	0.6572	0.5033	0.3671	0.2553	0.1691	0.1064	0.0632	0.0352
8	2	0.9999	0.9996	0.9987	0.9969	0.9942	0.9904	0.9853	0.9789	0.9711	0.9619	0.8948	0.7969	0.6785	0.5518	0.4278	0.3154	0.2201	0.1445
8	3	1.0000	1.0000	0.9999	0.9998	0.9996	0.9993	0.9987	0.9978	0.9966	0.9950	0.9786	0.9437	0.8862	0.8059	0.7064	0.5941	0.4770	0.3633
8	4	1.0000	1.0000	1.0000	1.0000	1.0000	1.0000	0.9999	0.9999	0.9997	0.9996	0.9971	0.9896	0.9727	0.9420	0.8939	0.8263	0.7396	0.6367
8	5	1.0000	1.0000	1.0000	1.0000	1.0000	1.0000	1.0000	1.0000	1.0000	1.0000	0.9998	0.9988	0.9958	0.9887	0.9747	0.9502	0.9115	0.8555
8	6	1.0000	1.0000	1.0000	1.0000	1.0000	1.0000	1.0000	1.0000	1.0000	1.0000	1.0000	0.9999	0.9996	0.9987	0.9964	0.9915	0.9819	0.9648
8	7	1.0000	1.0000	1.0000	1.0000	1.0000	1.0000	1.0000	1.0000	1.0000	1.0000	1.0000	1.0000	1.0000	0.9999	0.9998	0.9993	0.9983	0.9961
8	8	1.0000	1.0000	1.0000	1.0000	1.0000	1.0000	1.0000	1.0000	1.0000	1.0000	1.0000	1.0000	1.0000	1.0000	1.0000	1.0000	1.0000	1.0000

n	x	p = 0.55	0.60	0.65	0.70	0.75	0.80	0.85	0.90	0.91	0.92	0.93	0.94	0.95	0.96	0.97	0.98	0.99	1.00
8	0	0.0017	0.0007	0.0002	0.0001	0.0000	0.0000	0.0000	0.0000	0.0000	0.0000	0.0000	0.0000	0.0000	0.0000	0.0000	0.0000	0.0000	0.0000
8	1	0.0181	0.0085	0.0036	0.0013	0.0004	0.0001	0.0000	0.0000	0.0000	0.0000	0.0000	0.0000	0.0000	0.0000	0.0000	0.0000	0.0000	0.0000
8	2	0.0885	0.0498	0.0253	0.0113	0.0042	0.0012	0.0002	0.0000	0.0000	0.0000	0.0000	0.0000	0.0000	0.0000	0.0000	0.0000	0.0000	0.0000
8	3	0.2604	0.1737	0.1061	0.0580	0.0273	0.0104	0.0029	0.0004	0.0003	0.0001	0.0001	0.0000	0.0000	0.0000	0.0000	0.0000	0.0000	0.0000
8	4	0.5230	0.4059	0.2936	0.1941	0.1138	0.0563	0.0214	0.0050	0.0034	0.0022	0.0013	0.0007	0.0004	0.0002	0.0001	0.0000	0.0000	0.0000
8	5	0.7799	0.6846	0.5722	0.4482	0.3215	0.2031	0.1052	0.0381	0.0289	0.0211	0.0147	0.0096	0.0058	0.0031	0.0013	0.0004	0.0001	0.0000
8	6	0.9368	0.8936	0.8309	0.7447	0.6329	0.4967	0.3428	0.1869	0.1577	0.1298	0.1035	0.0792	0.0572	0.0381	0.0223	0.0103	0.0027	0.0000
8	7	0.9916	0.9832	0.9681	0.9424	0.8999	0.8322	0.7275	0.5695	0.5297	0.4868	0.4404	0.3904	0.3366	0.2786	0.2163	0.1492	0.0773	0.0000
8	8	1.0000	1.0000	1.0000	1.0000	1.0000	1.0000	1.0000	1.0000	1.0000	1.0000	1.0000	1.0000	1.0000	1.0000	1.0000	1.0000	1.0000	1.0000

n	x	p=0.01	0.02	0.03	0.04	0.05	0.06	0.07	0.08	0.09	0.10	0.15	0.20	0.25	0.30	0.35	0.40	0.45	0.50
9	0	0.9135	0.8337	0.7602	0.6925	0.6302	0.5730	0.5204	0.4722	0.4279	0.3874	0.2316	0.1342	0.0751	0.0404	0.0207	0.0101	0.0046	0.0020
	1	0.9966	0.9869	0.9718	0.9522	0.9288	0.9022	0.8729	0.8417	0.8088	0.7748	0.5995	0.4362	0.3003	0.1960	0.1211	0.0705	0.0385	0.0195
	2	0.9999	0.9994	0.9980	0.9955	0.9916	0.9862	0.9791	0.9702	0.9595	0.9470	0.8591	0.7382	0.6007	0.4628	0.3373	0.2318	0.1495	0.0898
	3	1.0000	1.0000	0.9999	0.9997	0.9994	0.9987	0.9977	0.9963	0.9943	0.9917	0.9661	0.9144	0.8343	0.7297	0.6089	0.4826	0.3614	0.2539
	4	1.0000	1.0000	1.0000	1.0000	1.0000	0.9999	0.9998	0.9997	0.9995	0.9991	0.9944	0.9804	0.9511	0.9012	0.8283	0.7334	0.6214	0.5000
	5	1.0000	1.0000	1.0000	1.0000	1.0000	1.0000	1.0000	1.0000	1.0000	0.9999	0.9994	0.9969	0.9900	0.9747	0.9464	0.9006	0.8342	0.7461
	6	1.0000	1.0000	1.0000	1.0000	1.0000	1.0000	1.0000	1.0000	1.0000	1.0000	1.0000	0.9997	0.9987	0.9957	0.9888	0.9750	0.9502	0.9102
	7	1.0000	1.0000	1.0000	1.0000	1.0000	1.0000	1.0000	1.0000	1.0000	1.0000	1.0000	1.0000	0.9999	0.9996	0.9986	0.9962	0.9909	0.9805
	8	1.0000	1.0000	1.0000	1.0000	1.0000	1.0000	1.0000	1.0000	1.0000	1.0000	1.0000	1.0000	1.0000	1.0000	0.9999	0.9997	0.9992	0.9980
	9	1.0000	1.0000	1.0000	1.0000	1.0000	1.0000	1.0000	1.0000	1.0000	1.0000	1.0000	1.0000	1.0000	1.0000	1.0000	1.0000	1.0000	1.0000

n	x	p=0.55	0.60	0.65	0.70	0.75	0.80	0.85	0.90	0.91	0.92	0.93	0.94	0.95	0.96	0.97	0.98	0.99	1.00
9	0	0.0008	0.0003	0.0001	0.0000	0.0000	0.0000	0.0000	0.0000	0.0000	0.0000	0.0000	0.0000	0.0000	0.0000	0.0000	0.0000	0.0000	0.0000
	1	0.0091	0.0038	0.0014	0.0004	0.0001	0.0000	0.0000	0.0000	0.0000	0.0000	0.0000	0.0000	0.0000	0.0000	0.0000	0.0000	0.0000	0.0000
	2	0.0498	0.0250	0.0112	0.0043	0.0013	0.0003	0.0000	0.0000	0.0000	0.0000	0.0000	0.0000	0.0000	0.0000	0.0000	0.0000	0.0000	0.0000
	3	0.1658	0.0994	0.0536	0.0253	0.0100	0.0031	0.0006	0.0001	0.0000	0.0000	0.0000	0.0000	0.0000	0.0000	0.0000	0.0000	0.0000	0.0000
	4	0.3786	0.2666	0.1717	0.0988	0.0489	0.0196	0.0056	0.0009	0.0005	0.0003	0.0002	0.0001	0.0000	0.0000	0.0000	0.0000	0.0000	0.0000
	5	0.6386	0.5174	0.3911	0.2703	0.1657	0.0856	0.0339	0.0083	0.0057	0.0037	0.0023	0.0013	0.0006	0.0003	0.0001	0.0000	0.0000	0.0000
	6	0.8505	0.7682	0.6627	0.5372	0.3993	0.2618	0.1409	0.0530	0.0405	0.0298	0.0209	0.0138	0.0084	0.0045	0.0020	0.0006	0.0001	0.0000
	7	0.9615	0.9295	0.8789	0.8040	0.6997	0.5638	0.4005	0.2252	0.1912	0.1583	0.1271	0.0978	0.0712	0.0478	0.0282	0.0131	0.0034	0.0000
	8	0.9954	0.9899	0.9793	0.9596	0.9249	0.8658	0.7684	0.6126	0.5721	0.5278	0.4796	0.4270	0.3698	0.3075	0.2398	0.1663	0.0865	0.0000
	9	1.0000	1.0000	1.0000	1.0000	1.0000	1.0000	1.0000	1.0000	1.0000	1.0000	1.0000	1.0000	1.0000	1.0000	1.0000	1.0000	1.0000	1.0000

(*Continued*)

(Continued)

n	x	p=0.01	0.02	0.03	0.04	0.05	0.06	0.07	0.08	0.09	0.10	0.15	0.20	0.25	0.30	0.35	0.40	0.45	0.50
10	0	0.9044	0.8171	0.7374	0.6648	0.5987	0.5386	0.4840	0.4344	0.3894	0.3487	0.1969	0.1074	0.0563	0.0282	0.0135	0.0060	0.0025	0.0010
10	1	0.9957	0.9838	0.9655	0.9418	0.9139	0.8824	0.8483	0.8121	0.7746	0.7361	0.5443	0.3758	0.2440	0.1493	0.0860	0.0464	0.0233	0.0107
10	2	0.9999	0.9991	0.9972	0.9938	0.9885	0.9812	0.9717	0.9599	0.9460	0.9298	0.8202	0.6778	0.5256	0.3828	0.2616	0.1673	0.0996	0.0547
10	3	1.0000	1.0000	0.9999	0.9996	0.9990	0.9980	0.9964	0.9942	0.9912	0.9872	0.9500	0.8791	0.7759	0.6496	0.5138	0.3823	0.2660	0.1719
10	4	1.0000	1.0000	1.0000	1.0000	0.9999	0.9998	0.9997	0.9994	0.9990	0.9984	0.9901	0.9672	0.9219	0.8497	0.7515	0.6331	0.5044	0.3770
10	5	1.0000	1.0000	1.0000	1.0000	1.0000	1.0000	1.0000	1.0000	0.9999	0.9999	0.9986	0.9936	0.9803	0.9527	0.9051	0.8338	0.7384	0.6230
10	6	1.0000	1.0000	1.0000	1.0000	1.0000	1.0000	1.0000	1.0000	1.0000	1.0000	0.9999	0.9991	0.9965	0.9894	0.9740	0.9452	0.8980	0.8281
10	7	1.0000	1.0000	1.0000	1.0000	1.0000	1.0000	1.0000	1.0000	1.0000	1.0000	1.0000	0.9999	0.9996	0.9984	0.9952	0.9877	0.9726	0.9453
10	8	1.0000	1.0000	1.0000	1.0000	1.0000	1.0000	1.0000	1.0000	1.0000	1.0000	1.0000	1.0000	1.0000	0.9999	0.9995	0.9983	0.9955	0.9893
10	9	1.0000	1.0000	1.0000	1.0000	1.0000	1.0000	1.0000	1.0000	1.0000	1.0000	1.0000	1.0000	1.0000	1.0000	1.0000	0.9999	0.9997	0.9990
10	10	1.0000	1.0000	1.0000	1.0000	1.0000	1.0000	1.0000	1.0000	1.0000	1.0000	1.0000	1.0000	1.0000	1.0000	1.0000	1.0000	1.0000	1.0000

n	x	p=0.55	0.60	0.65	0.70	0.75	0.80	0.85	0.90	0.91	0.92	0.93	0.94	0.95	0.96	0.97	0.98	0.99	1.00
10	0	0.0003	0.0001	0.0000	0.0000	0.0000	0.0000	0.0000	0.0000	0.0000	0.0000	0.0000	0.0000	0.0000	0.0000	0.0000	0.0000	0.0000	0.0000
10	1	0.0045	0.0017	0.0005	0.0001	0.0000	0.0000	0.0000	0.0000	0.0000	0.0000	0.0000	0.0000	0.0000	0.0000	0.0000	0.0000	0.0000	0.0000
10	2	0.0274	0.0123	0.0048	0.0016	0.0004	0.0001	0.0000	0.0000	0.0000	0.0000	0.0000	0.0000	0.0000	0.0000	0.0000	0.0000	0.0000	0.0000
10	3	0.1020	0.0548	0.0260	0.0106	0.0035	0.0009	0.0001	0.0000	0.0000	0.0000	0.0000	0.0000	0.0000	0.0000	0.0000	0.0000	0.0000	0.0000
10	4	0.2616	0.1662	0.0949	0.0473	0.0197	0.0064	0.0014	0.0001	0.0001	0.0000	0.0000	0.0000	0.0000	0.0000	0.0000	0.0000	0.0000	0.0000
10	5	0.4956	0.3669	0.2485	0.1503	0.0781	0.0328	0.0099	0.0016	0.0010	0.0006	0.0003	0.0002	0.0001	0.0000	0.0000	0.0000	0.0000	0.0000
10	6	0.7340	0.6177	0.4862	0.3504	0.2241	0.1209	0.0500	0.0128	0.0088	0.0058	0.0036	0.0020	0.0010	0.0004	0.0001	0.0000	0.0000	0.0000
10	7	0.9004	0.8327	0.7384	0.6172	0.4744	0.3222	0.1798	0.0702	0.0540	0.0401	0.0283	0.0188	0.0115	0.0062	0.0028	0.0009	0.0001	0.0000
10	8	0.9767	0.9536	0.9140	0.8507	0.7560	0.6242	0.4557	0.2639	0.2254	0.1879	0.1517	0.1176	0.0861	0.0582	0.0345	0.0162	0.0043	0.0000
10	9	0.9975	0.9940	0.9865	0.9718	0.9437	0.8926	0.8031	0.6513	0.6106	0.5656	0.5160	0.4614	0.4013	0.3352	0.2626	0.1829	0.0956	0.0000
10	10	1.0000	1.0000	1.0000	1.0000	1.0000	1.0000	1.0000	1.0000	1.0000	1.0000	1.0000	1.0000	1.0000	1.0000	1.0000	1.0000	1.0000	1.0000

n	x	p = 0.01	0.02	0.03	0.04	0.05	0.06	0.07	0.08	0.09	0.10	0.15	0.20	0.25	0.30	0.35	0.40	0.45	0.50
11	0	0.8953	0.8007	0.7153	0.6382	0.5688	0.5063	0.4501	0.3996	0.3544	0.3138	0.1673	0.0859	0.0422	0.0198	0.0088	0.0036	0.0014	0.0005
11	1	0.9948	0.9805	0.9587	0.9308	0.8981	0.8618	0.8228	0.7819	0.7399	0.6974	0.4922	0.3221	0.1971	0.1130	0.0606	0.0302	0.0139	0.0059
11	2	0.9998	0.9988	0.9963	0.9917	0.9848	0.9752	0.9630	0.9481	0.9305	0.9104	0.7788	0.6174	0.4552	0.3127	0.2001	0.1189	0.0652	0.0327
11	3	1.0000	1.0000	0.9998	0.9993	0.9984	0.9970	0.9947	0.9915	0.9871	0.9815	0.9306	0.8389	0.7133	0.5696	0.4256	0.2963	0.1911	0.1133
11	4	1.0000	1.0000	1.0000	1.0000	0.9999	0.9997	0.9995	0.9990	0.9983	0.9972	0.9841	0.9496	0.8854	0.7897	0.6683	0.5328	0.3971	0.2744
11	5	1.0000	1.0000	1.0000	1.0000	1.0000	1.0000	1.0000	0.9999	0.9998	0.9997	0.9973	0.9883	0.9657	0.9218	0.8513	0.7535	0.6331	0.5000
11	6	1.0000	1.0000	1.0000	1.0000	1.0000	1.0000	1.0000	1.0000	1.0000	1.0000	0.9997	0.9980	0.9924	0.9784	0.9499	0.9006	0.8262	0.7256
11	7	1.0000	1.0000	1.0000	1.0000	1.0000	1.0000	1.0000	1.0000	1.0000	1.0000	1.0000	0.9998	0.9988	0.9957	0.9878	0.9707	0.9390	0.8867
11	8	1.0000	1.0000	1.0000	1.0000	1.0000	1.0000	1.0000	1.0000	1.0000	1.0000	1.0000	1.0000	0.9999	0.9994	0.9980	0.9941	0.9852	0.9673
11	9	1.0000	1.0000	1.0000	1.0000	1.0000	1.0000	1.0000	1.0000	1.0000	1.0000	1.0000	1.0000	1.0000	1.0000	0.9998	0.9993	0.9978	0.9941
11	10	1.0000	1.0000	1.0000	1.0000	1.0000	1.0000	1.0000	1.0000	1.0000	1.0000	1.0000	1.0000	1.0000	1.0000	1.0000	1.0000	0.9998	0.9995
11	11	1.0000	1.0000	1.0000	1.0000	1.0000	1.0000	1.0000	1.0000	1.0000	1.0000	1.0000	1.0000	1.0000	1.0000	1.0000	1.0000	1.0000	1.0000

(Continued)

(*Continued*)

n	x	p = 0.55	0.60	0.65	0.70	0.75	0.80	0.85	0.90	0.91	0.92	0.93	0.94	0.95	0.96	0.97	0.98	0.99	1.00
11	0	0.0002	0.0000	0.0000	0.0000	0.0000	0.0000	0.0000	0.0000	0.0000	0.0000	0.0000	0.0000	0.0000	0.0000	0.0000	0.0000	0.0000	0.0000
11	1	0.0022	0.0007	0.0002	0.0000	0.0000	0.0000	0.0000	0.0000	0.0000	0.0000	0.0000	0.0000	0.0000	0.0000	0.0000	0.0000	0.0000	0.0000
11	2	0.0148	0.0059	0.0020	0.0006	0.0001	0.0000	0.0000	0.0000	0.0000	0.0000	0.0000	0.0000	0.0000	0.0000	0.0000	0.0000	0.0000	0.0000
11	3	0.0610	0.0293	0.0122	0.0043	0.0012	0.0002	0.0000	0.0000	0.0000	0.0000	0.0000	0.0000	0.0000	0.0000	0.0000	0.0000	0.0000	0.0000
11	4	0.1738	0.0994	0.0501	0.0216	0.0076	0.0020	0.0003	0.0000	0.0000	0.0000	0.0000	0.0000	0.0000	0.0000	0.0000	0.0000	0.0000	0.0000
11	5	0.3669	0.2465	0.1487	0.0782	0.0343	0.0117	0.0027	0.0003	0.0002	0.0001	0.0000	0.0000	0.0000	0.0000	0.0000	0.0000	0.0000	0.0000
11	6	0.6029	0.4672	0.3317	0.2103	0.1146	0.0504	0.0159	0.0028	0.0017	0.0010	0.0005	0.0003	0.0001	0.0000	0.0000	0.0000	0.0000	0.0000
11	7	0.8089	0.7037	0.5744	0.4304	0.2867	0.1611	0.0694	0.0185	0.0129	0.0085	0.0053	0.0030	0.0016	0.0007	0.0002	0.0000	0.0000	0.0000
11	8	0.9348	0.8811	0.7999	0.6873	0.5448	0.3826	0.2212	0.0896	0.0695	0.0519	0.0370	0.0248	0.0152	0.0083	0.0037	0.0012	0.0002	0.0000
11	9	0.9861	0.9698	0.9394	0.8870	0.8029	0.6779	0.5078	0.3026	0.2601	0.2181	0.1772	0.1382	0.1019	0.0692	0.0413	0.0195	0.0052	0.0000
11	10	0.9986	0.9964	0.9912	0.9802	0.9578	0.9141	0.8327	0.6862	0.6456	0.6004	0.5499	0.4937	0.4312	0.3618	0.2847	0.1993	0.1047	0.0000
11	11	1.0000	1.0000	1.0000	1.0000	1.0000	1.0000	1.0000	1.0000	1.0000	1.0000	1.0000	1.0000	1.0000	1.0000	1.0000	1.0000	1.0000	1.0000

n	x	p = 0.01	0.02	0.03	0.04	0.05	0.06	0.07	0.08	0.09	0.10	0.15	0.20	0.25	0.30	0.35	0.40	0.45	0.50
12	0	0.8864	0.7847	0.6938	0.6127	0.5404	0.4759	0.4186	0.3677	0.3225	0.2824	0.1422	0.0687	0.0317	0.0138	0.0057	0.0022	0.0008	0.0002
12	1	0.9938	0.9769	0.9514	0.9191	0.8816	0.8405	0.7967	0.7513	0.7052	0.6590	0.4435	0.2749	0.1584	0.0850	0.0424	0.0196	0.0083	0.0032
12	2	0.9998	0.9985	0.9952	0.9893	0.9804	0.9684	0.9532	0.9348	0.9134	0.8891	0.7358	0.5583	0.3907	0.2528	0.1513	0.0834	0.0421	0.0193
12	3	1.0000	0.9999	0.9997	0.9990	0.9978	0.9957	0.9925	0.9880	0.9820	0.9744	0.9078	0.7946	0.6488	0.4925	0.3467	0.2253	0.1345	0.0730
12	4	1.0000	1.0000	1.0000	0.9999	0.9998	0.9996	0.9991	0.9984	0.9973	0.9957	0.9761	0.9274	0.8424	0.7237	0.5833	0.4382	0.3044	0.1938
12	5	1.0000	1.0000	1.0000	1.0000	1.0000	1.0000	0.9999	0.9998	0.9997	0.9995	0.9954	0.9806	0.9456	0.8822	0.7873	0.6652	0.5269	0.3872
12	6	1.0000	1.0000	1.0000	1.0000	1.0000	1.0000	1.0000	1.0000	1.0000	0.9999	0.9993	0.9961	0.9857	0.9614	0.9154	0.8418	0.7393	0.6128
12	7	1.0000	1.0000	1.0000	1.0000	1.0000	1.0000	1.0000	1.0000	1.0000	1.0000	0.9999	0.9994	0.9972	0.9905	0.9745	0.9427	0.8883	0.8062
12	8	1.0000	1.0000	1.0000	1.0000	1.0000	1.0000	1.0000	1.0000	1.0000	1.0000	1.0000	0.9999	0.9996	0.9983	0.9944	0.9847	0.9644	0.9270
12	9	1.0000	1.0000	1.0000	1.0000	1.0000	1.0000	1.0000	1.0000	1.0000	1.0000	1.0000	1.0000	1.0000	0.9998	0.9992	0.9972	0.9921	0.9807
12	10	1.0000	1.0000	1.0000	1.0000	1.0000	1.0000	1.0000	1.0000	1.0000	1.0000	1.0000	1.0000	1.0000	1.0000	0.9999	0.9997	0.9989	0.9968
12	11	1.0000	1.0000	1.0000	1.0000	1.0000	1.0000	1.0000	1.0000	1.0000	1.0000	1.0000	1.0000	1.0000	1.0000	1.0000	1.0000	0.9999	0.9998
12	12	1.0000	1.0000	1.0000	1.0000	1.0000	1.0000	1.0000	1.0000	1.0000	1.0000	1.0000	1.0000	1.0000	1.0000	1.0000	1.0000	1.0000	1.0000

(Continued)

(Continued)

n	x	p = 0.55	0.60	0.65	0.70	0.75	0.80	0.85	0.90	0.91	0.92	0.93	0.94	0.95	0.96	0.97	0.98	0.99	1.00
12	0	0.0001	0.0000	0.0000	0.0000	0.0000	0.0000	0.0000	0.0000	0.0000	0.0000	0.0000	0.0000	0.0000	0.0000	0.0000	0.0000	0.0000	0.0000
12	1	0.0011	0.0003	0.0001	0.0000	0.0000	0.0000	0.0000	0.0000	0.0000	0.0000	0.0000	0.0000	0.0000	0.0000	0.0000	0.0000	0.0000	0.0000
12	2	0.0079	0.0028	0.0008	0.0002	0.0000	0.0000	0.0000	0.0000	0.0000	0.0000	0.0000	0.0000	0.0000	0.0000	0.0000	0.0000	0.0000	0.0000
12	3	0.0356	0.0153	0.0056	0.0017	0.0004	0.0001	0.0000	0.0000	0.0000	0.0000	0.0000	0.0000	0.0000	0.0000	0.0000	0.0000	0.0000	0.0000
12	4	0.1117	0.0573	0.0255	0.0095	0.0028	0.0006	0.0001	0.0000	0.0000	0.0000	0.0000	0.0000	0.0000	0.0000	0.0000	0.0000	0.0000	0.0000
12	5	0.2607	0.1582	0.0846	0.0386	0.0143	0.0039	0.0007	0.0001	0.0000	0.0000	0.0000	0.0000	0.0000	0.0000	0.0000	0.0000	0.0000	0.0000
12	6	0.4731	0.3348	0.2127	0.1178	0.0544	0.0194	0.0046	0.0005	0.0003	0.0002	0.0001	0.0000	0.0000	0.0000	0.0000	0.0000	0.0000	0.0000
12	7	0.6956	0.5618	0.4167	0.2763	0.1576	0.0726	0.0239	0.0043	0.0027	0.0016	0.0009	0.0004	0.0002	0.0001	0.0000	0.0000	0.0000	0.0000
12	8	0.8655	0.7747	0.6533	0.5075	0.3512	0.2054	0.0922	0.0256	0.0180	0.0120	0.0075	0.0043	0.0022	0.0010	0.0003	0.0001	0.0000	0.0000
12	9	0.9579	0.9166	0.8487	0.7472	0.6093	0.4417	0.2642	0.1109	0.0866	0.0652	0.0468	0.0316	0.0196	0.0107	0.0048	0.0015	0.0002	0.0000
12	10	0.9917	0.9804	0.9576	0.9150	0.8416	0.7251	0.5565	0.3410	0.2948	0.2487	0.2033	0.1595	0.1184	0.0809	0.0486	0.0231	0.0062	0.0000
12	11	0.9992	0.9978	0.9943	0.9862	0.9683	0.9313	0.8578	0.7176	0.6775	0.6323	0.5814	0.5241	0.4596	0.3873	0.3062	0.2153	0.1136	0.0000
12	12	1.0000	1.0000	1.0000	1.0000	1.0000	1.0000	1.0000	1.0000	1.0000	1.0000	1.0000	1.0000	1.0000	1.0000	1.0000	1.0000	1.0000	1.0000

n	x	p = 0.01	0.02	0.03	0.04	0.05	0.06	0.07	0.08	0.09	0.10	0.15	0.20	0.25	0.30	0.35	0.40	0.45	0.50
13	0	0.8775	0.7690	0.6730	0.5882	0.5133	0.4474	0.3893	0.3383	0.2935	0.2542	0.1209	0.0550	0.0238	0.0097	0.0037	0.0013	0.0004	0.0001
13	1	0.9928	0.9730	0.9436	0.9068	0.8646	0.8186	0.7702	0.7206	0.6707	0.6213	0.3983	0.2336	0.1267	0.0637	0.0296	0.0126	0.0049	0.0017
13	2	0.9997	0.9980	0.9938	0.9865	0.9755	0.9608	0.9422	0.9201	0.8946	0.8661	0.6920	0.5017	0.3326	0.2025	0.1132	0.0579	0.0269	0.0112
13	3	1.0000	0.9999	0.9995	0.9986	0.9969	0.9940	0.9897	0.9837	0.9758	0.9658	0.8820	0.7473	0.5843	0.4206	0.2783	0.1686	0.0929	0.0461
13	4	1.0000	1.0000	1.0000	0.9999	0.9997	0.9993	0.9987	0.9976	0.9959	0.9935	0.9658	0.9009	0.7940	0.6543	0.5005	0.3530	0.2279	0.1334
13	5	1.0000	1.0000	1.0000	1.0000	1.0000	0.9999	0.9999	0.9997	0.9995	0.9991	0.9925	0.9700	0.9198	0.8346	0.7159	0.5744	0.4268	0.2905
13	6	1.0000	1.0000	1.0000	1.0000	1.0000	1.0000	1.0000	1.0000	0.9999	0.9999	0.9987	0.9930	0.9757	0.9376	0.8705	0.7712	0.6437	0.5000
13	7	1.0000	1.0000	1.0000	1.0000	1.0000	1.0000	1.0000	1.0000	1.0000	1.0000	0.9998	0.9988	0.9944	0.9818	0.9538	0.9023	0.8212	0.7095
13	8	1.0000	1.0000	1.0000	1.0000	1.0000	1.0000	1.0000	1.0000	1.0000	1.0000	1.0000	0.9998	0.9990	0.9960	0.9874	0.9679	0.9302	0.8666
13	9	1.0000	1.0000	1.0000	1.0000	1.0000	1.0000	1.0000	1.0000	1.0000	1.0000	1.0000	1.0000	0.9999	0.9993	0.9975	0.9922	0.9797	0.9539
13	10	1.0000	1.0000	1.0000	1.0000	1.0000	1.0000	1.0000	1.0000	1.0000	1.0000	1.0000	1.0000	1.0000	0.9999	0.9997	0.9987	0.9959	0.9888
13	11	1.0000	1.0000	1.0000	1.0000	1.0000	1.0000	1.0000	1.0000	1.0000	1.0000	1.0000	1.0000	1.0000	1.0000	1.0000	0.9999	0.9995	0.9983
13	12	1.0000	1.0000	1.0000	1.0000	1.0000	1.0000	1.0000	1.0000	1.0000	1.0000	1.0000	1.0000	1.0000	1.0000	1.0000	1.0000	1.0000	0.9999
13	13	1.0000	1.0000	1.0000	1.0000	1.0000	1.0000	1.0000	1.0000	1.0000	1.0000	1.0000	1.0000	1.0000	1.0000	1.0000	1.0000	1.0000	1.0000

(Continued)

(Continued)

n	x	p = 0.55	0.60	0.65	0.70	0.75	0.80	0.85	0.90	0.91	0.92	0.93	0.94	0.95	0.96	0.97	0.98	0.99	1.00
13	0	0.0000	0.0000	0.0000	0.0000	0.0000	0.0000	0.0000	0.0000	0.0000	0.0000	0.0000	0.0000	0.0000	0.0000	0.0000	0.0000	0.0000	0.0000
13	1	0.0005	0.0001	0.0000	0.0000	0.0000	0.0000	0.0000	0.0000	0.0000	0.0000	0.0000	0.0000	0.0000	0.0000	0.0000	0.0000	0.0000	0.0000
13	2	0.0041	0.0013	0.0003	0.0001	0.0000	0.0000	0.0000	0.0000	0.0000	0.0000	0.0000	0.0000	0.0000	0.0000	0.0000	0.0000	0.0000	0.0000
13	3	0.0203	0.0078	0.0025	0.0007	0.0001	0.0000	0.0000	0.0000	0.0000	0.0000	0.0000	0.0000	0.0000	0.0000	0.0000	0.0000	0.0000	0.0000
13	4	0.0698	0.0321	0.0126	0.0040	0.0010	0.0002	0.0000	0.0000	0.0000	0.0000	0.0000	0.0000	0.0000	0.0000	0.0000	0.0000	0.0000	0.0000
13	5	0.1788	0.0977	0.0462	0.0182	0.0056	0.0012	0.0002	0.0000	0.0000	0.0000	0.0000	0.0000	0.0000	0.0000	0.0000	0.0000	0.0000	0.0000
13	6	0.3563	0.2288	0.1295	0.0624	0.0243	0.0070	0.0013	0.0001	0.0000	0.0000	0.0000	0.0000	0.0000	0.0000	0.0000	0.0000	0.0000	0.0000
13	7	0.5732	0.4256	0.2841	0.1654	0.0802	0.0300	0.0075	0.0009	0.0005	0.0003	0.0001	0.0001	0.0000	0.0000	0.0000	0.0000	0.0000	0.0000
13	8	0.7721	0.6470	0.4995	0.3457	0.2060	0.0991	0.0342	0.0065	0.0041	0.0024	0.0013	0.0007	0.0003	0.0001	0.0000	0.0000	0.0000	0.0000
13	9	0.9071	0.8314	0.7217	0.5794	0.4157	0.2527	0.1180	0.0342	0.0242	0.0163	0.0103	0.0060	0.0031	0.0014	0.0005	0.0001	0.0000	0.0000
13	10	0.9731	0.9421	0.8868	0.7975	0.6674	0.4983	0.3080	0.1339	0.1054	0.0799	0.0578	0.0392	0.0245	0.0135	0.0062	0.0020	0.0003	0.0000
13	11	0.9951	0.9874	0.9704	0.9363	0.8733	0.7664	0.6017	0.3787	0.3293	0.2794	0.2298	0.1814	0.1354	0.0932	0.0564	0.0270	0.0072	0.0000
13	12	0.9996	0.9987	0.9963	0.9903	0.9762	0.9450	0.8791	0.7458	0.7065	0.6617	0.6107	0.5526	0.4867	0.4118	0.3270	0.2310	0.1225	0.0000
13	13	1.0000	1.0000	1.0000	1.0000	1.0000	1.0000	1.0000	1.0000	1.0000	1.0000	1.0000	1.0000	1.0000	1.0000	1.0000	1.0000	1.0000	1.0000

n	x	p = 0.01	0.02	0.03	0.04	0.05	0.06	0.07	0.08	0.09	0.10	0.15	0.20	0.25	0.30	0.35	0.40	0.45	0.50
14	0	0.8687	0.7536	0.6528	0.5647	0.4877	0.4205	0.3620	0.3112	0.2670	0.2288	0.1028	0.0440	0.0178	0.0068	0.0024	0.0008	0.0002	0.0001
14	1	0.9916	0.9690	0.9355	0.8941	0.8470	0.7963	0.7436	0.6900	0.6368	0.5846	0.3567	0.1979	0.1010	0.0475	0.0205	0.0081	0.0029	0.0009
14	2	0.9997	0.9975	0.9923	0.9833	0.9699	0.9522	0.9302	0.9042	0.8745	0.8416	0.6479	0.4481	0.2811	0.1608	0.0839	0.0398	0.0170	0.0065
14	3	1.0000	0.9999	0.9994	0.9981	0.9958	0.9920	0.9864	0.9786	0.9685	0.9559	0.8535	0.6982	0.5213	0.3552	0.2205	0.1243	0.0632	0.0287
14	4	1.0000	1.0000	1.0000	0.9998	0.9996	0.9990	0.9980	0.9965	0.9941	0.9908	0.9533	0.8702	0.7415	0.5842	0.4227	0.2793	0.1672	0.0898
14	5	1.0000	1.0000	1.0000	1.0000	1.0000	0.9999	0.9998	0.9996	0.9992	0.9985	0.9885	0.9561	0.8883	0.7805	0.6405	0.4859	0.3373	0.2120
14	6	1.0000	1.0000	1.0000	1.0000	1.0000	1.0000	1.0000	1.0000	0.9999	0.9998	0.9978	0.9884	0.9617	0.9067	0.8164	0.6925	0.5461	0.3953
14	7	1.0000	1.0000	1.0000	1.0000	1.0000	1.0000	1.0000	1.0000	1.0000	1.0000	0.9997	0.9976	0.9897	0.9685	0.9247	0.8499	0.7414	0.6047
14	8	1.0000	1.0000	1.0000	1.0000	1.0000	1.0000	1.0000	1.0000	1.0000	1.0000	1.0000	0.9996	0.9978	0.9917	0.9757	0.9417	0.8811	0.7880
14	9	1.0000	1.0000	1.0000	1.0000	1.0000	1.0000	1.0000	1.0000	1.0000	1.0000	1.0000	1.0000	0.9997	0.9983	0.9940	0.9825	0.9574	0.9102
14	10	1.0000	1.0000	1.0000	1.0000	1.0000	1.0000	1.0000	1.0000	1.0000	1.0000	1.0000	1.0000	1.0000	0.9998	0.9989	0.9961	0.9886	0.9713
14	11	1.0000	1.0000	1.0000	1.0000	1.0000	1.0000	1.0000	1.0000	1.0000	1.0000	1.0000	1.0000	1.0000	1.0000	0.9999	0.9994	0.9978	0.9935
14	12	1.0000	1.0000	1.0000	1.0000	1.0000	1.0000	1.0000	1.0000	1.0000	1.0000	1.0000	1.0000	1.0000	1.0000	1.0000	0.9999	0.9997	0.9991
14	13	1.0000	1.0000	1.0000	1.0000	1.0000	1.0000	1.0000	1.0000	1.0000	1.0000	1.0000	1.0000	1.0000	1.0000	1.0000	1.0000	1.0000	0.9999
14	14	1.0000	1.0000	1.0000	1.0000	1.0000	1.0000	1.0000	1.0000	1.0000	1.0000	1.0000	1.0000	1.0000	1.0000	1.0000	1.0000	1.0000	1.0000

(Continued)

(*Continued*)

n	x	p = 0.55	0.60	0.65	0.70	0.75	0.80	0.85	0.90	0.91	0.92	0.93	0.94	0.95	0.96	0.97	0.98	0.99	1.00
14	0	0.0000	0.0000	0.0000	0.0000	0.0000	0.0000	0.0000	0.0000	0.0000	0.0000	0.0000	0.0000	0.0000	0.0000	0.0000	0.0000	0.0000	0.0000
14	1	0.0003	0.0001	0.0000	0.0000	0.0000	0.0000	0.0000	0.0000	0.0000	0.0000	0.0000	0.0000	0.0000	0.0000	0.0000	0.0000	0.0000	0.0000
14	2	0.0022	0.0006	0.0001	0.0000	0.0000	0.0000	0.0000	0.0000	0.0000	0.0000	0.0000	0.0000	0.0000	0.0000	0.0000	0.0000	0.0000	0.0000
14	3	0.0114	0.0039	0.0011	0.0002	0.0000	0.0000	0.0000	0.0000	0.0000	0.0000	0.0000	0.0000	0.0000	0.0000	0.0000	0.0000	0.0000	0.0000
14	4	0.0426	0.0175	0.0060	0.0017	0.0003	0.0000	0.0000	0.0000	0.0000	0.0000	0.0000	0.0000	0.0000	0.0000	0.0000	0.0000	0.0000	0.0000
14	5	0.1189	0.0583	0.0243	0.0083	0.0022	0.0004	0.0000	0.0000	0.0000	0.0000	0.0000	0.0000	0.0000	0.0000	0.0000	0.0000	0.0000	0.0000
14	6	0.2586	0.1501	0.0753	0.0315	0.0103	0.0024	0.0003	0.0000	0.0000	0.0000	0.0000	0.0000	0.0000	0.0000	0.0000	0.0000	0.0000	0.0000
14	7	0.4539	0.3075	0.1836	0.0933	0.0383	0.0116	0.0022	0.0002	0.0001	0.0000	0.0000	0.0000	0.0000	0.0000	0.0000	0.0000	0.0000	0.0000
14	8	0.6627	0.5141	0.3595	0.2195	0.1117	0.0439	0.0115	0.0015	0.0008	0.0004	0.0002	0.0001	0.0000	0.0000	0.0000	0.0000	0.0000	0.0000
14	9	0.8328	0.7207	0.5773	0.4158	0.2585	0.1298	0.0467	0.0092	0.0059	0.0035	0.0020	0.0010	0.0004	0.0002	0.0000	0.0000	0.0000	0.0000
14	10	0.9368	0.8757	0.7795	0.6448	0.4787	0.3018	0.1465	0.0441	0.0315	0.0214	0.0136	0.0080	0.0042	0.0019	0.0006	0.0001	0.0000	0.0000
14	11	0.9830	0.9602	0.9161	0.8392	0.7189	0.5519	0.3521	0.1584	0.1255	0.0958	0.0698	0.0478	0.0301	0.0167	0.0077	0.0025	0.0003	0.0000
14	12	0.9971	0.9919	0.9795	0.9525	0.8990	0.8021	0.6433	0.4154	0.3632	0.3100	0.2564	0.2037	0.1530	0.1059	0.0645	0.0310	0.0084	0.0000
14	13	0.9998	0.9992	0.9976	0.9932	0.9822	0.9560	0.8972	0.7712	0.7330	0.6888	0.6380	0.5795	0.5123	0.4353	0.3472	0.2464	0.1313	0.0000
14	14	1.0000	1.0000	1.0000	1.0000	1.0000	1.0000	1.0000	1.0000	1.0000	1.0000	1.0000	1.0000	1.0000	1.0000	1.0000	1.0000	1.0000	1.0000

n	x	p = 0.01	0.02	0.03	0.04	0.05	0.06	0.07	0.08	0.09	0.10	0.15	0.20	0.25	0.30	0.35	0.40	0.45	0.50
15	0	0.8601	0.7386	0.6333	0.5421	0.4633	0.3953	0.3367	0.2863	0.2430	0.2059	0.0874	0.0352	0.0134	0.0047	0.0016	0.0005	0.0001	0.0000
15	1	0.9904	0.9647	0.9270	0.8809	0.8290	0.7738	0.7168	0.6597	0.6035	0.5490	0.3186	0.1671	0.0802	0.0353	0.0142	0.0052	0.0017	0.0005
15	2	0.9996	0.9970	0.9906	0.9797	0.9638	0.9429	0.9171	0.8870	0.8531	0.8159	0.6042	0.3980	0.2361	0.1268	0.0617	0.0271	0.0107	0.0037
15	3	1.0000	0.9998	0.9992	0.9976	0.9945	0.9896	0.9825	0.9727	0.9601	0.9444	0.8227	0.6482	0.4613	0.2969	0.1727	0.0905	0.0424	0.0176
15	4	1.0000	1.0000	0.9999	0.9998	0.9994	0.9986	0.9972	0.9950	0.9918	0.9873	0.9383	0.8358	0.6865	0.5155	0.3519	0.2173	0.1204	0.0592
15	5	1.0000	1.0000	1.0000	1.0000	0.9999	0.9999	0.9997	0.9993	0.9987	0.9978	0.9832	0.9389	0.8516	0.7216	0.5643	0.4032	0.2608	0.1509
15	6	1.0000	1.0000	1.0000	1.0000	1.0000	1.0000	1.0000	0.9999	0.9998	0.9997	0.9964	0.9819	0.9434	0.8689	0.7548	0.6098	0.4522	0.3036
15	7	1.0000	1.0000	1.0000	1.0000	1.0000	1.0000	1.0000	1.0000	1.0000	1.0000	0.9994	0.9958	0.9827	0.9500	0.8868	0.7869	0.6535	0.5000
15	8	1.0000	1.0000	1.0000	1.0000	1.0000	1.0000	1.0000	1.0000	1.0000	1.0000	0.9999	0.9992	0.9958	0.9848	0.9578	0.9050	0.8182	0.6964
15	9	1.0000	1.0000	1.0000	1.0000	1.0000	1.0000	1.0000	1.0000	1.0000	1.0000	1.0000	0.9999	0.9992	0.9963	0.9876	0.9662	0.9231	0.8491
15	10	1.0000	1.0000	1.0000	1.0000	1.0000	1.0000	1.0000	1.0000	1.0000	1.0000	1.0000	1.0000	0.9999	0.9993	0.9972	0.9907	0.9745	0.9408
15	11	1.0000	1.0000	1.0000	1.0000	1.0000	1.0000	1.0000	1.0000	1.0000	1.0000	1.0000	1.0000	1.0000	0.9999	0.9995	0.9981	0.9937	0.9824
15	12	1.0000	1.0000	1.0000	1.0000	1.0000	1.0000	1.0000	1.0000	1.0000	1.0000	1.0000	1.0000	1.0000	1.0000	0.9999	0.9997	0.9989	0.9963
15	13	1.0000	1.0000	1.0000	1.0000	1.0000	1.0000	1.0000	1.0000	1.0000	1.0000	1.0000	1.0000	1.0000	1.0000	1.0000	1.0000	0.9999	0.9995
15	14	1.0000	1.0000	1.0000	1.0000	1.0000	1.0000	1.0000	1.0000	1.0000	1.0000	1.0000	1.0000	1.0000	1.0000	1.0000	1.0000	1.0000	1.0000
15	15	1.0000	1.0000	1.0000	1.0000	1.0000	1.0000	1.0000	1.0000	1.0000	1.0000	1.0000	1.0000	1.0000	1.0000	1.0000	1.0000	1.0000	1.0000

(Continued)

(Continued)

n	x	p = 0.55	0.60	0.65	0.70	0.75	0.80	0.85	0.90	0.91	0.92	0.93	0.94	0.95	0.96	0.97	0.98	0.99	1.00
15	0	0.0000	0.0000	0.0000	0.0000	0.0000	0.0000	0.0000	0.0000	0.0000	0.0000	0.0000	0.0000	0.0000	0.0000	0.0000	0.0000	0.0000	0.0000
15	1	0.0001	0.0000	0.0000	0.0000	0.0000	0.0000	0.0000	0.0000	0.0000	0.0000	0.0000	0.0000	0.0000	0.0000	0.0000	0.0000	0.0000	0.0000
15	2	0.0011	0.0003	0.0001	0.0000	0.0000	0.0000	0.0000	0.0000	0.0000	0.0000	0.0000	0.0000	0.0000	0.0000	0.0000	0.0000	0.0000	0.0000
15	3	0.0063	0.0019	0.0005	0.0001	0.0000	0.0000	0.0000	0.0000	0.0000	0.0000	0.0000	0.0000	0.0000	0.0000	0.0000	0.0000	0.0000	0.0000
15	4	0.0255	0.0093	0.0028	0.0007	0.0001	0.0000	0.0000	0.0000	0.0000	0.0000	0.0000	0.0000	0.0000	0.0000	0.0000	0.0000	0.0000	0.0000
15	5	0.0769	0.0338	0.0124	0.0037	0.0008	0.0001	0.0000	0.0000	0.0000	0.0000	0.0000	0.0000	0.0000	0.0000	0.0000	0.0000	0.0000	0.0000
15	6	0.1818	0.0950	0.0422	0.0152	0.0042	0.0008	0.0001	0.0000	0.0000	0.0000	0.0000	0.0000	0.0000	0.0000	0.0000	0.0000	0.0000	0.0000
15	7	0.3465	0.2131	0.1132	0.0500	0.0173	0.0042	0.0006	0.0000	0.0000	0.0000	0.0000	0.0000	0.0000	0.0000	0.0000	0.0000	0.0000	0.0000
15	8	0.5478	0.3902	0.2452	0.1311	0.0566	0.0181	0.0036	0.0003	0.0002	0.0001	0.0000	0.0000	0.0000	0.0000	0.0000	0.0000	0.0000	0.0000
15	9	0.7392	0.5968	0.4357	0.2784	0.1484	0.0611	0.0168	0.0022	0.0013	0.0007	0.0003	0.0001	0.0001	0.0000	0.0000	0.0000	0.0000	0.0000
15	10	0.8796	0.7827	0.6481	0.4845	0.3135	0.1642	0.0617	0.0127	0.0082	0.0050	0.0028	0.0014	0.0006	0.0002	0.0001	0.0000	0.0000	0.0000
15	11	0.9576	0.9095	0.8273	0.7031	0.5387	0.3518	0.1773	0.0556	0.0399	0.0273	0.0175	0.0104	0.0055	0.0024	0.0008	0.0002	0.0000	0.0000
15	12	0.9893	0.9729	0.9383	0.8732	0.7639	0.6020	0.3958	0.1841	0.1469	0.1130	0.0829	0.0571	0.0362	0.0203	0.0094	0.0030	0.0004	0.0000
15	13	0.9983	0.9948	0.9858	0.9647	0.9198	0.8329	0.6814	0.4510	0.3965	0.3403	0.2832	0.2262	0.1710	0.1191	0.0730	0.0353	0.0096	0.0000
15	14	0.9999	0.9995	0.9984	0.9953	0.9866	0.9648	0.9126	0.7941	0.7570	0.7137	0.6633	0.6047	0.5367	0.4579	0.3667	0.2614	0.1399	0.0000
15	15	1.0000	1.0000	1.0000	1.0000	1.0000	1.0000	1.0000	1.0000	1.0000	1.0000	1.0000	1.0000	1.0000	1.0000	1.0000	1.0000	1.0000	1.0000

n	x	p = 0.01	0.02	0.03	0.04	0.05	0.06	0.07	0.08	0.09	0.10	0.15	0.20	0.25	0.30	0.35	0.40	0.45	0.50
20	0	0.8179	0.6676	0.5438	0.4420	0.3585	0.2901	0.2342	0.1887	0.1516	0.1216	0.0388	0.0115	0.0032	0.0008	0.0002	0.0000	0.0000	0.0000
20	1	0.9831	0.9401	0.8802	0.8103	0.7358	0.6605	0.5869	0.5169	0.4516	0.3917	0.1756	0.0692	0.0243	0.0076	0.0021	0.0005	0.0001	0.0000
20	2	0.9990	0.9929	0.9790	0.9561	0.9245	0.8850	0.8390	0.7879	0.7334	0.6769	0.4049	0.2061	0.0913	0.0355	0.0121	0.0036	0.0009	0.0002
20	3	1.0000	0.9994	0.9973	0.9926	0.9841	0.9710	0.9529	0.9294	0.9007	0.8670	0.6477	0.4114	0.2252	0.1071	0.0444	0.0160	0.0049	0.0013
20	4	1.0000	1.0000	0.9997	0.9990	0.9974	0.9944	0.9893	0.9817	0.9710	0.9568	0.8298	0.6296	0.4148	0.2375	0.1182	0.0510	0.0189	0.0059
20	5	1.0000	1.0000	1.0000	0.9999	0.9997	0.9991	0.9981	0.9962	0.9932	0.9887	0.9327	0.8042	0.6172	0.4164	0.2454	0.1256	0.0553	0.0207
20	6	1.0000	1.0000	1.0000	1.0000	1.0000	0.9999	0.9997	0.9994	0.9987	0.9976	0.9781	0.9133	0.7858	0.6080	0.4166	0.2500	0.1299	0.0577
20	7	1.0000	1.0000	1.0000	1.0000	1.0000	1.0000	1.0000	0.9999	0.9998	0.9996	0.9941	0.9679	0.8982	0.7723	0.6010	0.4159	0.2520	0.1316
20	8	1.0000	1.0000	1.0000	1.0000	1.0000	1.0000	1.0000	1.0000	1.0000	0.9999	0.9987	0.9900	0.9591	0.8867	0.7624	0.5956	0.4143	0.2517
20	9	1.0000	1.0000	1.0000	1.0000	1.0000	1.0000	1.0000	1.0000	1.0000	1.0000	0.9998	0.9974	0.9861	0.9520	0.8782	0.7553	0.5914	0.4119
20	10	1.0000	1.0000	1.0000	1.0000	1.0000	1.0000	1.0000	1.0000	1.0000	1.0000	1.0000	0.9994	0.9961	0.9829	0.9468	0.8725	0.7507	0.5881
20	11	1.0000	1.0000	1.0000	1.0000	1.0000	1.0000	1.0000	1.0000	1.0000	1.0000	1.0000	0.9999	0.9991	0.9949	0.9804	0.9435	0.8692	0.7483
20	12	1.0000	1.0000	1.0000	1.0000	1.0000	1.0000	1.0000	1.0000	1.0000	1.0000	1.0000	1.0000	0.9998	0.9987	0.9940	0.9790	0.9420	0.8684
20	13	1.0000	1.0000	1.0000	1.0000	1.0000	1.0000	1.0000	1.0000	1.0000	1.0000	1.0000	1.0000	1.0000	0.9997	0.9985	0.9935	0.9786	0.9423
20	14	1.0000	1.0000	1.0000	1.0000	1.0000	1.0000	1.0000	1.0000	1.0000	1.0000	1.0000	1.0000	1.0000	1.0000	0.9997	0.9984	0.9936	0.9793
20	15	1.0000	1.0000	1.0000	1.0000	1.0000	1.0000	1.0000	1.0000	1.0000	1.0000	1.0000	1.0000	1.0000	1.0000	1.0000	0.9997	0.9985	0.9941
20	16	1.0000	1.0000	1.0000	1.0000	1.0000	1.0000	1.0000	1.0000	1.0000	1.0000	1.0000	1.0000	1.0000	1.0000	1.0000	1.0000	0.9997	0.9987
20	17	1.0000	1.0000	1.0000	1.0000	1.0000	1.0000	1.0000	1.0000	1.0000	1.0000	1.0000	1.0000	1.0000	1.0000	1.0000	1.0000	1.0000	0.9998
20	18	1.0000	1.0000	1.0000	1.0000	1.0000	1.0000	1.0000	1.0000	1.0000	1.0000	1.0000	1.0000	1.0000	1.0000	1.0000	1.0000	1.0000	1.0000
20	19	1.0000	1.0000	1.0000	1.0000	1.0000	1.0000	1.0000	1.0000	1.0000	1.0000	1.0000	1.0000	1.0000	1.0000	1.0000	1.0000	1.0000	1.0000
20	20	1.0000	1.0000	1.0000	1.0000	1.0000	1.0000	1.0000	1.0000	1.0000	1.0000	1.0000	1.0000	1.0000	1.0000	1.0000	1.0000	1.0000	1.0000

(Continued)

(Continued)

n	x	p = 0.55	0.60	0.65	0.70	0.75	0.80	0.85	0.90	0.91	0.92	0.93	0.94	0.95	0.96	0.97	0.98	0.99	1.00
20	0	0.0000	0.0000	0.0000	0.0000	0.0000	0.0000	0.0000	0.0000	0.0000	0.0000	0.0000	0.0000	0.0000	0.0000	0.0000	0.0000	0.0000	0.0000
20	1	0.0000	0.0000	0.0000	0.0000	0.0000	0.0000	0.0000	0.0000	0.0000	0.0000	0.0000	0.0000	0.0000	0.0000	0.0000	0.0000	0.0000	0.0000
20	2	0.0000	0.0000	0.0000	0.0000	0.0000	0.0000	0.0000	0.0000	0.0000	0.0000	0.0000	0.0000	0.0000	0.0000	0.0000	0.0000	0.0000	0.0000
20	3	0.0003	0.0000	0.0000	0.0000	0.0000	0.0000	0.0000	0.0000	0.0000	0.0000	0.0000	0.0000	0.0000	0.0000	0.0000	0.0000	0.0000	0.0000
20	4	0.0015	0.0003	0.0000	0.0000	0.0000	0.0000	0.0000	0.0000	0.0000	0.0000	0.0000	0.0000	0.0000	0.0000	0.0000	0.0000	0.0000	0.0000
20	5	0.0064	0.0016	0.0003	0.0000	0.0000	0.0000	0.0000	0.0000	0.0000	0.0000	0.0000	0.0000	0.0000	0.0000	0.0000	0.0000	0.0000	0.0000
20	6	0.0214	0.0065	0.0015	0.0003	0.0000	0.0000	0.0000	0.0000	0.0000	0.0000	0.0000	0.0000	0.0000	0.0000	0.0000	0.0000	0.0000	0.0000
20	7	0.0580	0.0210	0.0060	0.0013	0.0002	0.0000	0.0000	0.0000	0.0000	0.0000	0.0000	0.0000	0.0000	0.0000	0.0000	0.0000	0.0000	0.0000
20	8	0.1308	0.0565	0.0196	0.0051	0.0009	0.0001	0.0000	0.0000	0.0000	0.0000	0.0000	0.0000	0.0000	0.0000	0.0000	0.0000	0.0000	0.0000
20	9	0.2493	0.1275	0.0532	0.0171	0.0039	0.0006	0.0000	0.0000	0.0000	0.0000	0.0000	0.0000	0.0000	0.0000	0.0000	0.0000	0.0000	0.0000
20	10	0.4086	0.2447	0.1218	0.0480	0.0139	0.0026	0.0002	0.0000	0.0000	0.0000	0.0000	0.0000	0.0000	0.0000	0.0000	0.0000	0.0000	0.0000
20	11	0.5857	0.4044	0.2376	0.1133	0.0409	0.0100	0.0013	0.0001	0.0000	0.0000	0.0000	0.0000	0.0000	0.0000	0.0000	0.0000	0.0000	0.0000
20	12	0.7480	0.5841	0.3990	0.2277	0.1018	0.0321	0.0059	0.0004	0.0002	0.0001	0.0000	0.0000	0.0000	0.0000	0.0000	0.0000	0.0000	0.0000
20	13	0.8701	0.7500	0.5834	0.3920	0.2142	0.0867	0.0219	0.0024	0.0013	0.0006	0.0003	0.0001	0.0000	0.0000	0.0000	0.0000	0.0000	0.0000
20	14	0.9447	0.8744	0.7546	0.5836	0.3828	0.1958	0.0673	0.0113	0.0068	0.0038	0.0019	0.0009	0.0003	0.0001	0.0000	0.0000	0.0000	0.0000
20	15	0.9811	0.9490	0.8818	0.7625	0.5852	0.3704	0.1702	0.0432	0.0290	0.0183	0.0107	0.0056	0.0026	0.0010	0.0003	0.0000	0.0000	0.0000
20	16	0.9951	0.9840	0.9556	0.8929	0.7748	0.5886	0.3523	0.1330	0.0993	0.0706	0.0471	0.0290	0.0159	0.0074	0.0027	0.0006	0.0000	0.0000
20	17	0.9991	0.9964	0.9879	0.9645	0.9087	0.7939	0.5951	0.3231	0.2666	0.2121	0.1610	0.1150	0.0755	0.0439	0.0210	0.0071	0.0010	0.0000
20	18	0.9999	0.9995	0.9979	0.9924	0.9757	0.9308	0.8244	0.6083	0.5484	0.4831	0.4131	0.3395	0.2642	0.1897	0.1198	0.0599	0.0169	0.0000
20	19	1.0000	1.0000	0.9998	0.9992	0.9968	0.9885	0.9612	0.8784	0.8484	0.8113	0.7658	0.7099	0.6415	0.5580	0.4562	0.3324	0.1821	0.0000
20	20	1.0000	1.0000	1.0000	1.0000	1.0000	1.0000	1.0000	1.0000	1.0000	1.0000	1.0000	1.0000	1.0000	1.0000	1.0000	1.0000	1.0000	1.0000

n	x	p = 0.01	0.02	0.03	0.04	0.05	0.06	0.07	0.08	0.09	0.10	0.15	0.20	0.25	0.30	0.35	0.40	0.45	0.50
25	0	0.7778	0.6035	0.4670	0.3604	0.2774	0.2129	0.1630	0.1244	0.0946	0.0718	0.0172	0.0038	0.0008	0.0001	0.0000	0.0000	0.0000	0.0000
25	1	0.9742	0.9114	0.8280	0.7358	0.6424	0.5527	0.4696	0.3947	0.3286	0.2712	0.0931	0.0274	0.0070	0.0016	0.0003	0.0001	0.0000	0.0000
25	2	0.9980	0.9868	0.9620	0.9235	0.8729	0.8129	0.7466	0.6768	0.6063	0.5371	0.2537	0.0982	0.0321	0.0090	0.0021	0.0004	0.0001	0.0000
25	3	0.9999	0.9986	0.9938	0.9835	0.9659	0.9402	0.9064	0.8649	0.8169	0.7636	0.4711	0.2340	0.0962	0.0332	0.0097	0.0024	0.0005	0.0001
25	4	1.0000	0.9999	0.9992	0.9972	0.9928	0.9850	0.9726	0.9549	0.9314	0.9020	0.6821	0.4207	0.2137	0.0905	0.0320	0.0095	0.0023	0.0005
25	5	1.0000	1.0000	0.9999	0.9996	0.9988	0.9969	0.9935	0.9877	0.9790	0.9666	0.8385	0.6167	0.3783	0.1935	0.0826	0.0294	0.0086	0.0020
25	6	1.0000	1.0000	1.0000	1.0000	0.9998	0.9995	0.9987	0.9972	0.9946	0.9905	0.9305	0.7800	0.5611	0.3407	0.1734	0.0736	0.0258	0.0073
25	7	1.0000	1.0000	1.0000	1.0000	1.0000	0.9999	0.9998	0.9995	0.9989	0.9977	0.9745	0.8909	0.7265	0.5118	0.3061	0.1536	0.0639	0.0216
25	8	1.0000	1.0000	1.0000	1.0000	1.0000	1.0000	1.0000	0.9999	0.9998	0.9995	0.9920	0.9532	0.8506	0.6769	0.4668	0.2735	0.1340	0.0539
25	9	1.0000	1.0000	1.0000	1.0000	1.0000	1.0000	1.0000	1.0000	1.0000	0.9999	0.9979	0.9827	0.9287	0.8106	0.6303	0.4246	0.2424	0.1148
25	10	1.0000	1.0000	1.0000	1.0000	1.0000	1.0000	1.0000	1.0000	1.0000	1.0000	0.9995	0.9944	0.9703	0.9022	0.7712	0.5858	0.3843	0.2122
25	11	1.0000	1.0000	1.0000	1.0000	1.0000	1.0000	1.0000	1.0000	1.0000	1.0000	0.9999	0.9985	0.9893	0.9558	0.8746	0.7323	0.5426	0.3450
25	12	1.0000	1.0000	1.0000	1.0000	1.0000	1.0000	1.0000	1.0000	1.0000	1.0000	1.0000	0.9996	0.9966	0.9825	0.9396	0.8462	0.6937	0.5000
25	13	1.0000	1.0000	1.0000	1.0000	1.0000	1.0000	1.0000	1.0000	1.0000	1.0000	1.0000	0.9999	0.9991	0.9940	0.9745	0.9222	0.8173	0.6550
25	14	1.0000	1.0000	1.0000	1.0000	1.0000	1.0000	1.0000	1.0000	1.0000	1.0000	1.0000	1.0000	0.9998	0.9982	0.9907	0.9656	0.9040	0.7878
25	15	1.0000	1.0000	1.0000	1.0000	1.0000	1.0000	1.0000	1.0000	1.0000	1.0000	1.0000	1.0000	1.0000	0.9995	0.9971	0.9868	0.9560	0.8852
25	16	1.0000	1.0000	1.0000	1.0000	1.0000	1.0000	1.0000	1.0000	1.0000	1.0000	1.0000	1.0000	1.0000	0.9999	0.9992	0.9957	0.9826	0.9461
25	17	1.0000	1.0000	1.0000	1.0000	1.0000	1.0000	1.0000	1.0000	1.0000	1.0000	1.0000	1.0000	1.0000	1.0000	0.9998	0.9988	0.9942	0.9784
25	18	1.0000	1.0000	1.0000	1.0000	1.0000	1.0000	1.0000	1.0000	1.0000	1.0000	1.0000	1.0000	1.0000	1.0000	1.0000	0.9997	0.9984	0.9927
25	19	1.0000	1.0000	1.0000	1.0000	1.0000	1.0000	1.0000	1.0000	1.0000	1.0000	1.0000	1.0000	1.0000	1.0000	1.0000	0.9999	0.9996	0.9980
25	20	1.0000	1.0000	1.0000	1.0000	1.0000	1.0000	1.0000	1.0000	1.0000	1.0000	1.0000	1.0000	1.0000	1.0000	1.0000	1.0000	0.9999	0.9995
25	21	1.0000	1.0000	1.0000	1.0000	1.0000	1.0000	1.0000	1.0000	1.0000	1.0000	1.0000	1.0000	1.0000	1.0000	1.0000	1.0000	1.0000	0.9999
25	22	1.0000	1.0000	1.0000	1.0000	1.0000	1.0000	1.0000	1.0000	1.0000	1.0000	1.0000	1.0000	1.0000	1.0000	1.0000	1.0000	1.0000	1.0000
25	23	1.0000	1.0000	1.0000	1.0000	1.0000	1.0000	1.0000	1.0000	1.0000	1.0000	1.0000	1.0000	1.0000	1.0000	1.0000	1.0000	1.0000	1.0000
25	24	1.0000	1.0000	1.0000	1.0000	1.0000	1.0000	1.0000	1.0000	1.0000	1.0000	1.0000	1.0000	1.0000	1.0000	1.0000	1.0000	1.0000	1.0000
25	25	1.0000	1.0000	1.0000	1.0000	1.0000	1.0000	1.0000	1.0000	1.0000	1.0000	1.0000	1.0000	1.0000	1.0000	1.0000	1.0000	1.0000	1.0000

(Continued)

(Continued)

n	x	p = 0.55	0.60	0.65	0.70	0.75	0.80	0.85	0.90	0.91	0.92	0.93	0.94	0.95	0.96	0.97	0.98	0.99	1.00
25	0	0.0000	0.0000	0.0000	0.0000	0.0000	0.0000	0.0000	0.0000	0.0000	0.0000	0.0000	0.0000	0.0000	0.0000	0.0000	0.0000	0.0000	0.0000
25	1	0.0000	0.0000	0.0000	0.0000	0.0000	0.0000	0.0000	0.0000	0.0000	0.0000	0.0000	0.0000	0.0000	0.0000	0.0000	0.0000	0.0000	0.0000
25	2	0.0000	0.0000	0.0000	0.0000	0.0000	0.0000	0.0000	0.0000	0.0000	0.0000	0.0000	0.0000	0.0000	0.0000	0.0000	0.0000	0.0000	0.0000
25	3	0.0000	0.0000	0.0000	0.0000	0.0000	0.0000	0.0000	0.0000	0.0000	0.0000	0.0000	0.0000	0.0000	0.0000	0.0000	0.0000	0.0000	0.0000
25	4	0.0001	0.0000	0.0000	0.0000	0.0000	0.0000	0.0000	0.0000	0.0000	0.0000	0.0000	0.0000	0.0000	0.0000	0.0000	0.0000	0.0000	0.0000
25	5	0.0004	0.0001	0.0000	0.0000	0.0000	0.0000	0.0000	0.0000	0.0000	0.0000	0.0000	0.0000	0.0000	0.0000	0.0000	0.0000	0.0000	0.0000
25	6	0.0016	0.0003	0.0000	0.0000	0.0000	0.0000	0.0000	0.0000	0.0000	0.0000	0.0000	0.0000	0.0000	0.0000	0.0000	0.0000	0.0000	0.0000
25	7	0.0058	0.0012	0.0002	0.0000	0.0000	0.0000	0.0000	0.0000	0.0000	0.0000	0.0000	0.0000	0.0000	0.0000	0.0000	0.0000	0.0000	0.0000
25	8	0.0174	0.0043	0.0008	0.0001	0.0000	0.0000	0.0000	0.0000	0.0000	0.0000	0.0000	0.0000	0.0000	0.0000	0.0000	0.0000	0.0000	0.0000
25	9	0.0440	0.0132	0.0029	0.0005	0.0000	0.0000	0.0000	0.0000	0.0000	0.0000	0.0000	0.0000	0.0000	0.0000	0.0000	0.0000	0.0000	0.0000
25	10	0.0960	0.0344	0.0093	0.0018	0.0002	0.0000	0.0000	0.0000	0.0000	0.0000	0.0000	0.0000	0.0000	0.0000	0.0000	0.0000	0.0000	0.0000
25	11	0.1827	0.0778	0.0255	0.0060	0.0009	0.0001	0.0000	0.0000	0.0000	0.0000	0.0000	0.0000	0.0000	0.0000	0.0000	0.0000	0.0000	0.0000
25	12	0.3063	0.1538	0.0604	0.0175	0.0034	0.0004	0.0000	0.0000	0.0000	0.0000	0.0000	0.0000	0.0000	0.0000	0.0000	0.0000	0.0000	0.0000
25	13	0.4574	0.2677	0.1254	0.0442	0.0107	0.0015	0.0001	0.0000	0.0000	0.0000	0.0000	0.0000	0.0000	0.0000	0.0000	0.0000	0.0000	0.0000
25	14	0.6157	0.4142	0.2288	0.0978	0.0297	0.0056	0.0005	0.0000	0.0000	0.0000	0.0000	0.0000	0.0000	0.0000	0.0000	0.0000	0.0000	0.0000
25	15	0.7576	0.5754	0.3697	0.1894	0.0713	0.0173	0.0021	0.0001	0.0000	0.0000	0.0000	0.0000	0.0000	0.0000	0.0000	0.0000	0.0000	0.0000
25	16	0.8660	0.7265	0.5332	0.3231	0.1494	0.0468	0.0080	0.0005	0.0002	0.0000	0.0000	0.0000	0.0000	0.0000	0.0000	0.0000	0.0000	0.0000
25	17	0.9361	0.8464	0.6939	0.4882	0.2735	0.1091	0.0255	0.0023	0.0011	0.0005	0.0002	0.0001	0.0000	0.0000	0.0000	0.0000	0.0000	0.0000
25	18	0.9742	0.9264	0.8266	0.6593	0.4389	0.2200	0.0695	0.0095	0.0054	0.0028	0.0013	0.0005	0.0002	0.0000	0.0000	0.0000	0.0000	0.0000
25	19	0.9914	0.9706	0.9174	0.8065	0.6217	0.3833	0.1615	0.0334	0.0210	0.0123	0.0065	0.0031	0.0012	0.0004	0.0001	0.0000	0.0000	0.0000
25	20	0.9977	0.9905	0.9680	0.9095	0.7863	0.5793	0.3179	0.0980	0.0686	0.0451	0.0274	0.0150	0.0072	0.0028	0.0008	0.0001	0.0000	0.0000
25	21	0.9995	0.9976	0.9903	0.9668	0.9038	0.7660	0.5289	0.2364	0.1831	0.1351	0.0936	0.0598	0.0341	0.0165	0.0062	0.0014	0.0001	0.0000
25	22	0.9999	0.9996	0.9979	0.9910	0.9679	0.9018	0.7463	0.4629	0.3937	0.3232	0.2534	0.1871	0.1271	0.0765	0.0380	0.0132	0.0020	0.0000
25	23	1.0000	0.9999	0.9997	0.9984	0.9930	0.9726	0.9069	0.7288	0.6714	0.6053	0.5304	0.4473	0.3576	0.2642	0.1720	0.0886	0.0258	0.0000
25	24	1.0000	1.0000	1.0000	0.9999	0.9992	0.9962	0.9828	0.9282	0.9054	0.8756	0.8370	0.7871	0.7226	0.6396	0.5330	0.3965	0.2222	0.0000
25	25	1.0000	1.0000	1.0000	1.0000	1.0000	1.0000	1.0000	1.0000	1.0000	1.0000	1.0000	1.0000	1.0000	1.0000	1.0000	1.0000	1.0000	1.0000

CONFIDENCE INTERVAL FOR POPULATION PROPORTION: SMALL SAMPLE

This table was developed using the following equations and Excel.

$$P_L = \cfrac{1}{1 + \left(\cfrac{n-x+1}{x}\right) F_{((1-\alpha/2);2(n-x)+2;2x)}}$$

$$P_U = \cfrac{1}{1 + \left(\cfrac{n-x}{x+1}\right) F_{\left(\frac{\alpha}{2};2(n-x)+2;2x+2\right)}}$$

n = 5	α = 0.01		α = 0.05		α = 0.10	
x	P_L	P_U	P_L	P_U	P_L	P_U
1	0.001	0.815	0.005	0.716	0.010	0.657
2	0.023	0.917	0.053	0.853	0.076	0.811
3	0.083	0.977	0.147	0.947	0.189	0.924
4	0.185	0.999	0.284	0.995	0.343	0.990

n = 6	α = 0.01		α = 0.05		α = 0.10	
x	P_L	P_U	P_L	P_U	P_L	P_U
1	0.001	0.746	0.004	0.641	0.009	0.582
2	0.019	0.856	0.043	0.777	0.063	0.729
3	0.066	0.934	0.118	0.882	0.153	0.847
4	0.144	0.981	0.223	0.957	0.271	0.937
5	0.254	0.999	0.359	0.996	0.418	0.991

(Continued)

(*Continued*)

n = 7	$\alpha = 0.01$		$\alpha = 0.05$		$\alpha = 0.10$	
x	P_L	P_U	P_L	P_U	P_L	P_U
1	0.001	0.685	0.004	0.579	0.007	0.521
2	0.016	0.797	0.037	0.710	0.053	0.659
3	0.055	0.882	0.099	0.816	0.129	0.775
4	0.118	0.945	0.184	0.901	0.225	0.871
5	0.203	0.984	0.290	0.963	0.341	0.947
6	0.315	0.999	0.421	0.996	0.479	0.993

n = 8	$\alpha = 0.01$		$\alpha = 0.05$		$\alpha = 0.10$	
x	P_L	P_U	P_L	P_U	P_L	P_U
1	0.001	0.632	0.003	0.527	0.006	0.471
2	0.014	0.742	0.032	0.651	0.046	0.600
3	0.047	0.830	0.085	0.755	0.111	0.711
4	0.100	0.900	0.157	0.843	0.193	0.807
5	0.170	0.953	0.245	0.915	0.289	0.889
6	0.258	0.986	0.349	0.968	0.400	0.954
7	0.368	0.999	0.473	0.997	0.529	0.994

n = 9	$\alpha = 0.01$		$\alpha = 0.05$		$\alpha = 0.10$	
x	P_L	P_U	P_L	P_U	P_L	P_U
1	0.001	0.585	0.003	0.482	0.006	0.429
2	0.012	0.693	0.028	0.600	0.041	0.550
3	0.042	0.781	0.075	0.701	0.098	0.655
4	0.087	0.854	0.137	0.788	0.169	0.749
5	0.146	0.913	0.212	0.863	0.251	0.831
6	0.219	0.958	0.299	0.925	0.345	0.902
7	0.307	0.988	0.400	0.972	0.450	0.959
8	0.415	0.999	0.518	0.997	0.571	0.994

n = 10	$\alpha = 0.01$		$\alpha = 0.05$		$\alpha = 0.10$	
x	P_L	P_U	P_L	P_U	P_L	P_U
1	0.001	0.544	0.003	0.445	0.005	0.394
2	0.011	0.648	0.025	0.556	0.037	0.507

(*Continued*)

(*Continued*)

3	0.037	0.735	0.067	0.652	0.087	0.607
4	0.077	0.809	0.122	0.738	0.150	0.696
5	0.128	0.872	0.187	0.813	0.222	0.778
6	0.191	0.923	0.262	0.878	0.304	0.850
7	0.265	0.963	0.348	0.933	0.393	0.913
8	0.352	0.989	0.444	0.975	0.493	0.963
9	0.456	0.999	0.555	0.997	0.606	0.995

$n = 11$	$\alpha = 0.01$		$\alpha = 0.05$		$\alpha = 0.10$	
x	P_L	P_U	P_L	P_U	P_L	P_U
1	0.000	0.509	0.002	0.413	0.005	0.364
2	0.010	0.608	0.023	0.518	0.033	0.470
3	0.033	0.693	0.060	0.610	0.079	0.564
4	0.069	0.767	0.109	0.692	0.135	0.650
5	0.114	0.831	0.167	0.766	0.200	0.729
6	0.169	0.886	0.234	0.833	0.271	0.800
7	0.233	0.931	0.308	0.891	0.350	0.865
8	0.307	0.967	0.390	0.940	0.436	0.921
9	0.392	0.990	0.482	0.977	0.530	0.967
10	0.491	1.000	0.587	0.998	0.636	0.995

$n = 12$	$\alpha = 0.01$	$\alpha = 0.05$	$\alpha = 0.10$			
x	P_L	P_U	P_L	P_U	P_L	P_U
1	0.000	0.477	0.002	0.385	0.004	0.339
2	0.009	0.573	0.021	0.484	0.030	0.438
3	0.030	0.655	0.055	0.572	0.072	0.527
4	0.062	0.728	0.099	0.651	0.123	0.609
5	0.103	0.791	0.152	0.723	0.181	0.685
6	0.152	0.848	0.211	0.789	0.245	0.755
7	0.209	0.897	0.277	0.848	0.315	0.819
8	0.272	0.938	0.349	0.901	0.391	0.877
9	0.345	0.970	0.428	0.945	0.473	0.928
10	0.427	0.991	0.516	0.979	0.562	0.970
11	0.523	1.000	0.615	0.998	0.661	0.996

(*Continued*)

(*Continued*)

$n = 13$	$\alpha = 0.01$		$\alpha = 0.05$		$\alpha = 0.10$	
x	P_L	P_U	P_L	P_U	P_L	P_U
1	0.000	0.449	0.002	0.360	0.004	0.316
2	0.008	0.541	0.019	0.454	0.028	0.410
3	0.028	0.621	0.050	0.538	0.066	0.495
4	0.057	0.691	0.091	0.614	0.113	0.573
5	0.094	0.755	0.139	0.684	0.166	0.645
6	0.138	0.811	0.192	0.749	0.224	0.713
7	0.189	0.862	0.251	0.808	0.287	0.776
8	0.245	0.906	0.316	0.861	0.355	0.834
9	0.309	0.943	0.386	0.909	0.427	0.887
10	0.379	0.972	0.462	0.950	0.505	0.934
11	0.459	0.992	0.546	0.981	0.590	0.972
12	0.551	1.000	0.640	0.998	0.684	0.996

$n = 14$	$\alpha = 0.01$		$\alpha = 0.05$		$\alpha = 0.10$	
x	P_L	P_U	P_L	P_U	P_L	P_U
1	0.000	0.424	0.002	0.339	0.004	0.297
2	0.008	0.512	0.018	0.428	0.026	0.385
3	0.026	0.589	0.047	0.508	0.061	0.466
4	0.053	0.658	0.084	0.581	0.104	0.540
5	0.087	0.720	0.128	0.649	0.153	0.610
6	0.127	0.777	0.177	0.711	0.206	0.675
7	0.172	0.828	0.230	0.770	0.264	0.736
8	0.223	0.873	0.289	0.823	0.325	0.794
9	0.280	0.913	0.351	0.872	0.390	0.847
10	0.342	0.947	0.419	0.916	0.460	0.896
11	0.411	0.974	0.492	0.953	0.534	0.939
12	0.488	0.992	0.572	0.982	0.615	0.974
13	0.576	1.000	0.661	0.998	0.703	0.996

$n = 15$	$\alpha = 0.01$		$\alpha = 0.05$		$\alpha = 0.10$	
x	P_L	P_U	P_L	P_U	P_L	P_U
1	0.000	0.402	0.002	0.319	0.003	0.279
2	0.007	0.486	0.017	0.405	0.024	0.363

(*Continued*)

(*Continued*)

3	0.024	0.561	0.043	0.481	0.057	0.440
4	0.049	0.627	0.078	0.551	0.097	0.511
5	0.080	0.688	0.118	0.616	0.142	0.577
6	0.117	0.744	0.163	0.677	0.191	0.640
7	0.159	0.795	0.213	0.734	0.244	0.700
8	0.205	0.841	0.266	0.787	0.300	0.756
9	0.256	0.883	0.323	0.837	0.360	0.809
10	0.312	0.920	0.384	0.882	0.423	0.858
11	0.373	0.951	0.449	0.922	0.489	0.903
12	0.439	0.976	0.519	0.957	0.560	0.943
13	0.514	0.993	0.595	0.983	0.637	0.976
14	0.598	1.000	0.681	0.998	0.721	0.997

$n = 16$	$\alpha = 0.01$		$\alpha = 0.05$		$\alpha = 0.10$	
x	P_L	P_U	P_L	P_U	P_L	P_U
1	0.000	0.381	0.002	0.302	0.003	0.264
2	0.007	0.463	0.016	0.383	0.023	0.344
3	0.022	0.534	0.040	0.456	0.053	0.417
4	0.045	0.599	0.073	0.524	0.090	0.484
5	0.075	0.658	0.110	0.587	0.132	0.548
6	0.109	0.713	0.152	0.646	0.178	0.609
7	0.147	0.764	0.198	0.701	0.227	0.667
8	0.190	0.810	0.247	0.753	0.279	0.721
9	0.236	0.853	0.299	0.802	0.333	0.773
10	0.287	0.891	0.354	0.848	0.391	0.822
11	0.342	0.925	0.413	0.890	0.452	0.868
12	0.401	0.955	0.476	0.927	0.516	0.910
13	0.466	0.978	0.544	0.960	0.583	0.947
14	0.537	0.993	0.617	0.984	0.656	0.977
15	0.619	1.000	0.698	0.998	0.736	0.997

$n = 17$	$\alpha = 0.01$		$\alpha = 0.05$		$\alpha = 0.10$	
x	P_L	P_U	P_L	P_U	P_L	P_U
1	0.000	0.363	0.001	0.287	0.003	0.250
2	0.006	0.441	0.015	0.364	0.021	0.326

(*Continued*)

(*Continued*)

3	0.021	0.510	0.038	0.434	0.050	0.396
4	0.043	0.573	0.068	0.499	0.085	0.461
5	0.070	0.631	0.103	0.560	0.124	0.522
6	0.101	0.685	0.142	0.617	0.166	0.580
7	0.137	0.734	0.184	0.671	0.212	0.636
8	0.176	0.781	0.230	0.722	0.260	0.689
9	0.219	0.824	0.278	0.770	0.311	0.740
10	0.266	0.863	0.329	0.816	0.364	0.788
11	0.315	0.899	0.383	0.858	0.420	0.834
12	0.369	0.930	0.440	0.897	0.478	0.876
13	0.427	0.957	0.501	0.932	0.539	0.915
14	0.490	0.979	0.566	0.962	0.604	0.950
15	0.559	0.994	0.636	0.985	0.674	0.979
16	0.637	1.000	0.713	0.999	0.750	0.997

$n = 18$	$\alpha = 0.01$		$\alpha = 0.05$		$\alpha = 0.10$	
x	P_L	P_U	P_L	P_U	P_L	P_U
1	0.000	0.346	0.001	0.273	0.003	0.238
2	0.006	0.422	0.014	0.347	0.020	0.310
3	0.020	0.488	0.036	0.414	0.047	0.377
4	0.040	0.549	0.064	0.476	0.080	0.439
5	0.065	0.605	0.097	0.535	0.116	0.498
6	0.095	0.658	0.133	0.590	0.156	0.554
7	0.128	0.707	0.173	0.643	0.199	0.608
8	0.165	0.753	0.215	0.692	0.244	0.659
9	0.205	0.795	0.260	0.740	0.291	0.709
10	0.247	0.835	0.308	0.785	0.341	0.756
11	0.293	0.872	0.357	0.827	0.392	0.801
12	0.342	0.905	0.410	0.867	0.446	0.844
13	0.395	0.935	0.465	0.903	0.502	0.884
14	0.451	0.960	0.524	0.936	0.561	0.920
15	0.512	0.980	0.586	0.964	0.623	0.953
16	0.578	0.994	0.653	0.986	0.690	0.980
17	0.654	1.000	0.727	0.999	0.762	0.997

(*Continued*)

(*Continued*)

$n = 19$	$\alpha = 0.01$		$\alpha = 0.05$		$\alpha = 0.10$	
x	P_L	P_U	P_L	P_U	P_L	P_U
1	0.000	0.331	0.001	0.260	0.003	0.226
2	0.006	0.404	0.013	0.331	0.019	0.296
3	0.019	0.468	0.034	0.396	0.044	0.359
4	0.038	0.527	0.061	0.456	0.075	0.419
5	0.062	0.582	0.091	0.512	0.110	0.476
6	0.090	0.633	0.126	0.566	0.147	0.530
7	0.121	0.681	0.163	0.616	0.188	0.582
8	0.155	0.726	0.203	0.665	0.230	0.632
9	0.192	0.768	0.244	0.711	0.274	0.680
10	0.232	0.808	0.289	0.756	0.320	0.726
11	0.274	0.845	0.335	0.797	0.368	0.770
12	0.319	0.879	0.384	0.837	0.418	0.812
13	0.367	0.910	0.434	0.874	0.470	0.853
14	0.418	0.938	0.488	0.909	0.524	0.890
15	0.473	0.962	0.544	0.939	0.581	0.925
16	0.532	0.981	0.604	0.966	0.641	0.956
17	0.596	0.994	0.669	0.987	0.704	0.981
18	0.669	1.000	0.740	0.999	0.774	0.997

$n = 20$	$\alpha = 0.01$		$\alpha = 0.05$		$\alpha = 0.10$	
x	P_L	P_U	P_L	P_U	P_L	P_U
1	0.000	0.317	0.001	0.249	0.003	0.216
2	0.005	0.387	0.012	0.317	0.018	0.283
3	0.018	0.449	0.032	0.379	0.042	0.344
4	0.036	0.507	0.057	0.437	0.071	0.401
5	0.058	0.560	0.087	0.491	0.104	0.456
6	0.085	0.610	0.119	0.543	0.140	0.508
7	0.114	0.657	0.154	0.592	0.177	0.558
8	0.146	0.701	0.191	0.639	0.217	0.606
9	0.181	0.743	0.231	0.685	0.259	0.653
10	0.218	0.782	0.272	0.728	0.302	0.698
11	0.257	0.819	0.315	0.769	0.347	0.741

(*Continued*)

(*Continued*)

12	0.299	0.854	0.361	0.809	0.394	0.783
13	0.343	0.886	0.408	0.846	0.442	0.823
14	0.390	0.915	0.457	0.881	0.492	0.860
15	0.440	0.942	0.509	0.913	0.544	0.896
16	0.493	0.964	0.563	0.943	0.599	0.929
17	0.551	0.982	0.621	0.968	0.656	0.958
18	0.613	0.995	0.683	0.988	0.717	0.982
19	0.683	1.000	0.751	0.999	0.784	0.997

SCORECARD FOR PERFORMANCE REPORTING

Lean Six Sigma scorecard for performance reporting Green and Black belt

Define	Measure	Analyze	Improve	Control	Cumulative Percent Complete
Identify Process Owner	Develop Existing Process Flow (Deployment Flow Chart)	Validate Project Scope and Process Flow	Identify/Prioritize Root Causes	Develop advanced Process Control Charting Concepts	10%
Develop Project Charter and Problem Statement	Implement Kaizen an 5 S Housekeeping	Identify Cause of Variation	Screen Potential Causes/Design of Experiments (DOE)	Develop and Implement Control Plan	20%
Develop Stakeholder Analysis	Identify Areas of the 7 Wsate Streams	Cause & Effect Diagram	Taguchi/Full factorial Design	Issue Revised Procedures	30%
Project Selection/Prioritization (Quality Function Deployment)	Develop Data Collection Plan/ Gage Capability Study	Analysis of Mean (ANOM)/Box Plot of X and/ or Y Variables	Identify and Prioritize Improvement Actions	Revalidate Cause and Effect of X's to Achieve Y's	40%
Establish CTQ and SIPOC Diagram	Measurement Systems Analysis (MSA)	Multi-Vari Analysis	Develop Implementation Plan	Implement Mistake Proofing - Poka-Yoke	50%
Establish Project Metrics and Desired Baseline Performance	Plot Output (Y) Data Over Time [Box Plots and Time Series]	Histogram of Input (X) Variables	Cost Benefit Analysis/Quality and Delivery Analysis	Monitor Process Performance (Run Chart/Control Chart/	60%
Establish Goals (percent or actuals)	Pareto High Impact Cause/ Process Input	Regression Analysis/Scatter Diagram/Correlation	Risk Assessment/Failure Mode Effect Analysis	Develop Before/After Summary Table of Performance/Metrics	70%
Develop Value Stream Map	Establish Hitogram of Existing Data/Applv descriptive Stat	Scale and Probability Analysis: Discrete and Continuous	Establish Operating Tolerance for Input Variables	Identify Opportunities to Leverage Improvements	80%
Develop Value Added Analysis	Establish Process Capability of Existing Data	Hypothesis Testing Concept and Chi Square	Develop New Process Flow (Deployment Flowchart)	Process and Capability Study (CP, Cpk PP, Ppk)	90%
Establish Potential Project Cost Savings/Avoidance	Determine Sigma Level Baseline of Existing Process	Analysis of Variance (ANOVA)	Portter's Five Forces	Risk Analysis	100%
Review with Sponsor	Review with Sponsor	Review with Sponsor	Review with Sponsor	Review with Sponsor	

Overall Project Completion Percentage %:

Sponsor Signature	Sponsor Signature	Sponsor Signature	Sponsor Signature	Sponsor Signature

BIBLIOGRAPHY

1. ASQ. (2003). *Six Sigma Forum Magazine*.
2. Boyle, K. (2003, April). *Six Sigma Black Belt Lecture Notes*, The University of San Diego *Six Sigma* Master Black Belt Program.
3. Box, G.E.P. and Wilson, K.B. (1951). On the Experimental Attainment of Optimum Conditions, *Journal of the Royal Statistical Society*, B, *13*, 1–45.
4. Ellis, K. (2001, December). *Mastering Six Sigma*, Training.
5. http://epa.gov/lean/thinking/kanban.htm, (2005).
6. http://nationalatlas.gov/articles/mapping/a_statistics.html
7. http://www.*six-sigma*-material.com/
8. http://en.wikipedia.org/wiki/Six_Sigma
9. http://www.isixsigma.com/
10. http://asq.org/index.aspx
11. http://www.sixsigmaonline.org/index.html
12. http://www.qualitycouncil.com/cssbb.asp
13. http://www.ge.com/sixsigma/sixsigstrategy.html
14. http://syque.com/quality_tools/tools/TOOLS12.htm
15. http://www.mikeljharry.com/story.php?cid=13
16. http://www.epa.gov/lean/environment/methods/kaizen.htm
17. http://www.villanovau.com/online-courses/lean-sigma-master-black-belt.aspx
18. Groebner, D.F., Shannon, P.W., Fry, P.C., Smith, K.D. (2005). *Business Statistics*, Printice Hall, Englewood Cliffs, NJ.
19. Harry, M.J., Mann, P.J., Hodgins, O.D., Hulbert, R.L., Lacke, C.J. (2010). *Practitioner's Guide for Statistics and Lean Six Sigma for Process Improvement*, Wiley, New York.
20. Kvanli, A.H., Pavur, R. J., Guynes, C. S. (2000). *Introduction to Business Statistics*, 5th ed., South-Western College Publishing.
21. Lunau, S., John, A., Meran, R., Roenpage, O., Staudter, C., *Six Sigma + Lean Toolset*, Springer, Frankfurt.
22. *NIST/SEMATECH e-Handbook of Statistical Methods (2005)*. http://www.itl.nist.gov/div898/handbook/.
23. Schmit, S.R. Launsby, R.G. (1994). *Understanding Industrial Designed Experiments*, Air Academy Press, Colorado Spring, CO.
24. Snee, R.D., Hoerl, R.W. (2003). *Leading Six Sigma*, FT Prentice Hall, New York.
25. Taghuchi, G., Chowdhury, Taguchi, S. (2000). *Robust Engineering*, McGraw Hill, New York.
26. Taghizadegan, S. (2006). *Essentials of Lean Six Sigma*, Elsevier, New York.
27. Taghizadegan, S. (2006). *Injection Molding: Experimental Modelling, Simulation, and Statistical Process Control*, Ph.D. Dissertation, University of Louisville, Louisville, KY.
28. Taghizadegan, S., Harper, D. O. (1996). *Statistical Process Control of Injection Molding Simulation Based on An Experimental Study*, SPE ANTEC Technical Papers, 42, 598–602.
29. Taghizadegan, S. (2013, April) *Six Sigma Master Black Belt Lecture Notes*, The California State University Six Sigma Master Black Belt Program.

INDEX

THIS TITLE IS FROM OUR MANUFACTURING COLLECTION.
OTHER TITLES OF INTEREST MIGHT BE...

Alarm Management for Process Control: A Best-Practice Guide for Design,
Implementation, and Use of Industrial Alarm Systems,
By Douglas H. Rothenberg

Process Control Case Histories: An Insightful and Humorous Perspective
from the Control Room,
By Gregory K. McMillan

Protecting Industrial Control Systems from Electronic Threats,
By Joseph Weiss

Industrial Resource Utilization and Productivity: Understanding the Linkages,
By Anil Mital, Arun Pennathur

Robust Control System Networks: How to Achieve Reliable Control After Stuxnet,
By Ralph Langner

Quality Recognition & Prediction: Smarter Pattern Technology with
the Mahalanobis-Taguchi System,
By Shoichi Teshima, Yoshiko Hasegawa, Kazuo Tatebayashi

Variable Speed Drives: Principles and Applications for Energy Cost Savings,
By David W. Spitzer

Textile Processes: Quality Control and Design of Experiments,
By Georgi Damyanov

Plant IT Integrating Information Technology into Automated Manufacturing,
By Dennis L. Brandl, Donald E. Brandl

Announcing Digital Content Crafted by Librarians

Momentum Press offers digital content as authoritative treatments of advanced engineering topics, by leaders in their fields. Hosted on ebrary, MP provides practitioners, researchers, faculty and students in engineering, science and industry with innovative electronic content in sensors and controls engineering, advanced energy engineering, manufacturing, and materials science. **Momentum Press offers library-friendly terms:**

- perpetual access for a one-time fee
- no subscriptions or access fees required
- unlimited concurrent usage permitted
- downloadable PDFs provided
- free MARC records included
- free trials

The **Momentum Press** digital library is very affordable, with no obligation to buy in future years.

For more information, please visit **www.momentumpress.net/library** or to set up a trial in the US, please contact **mpsales@globalepress.com**.

CPSIA information can be obtained at www.ICGtesting.com
Printed in the USA
BVOW06*1618230713

326458BV00005B/6/P

9 781606 504048

.